The Origins of Modern Humans
A World Survey of the Fossil Evidence

The Origins of Modern Humans
A World Survey of the Fossil Evidence

Editors

Fred H. Smith
Department of Anthropology
University of Tennessee
Knoxville, Tennessee

Frank Spencer
Department of Anthropology
Queens College of the City University of New York
Flushing, New York

Alan R. Liss, Inc., New York

Address all Inquiries to the Publisher
Alan R. Liss, Inc., 150 Fifth Avenue, New York, NY 10011

Library of Congress Cataloging in Publication Data
Main entry under title:

The Origins of modern humans.

    Includes bibliographies and index.
    1. Human evolution—Addresses, essays, lectures.
I. Smith, Fred H. II. Spencer, Frank.
GN281.075 1984   573.2   84-859
ISBN 0-8451-0233-8

Contents

Contributors

C.L. Brace, Museum of Anthropology, University of Michigan, Ann Arbor, MI 48109 [485]

Günter Bräuer, Anthropologisches Institut, Universität Hamburg, 2000 Hamburg 13, Federal Republic of Germany [327]

David W. Frayer, Department of Anthropology, University of Kansas, Lawrence, KS 66045 [211]

F. Clark Howell, Laboratory of Human Evolutionary Studies, University of California, Berkeley, Berkeley, CA 94720 [xiii]

J.J. Hublin, Laboratoire de Paléontologie des Vertébrés et de Paléontologie Humaine, 75230 Paris, Cedex 05, France [51]

Roger C. Owen, Department of Anthropology, Queens College of the City University of New York, Flushing, NY 11367 [517]

G. Philip Rightmire, Department of Anthropology, State University of New York, Binghamton, NY 13901 [295]

Fred H. Smith, Department of Anthropology, University of Tennessee, Knoxville, TN 37916 [ix,137]

Frank Spencer, Department of Anthropology, Queens College of the City University of New York, Flushing, NY 11367 [ix,1]

C.B. Stringer, Department of Paleontology, British Museum (Natural History), London SW7 5BD, United Kingdom [51]

Alan G. Thorne, Department of Prehistory, Research School of Pacific Studies, Australian National University, Canberra, Australia [411]

Erik Trinkaus, Department of Anthropology, University of New Mexico, Albuquerque, NM 87131 [251]

B. Vandermeersch, Laboratoire d'Anthropologie, Université de Bordeaux, Talence, France [51]

Milford H. Wolpoff, Department of Anthropology, University of Michigan, Ann Arbor, MI 48104 [411]

Shao Xiang-qing, Anthropology Section, Department of Biology, Fudan University, Shanghai, People's Republic of China [485]

Zhang Zhen-biao, Institute of Vertebrate Paleontology and Paleoanthropology, Beijing, People's Republic of China [485]

Wu Xin Zhi, Institute of Vertebrate Paleontology and Paleoanthropology, Beijing, People's Republic of China [411]

The number in brackets is the opening page number of the contributor's article.

Preface

The idea for this book emerged from our belief that a need existed for a detailed consideration of the "origin" of anatomically modern (a.m.) *Homo sapiens*, encompassing both the pertinent fossil material and theoretical perspectives currently employed to approach the issue of archaic/modern *H sapiens* relationships.

In the last decade or so, our knowledge of the pattern and course of human phylogeny has increased markedly. One of the many factors contributing to this has been the discovery of several new human fossil remains. Also, more sophisticated recovery techniques have improved the ability to determine chronological sequences and relationships, and an increased understanding of the underlying processes of biological evolution, all of which have greatly facilitated our knowledge of the types of information that *should* be gleaned from the hominid fossil record, as well as the methodologies to accomplish this. Finally, and perhaps most significantly, the geometric increase in the number of scientists actively engaged in paleoanthropological research over the past two decades has broadened the scope and increased the amount of such research.

However, during the period since the discovery of the first East African australopithecine in 1959, considerably more attention has been focused on the origin and diversification of early (Pliocene and Lower Pleistocene) hominids than on the more recent (Middle and Upper Pleistocene) stages of human evolution. This development is quite understandable given the unwritten law of paleoanthropology that "older is better," and the fact that prior to these discoveries in East Africa and elsewhere, information on the earlier phases of hominid evolution was extremely limited. As a consequence, advances in our knowledge of the origin of a.m. *H sapiens* and the role of the various geographically diverse groups of archaic *H sapiens* in that phenomenon have been overshadowed by the dramatic and often highly publicized developments in early hominid research.

Although several books or collections of articles have been published on the initial phases of hominid evolution in recent years, the most recent text with a (limited) worldwide perspective and specifically oriented toward a consideration of archaic *H sapiens* appeared in 1958, namely the volume edited by the late GHR von Koenigswald entitled "Hundert Jahre Neanderthaler 1856–1956," which was published by Kemink en Zoon, Utrecht, Netherlands, under the auspices of the Wenner-Gren Foundation for Anthropological Research. This particular work was based on the proceedings of an international symposium convened in Dusseldorf, Germany, in August 1956 to commemorate the discovery of the Feldhofer (Neandertal) remains in the Neander Valley a hundred years earlier. As the contents of this volume indicates, the fossil hominid record as it pertained to the origin of a.m. *H sapiens* was then still largely confined to a consideration of the European assemblages. Although the European Upper Pleistocene hominid fossil record is still more extensive than anywhere else, there has nevertheless been, during the intervening three decades, a decided shift from this previously Eurocentric bias in the hominid fossil record. To a large extent this change in perspective has come about through the fortuitous discovery of a number of important specimens in Africa, Southwest Asia, East Asia, and Australasia.

In conceiving this book, we decided to invite several paleoanthropologists to contribute a comprehensive assessment of Upper and, where appropriate, late Middle Pleistocene hominid remains in a specific geographic region of the world (namely, western Europe, central Europe, the Near East, Africa, East Asia, Australasia, and the New World) with which they had considerable experience. Following this, the respective contributors were asked to discuss the skeletal and other pertinent archeological data from a given region in the context of the question of modern human origins. No attempt was made to select contributors on the basis of their theoretical orientation. This we believe was necessary if an accurate presentation of the state-of-the-art was be achieved, for scientific opinion is still very much divided over the question of which model should be adopted to account for the emergence of modern *H sapiens*. As such, we have endeavored to reflect this diversity of perspective and refrained from exerting editorial pressure on authors who do not share our theoretical point of view.

The structure of the book is self-evident. It commences with a brief historical overview of the Neandertal debate and the emergence of the three major schools of thought on the origin of *H sapiens*, namely the Neandertal-phase hypothesis, the presapiens hypothesis, and the compromise model, the pre-Neandertal hypothesis. This is followed by nine chapters, each of which deals with a specific region, commencing with western Europe and then proceeding eastwards across the Old World and finishing with a consideration of the appearance of human populations in the New World.

There are a few editorial matters which we feel merit some explanation. First, we have uniformly followed modern German orthography in the spelling of "Neandertal." Although some anthropologists still prefer to use the old spelling ("Neanderthal"), most German workers utilize the new spelling. According to the International Rules of Zoological Nomenclature, Article 19, "The original orthography of a name is to be preserved unless an error of transcription, a *lapsus calami*, or a typographical error is evident." Hence, for this reason we have continued to use the old spelling when it is employed in the formal taxonomic name, ie, *H neanderthalensis* (first proposed by William King at a meeting of the British Association for the Advancement of Science in 1863) or *H sapiens neanderthalensis*. The new spelling is reserved for colloquial or adjectival forms of the name. Next, throughout the text, we have used the abbreviation "ky" to denote thousands of years. Thus, dates are written, for example, as 43 ky BP rather than 43,000 years BP. Finally, the publisher has adopted the style, endorsed by the American Medical Association, which deletes the period after initials. Names are thus written as EO Anthropus rather than E.O. Anthropus, and the taxon *H sapiens* rather than *H. sapiens*.

We are greatly indebted to several individuals for various forms of assistance with this volume. We are especially grateful to the contributing authors, without whose patience and cooperative spirit this work would never have been realized. We would particularly like to express our appreciation to F. Clark Howell for his forebearance in waiting for the completion of all of the papers in order to provide us on short notice with his excellent introduction. In addition, FHS would like to thank the Alexander von Humboldt Foundation (West Germany) for fellowship support during part of the period this volume was being developed. Similarly, FS thanks the Wenner-Gren Foundation for Anthropological Research for a grant-in-aid that supported his (ongoing) researches into the history of physical anthropology. We also thank M.E. Peters for her assistance in the compilation of the author index. Finally, we would like to thank Mr. Alan R. Liss for his enthusiastic support, and his staff for their energetic commitment to quality throughout the production stage of this volume. In particular we wish to thank Ms. Rainelle Peters (production editor) for her patience and attention to detail.

Goethe once remarked, "Of all things the most interesting to man is man." As anthropologists, we are inclined to agree. Certainly few facets of the study of human history have so stirred the imagination or stimulated so much controversy as the questions of our initial origins and the emergence of early humans of a modern aspect. Although this work does not pretend to provide an unequivocal answer to the latter question, it is our hope that it will contribute to its eventual resolution.

Fred H. Smith and Frank Spencer

Introduction

F. Clark Howell

Laboratory of Human Evolutionary Studies, University of California, Berkeley,
Berkeley, CA 94720

In *The Origins of Modern Humans* sixteen authors from six countries consider the nature and phylogenetic implications of the available human skeletal evidence relevant to the origins of *Homo sapiens sapiens* (anatomically modern humans). This evidence is now both more substantial and more geographically representative than ever before. The authors are all anthropologists with skills and experience in human skeletal biology and immediate personal familiarity with the original collections that afford the database(s) for this conjoint effort. Consequently, this is an important and timely contribution to a subject of longstanding anthropological and biological interest—in fact to any and all people concerned with their own biological roots.

There is now a near consensus among students of human evolutionary biology that the origins of our own species, *Homo sapiens*, is somehow intimately linked with the first intercontinental ancient hominid, *Homo erectus*. However, neither the transformation of *erectus* to *sapiens* nor the transformation of ancient (archaic) populations of *Homo sapiens* to their anatomically modern succedents (*H s sapiens*) are matters of agreement in this scientific fraternity. Undoubtedly, there are many factors that make this the case, and any reader of this volume will discern some of those that are most obvious. In fact, there is no consensus among the authors represented in this volume, although the major issues are generally well delineated, and the limitations of the diverse and often disparate lines of evidence are usually apparent. My purpose here is to offer some comments on the limitations of some evidence, and on what I regard as (some) central issues that, hopefully, future research will further elucidate.

Homo erectus has its origins, so far as is now known, well back in the Lower Pleistocene of Africa. The species was both coexistent and, by all evidence, broadly (sometimes) sympatric with a derivative species of *Australopithecus* (*A boisei* and, probably, *A robustus*), although the ecological (including behavioral) factors that enabled (or limited) such coexistence are at best still a matter for speculation. The first demonstrable occurrence of the species outside Africa is hundreds of thousands of years subsequently, late in the Lower Pleistocene (upper Matuyama chron), and thousands of kilometers distant, both on the Sunda Shelf (Java) and in continental eastern Asia (China). The nature and timing of this extra-African dispersal is now a central problem in paleoanthropology, and the general deficiency of Lower Pleistocene human fossils and/or archeological occurrences throughout western and southern Asia severely limits any efforts toward its resolution. The now well-documented occurrence of successive ancient human occupations at 'Ubeidiya, Jordan Valley (Israel), in a mid-Lower Pleistocene context, is of particular significance in this regard (although skeletal remains of any hominid are unknown from this unique locality). Nonetheless, the affinities of the several lithic assemblages at 'Ubeidiya are definitely with (eastern) Africa.

Throughout Eurasia in fact the most adequate documentation of early hominid presence is largely a Middle Pleistocene phenomenon (within the Brunhes (N) chron). This is now known to be true for the extensive sample from Zhoukoudian-1 (China), as well as that from localities sampling the Kabuh Formation in central Java. Moreover, not only are these and other such Asian occurrences both relatively (geologically) and absolutely younger than has been commonly supposed, but also they sample a shorter overall time span and occur geologically more recently than is often envisioned. These are important matters for consideration with respect either to phylogenetic interpretation or to model-building of *erectus–sapiens* relationships.

Wolpoff, Wu, and Thorne discuss the Asian evidence relevant to modern *Homo sapiens* origins. These authors conclude that "all across east Asia, hominid evolution led to the appearance of modern populations without either a speciation *event* or a dramatic influx of populations from elsewhere," and thus that "the evolution of the genus *Homo* in east Asia was gradual (continuous, although not necessarily at a constant rate), and not punctuational." Hence, they consider "the fossil evidence from east Asia as disproving the hypothesis of a simple migratory origin for the modern populations of the region." Moreover, they do not find either useful or appropriate the concept of grades (whether dubbed subspecies or not) within *erectus* or *sapiens* (the former taxon they would in fact prefer to sink), and also "particularly question whether there are any morphological criteria that can be applied to this archaic/modern [human] boundary on a worldwide basis."

They consider local regional continuity (such that human populations have substantial time depths in terms of ancestral/descendant populational relationships) to be the most appropriate explanatory model. Ancestral (founder) populations within a dispersing genus *Homo* they regard as having had decreased genetic and morphological variability (for several reasons), with subsequent adaptive divergence in particular geographic/environmental situations resulting in genetic differences between peripheral populations. For them *"clines will invariably form in a polytypic species,"* and "there were two distinct although not geographically discontinuous morphological clades at the north [=China] and south [=Sunda Shelf] ends of the east [=Asian] periphery."

It should be noted that there is substantially more and better evidence in eastern Asia for *Homo erectus* (as usually defined), known from some ten localities,[1] than for succeedent forms, usually referred to as an archaic form of *Homo sapiens*. The

---

[1] Yuanmou (early Brunhes (N) chron); Lantian (Gondwangling, uppermost Mutayama chron); Chenjiawo (earlier Brunhes (N) chron); Zhoukoudian-1 (mid-Brunhes (N) chron, ca. 0.46–0.23 Ma.); Hexian (ca. 0.25–0.30 Ma.); and the broadly mid-Pleistocene localities of Jian Shi (perhaps the oldest), Yunxian, Yunxi, Xinghua, and Xichuan (all of which yield only teeth).

youngest occurrences are presumably on the order of, perhaps, a quarter of a million years (or rather less), as in the case of upper ZKD-1 (Bed 3, Skull 5) and the recent Hexian specimens.

Archaic *Homo sapiens* specimens are known from only six localities,[2] the most complete remains being the well-preserved cranium from Dali, and the partial anterior cranium from Maba; some teeth and a number of cranial fragments derive from Xujiayao.

There is inadequate relative temporal and no geographical control on these hominid fossils. The oldest appears to be the Dali specimen which, based on biostratigraphic grounds (perhaps to the final Middle Pleistocene), is a hundred thousand years or so younger than the youngest representatives of *Homo erectus* found elsewhere.

All other such remains are of largely undetermined Upper Pleistocene age. All have been treated as archaic *Homo sapiens* (the Dali specimen having been referred to as *H sapiens daliensis*). The best preserved specimens (Dali, Maba) show an overall morphology substantially divergent from that of antecedent east Asian *H erectus*, in spite of the fact that there are some shared, presumably plesiomorphic features (and others that seem to have definite Mongoloid resemblances). Thus, there is a substantive change in cranial morphology in the final phases of the Middle Pleistocene, and from the still scant available evidence (particularly Xujiayao cranial parts) this pattern continues into the (earlier) Upper Pleistocene. It is also noteworthy that there are substantial differences in cranial morphology between these Asian hominids and those of penecontemporaneous (or greater) age in Europe, that is the Neandertals and their Middle Pleistocene predecessors. I consider that the evidence is both so scant and so incomplete as to thwart efforts, however thoughtful and elaborated, to seek to understand an *erectus–sapiens* transition, however conceived, in this instance. Moreover, there is almost a complete absence of subsequent Upper Pleistocene fossil documentation, which would afford a background for the evaluation of the terminal Pleistocene/post-Pleistocene human crania, which are admittedly strongly Mongoloid in overall morphology (those from the Upper Cave, Ziyang, Liujiang, and Chilinshan).

There has been remarkable progress in documenting the earlier archeological record and attendant human populations on the Sahūl shelf in Australia. There is, however, an utter dearth of comparable and antecedent evidence from the presumed source area of such human populations on the Sunda Shelf. There is a near consensus that the Ngandong sample represents a derivative subspecies of an earlier *Homo erectus* lineage. However, the subsequent evolutionary history of the lineage is undocumented and hence a mystery.

Africa was apparently the source of Hominidae in the Mio-Pliocene, and its rich Pliocene hominid fossil record is now *the* basis for seeking to understand the radiation of *Australopithecus* and the emergence of the genus *Homo*. The Pleistocene hominid fossil record, with reference to the origins of *Homo sapiens* and particularly anatomically modern representatives of that species, is discussed here by Bräuer and

---

[2]Dingcun, Tongzi, or Yanhui (2 teeth), Xujiayao, Maba, and Dali.

by Rightmire. Over the past quarter century there has been remarkable progress in expanding both the prehistoric archeological and human paleontological database in several areas of the continent such that these analytical and comparative studies are particularly timely.

Homo erectus has its earliest known documentation in eastern Africa, early in the Lower Pleistocene. Overall, this is an important sample as it comprises not only cranial and jaw-parts, but also varied, and sometimes associated portions of the postcranial skeleton. Subsequently, presence of the species is well attested to at several localities of later Lower Pleistocene age and occurs also in the Maghreb at least by the Middle Pleistocene (Ternifine). At least some occurrences are directly associated with the Acheulian industrial complex, and others might reasonably be considered (parsimoniously) to have been so. There are fundamental overall resemblances in cranial morphology between such specimens from Africa and those (from the Kabuh Formation) on the Sunda Shelf. This writer considers that the Bodo (Middle Awash) cranium is of fundamentally Homo erectus morphology.

In both the Maghreb and in eastern Africa some human cranial remains of upper Middle Pleistocene age reveal varying degrees of morphological divergence from the antecedent Homo erectus morphology. There is, unfortunately, still inadequate stratigraphic and geochronological control to fix at all closely either the relative or the absolute ages of these (and other, more distinctly Homo erectus) occurrences. In the Maghreb all apparently fall within the several hundred thousand year span of the continental Tensiftian, which has afforded such remains (Salé, and the geologically younger Kebibat sample), as well as penecontemporaneous, if not actually older specimens of Homo erectus aspect (Littorina Cave, Thomas Quarry sites). A very tentative age estimate might be between 0.30 and 0.20 Ma. In sub-Saharan Africa comparable examples are known from Ndutu (Tanzania) (perhaps between 0.40 and 0.20 Ma.) and from Kabwe (Zambia) and from Elandsfontein (Cape) in situations where comparative biostratigraphy suggests only a broadly upper Middle Pleistocene age.[3] The Maghreb and Ndutu samples reveal a mosaic of plesiomorphic (erectus-like) features associated with divergent, derived (apomorphic) features, not only related to changes in robusticity, which approach a (archaic) sapiens morphological pattern. This transformation is even more fully evidenced in the Kabwe cranium, in which both cranial vault and facial morphology are excellently preserved. Bräuer is apparently receptive to the view that this was an autochthonous, relatively gradual and presumably continuous transformation, whereas Rightmire is less strongly (if at all) committed to such a position.

Bräuer recognizes here three major "grades" of Homo sapiens evolution in Africa, whereas Rightmire has sidestepped such a semiformal procedure. Both authors consider that there is a demonstrably substantial, early Upper Pleistocene antiquity for (somewhat archaic/primitive) Homo sapiens sapiens in sub-Saharan Africa. The examples include cranial and/or jaw specimens from Ethiopia (Omo Kibish Fm.), Tanzania (Ngaloba Beds, Laetoli), and South Africa (Florisbad; Border Cave; Klasies

---

[3]It is unclear at this point how the Kapthurin (Baringo basin) mandible, and partial cranium from Garba-3 (Melka-Konturé) relate, if at all, to this sample.

River Mouth). The morphological evidence afforded by this rather variable sample is convincing of their near anatomically modern affinities. Their age attributions have varying degrees of reliability, those from Laetoli and Klasies apparently the most reliable, and the others to some extent open to question. Undoubtedly, it is only a matter of time before the recovery of additional human remains in suitably datable, and reliably decipherable stratigraphic contexts will afford the necessary solid evidence to resolve this question. Nonetheless, even at this juncture, the evidence is strongly in favor of an early differentiation of near anatomically modern *Homo sapiens* populations in sub-Saharan Africa. The implications of that situation are profound for the unraveling of the later stages of human evolution in the adjacent regions of Asia as well as Europe.

Europe was long the principal focus of studies, initially by informed amateurs whose professional pursuits lay elsewhere, directed toward the elucidation of human antiquity within the emerging framework of earth history in the nineteenth century. Although the significance of earlier human evolutionary studies is not really considered by the majority of the authors in this volume, Spencer, in his chapter, treats in detail the impact and persistent influence on conceptions of human evolution in Europe following the discovery and recognition of Neandertal peoples, as well as the pernicious effects of belief in the Piltdown fossils before they were proved a hoax.

In Europe there is an increasingly improved documentation of human skeletal remains within the Middle Pleistocene. It should be recognized, however, that for the most part this documentation largely represents only the latter two-thirds of this approximately 600 ky time span. Although a consensus still cannot be said to have been reached, some investigators consider that these specimens sample populations that are distinct, in a number of ways, from (Asian/African) *Homo erectus*, and some even consider (largely on the basis of parsimony) that they already constitute an emergent (archaic) form of *Homo sapiens*. Here, and elsewhere, various authors have noted the divergences in cranial and/or mandibular morphology of such remains (Mauer, Montmaurin, Arago, Petralona, Vértesszöllös, Swanscombe, Steinheim, Bilzingsleben) from Afro-Asian *H erectus*. If, as now appears probable, there is a temporal overlap between late representatives of *H erectus* in eastern (and southern) Asia and this European hominid sample, then the question of a straightforward and gradual transformation of the former to a *sapiens*-grade is thrown into an entirely different perspective.

Substantive discussions of the Upper Pleistocene human fossil record in western Europe, central Europe, and western Asia are provided here by Stringer, Hublin, and Vandermeersch; by Smith; and by Trinkaus, respectively. Similarly, Bräuer treats the available evidence from northern Africa. There is a limited, but at best only partial consensus among these contributors to this volume. All recognize the skeletal distinctiveness of Neandertal populations whose roots are discernible at least in the European late Middle Pleistocene, and very probably earlier, and whose geographic distribution extended over a vast area from western Europe to central Asia and both north and south of the Mediterranean basin, in the latter instance at least to Cyrenaica, and even possibly to the Maghreb. At the minimum, a time span in excess of 0.20 Ma. is indicated for such peoples.

It is the fate of these successful populations about which a consensus has not been reached and over which debate continues. However, in at least two areas of their range the evidence is now very strongly suggestive, indeed almost overwhelming (to this writer) that these peoples were displaced and ultimately replaced by dissimilar populations of anatomically modern, *Homo sapiens sapiens* affinities (without addressing the question of potential or actual hybridization, which in fact Bräuer favors). In each instance the absence of very refined stratigraphic and geochronological control severely limits a fine analysis of this process. (Ultimately, the latter will undoubtedly be afforded by extensive application of a refinement of the radiocarbon dating method employing an accelerator as a mass spectrometer directly to count ^{14}C atoms).

Trinkaus considers that in western Asia "the morphological changes associated with the archaic [western Asian Neandertal] to anatomically modern [Skhūl-Qafzeh samples] human transition suggest a more complex process, with significant elevations of interregional genetic exchange." He stresses that "the total amount of change was of a different order of magnitude than that which is evident within either the archaic [western Asian Neandertal] or the anatomically modern *H sapiens* samples." He considers that there was possibly "an interval of about 5,000–10,000 years in which the transition could have taken place." This is an admittedly very brief span of time, but he considers such an *in situ* transformation to be possible, given certain conditions. However, if there is a temporal overlap between these human population samples, as some workers in fact maintain, then the validity of such an interpretation falls away. Clearly, this is a problem seriously in need of resolution; were the absolute ages of important Neandertal samples and those from Skhūl and Qafzeh more precisely fixed, then the problem might well be viewed in an entirely different perspective.

The fossil record from the south Mediterranean and Atlantic littoral of Morocco is not helpful in this regard. Fragmentary jaw parts from Haua Fteah (Cyrenaica), which have an age of about 0.050–0.045 Ma., appear to resemble Neandertal counterparts in western Asia. The oldest Upper Pleistocene human sample in the Maghreb, that from J'bel Irhoud, is of earlier Soltanian (I) age, broadly correlative with the earlier third of the Last Glaciation. These specimens (two adult crania and a juvenile mandible) reveal a mosaic of both (plesiomorphic) Neandertal-like and more anatomically modern cranial vault features, with a facial structure and proportions (short and without mid-facial prognathism, for example) verging on an anatomically modern form. Trinkaus notes that such populations might "have provided a source for gene flow into western Asia, although this does not necessarily mean that they did." Bräuer believes that there "certainly . . . were connections with the Near East," although he does not seek to explore further their nature, extent, and potential significance (apparently he considers hybridization probable).

In western (France) and southwestern (Spain, Italy) Europe Neandertal peoples occur occasionally in the initial, anaglacial phases of the buildup to the Last Glaciation (broadly correlative with ^{18}O stage 5d–5a), but predominately occur in the initial cold (Lower and early Middle Pleniglacial) substage (equivalent to ^{18}O stage 4) of that glaciation. In northern, southwestern (Périgord, Charente), and parts of southeastern France, as well as in Cantabrian Spain, there is a refined stratigraphy for this interval in which "facies" of the Mousterian industrial complex are ubiquitous. This complex occurs up to the Hengelo interstadial (~ 38 ky BP) and, at least in a few sites in some parts of southeastern France, into the subsequent mid-Middle Pleniglacial (pre-Denekamp interstadial), the latter interstadial having an age of ~ 30 ky BP Radiocarbon age control on most Mousterian occurrences is still far from adequate; some of the more recent occurrences seem to have ages between approximately 38 and 35 ky BP

Throughout western/southwestern Europe the Aurignacian industry (whether Aurignacian 0, or more commonly Aurignacian 1) is represented in the mid-Middle Pleniglacial, initially at approximately 35–36 ky BP (in southern Germany) and approximately 33.5 ky BP (in southwestern France). Whenever there are also associated human skeletal remains, these are of anatomically modern, Cro-Magnon peoples. In all known archeological sites in (greater) western Europe, occurrences of the Mousterian industry, if overlain by subsequent occupations, are succeeded by an Aurignacian industry (or a subsequent Upper Paleolithic assemblage) or, in other instances, in the Franco-Cantabrian region by the Châtelperronian (=Lower Périgordian of D. Peyrony and other authors). There are no known instances in which the latter is overlain by Mousterian.

The integrity or affinities, or both, of the Châtelperronian have been repeatedly questioned since H. Breuil introduced the term Châtelperronian over 70 years ago. This is no longer the case, and the distinctive Châtelperronian lithic assemblage, recorded at a total of some 35 (and conceivably as many as 66) sites (at least 17 of which are demonstrably uncontaminated with reference to adjacent stratified assemblages, two in Cantabria and the remainder in France), merits all due consideration in any discussion of the fate of the Neandertals in southwestern Europe. The Châtelperronian occurs in the Périgord and Charente between about 35 and 33 ky BP, and at a broadly comparable time in Catabrian Spain; in north-central France (Grotte du Renne, Arcy) it persists well into the Middle Pleniglacial (to the Arcy = Denekamp interstadial). There is some evidence (in France) to suggest a progressive northwestward expansion of Aurignacian occupation and a corresponding north to northeastward (and attendant diminution) of Châtelperronian occupation through this ca. 4,000-year time span.

At least four such sites in France, three of them open sites, have single Châtelperronian occupations, and lack antecedent or succedent occupations. At least six other sites have multiple (2 to 4) Châtelperronian occupations. Nine sites have Châtelperronian succeeding Mousterian occupation(s), and two sites (Le Piage in Périgord, and El Pendo in Cantabria) have Châtelperronian occupation interstratified between Aurignacian 0 (below) and Aurignacian 1 (above). At least two sites (Saint-Césaire in Charente, and Cueva Morin in Cantabria) have Aurignacian 0 and 1 occupations after the Châtelperronian industrial level(s).

Human remains certainly associated with the Châtelperronian industry long remained unfound. Some workers (largely following D. Peyrony) presumed that the human skeletal interment at Roc de Combe Capelle (Périgord) probably represented such an association, although this was never demonstrated, and an Aurignacian association is equally or even more probable. However, a series of (eight) human teeth from two Châtelperronian ocupations at Grotte du Renne (Arcy-sur-Cure) have decidedly Neandertal features. Most important, much of a Neandertal skeleton recently found in a Châtelperronian occupation context at Roche à Pierrot, Saint-

Césaire (Charente-Maritime) affords an incontrovertible association of such people with that industry. Only modern *H sapiens* has been found associated with the Aurignacian, as in the case of the Cro-Magnon individuals from that eponymous shelter (and including the pseudomorph burials of tall stature at Cueva Morín in an Aurignacian 0 context). Although Aurignacian (I) has never certainly been found stratigraphically *beneath* the Châtelperronian, an older Aurignacian (0) does so occur in at least three known instances. This is not only indicative of penecontemporaneity of such human populations but also is at least suggestive of alternate use of the same sites by culturally (and biologically?) different groups (and, for what it is worth, the Châtelperronian and such Aurignacian assemblages are quite readily distinguishable on typological grounds).

Thus, the question of population impingement, competition, displacement, and (eventually) replacement in southwestern Europe between 35 and 30 ky ago becomes again, and in a different form, a hypothesis worthy of further examination and testing. This alternative to gradualistic hypotheses does not yet merit out-of-hand dismissal.

What have other areas of Europe to contribute to an understanding of this fascinating problem? Smith contributes a fundamental chapter on central/eastern Europe in which he seeks to clarify and advance such an understanding. His points are well taken: first, that whatever has been found there is often ignored in the "West," and, second, that a regionally broader documentation is necessary if human evolution (or paleocultural adaptations and extra-regional relationships) are to be adequately apprehended.

Smith concludes that "morphological continuity between Neandertals and the EMH [early modern human] sample in central Europe is clearly documented by the available information. . . ." He notes that the earlier of the Neandertals from this region (eg., Ganovce, Krapina, Ochoz, Salzgitter-Lebenstedt, and Subalyuk) "exhibit a total morphological pattern that differs only slightly from that of western European 'classic' Neandertals." Their successors (eg, Kůlna, Šipka, the Vindija G₃ sample and, perhaps, Sala), some of which quite definitely antedate the Podhradem (Hengelo-equivalent) interstadial, also "unquestionably qualify as archaic *H sapiens* on the basis of their total morphological pattern," although he sees a number of features in this sample whereby it "consistently approaches the early modern *H sapiens* (EMH) condition" more so than do its antecedents. These features are reflected particularly in supraorbital morphology, form of cranial vault (frontal squama and occipital region), facial skeleton, dental reduction, and form and proportions of the mandibular symphysis.

The writer believes that this evidence is nonetheless equivocal. Although certain differences in size and in the expression of cranial (and dental) morphology are evidenced between earlier and later Neandertals of this region, as Smith has clearly shown (both here and elsewhere), the fundamental, overall morphological pattern is Neandertal and not anatomically modern human. This is in spite of the admittedly fragmentary nature of most specimens, which militates against a fuller appreciation of the total cranial morphology.

There are some ten occurrences of human skeletal parts in this region that are considered to be demonstrably or presumably associated with an Aurignacian industry. The occurrences for which ages are best established (by radiocarbon dating) are those from Vogelherd, Württemberg (31.9 ± 1.1 ky BP) and from Velika Pećina, Yogoslavia (below a level dated 33.9 ± 0.52 ky BP) Thus, these specimens fall within, or just subsequent to the Podhradem (Hengelo)[4] interstadial. They are fully anatomically modern in cranial morphology, and in this respect might well be subsumed within the Cro-Magnon population. The most extensive sample is that from the cave of Backova dira, near Mladeč (Lautsch), Moravia; however, here, although a diagnostic (typical) Aurignacian assemblage is known, the immediate relationships of the industry and the human remains cannot be precisely determined (the enclosing sediments are derived from a talus cone from an aven), and any presumed age attribution (for example to the Podhradem interstadial) is purely inferential. In any case, the total morphological pattern of cranial and postcranial morphology is again distinctively anatomically modern human.

Other specimens that also have a distinctively anatomically modern human morphology are of uncertain stratigraphic relationship and/or age (Podbaha, Cioclinova). In contrast, the anatomically modern Zlatý Kun cranial/mandibular parts are of younger, later Middle Pleniglacial age (as is also the case for a juvenile mandible fragment from Miesslingtal), which is in keeping with what is known in western Europe.

Finally, two other occurrences are worthy of mention. The adult human frontal from Hanöfersand (Elbe valley) has an apparently secure radiocarbon age of 36.3 ± 0.60 ky BP and thus falls quite early in the Hengelo interstadial. The fundamentally anatomically modern features of this specimen are not denied by either Bräuer (who initially described it) or by Smith (who remarks that it "is unquestionably a modern *H sapiens*"), although each would see some "primitive" (underived) characters as well (and account for such in different ways, presumably). However, its fundamentally modern structure and very substantial antiquity, close to and perhaps even overlapping in time some known Neandertals (both in central Europe and in France), are directly relevant to the appropriateness of a hypothesis of population replacement versus one of gradual evolutionary change.

Lastly, there is the question of early Upper Paleolithic occupation in Bacho Kiro cave (central Bulgaria), a site first tested by Dorothy Garrod and recently extensively investigated jointly by Polish and Bulgarian workers. It contains four lower levels (14–12) with Mousterian occupations, a middle one having an age of > 47.5 ky BP (and corresponding to an earlier Last Glacial interstadial). Traces of another Mousterian (level 11a) are succeeded by four rich levels of an Aurignacian-like Upper Paleolithic (termed Bachokirian by J. K. Kozlowski) distinctive in raw materials

---

[4]It should be noted here that if the Šipka Neandertal fragment is indeed of Podhradem interstadial age, which is possible if not adequately substantiated, then there is a temporal overlap more or less with these occcurrences. Moreover, there are some, though not fully diagnostic Neandertal-like features in the few human fragments from Vindija G_1, which is considered to be of Podhradem interstadial age and, on tenuous grounds, to occur in an Aurignacian context. The few (3) permanent teeth from the overlying Aurignacian occupation (level F_d) similarly are inconclusive.

employed for lithic artifacts, in technology, and in typology; there is evidence also of bodily adornment, as well as of substantial hearths and some form of structure. This occupation has an age of > 43 ky BP Successive occupations (levels 9 and 8) contain a broadly similar industry, but also include more Aurignacian elements, including a split-base point (level 9) and a Mladeč point (level 8). A subsequent level (6b) has traces of human occcupation, but only a small lithic assemblage; it has a ^{14}C age of 32.7 ± 0.30 ky BP.

The lowest (level 11) Upper Paleolithic occupation here is older by far than any such occurrences elsewhere in Europe. The lithic assemblage is reminiscent in certain respects (including chronologically) of the "Pre-Aurignacian" of western Asia (as at Yabrud). The lithic-rich industry strongly resembles that small assemblage from the base of Istállöskö (Bükk mountains, Hungary), which, however, has a number of split-base points. The latter has been variably dated to 31.5 ± 0.60 ky BP (burned bone) or 44.3 ± 1.9 ky BP and 39.8 ± 0.90 ky BP (on bone). The older ages have been questioned by some workers but perhaps merit reconsideration. A single $M_{\overline{2}}$ germ recovered at Istállöskö appears to diverge in both size and morphology from the Neandertal condition, although such evidence is admittedly weak. At Bacho Kiro only a single worn $dm_{\overline{1}}$ in a mandible fragment is known from the older Bachokirian occupation, whereas four permanent teeth, a juvenile mandible fragment with $dm_{\overline{2}}$ and $M_{\overline{1}}$ and an adult parietal fragment is known from levels 6a–b/7. In general, the few available elements of the dentition are distinguished principally by their substantial size, often in the range of Krapina homologues (except $I_{\overline{1}}$ breadth, and p^2 length) and also Vindija (for $M_{\overline{1}}$ although not for C and $I_{\overline{2}}$. $M_{\overline{1}}$ has a smallanterior fovea, and a near +5 pattern. At this juncture it is difficult to attempt to judge possible affinities, but a fuller comparative study of the specimens is evidently warranted.

The greater Balkan area of southeastern Europe is still very poorly known in terms of its Pleistocene prehistory (as is also Turkey), and perhaps Bacho Kiro is a measure of the surprises that may be expected when further such studies are pursued there. This volume clearly indicates the substantial progress made in recent years toward a fuller understanding of the later stages of human evolution, and it also indicates, in a number of important ways, the need for continued and intensified efforts in this fascinating aspect of paleoanthropology.

The Origins of Modern Humans: A World Survey of the Fossil Evidence, pages 1–49
© 1984 Alan R. Liss, Inc., 150 Fifth Avenue, New York, NY 10011

The Neandertals and Their Evolutionary Significance: A Brief Historical Survey

Frank Spencer

Department of Anthropology, Queens College of the City University of New York, Flushing, New York 11367

Since the discovery of the Feldhofer Cave skeletal remains in the Neander Valley, Germany, in 1856, there has been an ongoing debate regarding the significance of these and similar fossils discovered in Europe and other parts of the Old World. The purpose of this chapter is to present a brief historical survey of this 125-year debate in order that some of the arguments and divergent opinions, as well as differences in theoretical approach, presented in this volume will be better understood, particularly by the nonspecialist reader.

THE DISCOVERY OF A FOSSIL MAN IN THE NEANDER VALLEY, GERMANY

During quarrying operations in the Neander Valley, Germany, in August 1856, a cave deposit was disturbed, revealing the remains of what appeared to be a fossilized human skeleton. The cave (Feldhofer Grotto) where these remains were found, was one of several limestone grottos in this valley named after Joachim Neander (1650–1680), a 17th-century poet and composer for the German Reformed Church, who apparently made frequent excursions to the region to contemplate and work. By 1856, however, the poet's "leafy nook" had been essentially destroyed by the relentless quarrying of limestone for the Prussian construction industry. Although the exact circumstances relating to this discovery have been obscured by time, it appears that the fossilized human remains, consisting of a skullcap and an assortment of postcranial bones (most of which were incomplete), were subsequently gathered up and delivered into the care of Carl Fuhlrott (1804–1877), a local mathematics teacher and amateur natural historian.

On examining the remains, Fuhlrott found a number of peculiar anatomical features that clearly placed the specimen outside the then-known limits of human variation. Of particular interest to him was the skullcap, which was unusually thick, with exaggerated eyebrow ridges and a low, sloping forehead. These and other aspects of the skull all testified to its owner's once rude and robust appearance. Excited by the prospect that the bones might in fact represent a primitive form of humanity that had once stalked the Neander Valley, Fuhlrott resisted the temptation of placing the bones in his own collection, and decided instead to seek a more informed scientific opinion, namely that of Hermann Schaaffhausen (1816–1893), then professor of anatomy at the University of Bonn. This decision was evidently influenced by the fact that only 3 years earlier, Schaaffhausen had publicly rejected the theoretical implications of Catastrophism, proclaiming that:

> . . . living plants and animals are not separated from the extinct by new creations, but are to be regarded as their descendants through continued reproduction (Schaaffhausen 1853, cited in Darwin, 1872 [1976]: xxiv).

Impressed by what he was shown, Schaaffhausen [1857a] made a preliminary report on the Feldhofer cranium at a meeting of the Lower Rhine Medical and Natural History Society of Bonn on February 4, 1857. Later that year Schaaffhausen [1857b] and Fuhlrott [1865, 1859] gave a full account of the discovery at the general meeting in Bonn of the Natural History Society of the Prussian Rhineland and Westphalia. Consistent with his evolutionary beliefs, Schaaffhausen argued that the large frontal eminences of the human fossil were not a pathological deformity, but rather a character that was in all probability typical of the "barbarous aboriginal race" to which it had once belonged. In summary, Schaaffhausen's conclusions [see also Fuhlrott, 1859, 1865] were:

> First, the extraordinary form of the skull was due to a natural conformation, hitherto not known to exist even in the most barbarous race. Second, these remarkable human remains belonged to a period antecedent to the time of the Celts and Germans, and were in all probability derived from one of the wild races of northwestern Europe, spoken of by Latin writers, and which were encountered as autochthones by the German immigrants. And third, it was beyond doubt that these human relics were traceable to a period at which the latest animals of the Diluvium still existed . . . [Schaaffhausen, 1859b (English translation by Busk, 1861: pp 171, 172)].

THE INCIPIENT DEBATE ON THE SIGNIFICANCE OF THE FELDHOFER FOSSIL

In the 1850s, particularly on the European Continent, the majority of workers with an interest in paleontology and geology belonged to the so-called school of Catastrophism, which had been essentially the brainchild of the French paleontologist Georges Cuvier (1769–1832). By contrast, across the English Channel, the Uniformitarian school, as articulated first by James Hutton (1726–1797) and then Charles Lyell (1797–1875), was far more popular. To some extent the apparent differences between these two schools of thought reveal a fundamental distinction between the religious and cultural milieux that had spawned them. By its very nature, British Protestantism was far more liberal than Continental (and particularly French) Roman Catholicism, a fact clearly manifested in the rearguard action of Cuvier's disciples against the work of such men as Boucher de Perthes (see below), and more particularly Darwin [Glick, 1974; Brace, 1981].

From his studies in geology as well as comparative anatomy and paleontology, it had seemed to Cuvier that there had been a succession of geological epochs, each populated by a unique suite of animals and plants. This, Cuvier claimed, was clearly demonstrated by the different fossils found in the various strata of the geological column. Cuvier interpreted this to mean that the earth's history had been punctuated by a series of violent terrestrial "revolutions," during which all living things were destroyed. Through divine intervention the succeeding epoch was supplied with a new inventory of plants and animals. Just as institutions of the old regime in France had been swept away by revolution and replaced by new ones, so it appeared to Cuvier that the same had occurred in nature. Most, if not all animal fossils were, so Cuvier contended, extinct forms; fossils of modern forms were comparatively rare, and considered to be of little or no importance. To Cuvier, each species was the result of a Special Creation and exclusive to its particular geologic epoch—which included man. As a consequence he was unwilling to accept the proposition of early evolutionary thinkers such as Jean Baptiste Lamarck (1744–1829) that a similarity in structure between fossil and living forms supported the idea of continuity and phylogenetic affinity. Constrained by the religious belief that anthropogenesis had occurred a little more than 6,000 years ago, Cuvier had viewed the proposition of antediluvian man with prudent scepticism [Cartialhac, 1884]. Apparently, what many of Cuvier's followers found particularly attractive about his nonevolutionary scheme, in which species were portrayed as immutable entities, was that on the one hand it fortified the Linnean doctrine of a fixed hierarchical order in nature, while satisfying on the other the eschatological requirements that were implicit in this old typological concept.

The British school of Uniformitarianism, whose influence began to burgeon after the publication of Lyell's book *Principles of Geology* (1830–1833), rejected the Cuverian notion of catastrophic cycles. According to this school of thought, the world had from its creation been subjected to gradual modification by the action of natural agencies. Largely indifferent to religious doctrine, the Uniformitarian geologists enthusiastically assigned a great antiquity to the earth (now measured in terms of millions of years, rather than thousands). But while these ideas played an important role in setting the stage for both the eventual arrival of the Darwinian synthesis and the "new geology," it should be noted that Progressionist thinkers tended to assert that the paleontological record displayed a trend toward ever-increasing biotic complexity, contrary to Lyell's position that all life forms, both simple and complex, had come into existence simultaneously. Early in his career, Lyell had rejected Lamarck's vitalistic theory of transformism, largely because it brought man too close to brute creation to suit his traditional, theological view of the uniqueness of man. In the realm of geology, however, he was instrumental in removing from the natural history of the earth the implications of Catastrophist supernaturalism. Furthermore, the Uniformitarian concepts of progressive change and continuity of process did much, as suggested above, to support the notion of gradual organic evolution that had already begun to surface in the minds of many naturalists such as Robert Chambers (1802–1871), who in 1844 published anonymously, under the title *The Vestiges of the Natural History of Creation*, a short treatise anticipating much of what Darwin was to propound in the next decade [Gillespie, 1979, pp 88–91].

Traditionally, the terms "diluvium" and "antediluvium" served to separate the Recent period from ancient times, but they lost their meaning as the findings of the new geology's study of glacial mechanics and chronology began to accumulate. In 1856, the significance of the new geology was only partially perceived, particularly on the continent where Cuvier's ideas still continued to dominate paleontological and geological thinking. As such, in the absence of reliable and accurate dating methods, even the division of "postdiluvian" archaeological artefacts into a simple three-stage sequential model, proposed in the mid 1830s by Danish antiquarians based on technico-cultural developments, namely Stone, Bronze, and Iron, conveyed no indication of the chronological age of these three cultural stages [see Daniel, 1943].

Prior to the discovery of the Feldhofer remains, evidence supporting the existence of antediluvian man had been based essentially on the discovery of crude stone implements in reportedly "diluvial" deposits made by Boucher de Perthes (1788–1868) in and around his home at Abbeville in northwestern France [Aufrère, 1936]. While initially shunned by the French scientific establishment, Boucher de (Crèvecoeur) Perthes' (1788–1868) work [1847] became subsequently a crucial armament in the evolutionists' arsenal.

As this indicates, the idea of prehistory preceded the establishment of the antiquity of man. The Danish three-stage model recognized the fact that human activity had antedated the historic period in Europe. But until the discovery of human fossils in a glacial context, the problem of human prehistory had been confined to a consideration of what was then generally perceived as the post-paleontological period or Recent geological epoch. Although the term "Pleistocene" had been introduced by Lyell as early as 1839 to describe the paleontological period following the Pliocene, it did not come into general use until the last quarter of the century. Up to then, British workers tended to favor the term "post-Pliocene," while in France the term generally used to describe the post-Pliocene period was "Quaternary." Eventually, the definition of this latter term was expanded to embrace both the Pleistocene and the Recent period. By contrast, the Germans favored the term "Diluvium," which they continued to use well into the 20th century.

Following the discovery in the Neander Valley, the French paleontologist Edouard Lartet (1801–1871), in 1860, attempted to subdivide the Pleistocene based on faunal changes: the early Mammoth Age, the late Mammoth Age, and the Reindeer Age. Shortly thereafter, Lartet attempted a cultural subdivision of the Pleistocene. Here he recognized three major cultural subdivisions: Mousterian, Solutrean, and Magdalenian. Later, in 1867, the French prehistorian Gabriel de Mortillet (1820–1893) elaborated on Lartet's cultural sequence, recognizing five major subdivisions, beginning with the Chellean, followed by the Mousterian, Aurignacian, Solutrean, and terminating with the Magdalenian. Finally, one other innovation important to the developing understanding of man's antiquity was Sir John Lubbock's (1834–1913) proposal, made in 1865, to expand the Danish system by dividing the "Stone Age" into the "Palaeolithic" and "Neolithic." Following the adoption of this proposal, the Paleolithic-Neolithic cultural boundary was generally considered by most workers to coincide with the geologic Pleistocene-Recent boundary.

Although there had been a number of claims of having discovered the skeletal remains of antediluvian man made prior to 1856, they had all been rejected as unworthy candidates. From all indications, the earliest such claim had been made at the beginning of the 18th century by the Swiss naturalist and physician Johann Scheuchzer (1672–1733), who in 1726 reported finding "A Human Witness of the Deluge and divine Messenger" (Homo Diluvii Testis et ΘΕΟΣΚΟΠΣ). It was subsequently shown, however, that Scheuchzer's diluvian fossil was nothing more than a giant Miocene salamander. During this same time period a fragment of a human cranium (consisting of an incomplete frontal bone and the anterior half of the right parietal) was seemingly discovered along with the remains of extinct mammalian fauna at Cannstatt, near Stuttgart, Germany, circa 1700[1]. Apparently Cuvier examined this fossil and rejected the specimen's possible antiquity [Boule, 1923, p 6]. Later, however, Quatrefages and Hamy [1882] selected the Cannstatt skull fragment as the prototype of their "early" fossil race (see later). Similarly, due to Cuvier's influence, the antiquity was also denied to the skeletal remains discovered by the Belgian paleontologist P.C. Schmerling (1791-1836) in the Engis and Engihoul caves, situated in the limestone cliffs that border the River Meuse in the province of Liege, Belgium [eg, Lyell, 1863, p 68].

While many nonevolutionists, such as Cuvier, did not deny the possible existence of antediluvian man, it is evident that their expectations, regarding the morphological appearance of such material, were significantly different from workers favorably disposed toward the concept of evolutionism. Among the early evolutionists, especially after the publication of Darwin's *Origin of Species* in 1859, there was an eagerness to demonstrate a phylogenetic relationship between fossils, such as the Feldhofer specimen, and the anthropoid apes, as well as an emerging notion that there was a general equation between morphology and chronology. The "pithecoid" or "simianlike" character of fossil man was both required and predicted by human evolutionary theory. But having said this, it is possible to find nonevolutionists, such as Lyell, employing terms such as "apelike" to describe human fossils, which to the unsuspecting reader could be misconstrued to imply a recognized phylogenetic relationship with the anthropoid apes. Such terms were employed by the nonevolutionist in a strictly descriptive sense only. Acknowledging the apparent fact that the acquisition of civilization had been preceded by a state of savagery and barbarism, terms such as "pithecoid," "apelike," and "simious" were essentially interchangeable with those

---

[1]The history of these human remains is obscure. For instance the first account of fossil material found at Cannstatt, Wurtemburg, made by David Špleiss in 1701 (Oedipus osteolithologicus sen Dissertatio historico-physica de cornibus et ossibus fossilibus canstadiensis, in duas partes divisa. Scaphusiensemi: Apud J. Rudolphum) makes no mention of the Cannstatt skull. However, as noted in the text, Cuvier (c 1812) is reported having examined human bones presumably excavated along with the remains of extinct mammals. Similarly, in 1839, the German paleontologist, Georg Jaeger reported on the human skull fragments evidently present in this collection. For further details on the history of the Cannstatt skull see Buchner [1894] and Obermaier [1905].

of "savagery," "barbarous," "inferior," and "primitive." Accordingly, the nonevolutionists' perception of an ancient, inferior race of men appears to have been a peculiar blend of contemporary ethnological accounts of "primitive" hunters and gatherers (such as the Hottentot bushmen or perhaps even the natives of Tierra del Fuego, so vividly described by Darwin [1839] in his narrative of the voyage of HMS Beagle) and the legend of "*Magnus homo agrestis*" ("large wild men"). In a virtually uninterrupted continuum, the idea of "*homines agrestes*" had been perpetuated from classical times through the Middle Ages to the close of the 18th century. Early references to this myth can be found in the works of Virgil and Ovid. For example, in the *Aeneid*, Virgil speaks of a "race of men born of tree trunks and hard oak who had neither a rule of life nor civilization," and "who fed themselves from trees and the rough fare of the huntsman." Similarly, Ovid in his *Ars Amatoria*, refers to these wild men as having "merae vires et rude corpus": "strength unabated and their bodies rough," which, so it was explained, rendered them fit to live in the woods, sleeping on beds of leaves and subsisting on herbs.

Although Schaaffhausen's characterization of the Feldhofer remains as "one of the wild races of northwestern Europe, spoken of by the Latin writers" [Schaaffhausen, 1858, p 453], does not appear to be far removed from the above theme, his allusions to the specimen's apelike affinities were clearly perceived from an evolutionary standpoint [Schaaffhausen, 1858, pp 458–459]. Unlike many of his contemporaries, who viewed animal species as static entities, forming an uninterrupted chain of anatomical development—from the simplest to the most complex (a concept known as the Scale of Nature or the Great Chain of Being), Schaaffhausen saw this scale in dynamic terms, being convinced that one form of life could give rise to another. In other words, the links in the chain were not fixed but capable of moving their position on the scale. Thus, given the close proximity of the anthropoid apes to man in this anatomical chain, Schaaffhausen had little difficulty in perceiving a gradual evolution of man from some ancient anthropoid ape. Applying this perspective to the examination of the Feldhofer specimen, Schaaffhausen felt that the "remarkable conformation of the forehead" closely approximated that of "the larger apes." In fact, "the only thing human about the skullcap," he later wrote, "was its size and cranial capacity" [Schaaffhausen, 1873, p 454]. And from all indications, it was this particular feature that gave him reason to pause and consider the remains as those of a "primitive race" rather than a "primitive species" of mankind.

In 1860, Charles Lyell visited the Feldhofer Cave in the company of Fuhlrott to gather firsthand information on the discovery for his book, *The*

Geological Evidences of Antiquity of Man, published in 1863. Having examined all of the available evidence, Lyell arrived at essentially the same conclusion as Fuhlrott [1859] and Schaaffhausen [1857a,b, 1858, 1861], namely that it was "probable" that the Neandertal remains were ancient. But he cautioned: "As no other animal remains were found with it, there is no proof that it may not be newer. Its position lends no countenance whatever to the supposition of its being more ancient" [Lyell, 1863, pp 77–78].

On returning to England, Lyell invited Thomas H. Huxley (1825–1895), an anatomist who later became Curator of the Hunterian Collection at the Royal College of Surgeons, London, to prepare a detailed study of the casts given him by Fuhlrott. Thus, Huxley's first detailed discussion of the Feldhofer remains appeared in Lyell's *Antiquity of Man* [1863, pp 80–89]. An expanded version of this paper is included in Huxley's own classic work: *Evidence as to Man's Place in Nature*, published that same year. But where Lyell's book is one long evasion of Darwinian evolution, Huxley's text was designed to demonstrate that, zoologically, man is a primate, and that the presumed chasm between man and the apes had been grossly exaggerated.

However, Huxley felt that the Feldhofer fossil could not by any stretch of the imagination be regarded as a phylogenetic link between "Man and the apes." Comparing the skullcap with the Engis skull and a series of Neolithic and modern crania, Huxley concluded:

> . . . though truly the most pithecoid of known skulls, the Neanderthal cranium is by no means so isolated as it appears to be at first, but forms, in reality, the extreme term of a series leading gradually from it to the highest and best developed of human crania. On the one hand, it is closely approached by the flattened Australian skulls . . . from which other Australian forms lead us gradually up to skulls having very much the type of the Engis cranium. And, on the other hand, it is even more closely affined to the skulls of certain ancient people [Borreby tumuli] who inhabited Denmark during the "stone period" [Huxley, 1863, pp 182–183].

To which he added prophetically:

> . . . I may say, that the fossil remains . . . do not take us appreciably nearer to that lower pithecoid form, by the modification of which he has, probably, become what he is . . . Where, then must we look for primeval Man? Was the oldest *Homo sapiens* pliocene or miocene, or yet more ancient? In still older strata do the fossilized bones of an Ape

more anthropoid, or a Man more pithecoid, than any yet known await the researches of some unborn palaeontologist? [Huxley, 1863, pp 183–184].

Concurring with this diagnosis, the zoologist and scion of German Naturphilosophie at the University of Jena, Ernst Haeckel (1834–1919), suggested, first in 1866 and then more specifically in 1868, that in all probability "Homo primigenius" (a term he used to denote primeval man) had evolved from a preglacial member of the anthropomorphous primates outside of Europe, somewhere in either "Lemuria" (south Asia) or Africa. "Considering the extraordinary resemblance between the lowest woolly-haired men, and the highest man-like apes," Haeckel said:

. . . it requires but a slight imagination to conceive an intermediate form connecting the two . . . The form of their skull was probably very long, with slanting teeth; their hair woolly; the colour of their skin dark; of brownish tint. The hair covering the whole body was probably thicker than that of any of the still living human species; their arms comparatively longer and stronger; their legs, on the other hand, knock-kneed, shorter and thinner, with entirely under-developed calves; their walk but half-erect [Haeckel, 1868, cited in McCown and Kennedy, 1972, p 143].

To this phantom-construct, a blend of pongid and hominid traits, Haeckel later gave the name *Pithecanthropus alalus*: speechless ape-man [Haeckel, 1870, p 590].

The relevance of Haeckel's prediction to the subject at hand is that it was to lead the Dutch physician Eugene Dubois (1858–1941) to search for and subsequently discover just such a fossil at Trinil, Java, in 1890–1891. By then the existence of a Pleistocene Neandertal race, characterized by a suite of distinct anatomical traits, had been fully recognized, and it was just a matter of time before the Strassburg anatomist Gustav Schwalbe (1844–1917), proposed a simple phyletic scheme in which the Neandertals were portrayed as an intermediary form between the newly discovered *Pithecanthropus* and modern *Homo sapiens*.

THE NEANDERTALS AND THE ARYAN MYTH

As it appeared to the majority of Huxley's contemporaries, the problem of the Feldhofer specimen was primarily one of determining its racial affinities and genealogical relationship with modern European racial types. Although there were a number of workers during the mid 1860s and early 1870s who considered the fossil as either a recent [eg, Wagner, 1865; Mayer, 1864] or pathological specimen [Blake, 1862; Mayer, 1864; Prunner-Bey, 1864; Lucae, 1864; Virchow, 1872], the consensus of those investigating this fossil during this period was that it represented an ancient, inferior race of *Homo sapiens*. At this juncture no one seems to have taken seriously the opinions of William King (1809–1886), then professor of geology at Queen's College, Galway, Ireland, that the features of the Feldhofer remains placed the specimen beyond the limits of *Homo sapiens*, and proposed that it be placed in a separate taxon labeled *Homo neanderthalensis* [King, 1864a,b].

As Holtzman [1970] has shown in considerable detail, the initial recognition and characterization of the Neandertal type was not made in an evolutionary context, but rather in a racial-succession paradigm developed primarily by antievolutionist French workers during the third quarter of the 19th century. This paradigm was based on the notion that the autochthonous populations of Europe had on successive occasions during the "post-Pliocene period" been either replaced or mixed with invading peoples (the so-called Aryans) originating from somewhere in Asia. Initial justification for such a proposition appears to have been based primarily on early comparative philological research by such workers as William Jones (1746–1794) and James Burnett (1714–1799), who, in demonstrating that Sanskrit belonged to the Indo-European linguistic stock, opened the door to the philological hypothesis of the existence of a primitive "Aryan" language from which all Indo-European languages had been derived. The general acceptance of this hypothesis, however, had no foundation other than such aphorisms as "ex oriente lux" put forward by August F. Pott (1802–1887), or "the irresistible impulse towards the west" invented by Jacob Grimm (1785–1863).

The transposition of the "Aryan myth" into the realm of physical anthropology appears to have been first made by the Swedish anatomist Anders A. Retzius (1796–1860), who, in 1842, argued that the autochthones of Europe had been brachycephals and spoke languages quite unrelated to any of the Indo-European languages. The probable descendants of these aboriginal paleolithic races were the Basques, Finns, and Lapps, all of whom, as Retzius well knew and demonstrated, spoke languages evidently not derived from the Indo-European linguistic stock and were brachycephalic. These ancient brachycephals, Retzius contended, preceded the invading dolichocephalic, Indo-European, or Aryan race, who brought with them ancestral European languages and the use of iron into the West.

During the 1860s and 1870s, Retzius' theory of "ancient brachycephals" was the subject of considerable debate, particularly in France among members of the newly founded Société d'Anthropologie de Paris. Here, opinion was divided between workers supporting the view that the autochthones of Europe had been brachycephalic and those who argued that the aboriginal cranial type had been dolichocephalic. The importance of this controversy is that it prompted many of these investigators, in an effort to support their respective views, to examine all of the available cranial material then attributed to ancient man in Europe (Table I). Thus, as the debate progressed these various fossils were shuffled and reshuffled to fit a particular theoretical position, which resulted in the subsequent association between the Feldhofer skullcap and the newly discovered Naulette jaw. This association was made in 1873, when Jean Louis Armand de Quatrefages (1810–1892) and Jules Hamy (1842–1908), his assistant at the Museum d'Histoire Naturelle, Paris, presented their argument for an autochthonous, dolichocephalic race of Europe. This race, dubbed the "Cannstatt race," and assigned to the "earliest age of the Quaternary epoch," was fabricated from a number of fossils, including the remains from the Feldhofer Cave, Gibraltar, Cannstatt, Brüx, Eguisheim, and Naulette. The anatomical features characterizing the race, namely prominent brow ridges and a low, sloping forehead, were considered to be typically represented in the Cannstatt specimen. By contrast, the Feldhofer specimen represented to Quatrefages

TABLE I. Some Skeletal Remains Important to the Recognition of the Neandertal Type (Listed in Chronological Order of Discovery)

Fossil	Location	Date discovered	Described by:
Cannstatt	Germany	c 1700	Jaeger [1839], Quatrefages [1973], Quatrefages and Hamy [1882]
Engis[a]	Belgium	1830–1832	Schmerling [1833–1834]
Gibraltar[a]	Rock of Gibraltar	1848	Busk and Falconer [1865]
Feldhofer	Germany	1856	Schaafhausen [1857a,b, 1858] Fuhlrott [1859]
Olmo	Italy	1863	Cocchi [1867]
Moulin Quignon	France	1863–1864	Quatrefages [1873]
Eguisheim	Alsace-Lorrain	1865	Faudel [1867]
Naulette[a]	Belgium	1866	Dupont [1866]
Crô-Magnon[a]	France	1868	Lartet and Christy [1865–1875]
Brüx	Bohemia (Czechoslovakia)	1871	von Luschan [1873]
Spy[a]	Belgium	1886	Fraipont and Lohest [1886, 1887]

[a]For further details and discussion of these specimens see Stringer et al, this volume.

an "exaggeration" of the racial type. The Brüx skullcap, on the other hand, was considered intermediate between that of the Cannstatt and Eguisheim specimens, while the Gibraltar fossil, being the only complete skull in the series, provided an idea of the face of the Cannstatt race, and unlike the Feldhofer specimen, was considered a less extreme representative of the race [see Quatrefages, 1873; Quatrefages and Hamy, 1882].

While seemingly well-received by the majority of French workers, in Germany the Quatrefages-Hamy synthesis was severely criticized by the eminent Berlin pathologist and anthropologist Rudolph Virchow (1821–1902). Unlike Quatrefages, who was a monogenist, Virchow tended to a polygenist point of view and the notion that major racial types were fixed entities. As such, Virchow could not see how Quatrefages and Hamy had been able to derive racial types from seemingly isolated and pathological specimens [Virchow, 1874, cited in Lartet and Chaplain-Duparc, 1876, pp 327–328]. And furthermore, he expressed some doubt about the claimed antiquity for the Feldhofer specimen, because of suture closure, which Virchow took to indicate an advanced age. He felt no one could live to such an old age in a hunting and gathering society, and therefore must have belonged to a sedentary group, and since a sedentary way of life was considered to be a comparatively recent event, it clearly invalidated the great age claimed for the specimen [Virchow, 1872, p 163]. Despite their differences on this issue and the all important question of human origins, both Virchow and Quatrefages were clearly united in opposition to the idea that humankind were derived from the apes.

RECOGNITION OF THE NEANDERTAL TYPE
AND SCHWALBE'S SYNTHESIS

So long as the Feldhofer Neandertal was the only example of its kind, its true evolutionary significance remained obscured and open to question. The primary issue, throughout the 1870s, was the great antiquity assigned to this specimen, as well as other remains constituting the race of Cannstatt. However, any doubts there may have been in this regard were quickly dispelled in 1886 by the discovery of two human skeletons at the mouth of a cave at Spy, in the Belgian Province of Namur. According to Marcel de Puydt and Max Lohest, both geologists, the Spy skeletons had been found by them in direct association with the remains of mid-Quaternary fauna and stone tools, which showed affinities with a Paleolithic industry known as the Mousterian [De Puydt and Lohest, 1886a,b; Fraipont and Lohest, 1886, pp 692, 741–784].

Beside providing incontrovertible evidence for the claimed antiquity of Paleolithic man, the discovery at Spy also led to the eventual recognition of the Feldhofer skullcap as the type specimen of a group of humanity, later known as the Neandertals, that had prevailed in Europe during the early Upper Pleistocene. In this regard, the Belgian anatomist, Julien Fraipont (1857–1910), who had been invited to examine the Spy remains, noted in his report to the Belgian Royal Academy that while their general appearance conformed to that of the Cannstatt type, he felt that in many specifics the crania were "eminently Neanderthaloid" [Fraipont and Lohest, 1887], and he went on to provide what appears to have been the first description of the European Neandertals. According to Fraipont, this primitive form of humanity was short of stature but powerfully built, with strong and curiously curved thigh bones, which seemingly must have required this "bestial form of humanity" to walk with a bend at the knees. Their skulls were long and depressed with very strong eyebrow ridges and a heavy face with a lower jaw of "brutal depth and solidity" that sloped away downward and backward from the teeth. But while emphasizing the brutish and one might say "pithecoid" features of the Spy specimens, Fraipont had to confess, just as Huxley had 20 years earlier, that this ancient form of humanity did little to narrow the distance between man and ape.

Another problem confronting late 19th century human evolutionists was the incipient argument for the relative stability of the human form. From accumulating skeletal evidence it appeared as if the modern human skeleton extended far back in time, an apparent fact which led many workers to either abandon or modify their views on human evolution. One such apostate was Alfred Russel Wallace (1823–1913). In 1887, Wallace examined the evidence for early man in the New World, and like the German anatomist Julian Kollman (1834–1918), who three years earlier had made a similar survey, found not only considerable evidence of antiquity for the available specimens, but also a continuity of type through time. In an effort to explain this, Wallace [1889, pp 454–461] suggested that once man had become morphologically differentiated from his apish kin (during the mid-Tertiary period), he had remained physically stable. To justify this opinion, he argued that with the emergence of the human brain, man, through culture, had been essentially partitioned from the vagaries of natural selection and was, thereby, a unique creation of the biotic realm.

This argument, however, lost some of its potency as well as a few supporters when news began circulating of the discovery of a remarkable hominid fossil in Java that displayed many of the features predicted by Haeckel. According to the first published accounts, this fossil, fittingly dubbed *Pithecanthropus erectus* by its discoverer Eugene Dubois, had been found in a geological stratum containing fauna placing it at the interface of the Pliocene and Pleistocene [Dubois, 1894, 1895, 1896].

Immediate differences of interpretation arose over whether Dubois' specimen was a fossil ape [Branco, 1898; Kate, 1894; Klaatsch, 1899b; Kollman, 1895; Virchow, 1895; Volz, 1896; Waldeyer, 1895], a transitional ape-man form [Cunningham, 1895; Haeckel, 1898; Manouvrier, 1895; Marsh, 1895; Schwalbe, 1899a; Sollas, 1895; Verneau, 1895], or even *H sapiens* [Keith, 1895; Turner, 1895]. If its geology was accepted along with the interpretation that it represented an incipient form of humanity, then this evidence placed a considerable burden upon those workers like Wallace and Virchow who claimed the great antiquity of the modern human form.

Under the stimulus of renewed interest in human paleontology awakened by Dubois' discovery, Gustav Schwalbe embarked on an elaborate study and re-evaluation of the fossils attributed to the Cannstatt-Neandertal race by Quatrefages and Hamy in the early 1870s. From this protracted study, Schwalbe [1897, 1899b, 1900, 1901, 1904, 1906a,b] concluded that the Feldhofer skullcap and Spy specimens represented the remnants of a hominid group so far removed from all existing varieties of humanity as to warrant the rank of a distinct species, named by him *"Homo primigenius"*; all other fossil material previously assigned to the same group were removed. This claim for a new taxon, Schwalbe contended, was supported by the numerous pithecoid features manifest in Neandertal crania. Unfortunately, Schwalbe made no effort to inform his readers of the sense in which he employed the term "species," a fact that subsequently led to considerable confusion and misunderstanding in the minds of most scholars of the period, [eg, Sollas, 1908; Keith, 1915; Boule, 1921; Osborn, 1915a,b, 1927a,b; Burkitt, 1921; MacCurdy, 1924; Elliot Smith, 1928]. This fact was recognized by Hrdlička, who later noted in his 1927 attack on the presapiens models that "all these opinions can probably be traced, directly or indirectly, to the authoritative notions arrived at . . . by one of the foremost students of Neanderthal man, Gustav Schwalbe" [Hrdlička, 1927, p 250]. Although Schwalbe wrote that the majority of these features were intermediate between those of modern humans and the anthropoid apes, he confessed that his efforts to determine the degree of this intermediateness had given unsatisfactory results [Schwalbe, 1900, pp 57–58]. But be this as it may, on the basis of this "evidence," Schwalbe proposed two possible arrangements of the then-known pithecanthropine *(H erectus)*, Neandertal, and modern *H sapiens* fossils [see Schwalbe, 1906a, p 14]. The first, for which he declared his preference, was a simple phyletic scheme in which the Neandertals were portrayed as an intermediary link between Dubois' *Pithecanthropus* and modern humans, while the second depicted the Neandertals and *Pithecanthropus* as offshoots from the human lineage. Later, in the light of Marcellin Boule's evaluation of the La Chapelle skeleton and the Piltdown discovery, Schwalbe [1913] expressed some doubt in his earlier contention that Neandertals had been the precursor of modern H *sapiens*.

THE EMERGENCE OF THE PRESAPIENS THEORY AS A THEORETICAL ALTERNATIVE TO SCHWALBE'S HYPOTHESIS

Initially, as might be expected, Schwalbe's unilineal scheme was greeted with some satisfaction by Haeckel, but elsewhere in Germany it gathered little or no support. Indeed, the Breslau anatomist Hermann Klaatsch (1863–1916) argued that any attempt to establish a specimen's (such as the Feldhofer skullcap) remote ape-ancestry by searching for pithecoid characters was a fool's errand, since even the earliest of these fossils were already strongly stamped with the modern human morphotype [Klaatsch, 1899a, p 356, 1900, pp 492–493]. Shortly thereafter, however, Klaatsch retracted this view by claiming that it was possible to trace the origin of modern human racial groups to a particular member of either the Asiatic or African great apes. According to his theory the modern African Negro had evolved from the European Neandertal, which in turn had been derived from an African "gorilloid" ape, while the Caucasian stock, represented in the fossil record by the Aurignacian skeletons of Combe Capelle, had in his opinion been derived from an "Aurignac invader" from Asia that had evolved from the orang-outang [Klaatsch, 1910, p 567]. This scheme was one of several polyphyletic theories (a modified polygenism) developed by a number of German workers such as Vogt [1865], Schaaffhausen [1866], and Virchow [1872] during the third quarter of the 19th century, and serves to characterize the distinctly Romantic ethos of German paleoanthropology prior to World War I.

In England, by contrast, Schwalbe's scheme was initially adopted with qualified approval by the Oxford geologist W.J. Sollas (1849–1936) [1908] and by Arthur Keith (1866–1955) [1911], then curator of the Hunterian Collection at the Royal College of Surgeons, London. Between 1910 and 1912, however, Keith had a change of heart. To some extent this shift in Keith's position was not entirely unexpected. From the outset he had been favorably disposed to the notion of the great antiquity of the modern human form [see Keith, 1911, pp 142–146]. He believed that anything so unique as the modern human skeletal configuration must have required an

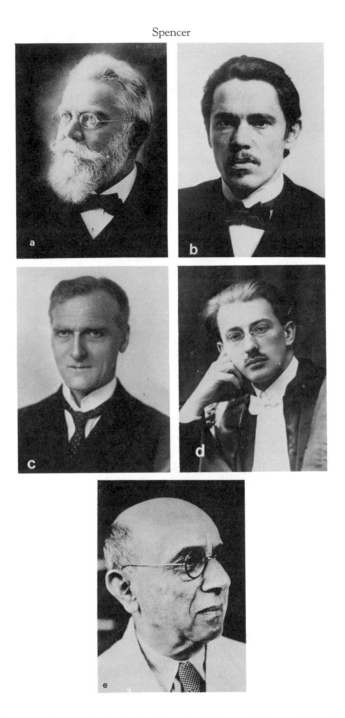

Fig. 1. Some prominent scientists in the Neandertal debate: (a) Gustav Schwalbe, (b) Aleš Hrdlička, (c) Sir Arthur Keith, (d) Henri Vallois, (e) Franz Weidenreich.

immense period of time to evolve. To accommodate this need, Keith proposed that the modern form had appeared at the commencement of the Holocene, some 200 ky ago by his estimation. His time-framework of the Pleistocene had had a duration of some 500 ky, a period dominated by the Neandertals. Thus, with an envisaged stretch of some 700 ky, Keith evidently felt reasonably comfortable in embracing Schwalbe's proposal that the Neandertals had been the precursors of modern human populations [Keith, 1911, pp 78–79, 99–100, 118–119, 139].

At this juncture it should be noted that, although it was generally accepted that the Pleistocene had involved a number of glacial cycles, the duration of these glaciations was still far from clear. According to the Scottish geologist James Geikie (1839–1915) in his work *The Great Ice Age* [1894], the European Pleistocene was characterized by four major glacial advances, punctuated by three interglacial phases. A similar conclusion had been reached by Albrecht Penck (1858–1945), director of the Berlin Institute for Oceanography and Geography. From his study of alpine glacial changes, Penck estimated that the duration of the Pleistocene had been between 500 ky and 1,500 ky [see Penck, 1908]. His estimates were not, however, generally accepted. For example, Sollas [1900, p 481] considered this to be too long. Based on the apparent fact that deposits laid down during the Pleistocene had a general depth of approximately 4,000 feet, and that this formation had proceeded at the rate of a foot per century, Sollas concluded that the collective deposits of the Pleistocene must have taken about 400 ky to form. Another problem was that it was still far from clear to what extent Penck's alpine glacial-sequences (namely the "Günzien, Mindelien, Rissien, and Würmien") could be correlated with geological formations elsewhere.

From all indications Keith's subsequent rejection of Schwalbe's Neandertal hypothesis was greatly influenced by the discovery, in 1911, of a human skeleton in Ipswich (Suffolk), England, in a geological deposit considered to have been laid down during or shortly after the "period of maximum glaciation" [Keith, 1915, p 218]. The problem raised by the so-called Ipswich skeleton was akin, as Keith so aptly put it, to finding "a modern aeroplane in a church crypt which had been bricked up since the days of Queen Elizabeth" [Keith, 1911, p 143]. According to Keith this skeleton was in almost every essential detail modern in form and had been found in geological circumstances that in his mind left little doubt as to its great antiquity [Keith, 1912a]. The importance of this lay in the fact that it confirmed, so Keith believed, the validity of the Galley Hill specimen and his long-held suspicion of the great antiquity of the modern human form.

The Galley Hill specimen had been found in 1888 at a site not far from Swanscombe (Kent), England [Newton, 1895, p 505], in circumstances that seemingly confirmed its great antiquity, despite the fact its modern morphology was clearly at variance with the succession of evolutionary stages demanded by Schwalbe's hypothesis. Thus, for a while at least, the Galley Hill specimen was regarded by supporters of Schwalbe's evolutionary synthesis with some suspicion. Towards the end of the first decade of the 20th century, however, the Galley Hill remains were re-examined by a number of European workers, namely M.A. Rutot [1909], H. Klaatsch [1909] and V. Giuffrida-Ruggeri [1910], all of whom endorsed the authenticity of the specimen and suggested that it represented a form that had been either contemporaneous with or preceded the Neandertals in Europe. Interestingly enough, when these remains were shown in Paris in 1909 [Corner and Raymond, 1909, p 487; Giuffrida-Ruggeri, 1910, p 259], Marcellin Boule dismissed them as merely "bric-a-brac" [Keith, 1912a, p 308]. Keith thought otherwise, particularly in the light of the Ipswich discovery. He was convinced this evidence proved beyond doubt that England, if not elsewhere in Europe, had been populated during Mousterian times by "men of the modern type—not Neanderthaloid" [Keith, 1912a, p 309; see also 1912b,c].

At that time the only hominid remains that could be said to antedate the Ipswich skeleton in Europe, with any certainty, was the Heidelberg mandible, which had been found in a sand pit located on the outskirts of the village of Mauer, some 10 km southeast of Heidelberg, Germany. On the basis of fossil bones of *Elephas antiquus* and other extinct fauna found in the stratigraphical sequences at the Mauer sand pit, Otto Schoetensack (1850–1912), a lecturer in geology at Heidelberg University, had determined that the mandible, or *"Homo heidelbergensis"* as he referred to it, belonged to the lower Pleistocene, or specifically to the warm period between the first and second glacial stage of the European Pleistocene [Schoetensack, 1908, pp 1–67]. The robust and primitive-looking mandible was considered by Schoetensack and others [eg, Keith, 1911, pp 78, 79] to represent a precursor of the Neandertal mandibular form (established by the finds at La Naulette, Šipka, Spy, and Krapina—see Tables I and II). In a nutshell, the Heidelberg specimen had all the hallmarks of being the European equivalent of Dubois' *Pithecanthropus* [see Hrdlička, 1914, pp 510, 511]. Thus, since, in Keith's opinion, there was apparently not sufficient time separating the Ipswich skeleton from the Heidelberg specimen for the latter to have been the former's precursor, it seemed to him highly improbable that either "Heidelberg Man" or the Neandertals could have been ancestral to modern man as he originally contended.

Another factor influencing Keith's capitulation had been Gioacchino Sera's [1910a,b] study of the Gibraltar skull[2] and Marcellin Boule's initial pronouncements on the La Chapelle-aux-Saints skeleton, which had been discovered in 1908 (see below). In both cases, these workers advocated the view that the European Neandertals stood much closer to the anthropoid apes than they did to modern humanity. Hence in a series of lectures presented at the Royal College of Surgeons, London, between February 28 and March 8, 1912, Keith prepared the ground for the eventual removal of the Neandertals from the ancestry of modern man by proposing the idea of two parallel forms of humanity present during the Pleistocene:

> On the evidence at present available, it must be inferred that two types of man were in existence in Europe during the Pleistocene epoch: (1) the Neanderthal type, represented by the Heidelberg mandible, near the beginning of that epoch, and the various skeletons found in Belgium and France near its end; and (2) the modern type, represented by the remains of many races belonging to the inferior, middle and superior formations of the Pleistocene epoch [Keith, 1912b, p 155; see also 1912c, pp 734–736].

Meanwhile, in France, Marcellin Boule (1861–1942) was busily arriving at the same conclusion. Essentially, Boule's reaction to Schwalbe's unilineal scheme emerged in a series of reports published between 1909 and 1912 on a Neandertal skeleton found at La Chapelle-aux-Saints in the Dordogne region of France in 1908.

In his first reports (based on his original presentation to the French Academy of Science, Paris in 1908), published between 1908 and 1909 in

---

[2]Sera's study gave particular attention to the abrupt downward bending of the skull's base near its middle—a structural peculiarity that had long been known to be more highly developed in humans than in other primates. In studying the base of the Gibraltar skull, Sera reconstructed the profile of the cranial floor and compared it with the corresponding profile in the skull of an Australian aborigine—a living race he considered closely approximating Neandertals. He found that the angle at which the floor was bent differed significantly in two specimens. In the Gibraltar skull the angle was about 140°, while in the Australian it was only 112°. Since the greatest angle measured by Sera in a modern human skull was only 132°, he considered the difference of 28° to be significant. Summarizing the results of his study, Sera felt the Gibraltar skull demonstrated the fact that the acquisition of modern anatomical characters in hominid evolution had appeared quite late in a stock that had marked simian peculiarities. Moreover, he considered these traits of the skull base pointed to the stock's nearness to the chimpanzee and gorilla. In this regard, he noted the close proximity of Gibraltar to Africa, the home of these two apes.

L'Anthropologie, Boule presented his initial findings and tentative conclusions regarding the La Chapelle skeleton and the Neandertals in general. He noted that the La Chapelle skeleton confirmed that the skeletal configuration of the Neandertals was not an anomalous modern form, but rather an archaic population characterized by a distinct suite of cranial and postcranial morphological features. With regard to the cranial features he drew attention to the retreating forehead, prominent supraorbital torus, the large flattened braincase, massive jaw, and nearly absent chin. As for the postcranial features, Boule noted the elongated, nonretroverted and nonbifurcated cervical spinal processes, as well as the short and massive character of the vertebral column. To this he added the extensive curvature of the femora and retroversion of the tibae heads. In sum this indicated to Boule that the Neandertals would have been incapable of assuming a totally orthograde posture. He noted further that the entire skeleton was dominated by numerous other simianlike characters. Although acknowledging that the average cranial capacity of the Neandertals was larger than that of modern humans, he said that this fact was offset by the poor development of the cerebral lobes and the presence of a number of other primitive features. Thus, while the Neandertals were undoubtedly superior in cerebral quantity, they were, so Boule contended, qualitatively inferior [Boule and Anthony, 1911, pp 129–196].

Between 1911 and 1912, Boule elaborated on these findings in a series of communications extending through the sixth, seventh, and eighth volumes of the *Annales de Paléontologie*, and culminating in their republication as a large independent monograph in 1913. Here he made a meticulous descriptive study of the skeleton, comparing it with the anatomy of the apes, other Neandertals, and modern man. From this empirical base, Boule argued that there was an enormous morphological hiatus between Neandertals and modern humans, and concluded that the former were not a reasonable antecedent of the latter. Furthermore, guided by his earlier work in 1906 on the skeletons from Grotte de Enfants at Grimaldi, Boule felt that the temporal gap between modern humans and Neandertals was relatively small. Indeed, he felt there was every reason to suppose that the two populations had been contemporaneous, and that the "sapiens" lineage could be traced to an earlier period, independently of Neandertals.

Summarizing this work in a paper presented in September 1912 at the XIVth International Congress of Prehistoric Anthropology and Archaeology in Geneva, Boule [1914] proclaimed that the Neandertals were an archaic and extinct species, and advocated their immediate removal from the human phylogenetic tree. At the same time he also axed Dubois'

Pithecanthropus, declaring it to be nothing more than a giant gibbon. But if the Neandertals were merely an evolutionary cul-de-sac, when, where, and from what had the "sapiens" lineage evolved? In the absence of a suitable candidate Boule was obliged, for the moment at least, to leave this question in abeyance. Two months later, however, the discovery of the now-infamous Piltdown remains was announced, providing Boule with a possible precursor of the "presapiens" lineage and justification of the opinion that there had not been one, but two "races of man" living in the lower Pleistocene—one being the "Piltdown race," and the other being the "Heidelberg race." Of these two races, Boule said: "The Piltdown race seems to us the probable ancestor in the direct line of recent species of man, *Homo sapiens;* while the Heidelberg race may be considered, until we have further knowledge, as a possible forerunner of *"Homo neanderthalensis"* [Boule, 1913, pp 245-246].

THE PILTDOWN REMAINS AND THE PRESAPIENS THEORY

Details of the Piltdown discovery were unveiled to scientific scrutiny at a meeting of the Geological Society of London on the evening of December 18th, 1912 [Dawson and Smith Woodward, 1913, 1914]. At this time, Charles Dawson, a lawyer with an avid interest in archeology, reported that during the autumn of 1911 he had been searching for paleolithic implements at a gravel pit located on the outskirts of the Sussex village of Piltdown and had quite by "accident" stumbled upon several fragments of a thick, fossilized human skull[3] which he subsequently handed to his friend Arthur Smith Woodward (1864-1944), Keeper of Geology at the Museum of Natural History, Kensington, London, for examination. Impressed by the fragments, Smith Woodward had apparently urged Dawson to resume his search at Piltdown. The following summer, Dawson found several more cranial fragments, along with the right half of an imperfect lower jaw.

From these fragments Smith Woodward reconstructed the skull whose general conformation was quite remarkable, since it coupled an essentially

---

[3]During the summer of 1983, I had an opportunity to examine the original Piltdown remains thanks to the generosity of Dr. Chris Stringer at the British Museum. The skullcap is extraordinarily thick and, as Dr. Stringer noted during our discussion on this point, the perpetrator of the forgery must have had access to pathological material. From all appearances the thickened Piltdown skullcap may in fact have come from an individual suffering from Cushing's syndrome, which is a primary pituitary disorder having secondary effects on the adrenals. A common clinical feature of this syndrome is osteoporosis. As far as I know this fact has not been used as a possible clue in the identification of Dawson's co-conspirator(s) in the Piltdown forgery [see Spencer, in preparation and footnote 5].

modern braincase with an ape-like jaw. From this reconstruction, the anatomist Grafton Elliot Smith (1872–1937) had made an endocranial cast, which he said was "vastly superior" to that of Dubois' *Pithecanthropus* [Elliot Smith, 1913a–c]. Compounding this finding with the apparent geological antiquity of the specimen, Elliot Smith said there was little doubt in his mind that the Piltdown skull could be regarded as "the most primitive and most simian brain so far recorded," and he endorsed Smith Woodward's appellation: "Eoanthropus"—the Dawn Man.

Keith [1914a,b, and particularly, 1915, pp 337–355] was strongly opposed to Smith Woodward's reconstruction of the skull, which he contended was too small and apelike. Keith argued that if the two parietal bone fragments were properly restored and made approximately symmetrical, this would elevate the cranial capacity and thereby conform more accurately to the skull's patent modern morphology. Smith Woodward refused to budge, contending that his reconstruction was in perfect harmony with the creature's apelike jaw. In fact, in reconstructing the jaw, he had predicted that it would be chinless and possess a prominent and projecting canine. He felt such a prediction was not unreasonable given the specimen's intermediary relationship between the apes and *"Homo heidelbergensis."* This prediction was later confirmed by the fortuitous discovery of just such a tooth in the summer of 1913 at the Piltdown site by a young aspiring French paleontologist, Pierre Tielhard de Chardin.

With the notable exception of Keith, there was by 1915 a consensus in British anatomical circles regarding the reconstruction of the Piltdown braincase, as well as the phylogenetic significance of the find. Keith, troubled by the fact that the remains had been found in geological circumstances that seemingly corresponded closely with the geology of both Ipswich and Galley Hill, was obliged to reject the view that it represented the precursor of the "sapiens" lineage [eg, Lankester, 1915; Moir, 1915; Elliot Smith, 1916, 1917; Sollas, 1915; Smith Woodward, 1915]. Instead he [1915, p 509] proposed that Piltdown, along with the Neandertals, were extinct offshoots from the main evolutionary stem that led to the emergence of modern *H sapiens* during the middle Pliocene.

In the decade immediately following the outbreak of World War I, there was a general retreat from Schwalbe's Neandertal hypothesis to what later became known as the presapiens theory. According to this theory, there had been an ancestral split in the human lineage which led to the early appearance of a relatively modern skeletal form alongside a more archaic hominid, represented in the fossil record by the Neandertals. Although support for this scheme was invariably based on the Piltdown fossil [eg,

Broom, 1918; Burkitt, 1921; Giuffrida-Ruggeri, 1918; Lull, 1917; MacCurdy, 1924; Osborn, 1916, 1927a,b; Sollas, 1924; Todd, 1914a,b], both Boule and Keith, who were probably the most influential in promulgating this multilineal view of human evolution, based their respective arguments on quite different evidence.

In the case of Boule [1921], his arguments were essentially a theoretical expectation derived from his understanding of the "specialized" anatomy of the Neandertals, whereas Keith [1915, 1925] based his arguments for a presapiens lineage on the actual identification of early modern "ancestral sapiens"—namely such specimens as Galley Hill and Ipswich (and after 1936, the Swanscombe fossil, see Keith and McCown [1937]). But irrespective of this, it is clear that the Piltdown fossil presented a formidable obstacle to those workers such as Aleš Hrdlička (1869–1943), Rene Verneau (1852–1927), Franz Weidenreich (1873–1948), and Hans Weinert (1877–19??) who still continued to support some form of a basically unilineal concept of hominid evolution.

HRDLIČKA'S ATTEMPT TO REMOVE THE PILTDOWN OBSTACLE AND REINSTATE NEANDERTALS

From the outset, Aleš Hrdlička, then Curator of the Division of Physical Anthropology at the United States National Museum of Natural History, Washington, DC, had viewed with increasing scepticism Smith Woodward's views on Piltdown and the proposition that it represented a "mid-Tertiary precursor" of H sapiens. Besides there being some doubt about the reliability of Piltdown fossil's geologic provenience, there was what Hrdlička called "the insurmountable problem" of equating the essentially modern skullcap with a patent apelike jaw. Had this mandible articulated with an equally primitive-looking skullcap, like that of Pithecanthropus, then this would have been an entirely different matter altogether. But as it was, Smith Woodward's "monstrous" hybrid not only offended Hrdlička's understanding of biomechanics, it also demanded the existence in Tertiary times of a hominid with a steep, smooth-browed forehead.

It was claimed by Smith Woodward that the skull form of the common ancestor of the higher primates (including man) had been "more rounded" [Woodward, cited in Dawson and Smith Woodward, 1913, p 139]. This assumption was based on the fact that during the fetal stage of development, the anthropoid apes (like humans) exhibited a "more rounded, high, and expanded forehead." This indicated, in Smith Woodward's estimation, that the descendants of a common ancestral stock in the course of their onto-genic development passed through structural stages that were reminiscent of the adult common ancestor. But for reasons unknown it appeared that where the hominid branch had essentially retained the archaic rounded form, the anthropoid apes had since late Tertiary times undergone progressive change. Thus, using this general argument, Smith Woodward was able to explain the retreating forehead and projecting brow ridges of the Neandertals as "degenerative gorilloid specializations," and thereby justify, as Boule had done, their exclusion from the ancestral line leading to modern H sapiens.

From Hrdlička's standpoint this theory was at variance with the evidence furnished by the fossil record, which, in his opinion, clearly demonstrated that the modern human form had been derived from a process of progressive differentiation (as had the modern anthropoid apes) from an as yet unknown "anthropoid precursor" [Hrdlička 1912a, 1920]. Also, unlike a good many of his contemporaries who saw this evolutionary process abruptly terminating at some undetermined point in the late Tertiary or lower Quaternary, Hrdlička was of the opinion that it was an ongoing process, and for this reason regarded human skeletal diversity (both contemporary and historic) as essentially the product of the interaction between the "germplasm" and the environment (see later). On the basis of these two propositions Hrdlička envisioned man as having undergone a gradual morphological metamorphosis: from a "more theroid form" to his present form. Hence by definition, Hrdlička's conception of a human precursor was the antithesis of Smith Woodward's and others who supported the presapiens theory.

Hrdlička's commitment to a fundamentally unilineal interpretation of human evolution can be traced to the influence of Léonce Manouvrier (1850–1927) under whom he had studied anthropology at the Ecole d'Anthropologie, Paris, in 1896 [Spencer, 1979]. The faculty of the Ecole, unlike its counterpart at the Museum de Histoire Naturelle, was more inclined to Darwinism and the evolutionary perspective.[4] As such, Manouvrier, in contrast to Boule, subscribed to the notion of the plasticity of the human organism that had undergone a slow, yet progressive transformation to its

---

[4]The Laboratoire d'Anthropolgie de L'Ecole pratique des Hautes Etudes was established in 1868 under the direction of Paul Broca. After Broca's death in 1880, the directorship of the Laboratoire passed to the histoanatomist Mathias Duval (1844–1907), an emigré from Strassburg, who, apart from entrusting Manouvrier with developing Broca's work on craniometry and anthropometry, also contributed to Manouvrier's pro-Darwinist posture. In fact, Duval was one of the few protagonists of Darwinian theory in France during the latter 19th

modern form during the Pleistocene—hence his early support of Dubois' interpretation of the *Pithecanthropus* [Manouvrier, 1896]. Furthermore, Hrdlička's developing views on the antiquity of man in the New World had demanded a Neandertal-like phase of human evolution and along with it the growing conviction that the initial appearance of modern *H sapiens* in the Old World had occurred during the Upper Pleistocene prior to their eventual spread into the New World at the Pleistocene-Holocene boundary [see Hrdlička, 1907, 1912a,b, 1914, 1921].

Hrdlička's views on Neandertals at the commencement of World War I are conveniently summarized in a letter he wrote to Henry Fairfield Osborn (1857–1935), the paleontologist and then director of the American Museum of Natural History, New York City:

> My conviction that the Neanderthal type is merely one phase in the more or less gradual process of evolution of man to his present form, is steadily growing stronger. It isn't only the skulls we have, but also other skeletal parts which in some respects are very important; and I find on the one hand, not withstanding many lapses, nothing but gradual evolution and involution, and on the other hand no substantial break in the line from at least the earliest Neanderthal specimen to the present day. We have today not one but many intermediary forms in both the skull and other bones between Neanderthal and present man. Take Spy

century. Indeed his book, *Le Darwinisme* [1886], illustrates particularly well this minority position. As inferred in my text the Europeans, and particularly the French, tended to regard transformism-evolution as an essentially French inspiration (namely Lamarck) and as such received Darwin's thesis with frigid indifference. The French (as indicated by Boule and others at the Museum de Histoire Naturelle in Paris) were generally more inclined toward a neo-Lamarckian viewpoint, an attitude governed not so much by their evident national chauvinism as a scientific reservation they held toward the theory of natural selection to account adequately for the indubitable evidence of evolution presented in the natural world.

Unable to equate the vitalistic implications of neo-Lamarckianism with his distinctly mechanistic viewpoint, Manouvrier forcefully rejected all notions of innate and inferior beings, and championed the plasticity of human nature and the dynamic influence that the social environment exercised upon the individual organism [See Manouvrier, 1893].

Also it should be noted that, in addition to Duval, Manouvrier was also greatly influenced by the physiologist Etienne Jules Marey (1830–1904). One of the founders of nonvitalistic physiology in France, Marey conducted his researches at an experimental station situated in the Bois de Boulogne, and it was in this laboratory that Manouvrier was offered a part-time research position shortly after receiving his doctorate in 1882. After Marey's death in 1904, Manouvrier became sub-director of this laboratory, a position he kept until his death in 1927.

No. 2 for instance; it is a most important specimen in this connection, or take the long bones of the skeletons of Krapina or La Ferrassie. Then take *Homo aurignacensis*, the Předmost material, and some of the Neolithic crania in Brussels and Warsaw. All these speak for the continuity of the race. Besides which we have the multiple evidence pointing in the same direction of the remains of the primitive and civilized modern man [Hrdlička to Osborne February 17, 1916, cited in Spencer, 1979, p 475].

Hrdlička, however, was not the only worker who balked at Smith Woodward's reconstruction of the Piltdown remains. Indeed, among the first of several American workers to publicly question the association of the Piltdown jaw and skullcap was Gerrit S. Miller (1869–1956), curator of the Division of Mammals at the Smithsonian Institution–a close friend and colleague of Hrdlička. In a paper entitled *The Jaw of the Piltdown Man*, published by the Smithsonian Institution in November, 1915, Miller declared that he could find no justification for the assertion that the jaw was human. In his opinion it was plainly "simian," and advocated that the name *Eoanthropus* be restricted to the cranial parts, while the mandible should be described as a fossil ape, and proposed the name *Pan vetus*.

To begin with, Miller's argument was endorsed by a group of workers associated with the American Museum in New York City, namely Henry Fairfield Osborn [1915a, pp 130–144, 1915b], William K. Gregory [1916], Richard S. Lull [1917, pp 681–682] and J.H. McGregor [1917, cited in Spencer, 1979]. Although a few European workers supported the Hrdlička-Miller viewpoint [eg, Waterston, 1913; Boule, 1917; Frassetto, 1918; Guiffrida-Ruggeri, 1918; Ramstrom, 1919; Mollison, 1921; Freidrichs, 1932], by and large the Europeans tended to support the association of the jaw and skull [eg, Sergi, 1914a,b, pp 192–194; Sera, 1917].

In the meantime the Washington position was seriously weakened by the fortuitous discovery of a second Piltdown skull! According to Smith Woodward, who made the announcement at a meeting of the Geological Society of London in February 1917, Charles Dawson had found in the summer of 1915 (just prior to his death) a fragment of a frontal and occipital bone, and a molar tooth in a field located some two miles from the original Piltdown site [Smith Woodward, 1917, p 6].

In the light of this new evidence several of Miller's American supporters were converted to the British point of view. This retreat was completed following the examination of the original specimens in London by McGregor and Osborn in 1921 and 1922, respectively [Spencer, 1979, p 522].

After the resumption of peace in Europe, Hrdlička made a detailed study of the Piltdown remains in an effort to discredit the association of the jaw and skull [Hrdlička, 1922, 1923a,b, 1924]. The purpose of this exercise was essentially to undermine the theoretical foundation of the presapiens movement. But while rejecting the antiquity of the Piltdown calotte, Hrdlička surprisingly accepted the jaw as an authentic fossil. Commenting on the jaw, he said it was a "truly remarkable specimen," whose morphology was commensurate with its suggested geological age. From this protracted study (made between 1922 and 1925), Hrdlička became increasingly convinced of a demonstrable relationship between the dentition of the European dryopthecines and the Piltdown jaw, a "finding" he believed gave credence to the proposition that Europe had been the "cradle and nursery" of the genus *Homo*.

Acutely aware of the fact that he was now almost alone in considering the Neandertals as the antecedents of modern *H sapiens* and that this was the linchpin of his hypothesis on the peopling of the New World [Hrdlička, 1921], Hrdlička decided to use the invitation to deliver the 1927 Huxley Memorial Lecture in London to summarize his recent survey of all the available evidence related to the Neandertal question (see Table II).

In his defense of a Neandertal phase of human evolution, Hrdlička [1927, also 1930, pp 319–349] presented his classic definition of Neandertal as "man of the Mousterian culture" and outlined their geographic distribution (Europe, Western Asia, and North Africa), as well as their archeological and geological context.

Since the details of Hrdlička's Huxley Memorial Lecture have been dealt with recently at length by a number of authors, most notably Brace [1964],

TABLE II. Major Neandertal Remains Discovered Between 1886 and 1927
(Listed in Chronological Order)

Fossil	Location	Date discovered	Described by:
Krapina	Croatia (Yugoslavia)	1899–1905	Gorjanović-Kramberger [1906]
Le Moustier	Dordogne (France)	1908	Klaatsch and Hauser [1909]
La Chapelle	Corrèze (France)	1908	Boule [1908]
La Quina	Charente (France)	1908	Martin [1911]
Ehringsdorf	Fischer's Quarry (Germany)	1908	Schwalbe [1914]
La Ferrassie	Dordogne (France)	1909 1910 1912 1916	Capitan and Peyrony [1909, 1912] Virchow [1920]
Ehringsdorf	Fischer's Quarry (Germany)	1925	Weidenreich [1927]
Galilee	Palestine	1925	Turville-Petre [1927]

Spencer [1979; in preparation], Spencer and Smith [1981], and Trinkaus [1982], only the salient features of his thesis will be presented here.

To begin with, Hrdlička noted that while it was not possible, at least with any certainty, to arrange the then-existing Neandertal fossil assemblage into a reliable chronological sequence, there was, however, evidence (contrary to popular scientific opinion) showing considerable morphological variation in the Neandertal phenotypic pattern. In this regard he noted that in the western sector of Europe the Neandertal fossils more often than not displayed a hyper-robust morphology, which he conjectured was a cold-adapted phenotype. By contrast, as one proceeded eastwards, this hyper-robusticity relaxed into a more refined morphology. Thus taken as a whole, Hrdlička claimed the European assemblage revealed evidence for not only specific climatic adapatational characters (superimposed on the more common grade features), but also a tendency, particularly in the Central European specimens (eg, Krapina), toward the modern H sapiens pattern. Considering this point, he said:

> Here is . . . evidently, a very noteworthy example of morphological instability, an instability, evidently of evolutionary nature, leading from old forms to more modern . . . Taking the . . . skulls, jaws, and bones attributed to the Neanderthal phase, it is seen that both the variability and the number of characters that tend in the direction of the later increase considerably. The Krapina series, by itself, is probably more variable from the evolutionary point of view than would be any similar series from one locality at the present [Hrdlička, 1927, pp 267, 268].

To which he added:

> During this period man is brought face to face with great changes of environment . . . which call for new adaptations and developments . . . such factors must inevitably have brought about, on the one hand, greater mental as well as physical exertion and, on the other, an intensification of natural selection, with the survival of only the more, and perishing of the less, fit. But . . . evolution would certainly differ from region to region, as the sum of the factors affecting man differed, reaching a more advanced grade where the conditions in general proved the most favorable, while to many of the less favored groups disease, famine and warfare would bring extinction . . . Here seems to be a relatively simple, natural explanation of the progressive evolution of Neanderthal man, and such an evolution would inevitably carry his most advanced form to those of primitive Homo sapiens [Hrdlička, 1927, pp 271, 272].

Turning to the notion frequently voiced by advocates of the presapiens theory that the Mousterian culture, along with its author the Neandertals, had been "swept away" by the coming of a distinct and superior group of hominids, namely the Aurignacians, Hrdlička contended that such an idea had no factual basis. To back his argument, he had made a careful comparative study of various cultural sites belonging to the European Paleolithic [Hrdlička, 1927, pp 256–259]. From this analysis, Hrdlička claimed that the Aurignacian was not an intrusive but an indigenous outgrowth from the Mousterian, a fact clearly shown, so he believed, by the work of the French paleoanthropologist, Henri Martin, on the Mousterian sequence at La Quina.

Another problem related to the issue of the "Aurignacian influx" and the supposed replacement of the European Neandertals, was why these people had seemingly waited until the height of the glacial period to invade Europe. This, Hrdlička said, violated what he called the "laws" governing the movement of human and animal populations [see Hrdlička, 1921, p 545]. According to these "laws," human populations always tended to move in the direction of least resistance (climate) and in the direction of better material prospects [Hrdlička, 1927, p 260].

Finally, and perhaps most important, the invasion-replacement hypothesis presupposed the existence of a non-Neandertal mother population, for which, as Hrdlička noted, there was no evidence from either Asia or Africa. Furthermore, even presuming that such a population had existed, the suggestion that this invasion may have been a "peaceful extension," would have led, Hrdlička argued, to an amalgamation with, rather than an extinction of, the European Neandertals [Hrdlička, 1927, p 260].

The immediate reaction to Hrdlička's Huxley Lecture can be summarized by Grafton Elliot Smith's comments published in the January 1928 issue of *Nature*:

> In his recent Huxley Lecture . . . Dr. Aleš Hrdlička has questioned the validity of the specific distinction of Neanderthal man, an issue which most anatomists imagined to have been definitely settled by the investigations of Schwalbe in 1899, and the corroboration afforded by the work of Boule and a host of other anatomists . . . It is the way of true science constantly to submit to scrutiny the foundations of its theoretical views—a discipline to which a restive anthropology is not always willing to submit. The only justification for re-opening the problem of the status of Neanderthal man would be afforded by new evidence or new views, either of a destructive or constructive nature. I do not think

Dr. Hrdlička has given any valid reason for rejecting the view that
Homo neanderthalensis is a species distinct from *H sapiens* [Elliot Smith,
1928, p 141].

As this indicates, the Huxley Lecture was not successful in turning the
tide of general scientific opinion in Hrdlička's favor. Furthermore, his
publication of a more detailed consideration of the issue, *The Skeletal
Remains of Early Man*, in 1930, was no more successful in this regard. The
presapiens theory was now firmly entrenched, and for the next two decades,
with the notable exception of the work of Hans Weinert [eg, 1944, 1947,
1951, 1953] and Franz Weidenreich (see below), Hrdlička's phylogenetic
interpretations were generally either avoided completely or mentioned as a
highly unlikely alternative to the presapiens view [eg, Black, 1929; Boule,
1929; Hooton, 1930, 1931; Keith, 1929; Osborn, 1930; MacCurdy, 1937].
In this intellectual context, new discoveries made during the 1920s and
1930s (see Table II) such as Chou-Kou-Tien [Black, 1929], Saccopastore
[Sergi, 1929; Breuil and Blanc, 1935], Steinheim [Berckheimer, 1933], Swanscombe [Marston, 1936], Skhūl and Tabūn [Keith and McCown, 1937;
McCown and Keith, 1939], and Circeo [Sergi, 1939] were simply assimilated
into existing phylogenetic schemes that almost without exception portrayed
Neandertals as an archaic and extinct hominid lineage.

WEIDENREICH'S DEVELOPMENT OF THE
NEANDERTAL HYPOTHESIS

Although Franz Weidenreich began his career in anatomy under the
tutelage of Gustav Schwalbe at the University of Strassburg in 1899, his
latent interest in human evolution did not surface until the mid 1920s,
when he published a detailed description and evaluation of the Ehringsdorf
skull found in 1925 [Weidenreich, 1926, 1928]. The reasons for this delay
are not altogether clear. A contributing factor may have been the war,
along with a simple lack of opportunity. Similarly, it is also possible that
during his tenure at the University of Strassburg from 1904–1918, he may
have been intimidated by both Schwalbe's omnipotent presence and international reputation. In this regard it is important to remember that Schwalbe
had not only been responsible for establishing the view that the Neandertals
were a separate species—an idea that had inadvertently fueled the merging
argument for a separate and independent presapiens lineage—but also, in
1913, under the influence of Boule's weighty scholarship, he had partially
retracted his former assertion that the Neandertals represented a reasonable
antecedent of modern humans [Schwalbe, 1913, p 602].

Weidenreich, however, rejected the view that there had been more than one hominid species in existence at any given time during the Pleistocene and accepted without reservation Hrdlička's unilineal approach to human evolution, which included a "Neanderthal phase of man" [Weidenreich, 1928, p 59, 1940, 1943a, 1947, 1949]. The similarities between Hrdlička's views on human phylogeny and those of Weidenreich are striking. According to Hrdlička:

> Evolutionary changes have not progressed and do not progress regularly in mankind as a whole, nor even in any of its divisions. Such changes may be thought of as a slowly-augmenting complex of zig-zags, with localized forward leaps, temporary haltings, retrogressions, and possibly even complete cessations. Thus it would not be unreasonable to expect that at any given date in the past or present all the branches or members of the human or protohuman family would be absolutely of uniform type [Hrdlička, 1912a, p 3].

Dealing with the same subject, Weidenreich wrote:

> If we admit that mankind of today, uniform regarding its general character but differing in special appearance, has developed from various regional stocks starting even from an earlier stage than that represented by the Prehominids, and if we assume, furthermore, that development was going on simultaneously everywhere but was accelerated in one place and retarded in another, perhaps as a consequence of local influences, then all the discrepancies between the morphologic and chronologic sequence of the known types of fossil man can be understood [Weidenreich, 1940].

An obvious difference between these two workers is that Hrdlička's synthesis had been confined essentially to an understanding of the European fossil record, whereas Weidenreich's concept of a Neandertal phase of human evolution embraced the entire Old World. This difference, however, is considered to be of minor importance, since Hrdlička was clearly hindered by the lack of evidence from outside of Europe. But as his comments on the Broken Hill cranium suggest [Hrdlička, 1930, pp 115–116], he may well have modified his patently Eurocentric model given the evidence subsequently placed at Weidenreich's disposal. As Davidson Black's (1884–1934) successor at the Peking Union Medical College and director of the work at Chou Kou Tien from 1935 on through to the outbreak of World War II, Weidenreich was provided with a unique opportunity to study firsthand "Sinanthropus" and other hominid fossils from China [Weidenreich, 1943b] and

Java [Weidenreich, 1945, 1951]. But far more important is the fact that both men were firmly committed to what Hrdlička called "the incomplete stability of the human organism" and the idea that hominid morphology being the product of an ongoing interaction between the "protoplasm and environment" would thereby "bear strongly against the persistence of the same type of man in any region from the Pleistocene or even older period to the present" [see Hrdlička, 1912a, pp 3–5].

Bolstered by many more fossils, many of which were from outside Europe, Weidenreich produced a sophisticated elaboration of Hrdlička's Neandertal hypothesis involving parallel evolutionary lineages in various regions of the Old World leading through separate Neandertaloid stages to the modern geographical variances of H sapiens. Aware that this might suggest to some that he was proposing a polyphyletic scheme of human evolution, Weidenreich cautioned his would-be critics by noting: "This is not what I have in mind. All facts so far available indicate that man branched off as a unit which split afterwards, within its already limited faculties into several lines . . . I regard all the hominids, living mankind included, as members of one species" [Weidenreich, 1949, p 157]. Succinctly, Weidenreich [1947, p 201] envisioned three major subdivisions of the evolutionary sequence: the "Archanthropinae" (pithecanthropines, ie, H erectus), "Paleoanthropinae" (Neandertals), and "Neoanthropinae" (modern H sapiens). These fossil grades were then sorted according to their morphologic and chronologic status, as well as their geographic location. In the latter category Weidenreich identified four major areal types: Australian, Mongolian, African, and Eurasian. While recognizing that environmental factors in different geographical areas could result in local differentiations, he was convinced these populations had not evolved in total genetic isolation and as such allowed for varying degrees of gene flow between adjacent lineages. Although Weidenreich argued that there was no justifiable reason to believe that Neandertal man in Europe could not have advanced to modern H sapiens [Weidenreich, 1943a, p 48], he nevertheless tended to the view that the actual evolutionary transition from Neandertals to modern Europeans occurred essentially outside of Europe, namely that they had most probably been derived from early west Asiatic Neandertal populations [Weidenreich, 1943a, 1949].

REMOVAL OF THE PILTDOWN OBSTACLE AND OTHER POST-WORLD WAR II DEVELOPMENTS

The 1940s and early 1950s witnessed a number of developments that significantly altered the paleoanthropological ethos. Crucial to this nascent change had been the formulation of a synthetic theory of evolution. This movement began in the 1930s under the auspices of such workers as Theodosius Dobzhansky [1937], Julian Huxley [1942], Ernst Mayr [1942] and George Gaylord Simpson [1944], and culminated in the highly influential conference on "Genetics, Paleontology and Evolution" at Princeton University in 1947 [Mayr and Provine, 1980]. Where previously paleontological thinking had been essentially descriptive and typological, the new synthesis propounded at Princeton demanded a distinctly evolutionary, genetic, and populational perspective. This "new" approach rapidly gained widespread acceptance in both the biological and anthropological communities and was responsible for the subsequent organization of the multidisciplinary symposium on the "Origin and Evolution of Man," held at Cold Spring Harbor, Long Island, New York in June 1950 [Warren, 1951]. The impact of this symposium on physical anthropologists, especially in the United States, was profound [see Spencer, 1982]. In particular it prompted not only the immediate appearance of populational thinking in paleoanthropological research, but also led to a more rational and systematic approach to hominid taxonomy and phylogeny.

Of particular interest here is the question raised at the Cold Spring Harbor Symposium regarding the taxonomic and phylogenetic status of the South African australopithecines. Although by this time, most workers accepted Raymond Dart's original description of *Australopithecus* in 1925 as an early apelike hominid [eg, Gregory, 1949; Le Gros Clark, 1947, 1948], few, if any, were willing to adopt Ernst Mayr's [1951] suggestion that this taxon be upgraded to *Homo*. A major reason for this reservation was the problem of equating the australopithecines with the Piltdown remains and vice versa. Although seemingly of comparable age, both presented quite divergent morphologies. This obstacle to the admission of *Australopithecus* to the human phylum, however, was dramatically removed in 1953 when the Piltdown remains were exposed as a fraud [Weiner et al, 1953; Weiner, 1955] that had been presumably perpetrated by Charles Dawson.[5]

---

[5]In spite of the overwhelming evidence incriminating Dawson as the culprit, there has been nevertheless in recent years a steady stream of theories suggesting either other individuals as the author of the hoax, or possible co-conspirators in the fraud [eg, Millar, 1972; Halstead, 1978; Gould, 1980]. In retrospect it seems incredible, especially if one examines the original specimens, that Piltdown had been so widely accepted as a genuine fossil and that it had evaded detection for so long. But as suggested earlier in the text, the remains had clearly fulfilled the paleontological expectations of the period, and it was not until after World War II that many of the ideas and theoretical assumptions responsible for the fabrication were finally called into question.

This development had the immediate effect of placing the Pleistocene hominids in a different and more favorable light. The Neandertals (along with the pithecanthropines, ie, *H erectus*) now appeared to be much closer to modern *H sapiens* than previously believed. Thus it was in this developing intellectual context that over the next decade many of the earlier misconceptions of Neandertal anatomy were gradually corrected [eg, Arambourg, 1955; Schultz, 1955; Straus and Cave, 1957; Stewart, 1962]. But while many of the supposedly distinctive and specialized traits were shown to be well within the range of variation for modern humans, there were nevertheless still a small number of traits remaining considered by some workers to be so significantly divergent from modern human morphology as to warrant continuing adherence to the view that the European Neandertals were a specialized and extinct branch of the human evolutionary tree.

PRESAPIENS THEORY POST 1950: NEW WINE IN AN OLD BOTTLE

During the 1950s the presapiens view was promoted largely through the efforts of Henri Vallois, Boule's successor at the Musée de l'Homme and Institut de Paléontologie Humaine in Paris. In his now-classic presentation of the presapiens point of view, Vallois [1954] noted three major objections to Hrdlička's unilineal interpretation of the European fossil record. First, he pointed to a number of skeletal specializations of Neandertals and the lack, in his opinion, of morphological continuity between Neandertals and the earliest modern *H sapiens* in Europe. Second, he offered objections to several of Hrdlička's arguments, including the idea of the interrelationship between the selective pressures exerted by the glacial period and the progression of Neandertals in the direction of modern humans. Here Vallois [1954, p 119] noted that from "pre-Neanderthals" to "Wurm Neanderthals" these hominids became more specialized in their own direction and not in the direction of modern *H sapiens*. Third, he presented as his most conclusive evidence the actual fossil remnants of the presapiens lineage, which he claimed either predated or existed contemporaneously with Neandertals. While pointing out that most of the specimens previously purported to be presapiens, such as Ipswich, Galley Hill, Foxhall, Moulin Quignon, and Piltdown, were demonstrably more recent in age, he revealed that there were, at that time, two unquestionable representatives of this ancient lineage: the Swanscombe skull, found in England in 1935–36 [Marston, 1936] and the Fontéchevade skull [Vallois, 1949].

Having thus outlined both the direct and indirect evidence against a Neanderthal phase of human evolution, Vallois then went on to state that the overwhelming consensus of the previous 40 years (albeit on the basis of completely different "presapiens" fossils) permitted him to conclude with confidence that:

> The European *Homo sapiens* is not derived from the Neanderthal men who preceded him. His stock was long distinct from, and, under the name of Praesapiens, had evolved in a parallel direction to theirs. Long debated, the Praesapiens forms are thus not a myth. They did exist. The few remains of them we possess are the tangible evidence of the great antiquity of the phylum that culminates in modern man [Vallois, 1954, p 128].

However, subsequent studies made on Vallois' presapiens evidence in the early 1960s revealed that neither Swanscombe nor Fontechévade differed significantly from other contemporaneous fossils. In the case of Swanscombe, Weiner and Campbell [1964] showed conclusively that it was nonmodern in form, and in all probability represented a hominid grade transitional between *H erectus* and modern *H sapiens*. Similarly, Sergi [1953] and Brace [1964] independently demonstrated the affinities of the Fontéchevade specimen with Neandertals, rather than with modern *H sapiens*.

During the mid 1960s, a fragment of a hominid occipital bone was found in a quarry at Vértesszöllös, Hungary, and was initially labeled as an early presapiens form [Thoma, 1966, 1969]. This diagnosis, however, was subsequently contested by Wolpoff [1971], who from his study of the Hungarian occipital fragment concluded that it was indistinguishable from *H erectus*.

But in spite of a continuing lack of evidence to support the existence of a European presapiens lineage, this scheme still retained support, albeit with diminished enthusiasm, during the next decade. For example, in 1953, Louis B. Leakey (1903–1972) advanced the hypothesis based on remains recovered from Kanjera, Kenya, that there had been two contemporaneous hominid lineages in Africa. According to Leakey, one of these lineages led, via Kanjera, to modern *H sapiens*, while the other led in the direction of the so-called archaic Rhodesian forms (eg, Broken Hill Man). Later, in 1972, he modified this view slightly by claiming that the archaic Rhodesian forms were hybrids of the sapiens-erectus lineage.

Currently there are a number of young workers such as Stringer [1974, 1978], Bräuer [1980], and Hublin [1978] who argue that there is a significant qualitative difference between Neandertals and early modern *H sapiens* and as such seriously question the proposition that the former can be the evolutionary antecedent of the latter. Since many of these workers tend to regard the European late Middle and early Upper Pleistocene fossil record as solely documenting Neandertal evolution, they automatically infer from this that early modern H sapiens must have evolved outside of Europe (see the contributions of Stringer et al and Bräuer, respectively, in this present volume).

As indicated above, there was after 1950 a general retreat from the presapiens scheme to what might be best described as a "compromise" position—namely the Pre-Neandertal hypothesis. This theoretical position slowly acquired increasing support during the late 1950s and early 1960s, and eventually emerged in the 1970s as the most popular perspective on the phylogenetic position of the European Neandertals.

THE PRENEANDERTAL THEORY AND RESUSCITATION OF HRDLIČKA'S NEANDERTAL HYPOTHESIS

The Pre-Neandertal hypothesis was first enunciated in the early 1920s when it was postulated that certain Neandertal fossils, considered to be less specialized (ie, "early" or "progressive"), were conceived of as being ancestral to both modern H sapiens and the more specialized ("classic") Neandertals, especially the western European Wurm samples. Though first appearing in the writings of Grafton Elliot Smith [1924], this idea did not become popular until the early 1950s, at which time it was developed by the Italian paleontologist Sergio Sergi (the son of Giuseppe Sergi) and the American physical anthropologist F. Clark Howell.

In his writings on the Neandertal issue, Sergi [1944, 1948] had considered a probable genetic connection between some Neandertal-like populations and early modern H sapiens—though he was not very explicit on this point. Later, in 1953, he made his position that much clearer when he wrote: "The men of Swanscombe and Fontéchevade present a mixture of characteristics of Phaneranthropi and Palaeanthropi." According to Sergi [1953, 1958] the Phaneranthropi or morphologically modern humans were derived from the Prophaneranthropi, which had made their appearance during the early Middle Pleistocene. The so-called Prophaneranthropi, represented by such fossils as Swanscombe, Steinheim, Saccopastore, Ehringsdorf, and Fontéch-evade, were considered by Sergi to be a peculiar blend of Neandertal (Palaeanthropi) and modern H sapiens features. In Sergi's opinion the Pro-phaneranthropi had been the "common forerunner" of both the more specialized Neandertals (Palaeanthropi) of the Wurm period, and modern humans. Sergi left unexplained the circumstances leading to the appearance and subsequent success of the Phaneranthropi [1958, p 49], though he did note that the "extraordinary development of the great phaneranthropic complex in the post-Pleistocene . . . coincided with the final extinction of the Palaeanthropi."

Similarly, using the then current view that the Mount Carmel assemblage (Tabūn and Skhūl) belonged to the last interglacial period, Howell [1951] proposed that these hominids, along with Steinheim, Ehringsdorf, Krapina, Saccopastore, and Teshik Tash, represented a widespread and early "progressive" Neandertal population. This "progressive" Neandertal or Pre-Neandertal population was considered by Howell to have been both the precursor of the late "classic" Neandertals of western Europe and modern humans. At this juncture, however, Howell was not very specific about the issue of the emergence of the latter, other than saying that "modern man was developing further to the East." But later, in 1957, employing the redating of the Mount Carmel remains to the last glaciation, Howell elected the Skhūl hominids as a likely ancestor for modern humans in Europe, suggesting that they were derived from a stock similar to the early "progressive" European Neandertals.

Following the lead of Sergi and Howell, a number of variations on this general Pre-Neandertal theme appeared in the literature [eg, Le Gros Clark, 1955; Breitinger, 1957; Weiner, 1958]. Almost without exception, it was agreed that the Middle Pleistocene hominids of Europe, represented principally by the Swanscombe and Steinheim fossils, were plausible ancestral forms for both the so-called classic Neandertals and modern humans.

It was in the context of burgeoning support of the Pre-Neandertal interpretations of the Pleistocene hominid fossil record that the Neandertal hypothesis was resuscitated in the early 1960s by C. Loring Brace [1962a, 1964, 1966]. In retrospect, it appears that Brace, like Carleton S. Coon [1962] and others, made a conscious effort to embrace the new evolutionary synthesis. But where such workers as Coon attempted to explain European Upper Pleistocene hominid evolution in terms of biological adaptation without specifying the exact mechanics involved, Brace [1962b, 1963, 1967] endeavored to do so. Essentially, Brace's hypothesis is an elaborate restatement of Hrdlička's [1911] suggestion that changes in masticatory function were an integral process in later patterns of hominid cranial evolution. Accordingly, Brace [1962a,b, 1963] argues that with the introduction of more specialized tools in the Middle and Upper Paleolithic to perform tasks previously handled by the anterior teeth there was a progressive relaxation in the amount of stress generated in the craniofacial skeleton, which ultimately precipitated a series of morphological changes that signalled the appearance of the modern cranial configuration.

From the late 1960s and on through to the present, this approach to Pleistocene hominid evolution has been vigorously pursued by Brace [eg,

1967, 1979, see also in this present volume] as well as several other workers. For example, Wolpoff [Brose and Wolpoff, 1971; and more particularly Wolpoff, 1980, pp 298–300] has emphasized the role of cultural change to support the view of a widespread Neandertal-like phase of human evolution throughout the Old World. According to Wolpoff's version of this hypothesis, the changes necessary for the archaic-to-modern *H sapiens* are essentially the same throughout the Old World and can be understood and documented in terms of worldwide changes in selection forces acting on already different local populations of archaic *H sapiens*. Similarly, studies of European Upper Pleistocene hominids have been used to document a claimed transformation of Neandertals into modern humans from the evidence of the dentition [Frayer, 1978, and this volume], face, jaws, and teeth [Brace, 1979, and this volume], and supraorbital region [Smith and Raynard, 1980]. Also, detailed studies of European hominids and some interpretations of the archeological record have also been advanced to document "local continuity" between Neandertal and early modern humans in Eastern Europe [Smith, 1982, and this volume]. The need to evoke gene flow from outside Europe to account for the appearance of modern *H sapiens* is questioned by these workers, as is the evidence for earlier representatives of morphologically modern *H sapiens* outside of this area.

PRESENT STATUS OF DEBATE

As the various contributions to this volume demonstrate, there is still a continuing lack of consensus among paleoanthropologists regarding the phylogenetic significance of Neandertals—particularly the so-called classic Neandertals of Western Europe. These particular hominids present to many workers a unique suite of morphological traits, which serve to reinforce their conviction that Neandertals were a specialized offshoot from the main line leading to the emergence of modern *H sapiens* [Howells, 1975; Trinkaus, 1976; Hublin, 1978; Santa Luca, 1978; Stringer, 1974, 1978; Trinkaus and Howells, 1979; Stringer and Trinkaus, 1981]. In recent years this position has been strengthened by a clearer understanding of hominid variation during the Upper Pleistocene, afforded by recent additions to the African and Asian fossil assemblages, and the distinct possibility that anatomically modern humans may have evolved earlier in Africa than elsewhere in the Old World [see Rightmire, 1979; Rightmire and Bräuer, respectively, in this volume]. At present, however, the dates assigned to these African specimens are contentious. But even if these dates prove to be correct there is still no reason to assume that a similar transition did not take place in Europe and

Asia. In fact the work of Fred Smith and others in Central Europe strongly supports such a possibility. Furthermore, the old argument that there was simply not enough time separating Neandertals from modern H sapiens for the former to have evolved into the latter requires some modification in the light of revisions taking place in the traditional chronology of the last glaciation in Europe [Dennell, 1983]. According to the traditional chronology, the Mousterian of the last glaciation was confined to a period of some 30 ky (ie, 75,000 to 30,000 yrs BP). Recent geological studies have resulted in a doubling of this figure. Although the results of this work have yet to be fully evaluated, it does suggest that our traditional understanding of the European Upper Pleistocene is still far from complete. In fact it is becoming increasingly evident that continued adherence to either of the three traditional theoretical schemes of hominid Pleistocene evolution is a "gross oversimplification" of what really occurred. But while most workers in the field today willingly acknowledge that this is undoubtedly true, it appears we are still some way from achieving a new synthetic model that deals realistically with the Neandertal problem.

LITERATURE CITED

Arambourg C (1955): Sur l'attitude, en station verticale, des Néanderthaliens. CR Acad Sci [D] (Paris) 240:804–806.

Aufrère L (1936): Essai sur les premières découvertes de Boucher de Perthes et les origines de l'archéologie primitive. Paris: Stande.

Berckheimer F (1933): Ein Menschen-Schädel aus den diluvialen Schottern von Steinheim a. d. Murr. Anthropol Anz 10:318–321.

Black D (1929): Sinanthropus pekinensis: The recovery of further fossil remains of this early hominid from the Chou Kou Tien deposits. Science 69:674–676.

Blake CC (1862): On the crania of the most ancient races of men. Geologist 5:205–233.

Boucher de Perthes J (1847): Antiquités celtiques et antédiluviennes. Mémoire sur l'industrie primitive et les arts à leur origine. Paris: Treuttel et Wurtz.

Boule M (1906): "Les Grottes de Grimaldi: Résumes et Conclusions des Etudes Géologiques." Monaco.

Boule M (1908): L'homme fossile de La Chapelle-aux-Saints. l'Anthropologie 19:519.

Boule M (1909): L'homme fossile de La Chapelle-aux-Saints. l'Anthropologie 20:260, 264.

Boule M (1911–1913): L'homme fossile de La Chapelle-aux-Saints. Extrait Ann Paléont VI:111–172; VII:21–192, 65–208; VIII:1–70, 209–278.

Boule M (1914): L'Homo Néanderthalensis et sa place dans la nature. Congr. Int Anthropol Archéol Préhist. 1912 II:392–395.

Boule M (1917): The jaw of Piltdown. l'Anthropologie 28:157–159.

Boule M (1921): "Les Hommes Fossiles: Elements de Paléontologie Humaine, 1st ed." Paris: Masson et Cie.

Boule M (1923): "Les Hommes Fossiles: Elements de Paléontologie Humaine, 2nd ed." Paris: Masson et Cie.

Boule M (1929): Le Sinanthrope. l'Anthropologie 39:455.

Boule M, Anthony R (1911): L'encèphale de l'homme fossile de La Chapelle-aux-Saints. l'Anthropologie 22:129–196.

Bongard JH (1835): "Wanderung zur Neandershöhle." Dusseldorf: Arnz.

Brace CL (1962a): Refocusing on the Neanderthal problem. Am Anthropol 64:729–741.

Brace CL (1962b): Cultural factors in the evolution of the human dentition. In Montagu MFA (ed): "Culture and the Evolution of Man." New York: Oxford University Press, pp 343–354.

Brace CL (1963): Structural reduction in evolution. Am Nat 97:39–49.

Brace CL (1964): The fate of the classic Neanderthals: A consideration of hominid catastrophism. Curr Anthropol 5:3–43.

Brace CL (1966): More on the fate of the "classic" Neanderthals: Reply. Curr Anthropol 7:210–214.

Brace CL (1967): Environment, tooth form, and size in the Pleistocene. J Dent Res 46:809–816.

Brace CL (1979): Krapina "classic" Neanderthals, and the evolution of the European face. J Hum Evol 8:527–550.

Brace CL (1981): Tales of the phylogenetic woods: The evolution and significance of evolutionary trees. Am J Phys Anthropol 56:411–429.

Branco W (1898): Die menschenahnlichen Zähne aus dem Bohnerz der schwabischen Alb. Jahreshefte Verhandl. Vaterland Naturk 54:1–44.

Bräuer G (1980): Die morphologischen Affinitäten des jungpleistozanen Stirnbeines aus dem Elbmündungsgebiet bei Hahnöfersand. Z Morphol Anthropol 71:1–42.

Breitinger E (1957): Zur phyletischen Evolution von Homo sapiens. Anthropol Anz 21:62–83.

Breuil H, Blanc AC (1935): Rinvenimento in situ di un nuovo cranio di Homo neanderthalensis nei giacimento di Saccopasstore (Roma). Atti Accad Naz Lincei, 22:166–169.

Broom R (1918): The evidence afforded by the Boskop skull of a new species of primitive man (Homo capensis). Anthropol Papers Am Mus Nat Hist (New York) 23:67–79.

Brose DS, Wolpoff MH (1971): Early Upper Paleolithic man and late Middle Paleolithic tools. Am Anthropol 73:1156–1194.

Büchner L (1894): "Man in the Past, Present, and Future. A Popular Account of the Results of Recent Scientific Research Regarding the Origin, Position, and Prospects of Mankind." New York: Eckler.

Burkitt MC (1921): "Prehistory." Cambridge: Cambridge University Press.

Busk G, Falconer H (1865): On the fossil contents of the Genista cave, Gibraltar. Q J Geol Soc 21:364–370.

Capitan L, Peyrony D (1909): Deux squelettes humains au milieu de foyers de l'époque moustérienne. Rev Anthropol 19:402–409.

Capitan L, Peyrony D (1912): Station préhistoriques de la Ferrassie. Rev Anthropol 22:76–99.

Cartailhac E (1884): Georges Cuvier et l'anciennete de l'Homme. Materiaux Hist Nat Primitive Homme 27:187.

Chambers R (1844): "Vestiges of the Natural History of Creation." London: Churchill.

Cocchi I (1867): L'uomo fossile nell'Italia centrale. Mem Soc Ital Sci Nat 2(7):1–80.

Coon CS (1962): "The Origin of Races." New York: Knopf.

Corner F, Raymond P (1909): Le crâne de Galley-Hill. Bull Mem Soc Anthropol Paris 4–5:487.

Cunningham DJ (1895): Dr. Dubois' so-called missing link. Nature 51:428–429.

Daniel GE (1943): "The Three Ages: An Essay on Archaeological Method." Cambridge (England): The University Press.

Darwin C (1839): "Narrative of the Surveying Voyages of HMS 'Adventure' and 'Beagle' Between 1826 and 1836, Vol 3." London: J. Murray.

Darwin C (1872):"On the Origin of Species by Means of Natural Selection, or the Preservation of Favoured Races in the Struggle for Life, 6th Ed." Connecticut: Easton Press, 1976.

Dawson C, Smith Woodward A (1913): On the discovery of a palaeolithic human skull and mandible in flint bearing gravel overlying the Wealden (Hastings Bed) at Piltdown, Fletching (Sussex). Q J Geol Soc Lond 69:117–144.

Dawson C, Smith Woodward A (1914): Supplementary note on the discovery of a paleolithic human skull and mandible at Piltdown (Sussex). Q J Geol Soc Lond 70:82–93.

Dennell R (1983): A new chronology for the Mousterian. Nature 301:199–200.

De Puydt M, Lohest M (1886a): Exploration de la grotte de Spy. Ann Soc Geol Belg Liege 13:34–39.

De Puydt M, Lohest M (1886b): L'homme contemporain du mammoth a Spy, province de Namur (Belgique). CR Cong Namur.

Dobzhansky T (1937): "Genetics and the Origin of Species." New York: Columbia University Press.

Dubois E (1894): "Pithecanthropus Erectus, Eine Menschenähnliche Übergangsform aus Java Batavia Landes Druckerei." New York: GE Stechert (reprinted).

Dubois E (1895): Pithecanthropus erectus, betracht als eine wirkliche Übergangsform und als Stammform des Menschen. Ethnol 27:723–740.

Dubois E (1896): On Pithecanthropus erectus: A transitional form between man and the apes. Sci Trans R Dublin Soc. 6:1–18.

Dupont E (1866): Etude sur les fouilles scientifiques executées pendant l'hiver de 1865–1866 dans les cavernes des bordes de la Lesse. BullAcad R Sci Lett Beaux Arts Belg 22:31–54.

Duval MM (1886): "Le Darwinisme: leçons professées à l'école d'anthropologie par M. Duval." Paris: Delahaye et Decrosnier.

Elliot Smith G (1913a): Preliminary report on the cranial cast [of Piltdown skull]. Q J Geol Soc Lond 69:145–147.

Elliot Smith G (1913b): The Piltdown skull. Nature 92:131.

Elliot Smith G (1913c): The controversies concerning the interpretation and meaning of the remains of the Dawn Man found near Piltdown. Nature 92:468–469.

Elliot Smith G (1916): New phases of the controversies concerning the Piltdown skull. Proc Manchester Lit Philos Soc. 60:xxviii–xxix.

Elliot Smith G (1917): The problem of the Piltdown jaw: Human or subhuman? Eugenics Rev 9:167.

Elliot Smith G (1924): "The Evolution of Man." London: Oxford University Press.

Elliot Smith G (1928): Neanderthal man as a distinct species. Nature 121:141.

Faudel CF (1867): Sur l'découverte d'ossements humains fossiles dans le lehm de la vallée du Rhin, à Eguisheim près de Comar. Bul Soc Geol Fr 24:36–44.

Fraipont J, Lohest M (1886): La race de Néanderthal ou de Cannstadt, en Belgique. Recherches ethnographiques sur des ossements humains découverts en dépôts quaternaires d'une grotte à Spy et détermination de leur âge géologique—Note Préliminaire. Bull Acad R Sci Lett Beaux Arts Belg 12:741–784.

Fraipont J, Lohest M (1887): Recherches ethnographiques sur des ossements humains découverts dans les dépôts quaternaires d'une grotte à Spy et détermination de leur âge géologique. Arch Biol 7:587–757.

Frassetto F (1918): "Lezioni di Antropologia, Vol 1." Milano: Hoepli.

Frayer DW (1978): Evolution of the dentition in upper paleolithic and mesolithic Europe. U Kansas Publ Antropol 10:1–201.

Friedrichs HF (1932): Schädel und Unterkiefer von Piltdown. (Eoanthropus dawsonii Woodward): Neuer Untersuchungen. Z Anat Entwickl XCVIII:199–262.

Fuhlrott JK (1859): Menschliche Uberreste aus einer Felsengrotte des Düsselthals: Ein Beitrag zur Frage über die Existenz fossiler Menschen. Verhandl Nat 16:131–153.

Fuhlrott JK (1865): "Der fossile Mensch aus dem Neanderthal und sein Verhaltnis zum Alter des Menschengeschlechtes." Duisburg: Falk and Vollmer.

Geike J (1894): "The Great Ice Age and Its Relation to the Antiquity of Man." London: Stanford.

Gesner J-A (1749): "Nachricht von dem Canstadter Sultzwasser." Stuttgart: Erhardt.

Gillespie NC (1979): "Charles Darwin and the Problem of Creation." Chicago: University of Chicago Press.

Glick TF (ed) (1974): "The Comparative Reception of Darwinism." Austin: University of Texas Press.

Giuffrida-Ruggeri V (1910): Nuove addizioni at tipo di Galley-Hill, e l'antichita della brachicefalia. Arch Antropol Etnol XL:255–263.

Giuffrida-Ruggeri V (1918): Unicita del philum umano con pluralita dei centri specifici. Rev Ital Paleontol (Perugia) 24:1–15.

Gorjanović-Kramberger K (1906): "Der Diluviale Mensch von Krapina in Kroatia. Ein Beitrag zur Paläoanthropologie." Weisbaden: Kridels Verlag.

Gould SJ (1980): The Piltdown controversy. Nat Hist 89(8):8–28.

Gregory WK (1916): Note on the molar teeth of the Piltdown mandible. Am Anthropol 18:384–387.

Gregory WK (1949): The bearing of the Australopithecinae upon the problem of man's place in nature. Am J Phys Anthropol 7:485–512.

Haeckel E (1866): "Generelle Morphologie der Organismen." Berlin: Reimer.

Haeckel E (1870): "Natürliche Schöpfungsgeschichte, 2nd ed." Berlin: Reimer.

Haeckel E (1898): On our present knowledge of the origin of man. Annu Rep Smithsonian Inst pp 461–480.

Halstead LB (1978): New light on the Piltdown hoax. Nature 276:11–13.

Holtzman SF (1970): "History of the Early Discoveries and Determination of the Neandertal Race." Ph.D. Thesis, Department of Anthropology, University of California, Berkeley.

Hooton EA (1930): "The Indians of Pecos Pueblo: A Study of the Their Skeletal Remains." New Haven: Yale University Press.

Hooton EA (1931): "Up From the Ape." New York: Macmillan.

Howell FC (1951): The place of Neanderthal man in human evolution. Am J Phys Anthropol 9:379–416.

Howell FC (1957): The evolutionary significance of variation and varieties of "Neanderthal" man. Q Rev Biol 32:330–410.

Howells WW (1975): Neanderthal man: Facts and figures. In RH Tuttle (ed): "Paleoanthropology: Morphology and Paleocology." Paris: Mouton, pp 389–407.

Hrdlička A (1907): Skeletal remains suggesting or attributed to early man in North America. Bull Bur Am Ethnol Smithsonian Inst, XXXIII. Washington: Government Printing Office.

Hrdlička A (1911): Human dentition from the evolutionary standpoint. Dominion Dent J 23:403–422.

Hrdlička A (1912a): Early Man in South America. Bull Bur Am Ethnol Smithsonian Inst, CII. Washington: Government Printing Office.

Hrdlička A (1912b): Remains in eastern Asia of the race that peopled America. Smithsonian Misc Coll No. 16, LX.

Hrdlička A (1914): The most ancient remains of man. Annu Rep Smithsonian Inst, 1913. Washington: Government Printing Office. pp 491-552.

Hrdlička A (1920): "Human Evolution." Lecture series held at American University, Washington D.C. Manuscript: Hrdlička Papers, National Anthropological Archives, Smithsonian Institution, Washington, D.C.

Hrdlička A (1921): The peopling of Asia. Proc Am Philos Soc Philadelphia LX:535-552.

Hrdlička A (1922): The Piltdown jaw. Am J Phys Anthropol V:337-347.

Hrdlička A (1923a): Dimensions of the first and second lower molars, with their bearing on the Piltdown jaw and man's phylogeny. Am J Phys Anthropol VI:195-216.

Hrdlička A (1923b): Variation in the dimension of lower molars in man and anthropoid apes. Am J Phys Anthropol VI:423-438.

Hrdlička A (1924): New data on the teeth of early man and certain fossil European apes. Am J Phys Anthropol VII:109-132.

Hrdlička A (1927): The Neanderthal phase of man. J R Anthropol Inst 67:249-269.

Hrdlička A (1930): "The Skeletal Remains of Early Man." Smithsonian Miscell. Coll. 83:1-379.

Hublin J-J (1978): Le Torus Occipital Transverse et les Structures Associées: Evolution Dans le Genre Homo. Thèse de Docteur, 3e Cycle, Université de Paris VI.

Huxley J (1942): "Evolution, the Modern Synthesis." London: Allen and Unwin.

Huxley TH (1863): "Evidence as to Man's Place in Nature." New York: Appleton.

Jaeger GF von (1839): Uber die Fossilen Saugethiere, welche in Würtemberg in aufgefunden worden sind, nebst Grognostischen Bemerkungen über diese Formationem. Stuttgart: Erhard.

Kate H Ten (1894): [Notice of Dubois' paper in Verslag van het mijnwezen over het 3e kwartaal 1892. Extra bijvoegsel der Javansche Courant No 10, Batavia, 1893, p 10] Nederl Koloniall Centraalblad, Amsterdam 1:82-83.

Keith A (1895): Pithecanthropus erectus—A brief review of human fossil remains. Sci Prog 3:348-369.

Keith A (1911): "Ancient Types of Man." New York: Harper.

Keith A (1912a): Recent discoveries of ancient man. Bedrock 1:295-311.

Keith A (1912b): The Neanderthal's place in nature. Nature 88:155.

Keith A (1912c): Certain phases in the evolution of man. Br Med J March 734-736.

Keith A (1914a): The significance of the discovery at Piltdown. Bedrock 2:435-453.

Keith A (1914b): The reconstruction of fossil human skulls. J R Zool Centralbl 6:217-235.

Keith A (1915): "The Antiquity of Man." London: Williams and Norgate.

Keith A (1925): "The Antiquity of Man, 2nd ed." Philadelphia: Lippincott.

Keith A (1929): The fossil men of Peking. Lancet September 683-684.

Keith A and McCown TD (1937): Mount Carmel man: His bearing on the ancestry of modern races. In MacCurdy GG (ed): "Early Man." Philadelphia: Lippincott, pp 41-52.

King W (1864a): On the reputed fossil of man of the Neanderthal. Q J Sci London 1:88-97.

King W (1864b): On the Neanderthal Skull, or reasons for believing it to belong to the Clydian Period and to a Species different from that represented by Man. Rep Br Assoc Adv Sci (1863), Notices and Abstracts, pp 81-82.

Klaatsch H (1899a): Die Stellung des Menschen in der Reihe de Saugetiere, speciell der Primaten, und der Modus seiner Heranbildung aus einer niederen Form. Globus 76:329–332, 354–357.

Klaatsch H (1899b): Der gegenwartige Stand der Pithecanthropus-Frage. Zool Centralbl 6:217–235.

Klaatsch H (1900): Die Fossilen Knochenreste des Menschen und ihre Bedeutung fur das Abstammungsproblem. Ergeb Anat Entwickel Wiesbaden 9:415–496.

Klaatsch H (1909): Die fossilen Menschenrassen und ihre Beziehungen zu den rezenten. Ges Anthropol Ethnol Urgesch XLI:537.

Klaatsch H (1910): Die Aurignac-Rasse und ihre Stellung im Stammbau der Menschheit. Z Ethnol 42:513–577.

Klaatsch H, Hauser O (1909): Homo mousteriensis Hauseri. Ein altdiluvialer Skelettfund im Departement Dordogne und seine Zugehörigkeit zum Neandertaltypus. Arch Anthropol 7:287–297.

Kollman J (1895): [Remarks on Pithecanthropus] Ethnol 27:740–744.

Lankester R (1915): "The Missing Link: Diversions of a naturalist." London: MacMillan.

Lartet E (1860): "Mémoire sur la Station Humaine d'Aurignac." Paris.

Lartet E (1861): Nouvelles recherches sur la coexistence de l'homme et des grands mammifères fossiles réputés caractéristiques de la dernière période géologique. Ann Sci Nat 4 Sér. 177–253.

Lartet E, Christy H (1865–1875): "Reliquae Acquitanicae: Being Contributions to the Archaeology and Palaeontology of Perigord and the Adjoining Provinces." London: Williams and Norgate.

Lartet L, Chaplain-Duparc M (1876): Sur une sépulture des anciens troglodytes des Pyrénées, superposée à un foyer contenant des débris humains associées à des dents sculptées de lions et d'ours. CR Int Cong Anthropol Arch Préhist 7th Sess, Stockholm 1874. pp 302–330.

Leakey LSB (1953): "Adam's Ancestors." London: Methuen.

Leakey LSB (1972): Homo sapiens in the Middle Pleistocene and the evidence of Homo sapiens' evolution. In Bordes F (ed): "The Origin of Homo sapiens." Paris: UNESCO, pp 25–29.

Le Gros Clark WE (1947): Anatomy of the fossil Australopithecinae. J Anat 81:300–333.

Le Gros Clark WE (1948): Observations on the anatomy of the fossil Australopithecinae. Yrbk Phys Anthropol 3:143–177.

Le Gros Clark WE (1955): "The Fossil Evidence for Human Evolution." Chicago: University of Chicago Press.

Lubbock J (1865): "Prehistoric Times." London: Murray.

Lucae JCG (1864): Zur Morphologie der Rassenschadel. Einleitende Bomerkungen und Beitrage. Zweite Abt. Abhandl. Senckenberg. Naturforsch Ges Frankfurt 5:1–50.

Lull RS (1917): "Organic Evolution." New York: Macmillan.

Luschan F von (1873): Die Funde von Brux. Mitt Anthropol Ges Wien 3:32–94.

Lyell C (1830–1833): "Principles of Geology." London: John Murray.

Lyell C (1863): "The Geological Evidences of the Antiquity of Man, With Remarks on Theories of the Origin of Species by Variation." London: John Murray.

MacCurdy GG (1924): "Human Origins: A Manual of Prehistory, Vol 1." New York: Appleton.

MacCurdy GG (ed) (1937): "Early Man, as Depicted by Leading Authorities at the International Symposium of the Academy of Natural Sciences, Philadelphia." Philadelphia: Lippincott.

Manouvrier L (1893): Etude sur la rétroversion de la tête du tibia et l'attitude humaine à l'époque quaternaire. Mém Soc Anthropol Paris 4:219–264.

Manouvrier L (1895): Discussion du "Pithecanthropus erectus" comme précurseur presume de l'homme. Bull Soc Anthropol Paris 6:12–47.

Marsh OC (1895): On the Pithecanthropus erectus. Am J Sci 4:213–234.

Marston AT (1936): Preliminary note on a new fossil human skull from Swanscombe, Kent. Nature 138:200–201.

Martin H (1911): Sur un squelette humain de l'époque mousterienne trouvé en Charente. CR Acad Sci [D] (Paris) 153:728.

Mayer A (1864): Uber die fossilen Uberreste eines menschlichen Schädels und Skelettes in einer Felsenhohle des Düssel-oder Neander-Thales. Arch Anat Physiol Med 1:1–26.

Mayr E (1942): "Systematics and the Origin of Species." New York: Columbia University Press.

Mayr E (1951): Taxonomic categories in fossil hominids. Cold Spring Harbor Symp Quant Biol 15:109–117.

Mayr E, Provine WB (eds) (1980): "The Evolutionary Synthesis: Perspectives on the Unification of Biology." Cambridge: Harvard University Press.

McCown TD, Keith A (1939): "The Stone Age of Mount Carmel, Vol 2: The Fossil Human Remains From the Levalloiso-Mousterian." Oxford: Clarendon Press.

McCown TD, Kennedy KAR (1972): "Climbing Man's Family Tree: A Collection of Major Writings on Human Phylogeny, 1699–1971." New Jersey: Prentice-Hall.

Millar R (1972): "The Piltdown Men." London: St. Martin's Press.

Miller GS (1915): The jaw of the Piltdown. Smithsonian Misc Coll 65:1–31.

Moir JR (1915): Pre-Palaeolithic man in England. Sci Prog 12:465–474.

Mollison T (1921): Abstammung des Menschen. Naturwissenschaften 9:128–140.

Newton ET (1895): On a human skull and limb bones found in the paleolithic terrace gravel at Galley Hill, Kent. Q J Geol Soc Lond 51:505.

Obermaier H (1905): Les restes humains quaternaires dans l'Europe centrale. l'Anthropologie 41:385–410.

Osborn HF (1915a): " Men of the Old Stone Age: Their Environment, Life and Art." New York: Scribner.

Osborn HF (1915b): Review of the Pleistocene of Europe, Asia, and North Africa. Ann NY Acad Sci 26:215–315.

Osborn HF (1916): "Men of the Old Stone Age, 2nd Ed." New York: Scribner.

Osborn HF (1927a): Man of the cave period. In "Man Rises to Parnassus, Critical Epochs in the Prehistory of Man." New Jersey: Princeton University Press.

Osborn HF (1927b): Recent discoveries relating to the origin and antiquity of man. Paleobiologica 1:189–202.

Osborn HF (1930): The discovery of Tertiary man. Science 71:1–7.

Penck A (1908): Das Alter des Menschenhgeslichtes. Z Ethnol XL:390.

Prunner-Bey IF (1864): Réplique à Davis au sjet du crâne de Néanderthal. Bull Soc Anthropol Paris 5:775–778.

Quatrefages de Breau JLA de, Hamy EJT (1882): "Crania Ethnica: Les Crânes des Races Humaines." Paris: Ballière et Fils.

Ramstrom M (1919): Der Piltdown-Fund. Bull Geol Inst Up 16:261.

Retzius A (1846): Sur la forme du crâne des inhabitants du Nord. Ann Sci Nat 6:133–172.

Rightmire GP (1979): Implications of Border Cave skeletal remains for later Pleistocene human evolution. Curr Anthropol 20:23–35.

Rutot MA (1909): L'age probable du squelette de Galley-Hill. Bull Soc Belge Geol XXIII:239.

Santa Luca AP (1978): A re-examination of presumed Neandertal-like fossils. J Hum Evol 7:619–636.

Schaaffhausen H (1857a): [On the Feldhofer Cave Skeleton] Sitzungsberichte der niederrheinischenGesselschaft fur Natur-und Heilkunde zu Bonn. Verhandl Naturh 14:xxxviii–xlii.

Schaaffhausen H (1857b): [On the Feldhofer Cave Skeleton] Verhandl Naturh 14:50–52.

Schaaffhausen H (1858): Zur Kentniss der ältesten Rassenschädel. Arc Verbindung Mehreren Gelehrten 453–488.

Schaaffhausen H (1861): Ont he crania of the most ancient races of man. (Trans. by G. Busk of Schaaffhausen's 1858 paper) Nat Hist Rev 1:155–176.

Schaaffhausen H (1866): Les questions anthropologiques de notre temps. Rev Sci 796–797.

Schaaffhausen H (1873): Sur l'anthropologie préhistorique. CR Int Anthropol Arch Prehist Bruxelles 1872:535–549.

Schmerling PC (1833–1834): "Recherches sur les Ossements Fossiles Découvertes dans les Cavernes de la Province de Liège, Tomes 1 and 2." Liège: Collardin.

Schoetensack O (1908): "Unter Kiefer des Homo heidelbergensis, aus den Sanden von Mauer bei Heidelberg." Leipzig: W. Englemann.

Schultz WL (1955): The position of the occipital condyles and of the face relative to the skull base in primates. Am J Phys Anthropol 13:97–120.

Schwalbe G (1897): Über die Schädelformen der ältesten Menschenrassen mit besonderer Berücksichtigung des Schädels von Eguisheim. Mitt Philos Ges Elsass Lothringen 3:72–85.

Schwalbe G (1899a): Studien über Pithecanthropus erectus Dubois. 1. Theil. 1. Abtheilung. Morphol Anthropol 1:16–228.

Schwalbe G (1899b): Uber die Schädelformen der ältesten Menschenrassen mit besonderer Berücksichtigung des Schädels von Eguisheim. Mitt Naturh Ges Colmar 4:119–135.

Schwalbe G (1900): Der Neanderthalschädel. Jahrb Ver Altertsfr Rheinlande 106:1–72.

Schwalbe G (1901): Über die specifischen Merkmale des Neanderthalschädels. Anat Anz 19:44–61.

Schwalbe G (1904): "Die Vorgeschichte des Menschen." Braunschweig: F. Vieweg und Sohn.

Schwalbe G (1906a): "Studien zur Vorgeschichte des Menschen: I. Zur Frage der Abstammung des Menschen." Stuttgart: E. Scheizerbart.

Schwalbe G (1906b): "Studien zur Vorgeschichte des Menschen. II. Das Schadelfragment von Brux und Verwandte Schädelform." Stuttgart: E. Scheizerbart.

Schwalbe G (1913): Kritische Besprechung von Boule's Werk: "L'homme fossile de la Chapelle-aux-Saints" mit eigenen Unterscuhungen. Z Morphol Anthropol 1:527–610.

Schealbe G (1914): Uber einen bein Ehringsdorf in der Nahe von Weimar gefundenen Unterkiefer des Homo primigenius. Anat Anz 47:337–345.

Sera GL (1910a): Nuove observazioni ed induzioni sul cranio di Gibraltar. Arch Antropol Etnol (Firenze) 39:151–212.

Sera GL (1910b): Di alcuni caratteri importanti sinoria non rivelati nel cranio di Gibraltar. Atti Soc Romana Antropol 15:197–208.

Sera GL (1917): Un preteso Hominida miocenico: *Sivapithecus indicus*. Natura 8:149–173.

Sergi G (1914a): La mandibola umana. Rev Antropol (Roma) 19:119–168.

Sergi G (1914b): "L'Evoluzione Organica e le Origini Umane. Induzioni paleontologische." Torino: Bocca.

Sergi S (1929): La scoperta di un cranio del tipo di Neandertal presso Roma. Riv. Antropol 28:457–462.

Sergi S (1939): Il cranio neandertaliano del Monte Circeo. Atti Accad Naz Lincei 29:672–685.

Sergi S (1944): Craniometria e craniografia del primo Paleantropo di Saccopastore. Ric Morfol 21:1–60.

Sergi S (1948): Il cranio del secondo Paleantropo di Saccopastore. Palaeontograph Ital 42:25–64.

Sergi S (1953): I Profanerantropi di Swanscombe e di Fontechevade. Rend Accad Naz Lincei 15:601–608. (This article also appears as "Morphological position of the "Prophaeranthropi (Swanscombe and Fontechevade)". In Howells WW (ed): "Ideas on Human Evolution 1949–1961." Cambridge: Harvard University Press, 1962, pp 507–520.

Simpson GG (1944): "Tempo and Mode in Evolution." New York: Columbia University Press.

Smith FH (1982): Upper Pleistocene hominid evolution in South-Central Europe: A review of the evidence and analysis of trends. Curr Anthropol 23:667–703.

Smith FH, Raynard GC (1980): Evolution of the supraorbital region in Upper Pleistocene fossil hominids from South-Central Europe. Am J Phys Anthropol 53:589–610.

Smith Woodward A (1915): "A Guide to the Fossil Remains of Man in the British Museum." London: British Museum.

Smith Woodward A (1917): Fourth note on the Piltdown gravel, with evidence of a second skull of Eoanthropus dawsoni. With an appendix by G. Elliot Smith. Q J Geol Soc Lond 73:1–10.

Sollas WJ (1895): "Pithecanthropus erectus" and the evolution of the human race. Nature 53:150–151.

Sollas WJ (1900): Evolutional geology. Nature LXII:481.

Sollas WJ (1908): On the cranial and facial characters of the Neandertal race. Proc R Soc Lond [Biol] 199:281–339.

Sollas WJ (1915): "Ancient Hunter and Their Modern Representatives." New York: Macmillan.

Sollas WJ (1924): "Ancient Hunter and Their Modern Representatives." 3rd ed. New York: Macmillan.

Spencer F, Hrdlička A (In preparation): "Aleš Hrdlička: An architect of American Physical Anthropology."

Spencer F (1979): "Aleš Hrdlička M.D. 1869–1943: A Chronicle of the Life and Work of an American Anthropologist." Ph.D. Thesis, Department of Anthropology, University of Michigan, Ann Arbor.

Spencer F (ed) (1982): "A History of American Physical Anthropology, 1930–1980." New York: Academic Press.

Spencer F, Smith FH (1981): The significance of Aleš Hrdlička's "Neanderthal Phase of Man": A historical and current assessment. Am J Phys Anthropol 56:435–459.

Stewart TD (1962): Neanderthal scapulae with special attention to the Shanidar Neanderthals from Iraq. Anthropos 57:779–800.

Straus WL, Cave AJE (1957): Pathology and the posture of Neanderthal man. Q Rev Biol 32:348–369.

Stringer CB (1974): Population relationships of late Pleistocene hominids. A multivariate study of available crania. J Archaeol Sci 1:317–342.

Stringer CB (1978): Some problems in Middle and Upper Pleistocene hominid relationships. In Chivers D, Joysey K (eds): "Recent Advances in Primatology, Vol 3." London: Academic Press, pp 395–418.

Todd TW (1914a): Neanderthal Man. Cleveland Med J 13:375–384.

Todd TW (1914b): The ancestry of Homo sapiens. Cleveland Med J 13:460–469.

Thoma A (1966): L'Occipital de l'homme mindelien de Vertesszöllös. Anthropologie 70:495–535.

Thoma A (1969): Biometrische Studie uber das Occipitale von Vertesszöllös. Z Morphol Anthropol 60:229–242.

Trinkaus E (1975): "A Functional Analysis of the Neandertal Foot." Ph.D. Thesis. Department of Anthropology, University of Pennsylvania, Philadelphia.

Trinkaus E (1976): The morphology of European and Southwest Asia Neandertal pubic bones. Am J Phys Anthropol 44:95–104.

Trinkaus E (1981): Neandertal limb proportions and cold adaptation. In Stringer CB (ed): "Aspects of Human Evolution." London: Taylor and Francis, pp 187–224.

Trinkaus E (1982): A history of Homo erectus and Homo sapiens paleontology in America. In Spencer F (ed): "A History of American Physical Anthropology 1930–1980." New York: Academic Press, pp 261–280.

Trinkaus E, Howells WW (1979): The Neandertals. Sci Am 241(6):118–133.

Turner W (1895): On M. Dubois' description of remains recently found in Java and named by him Pithecanthropus erectus. With remarks on the so-called transitional forms between apes and man. J Anat Physiol 9:424–445.

Turville-Petre F (1927): Researches in prehistoric Galilee, 1925–1926. In: "Researches in Prehistoric Galilee." London: British School of Archaeology in Jerusalem, pp 1–52.

Vallois HV (1949): The Fontechevade fossil man. Am J Phys Anthropol 7:339–362.

Vallois HV (1954): Neanderthals and praesapiens. J R Anthropol Inst 84:111–130.

Verneau R (1895): Encore le Pithecanthropus erectus. Anthropologie 6:725–726.

Virchow H (1920): "Die Menschenlichen Skeletreste aus dem Kampfe 'schen Bruch im Travertin von Ehringsdorf bei Weimar." Jena: Fischer.

Virchow R (1872): Untersuchung des Neanderthal-Schädels. Verhandl Berliner Ges Anthropol Ethnol 4:157–165.

Virchow R (1895): Pithecanthropus erectus Dubois. Z Ethnol 27:336–337.

Vogt KC (1865): "Lecons sur l'Homme, sa place dans la création et dans l'histoire de la terre." (Trans. J.J. Moulinié). Paris: Reinwald.

Volz W (1896): Ueber Pithecanthropus erectus Dub. Eine menschenahnliche Ubergangsform aus Jaya. Jahres Ber Ges Kultur Sitz Zool Bot 74:5–8.

Wagner R (1865): Über einige Sendungen von Schädeln, die in der letzten Zeit. . . . gemacht worden sind etc. Göttinger Nachrichten 1864, pp 87–99.

Waldeyer W (1895): [Remarks on Pithecanthropus] Z Ethnol 27:88.

Wallace AR (1887): Antiquity of man in America. The Nineteenth Cent. Nov. pp 667–679.

Wallace AR (1889): "Darwinism." London: Macmillan.

Warren KB (ed) (1951): Origin and evolution of man. Cold Spring Harbor Symp Quant Biol 15.

Waterston D (1913): The Piltdown mandible. Nature 92:319.

Weidenreich F (1926): Fundbericht uber ein in den Travertinbruchen von Weimar-Ehringsdorf gefundes Schadel-Fragment vom Neanderthal Typus. Verh Ges Phys Anthropol Freiburg 32.

Weidenreich F (1928): "Der Schädelfund von Weimar-Ehringsdorf." Jena: Fischer.

Weidenreich F (1940): Some problems dealing with ancient man. Am Anthropol 42:375–383.

Weidenreich F (1943a): The "Neanderthal Man" and the ancestors of "Homo sapiens." Am Anthropol 45:39–45.

Weidenreich F (1943b): The skull of Sinanthropus pekinensis: A comparative study on a primitive hominid skull. Paleontol Sinica 127:1–484.

Weidenreich F (1945): Giant early man from Java and South China. Anthropol Papers Am Mus Nat Hist (New York) 40:1–134.

Weidenreich F (1947): Facts and speculations concerning the origin of Homo sapiens. Am Anthropol 49:187–203.

Weidenreich F (1949): Interpretations of the fossil material. In Howells WW (ed): "Early Man in the Far East. Studies in Physical Anthropology, No. 1." AAPA. pp 149–157.

Weidenreich F (1951): Morphology of Solo man. Anthropol Papers Am Mus Nat Hist (New York) 43:205–290.

Weiner JS (1955): "The Piltdown Forgery." London: Oxford University Press.

Weiner JS (1958): The pattern of evolutionary development of the genus Homo. S Afr J Med Sci 23:111–120.

Weiner JS, Campbell BG (1964): The taxonomic status of the Swanscombe skull. In Ovey CD (ed): "The Swanscombe Skull." London: Royal Anthropol. Inst., pp 175–209.

Weiner JS, Oakley FP, Le Gros Clark WE (1953): The solution of the Piltdown problem. Bull Br Mus Nat Hist Geol 2(3):141–146.

Weinert H (1944): "L'Homme Préhistorique." Paris: Payot.

Weinert H (1947): "Menschen der Vorzeit: Ein Uberblick Uber die Altsteinzeitlichen Menschenreste, 2nd Ed." Stuttgart: F. Vieweg.

Weinert H (1951): "Stammesentwicklung der Menschheit." Stuttgart: F. Enke.

Weinert H (1953): Der Fossile Mensch. In Kroeber AL (ed): "Anthropology Today." Chicago: Chicago University Press, pp 101–119.

Wolpoff MH (1971): Vértesszöllös and the presapiens theory. Am J Phys Anthropol 35:209–215.

Wolpoff MH (1980): "Paleoanthropology." New York: Knopf.

The Origins of Modern Humans: A World Survey of the Fossil Evidence, pages 51–135
© 1984 Alan R. Liss, Inc., 150 Fifth Avenue, New York, NY 10011

The Origin of Anatomically Modern Humans in Western Europe

C.B. Stringer, J.J. Hublin, and B. Vandermeersch

Department of Palaeontology, British Museum (Natural History), London SW7 5BD, United Kingdom (C.B.S.), Laboratoire de Paléontologie des Vertébrés et de Paléontologie Humaine, 75230 Paris, Cedex 05, France (J.J.H.), and Laboratoire d'Anthropologie, Université de Bordeaux, Talence, France (B.V.)

INTRODUCTION

Only in the last 30 years has the focus of discussions about the origin of anatomically modern (a.m.) *Homo sapiens* shifted from western Europe. The reasons for the dominant position of the western European fossils are not difficult to find. First, much of the significant Upper Pleistocene fossil hominid material was discovered while paleoanthropological studies were in their infancy and evidence was still lacking from elsewhere. Second, the evidence was relatively complete and well dated by reference to a framework of glacial chronology, biostratigraphic sequences, and archeological successions. Third, the majority of influential workers in the field were from western Europe, although some, most notably Hrdlička and Weidenreich [see Spencer, this volume], produced much of their research while based in the United States. At the present time, it is easier to see the European hominids in their wider context as not always typical of the fossil hominid and archeological record of the Old World, and as more discoveries are made outside of Europe, it may turn out that even in the Middle and Upper Pleistocene, western Europe was perhaps sometimes peripheral to the main events in technological and hominid population change.

Various scenarios have been proposed to account for the European fossil record, from the extreme of the presapiens theory to a simple unilinear scheme in which all European fossil hominids could be considered as ancestors for a.m. Europeans [see Stringer, 1974; Wolpoff, 1980a; Spencer, this volume; Spencer and Smith, 1981]. The presapiens scheme of Boule [1923] and Vallois [1954] has lost much support recently, although variations of it still persist [eg, Leakey, 1972; Vlček, 1978; Saban, 1982]. Additionally, workers who recognize a long Neandertal lineage in Europe, terminated by the appearance of modern human characteristics derived from elsewhere, automatically imply the existence of a long and separate presapiens lineage, but in this case the separate lineage of modern humans would have developed outside of Europe [eg, see Hublin, 1982, 1983; Stringer, 1974, 1982].

The unilinear schemes of workers such as Schwalbe [1904], Hrdlička [1930], Weidenreich [1943a, 1947], Coon [1962] and Brace [1967] have been further developed in a European context [Smith, 1982] and more generally [Wolfpoff, 1980a], as other contributions in this volume show. The need to invoke gene flow from outside of Europe at the time of the transition to a.m. H sapiens or an early appearance of a.m. hominids elsewhere is seriously questioned by these workers.

Intermediate or compromise schemes have been proposed by various workers such as Sergi [1953], Le Gros Clark [1955], Breitinger [1957], and Howell [1960]. Here generalized earlier Pleistocene populations, often termed "early H sapiens," preneandertals, or early Neandertals, are envisaged as ancestral to both the Neandertals and modern humans, either in Europe or an adjoining area.

INTERPRETATIONS OF THE FOSSIL HOMINID RECORD

If there were a gradual evolutionary transition in western Europe from Neandertals to a.m. H sapiens, the disappearance of Neandertal characters and the appearance of modern characters should not be abrupt; that is, there should be local continuity of Neandertal characters, and some supposedly modern characters should make a precocious appearance in the preceding hominids of the area. If the transition period was long enough, there should be a record of "mosaics" showing a genuinely intermediate morphology during this period. The archeological record should also demonstrate local continuity rather than sudden change. If the so-called pre-Neandertal model is applicable to western Europe, we might expect the record to document a progressive loss of primitive (plesiomorphous) characters shared with a.m. H sapiens through Middle and Upper Pleistocene times, as Neandertal specializations (derived characters or apomorphies) become established instead. On the other hand, if the strict European presapiens scheme is correct we would expect to recognize genuine synapomorphies shared with modern humans in some European Middle Pleistocene fossils. Such fossils should be distinguishable from the contemporaneous lineage of hominids characterized by the presence of Neandertal apomorphies.

In those models denying late Neandertals a place in the evolution of a.m. *H sapiens*, a replacement phase is necessary where the western European record would be expected to document the appearance and persistence of Neandertal apomorphies right up to the time of their rapid replacement. There would be no morphological continuity between the late Neandertal and early modern groups, nor would evolutionary trends in Neandertals toward a modern morphology be expected—unless there was evolutionary parallelism. If replacement was not instantaneous (in terms of the Upper Pleistocene timescale), it might be possible to document the two morphologies (ie, Neandertal and early modern) as broadly contemporaneous. Supporting evidence for the replacement model, and the required accompanying population movement or massive gene flow, would also come from evidence of cultural discontinuity. Furthermore, evidence of novel skeletal or archeological features should be apparent at an earlier date ouside western Europe.

A model involving more moderate gene flow at the time of the appearance of a.m. *H sapiens* in western Europe would suggest a more complicated pattern of population change. Depending on the rate and amount of gene flow from outside the area, there could be indications of gradual morphological change, together with intrusive characters in the transitional populations. Moreover, the pattern could vary from one area to another with "gene swamping" virtually indistinguishable from total population replacement, and a slow infiltration of exotic genes being virtually indistinguishable from total local continuity. However, if the characters in question originated outside western Europe there should again be supporting evidence of their earlier existence elsewhere.

In order to discuss the fossil hominid evidence on a consistent basis it is important to first identify those characters that, in our opinion, serve to differentiate a.m. *H sapiens* from Neandertals. Table I lists some anatomical characters distinguishing modern *H sapiens*. For further discussion of characters used in Table I see Hublin [1978a,b, 1982], Santa Luca [1978], Trinkaus and Howells [1979], Stringer and Trinkaus [1981], Day and Stringer [1982], and Stringer [in press(b)]. Many of the characters that are used to distinguish Neandertals from modern humans are merely plesiomorphies retained in the former and lost in the latter. These features include: low frontal bone, short and flat parietal bone, poorly developed mental eminence, and a high degree of postcranial robusticity. Such characters should not be used to exclude Neandertals from the ancestry of modern humans, although evidence of an early disappearance of these plesiomorphies outside of Europe would provide evidence that European

TABLE I. Anatomical Characters of Modern *H sapiens*

Cranium, mandible, dentition	*Details of usual morphology*
Vault relatively short and high[a]	Basibregmatic height/glabello-occipital length (gol) = ≥0.70, vertex radius/gol = ≥0.64
High frontal bone[a]	Frontal angle = ≤134°
Weak supraorbital torus[a]	Divided into medial and lateral portions, lateral portion especially weak
Parietal bone long and well-curved in midsagittal plane[a]	Parietal angle = ≤138°
Parietal arch long and high, inferiorly narrow, superiorly broad (shape: "en maison")[a]	Bregma-asterion chord/biastrionic breadth = ≥1.19
Occipital bone long, narrow, and not markedly projecting[a]	Occipital angle ≥113°
True external occipital protruberance may be developed[a]	
Canine fossa present	Inferior perimeter of zygomatic process of maxilla retracted compared with superior (orbital) perimeter
Well-developed mental eminence (chin)[a]	
Small dentition, particularly anterior teeth	
Postcranial skeleton	
Gracile morphology[a]	Thin-walled long bones with relatively small articular surfaces. Muscle insertions not strongly marked
Scapula has a single ventral axillary sulcus (less commonly bisulcate) (?)	
Superior pubic ramus is short and stout (?)	

[a]Probable autapomorphy; (?) phylogenetic status uncertain.

Neandertals may not be the best candidates for an ancestral position. The identification in Neandertals of derived characters shared with modern humans (synapomorphies) would be important in demonstrating the close relationship of Neandertals to a.m. *H sapiens*. A large number of synapomorphies might imply a very recent common ancestor for the two groups, or that one group was ancestral to the other. Shared features such as a large brain, reduced lateral development of the supraorbital torus, dental reduction, and reduced facial prognathism may all be indicative of a close relationship between Neandertal and modern humans, unless they are parallelisms independently acquired since the Middle Pleistocene.

The identification of derived characters that are actually unique only to the Neandertals (autapomorphies), however, would provide important evidence for excluding them from the ancestry of modern humans. Proposed autapomorphies for Neandertals are listed in Table II.

TABLE II. Proposed Neandertal Autapomorphies and Common Characteristics

Cranium, mandible, dentition	Details of usual morphology
Cranial shape spherical or oval in *norma occipitalis* (shape: "en bombe")[a]	Maximum diameter mid or low parietal
Occipitomastoid crest usually large compared with mastoid process[a]	
Anterior mastoid tubercle often present[a]	
External auditory meatus positioned superiorly[a]	High relative to root of zygomatic process
Suprainaic fossa[a]	
Lambdoid flattening and occipital protrusion (Chignon or occipital bun)	Lambda-inion portion highly curved
Relatively large sphenoidal angle and flattened cranial base (?)	
Horizontal-oval pattern of mandibular foramen common	
Taurodontism of molars common	
Nasal aperture voluminous, with projecting nasal bones	Lowered and sloping nasal floor
Double arched supraorbital torus	
Extensive pneumatization of torus and maxilla (?)	Frontal sinus extends laterally, not superiorly. Zygomatic process of maxilla inflated
Mid-facial prognathism:[a]	
Low subspinale angle	$< 115°$
Dentition positioned anteriorly	Retromolar space, mental foramen usually under M_1
Low nasiofrontal angle	$< 141°$
Large value for difference between M_1 alveolus and zygomaxillare radii	> 18 mm
Postcranial skeleton	
Short distal limb segments as in cold adapted recent *H. sapiens* (?)	Brachial index < 78.0 Crural index < 81.0
Scapula has dorsal (less commonly bisulcate) axillary sulcus (?)	
Superior pubic ramus long and plate-like (?)	

[a]Probably autapomorphy; (?) phylogenetic status uncertain.

THE FOSSIL RECORD

In the following discussion, the Middle and Upper Pleistocene fossil record of western Europe will be reviewed for evidence of the earliest appearance of modern skeletal characters and the existence of evolutionary trends. A more detailed discussion of many of the sites in question can be found in Oakley et al [1971], H. de Lumley [1976], and Cook et al [1982]. Sites that we consider irrelevant to the question of modern human origins have been omitted.

THE BRITISH ISLES
Swanscombe (Fig. 1)

The three parts of the Swanscombe skull were found in 1935 (occipital), 1936 (left parietal), and 1955 (right parietal) in the upper middle gravels of the site near the present south bank of the river Thames in Kent. Large numbers of artifacts described as "Middle Acheulian" had previously been collected from this and other levels, while the lower gravels and lower loam had yielded "Clactonian" artifacts. Although often attributed to the Hoxnian interglacial, it is possible that the Swanscombe hominid postdates the Hoxnian *sensu stricto* and belongs in the late Middle Pleistocene [Conway and Waechter, 1977; Bridgland, 1980].

Shortly after the discovery, Le Gros Clark [1938b], studying the endocranial cast, estimated the total volume to be 1,325 cm^3 and remarked that the cast showed no features that could distinguish it from those of modern crania. Similarly, Morant [1938] reported that while other archaic hominid crania could readily be distinguished from modern specimens by various measurements, the only unusual features of the Swanscombe fossil were the very broad occipital, the thick cranial walls, and the inferred large size of the sphenoidal sinus [also reported by Le Gros Clark, 1938a]. However, Morant was aware of resemblances between the Swanscombe and Steinheim specimens in their extant parts, and concluded that "it appears that the group they represent was either in direct line of descent of *Homo sapiens*, or if not in the direct line at least closer to it than was *H neanderthalensis*."

This view was taken up more strongly by Henri Vallois [1954], who made the specimen, along with the Fontéchevade remains, a cornerstone of his version of the presapiens scheme which he had taken over from Marcellin Boule [see Spencer, this volume; Boule, 1923; Boule and Vallois, 1957]. In Vallois' view, the Swanscombe skull was morphologically much more modern than Neandertal specimens and was not similar to the Steinheim fossil, which he considered more archaic. The modern features of Swanscombe, in his view, included the general shape, size and endocranial form, while two typical Neandertal specializations—a marked occipital torus and "chignon"—were completely lacking. By analogy with the Fontéchevade remains, Vallois [1954] considered that the morphology of the Swanscombe frontal bone would have been of modern type, lacking a supraorbital torus.

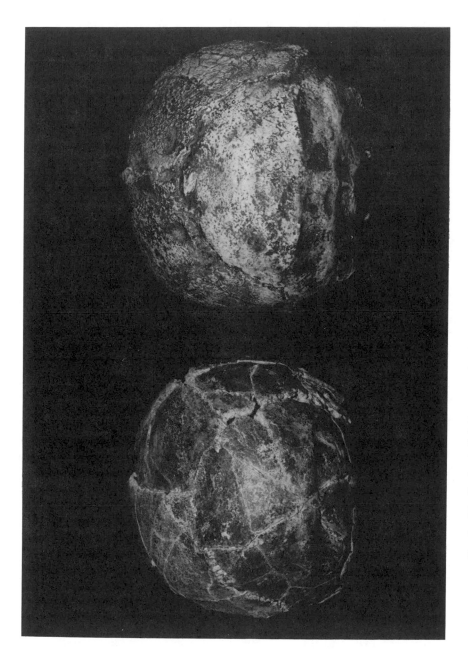

Fig. 1. The Biache skull (left) compared with a cast of the Swanscombe skull in occipital view.

Weidenreich [1943b] agreed that the Swanscombe specimen had "all the appearance of a modern human skull," although he added that there were also certain resemblances to his "Ehringsdorf group" of Neandertals, including the Steinheim skull. However, while conceding that it was "not entirely impossible that the missing frontal bone may have had supraorbitals like those of the Steinheim skull," he considered it doubtful.

Another group of workers such as Breitinger [1957] and Brace [1967] maintained that the Swanscombe fossil could be accommodated within an early Neandertal group. In fact, Breitinger [1964] proposed a reconstruction of the missing parts of the fossil cranium based on the Steinheim skull and like Sergi [1953], Howell [1957, 1960], and Le Gros Clark [1955], he suggested that the Swanscombe specimen belonged to an early unspecialized hominid group that could have been broadly ancestral both to Neandertals and modern humans. This group has been variously termed the "preneandertals," "early Neandertals," "progressive Neandertals," "early *H sapiens*," or "prophaneranthropi." Weiner and Campbell [1964] provided metrical data in support of this view.

The idea of a close relationship between Swanscombe and other archaic European hominids, including Neandertals, has gained much support recently. Metrical analyses by workers such as Stringer [1974, 1978] and Corruccini [1974] have provided supporting data, while detailed comparisons by Howell [1960], Wolpoff [1980b], and others have emphasized the specific resemblances between the cranial remains of Swanscombe and Steinheim. Also recent work has amplified Stewart's [1964] identification of a specific Neandertal character in the Swanscombe specimen by the recognition of further relevant features [eg, Hublin, 1978a,b, 1982, 1983; Santa Luca, 1978]. Hublin has suggested that the apparently modern cranial shape of the Swanscombe specimen in *norma occipitalis* may merely reflect the fact that this is basically a plesiomorphous condition. This is an important consideration, since the rounded occipital contour of the Swanscombe skull in lateral view also appears more modern than that of many Neandertals, but this too may have more to do with the lack of a Neandertal "chignon" than to the possession of a genuinely modern occiput. The Biache fossil (see page 89) illustrates how the accentuation of lambdoid flattening in specimens such as Swanscombe or Steinheim could produce a more typical Neandertal posterior vault morphology.

Pontnewydd

The first hominid specimen (a molar) from this site in North Wales was discovered circa 1870 and then lost! Recent excavations by Green and co-workers [1981] have recovered six hominid specimens. Four of these (PN1— probably a left adult M^2, PN4—a fragment of right maxilla containing dM^2 and M^1 from an individual aged c 8–9 years and two lower premolars) were recovered in a stratified context, while the others (PN2—a posterior fragment of the right mandible of an immature individual containing an unerupted molar, and PN3—a fragment of an adult thoracic vertebra) were found in disturbed strata. However, even the stratified material is in a derived context since a series of mudflows appear to have transported this hominid material, along with fauna and "Upper Acheulian" artifacts, from their original depositional context or contexts near the cave entrance. Dating of the Pontnewydd site by uranium series and thermoluminescence indicates that the stratified hominid finds were emplaced by c 200 ky. The unstratified finds are too fragmentary to be positively excluded from representing a.m. *H sapiens*, and therefore cannot be assigned to any other hominid taxon. The "stratified" molars closely match the size and morphology of certain Neandertal specimens, notably those from Krapina, particularly in their distinctive root morphology and taurodontism [see below, and Stringer, in press(a)]. While the unusual form of the teeth cannot be said to be outside the range for a.m. *H sapiens* [see Keith, 1925; Oakley et al, 1971, p 264], the only Pleistocene specimens closely resembling them are to be found in Neandertal samples. Although the Pontnewydd specimens do not provide clear evidence of the early emergence of any modern skeletal features, they can be used tentatively to reinforce the model of an early European emergence of characters later found (but in this case not exclusively) in Neandertals.

La Cotte de St. Brelade

Twelve hominid teeth and an incisor root were recovered from the cave of La Cotte de St. Brelade, Jersey, Channel Islands, during excavation of a hearth midden in 1910-1911. The molar teeth in this collection are characterized by their unusual prismatic root morphology and an accompanying apical expansion of the pulp cavities—a condition otherwise known by the term "taurodontism," which was coined by Sir Arthur Keith (1913). While known in other fossil hominids, it is highly developed and common in Neandertal samples and most notably those from Krapina [Kallay, 1963; Skinner and Sperber, 1982]. All the La Cotte teeth appear to be from a single (adult) individual, and appear to closely resemble other Neandertal specimens in morphology and dimensions [Stringer and Currant, in press]. A radiocarbon date of c 47 ky has been determined on charcoal extracted from the Mousterian levels of the cave, and in all probability represents a minimum age. But as it was taken from a section of the cave not associated with the hominid teeth, the relevance of the date is uncertain. Further hominid material was claimed on the basis of finds made in 1915 and 1954–1958 [Oakley et al, 1971]. However, only the 1915 occipital fragment can be reliably identified as hominid—but since it too was recovered from deposits outside the present cave entrance it is without clear faunal or archeological associations. Angel and Coon [1954] considered that it may have derived from a Neandertal child, but as their closest parallel was Skhūl 1 (now regarded by most workers as an a.m. *H sapiens* specimen), this view must be treated with caution. The fragmentary nature of the specimen and its uncertain antiquity preclude any firm attribution.

Picken's Hole

In 1961, two teeth were found at this cave site in Somerset which may represent the earliest known remains of a.m. *H sapiens* from the British Isles [Oakley et al, 1971]. The teeth comprise a moderately worn lower incisor and a heavily worn upper premolar, both small in size with no discernible characteristics by which they can be distinguished from those of modern humans. The associated last glaciation fauna incudes suslik, arctic fox, mammoth, and woolly rhinoceros, and although artifacts are present slightly lower in the sequence, hyaena activity appears to have contributed to the bone accumulations in the cave [Stuart, 1982]. A radiocarbon date on bone collagen from this layer (layer 3) of c 34 ky places these teeth amongst the oldest remains attributed to modern *H sapiens* in Europe. However, another collagen radiocarbon date from the underlying layer 5 inexplicably produced a younger age of circa 26 ky. This latter date has been questioned on the grounds of possible contamination, since it derives from a deposit that may represent interstadial conditions, and underlies artifacts that have been described as "Middle Palaeolithic" [Burleigh and Hewson, 1979]. If the older date for layer 3 can be substantiated, the importance of Picken's Hole from a paleoanthropological viewpoint would be in its chronological position close to the interface between Middle and Upper Paleolithic industries elsewhere in Europe. The meager hominid material gleaned from this site, however, does not allow further comment on the evolutionary significance of the specimens.

Kent's Cavern

This cave, first excavated extensively by Pengelly from 1865 to 1880, has provided the most complete archeological sequence from any single British site, spanning the Lower, Middle, and Upper Paleolithic and continuing on through into the Holocene. Unfortunately, most of the excavation work was not properly controlled or published. However, Pengelly's records, excellent for their time, have enabled workers to attempt a reconstruction of the vertical and horizontal distributions of finds from his excavations [Campbell, 1977]. A number of hominid bones have also been recovered from this site. A right maxilla (KC1), humeral fragment (KC2), and ulna [Oakley, 1980] found by Pengelly, and a cranium discovered in 1925 (KC3) and described by Keith [1931] appear to date from late Upper Paleolithic levels, while a more fragmentary right maxilla (KC4), discovered in 1926 [Keith, 1931], and an undescribed deciduous molar found in 1930 may be older. Campbell [1977] has suggested that the 1926 right maxilla might be associated with early Upper Paleolithic material at the site for which radiocarbon dates from faunal material give a possible age range from 28 to 38 ky. However, the 1926 maxilla closely resembles the supposedly younger KC1 specimen, and both fall completely within the morphological and metrical range of Upper Pleistocene and Holocene Europeans [Keith, 1931; Frayer, 1978; Stringer, personal observation].

Badger Hole

Between 1939 and 1945, excavations at this cave situated near the more famous site of Wookey Hole, Somerset, produced fragmentary remains of at least three individuals, possibly associated with "Proto-Solutrean" artifacts and charred animal bone fragments dated by a radiocarbon technique to > 18 ky. The first of these three individuals is represented by an abraded right mandibular ramus containing the roots and unerupted crowns of the permanent dentition of an individual aged approximately 8–9 years (BH1). The second individual (BH2) is also represented by a mandible, which though lacking the posterior portions, is nevertheless in a slightly better state of preservation than BH1. The age of BH2 has been estimated to be in the region of 4 to 5 years. The third hominid from Badger Hole (BH3) consists of a number of cranial fragments. Although still not described in any detail, the jaws have been illustrated from radiographs [Skinner and Sperber, 1982], and do not appear to display any archaic features. Association with the artifacts and dated faunal remains from the site is difficult to establish from the scanty (unpublished) excavation records, and because of the fact that dynamite was used to recover some of the material, including BH1 (Balch: unpublished records). Although post-Pleistocene material has also been recovered from this site, limited analyses failed to distinguish the relative date of the hominid material from a genuine Pleistocene *Crocuta* bone [Oakley, 1980].

Jacobi [1980] has suggested that the leaf points from the site (without accompanying Aurignacian tools) indicate at least an early Upper Paleolithic age. McBurney [1960], after additional excavations, felt able to propose that this site had been occupied only once early in the Upper Paleolithic. Clearly, until further excavations can be carried out at this site, or more sensitive dating techniques can be applied to the existing material, an early Upper Paleolithic age for the hominid fragments must remain unconfirmed. Morphologically they appear to be entirely modern and even gracile specimens.

Paviland

A human skeleton, probably representing an extended burial with red ochre, was excavated in 1822–1823 from the Paviland (Goat's Hole) cave in South Wales. These remains constitute the oldest moderately complete skeletal evidence of a.m. *H sapiens* from the British Isles. Unfortunately, the skull, vertebrae, and most of the bones of the right side had already been lost, probably by marine erosion. The remains of a second individual, represented only by a metatarsal and distal humeral fragment, were excavated in 1912 [Sollas, 1913]. The original skeleton, colloquially known as the "Red Lady," in fact represents a young adult male [Molleson, 1976] with a crural index typical of Upper Paleolithic and recent European values, rather than the lower values typical of Neandertals [Trinkaus, 1981]. A number of problems still surround the dating of the Paviland hominids [Jacobi, 1980; Oakley, 1980]. Archeological material, including ivory artifacts apparently associated with the burial, have been assigned to the "Proto-Solutrean" phase of the British Upper Paleolithic. However, direct radiocarbon dating of the skeletal material gave an age of c 18.5 ky—a date somewhat younger than expected in the light of the archeological diagnosis and close to the last glacial maximum in the area. Doubts over the relevance of the date to all the Paviland Upper Paleolithic material were reinforced by a radiocarbon date of c 27.5 ky on collagen from a bovine humerus derived from the 1912 excavations [Oakley, 1980]. Jacobi [1980] has suggested that there may have been a series of Upper Paleolithic occupations at the site, beginning with the manufacturers of leaf points, followed by groups producing tools and ivory artifacts of Aurignacian or Upper Perigordian affinities. The burial may have been associated with the latter groups. In this case, the direct date on the skeleton simply does not equate with the archeological evidence. Alternatively, the Paviland cave may have been used as a burial site at the extreme range of human activity in western Europe at the time of the last glacial maximum, long after regular human occupation of the area had ceased. This being the case, the main body of archeological evidence from the cave would be older, and unrelated to, the hominid burial [Molleson, 1976].

GERMANY

The German Pleistocene fossil hominid record probably covers the widest range in time and morphology of that of any European country. Unfortunately, the relative chronology of many of the fossils is not clearly established. However, the more complete Pleistocene specimens can be divided into three main groups. The first group (Mauer and Bilzingsleben) displays clear archaic characters and is difficult to relate to either Neandertal or a.m. *H sapiens*. The second group (Neandertal [Feldhofer specimen] and Salzgitter-Lebenstedt) displays clear Neandertal affinities, while the third and largest group contains fossils with an essentially modern morphology. There are also a few sites that have produced specimens of less certain affinities, namely those from Steinheim, Ehringsdorf and Hahnöfersand, which deserve special attention.

Mauer and Bilzingsleben (Fig. 2)

The mandible from the Mauer sand pit near Heidelberg is certainly the oldest known German hominid, and still may be the oldest known European hominid. Although a very robust specimen, it has a relatively small dentition, showing a moderate degree of taurodontism in the molars. Comparison of the Mauer mandible and Petralona cranium [Stringer, 1981] indicates that the Mauer face would have been more prognathic. The absence of retromolar spaces and the positions of the mental foramina (somewhat anterior to M_1), however, suggest that the prognathism may have been rather more total (ie, high nasion angle and low prosthion angle *sensu* Howells, 1973) than midfacial (low subspinale angle). The Mauer mandible has been classified as *H erectus* by some workers [eg, Campbell, 1964; Wolpoff, 1980a,b], but such an assessment is based on its overall robusticity and age rather than on specific characters. Since the mandible is also considered to share some characteristics with European fossil hominids not generally assigned to *H erectus* [Aguirrre and de Lumley, 1977; Wolpoff, 1982], its taxonomic position is unclear. Certainly its marked robusticity and lack of clear Neandertal characters (as mentioned above, together with an apparently "normal" morphology of the mandibular foramen) appear plesiomorphous amongst European hominids.

The hominid material so far recovered from the Steinrinne travertine quarry near Bilzingsleben (Halle) may represent the remains of a single

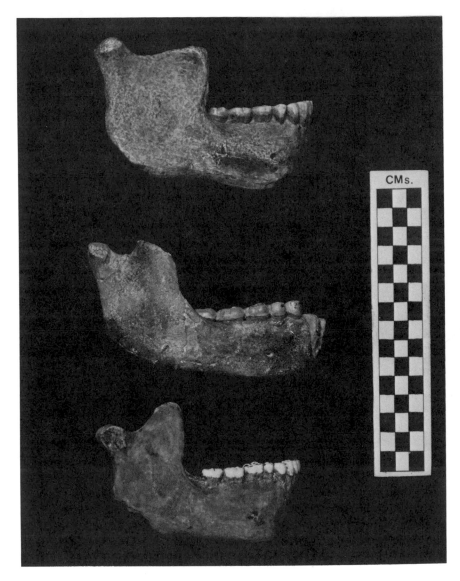

Fig. 2. Casts of Middle and Upper Pleistocene mandibles from Germany and France: Mauer (top), Arago 13 (middle), Régourdou (bottom). Note contrasts in morphology of ascending ramus, retromolar space, position of mental foramina, and symphysis.

individual dispersed by water. Consisting of anterior and posterior cranial fragments and an upper molar, the Bilzingsleben hominid remains have been placed in a new subspecies of H erectus by Vlček [1978]. This classification has been questioned on chronological [Wolpoff, 1980b] and morphological [Stringer, 1981] grounds. However, the anterior frontal fragment indicates a strongly built supraorbital torus with only moderate pneumatization, and a large interorbital breadth. The occipital is low, broad, thick and angulated, with a continuous transverse torus and a large distance between the endinion and external inion. All of these characters are typical of H erectus, yet the gracility of the occipital torus and general resemblances between comparable parts of the Bilzingsleben, Petralona, Arago, and Vértesszöllös specimens has led some workers to suggest that they should be classified as a unit, thereby making their allocation to H erectus more difficult [Stringer et al, 1979; Stringer, 1981]. From the available fragments, the Bilzingsleben remains are the most erectus-like and least modern or Neandertal-like of the group, since they lack the facial or mandibular parts which in the Arago and Petralona specimens appear to resemble Neandertals. Moreover, they lack the parietal portions that appear "progressive" in the same specimens. The fact that the Bilzingsleben remains appear more archaic than the Vértesszöllös occipital in the short and flat occipital plane and inferred lower cranial capacity, presents a problem for a European model of gradual hominid evolution. The Bilzingsleben material is evidently younger than the Vértesszöllös occipital given the usual faunal relative dating (Holsteinian versus Biharian), but Schwarcz [1982] has obtained uranium series ages suggesting that Bilzingsleben could be considerably the older. Although from the hominid remains alone these would be acceptable results, they are difficult to reconcile with the faunal and other data [Cook et al 1982]. But whatever the relative age of the Bilzingsleben material, it only provides evidence for the absence of Neandertal or modern characters in the population it represents.

Steinheim and Ehringsdorf

The Steinheim cranium was discovered in the Sigrist quarry at Steinheim (Würtemberg) in 1933. It has been dated as "Riss" or "Mindel-Riss," with the presumed associated fauna favoring assignment to a late Middle Pleistocene interglacial period comparable with that of the Swanscombe hominid. Two recent discussions of the specimen by Howell [1960] and Wolpoff [1980b] have emphasized resemblances to the Swanscombe specimen, as noted by authors such as Morant as long ago as 1938. These two specimens have usually been assigned to a primitive form of H sapiens (eg, H s *steinheimensis* [Campbell, 1964]) that for some workers might represent the ancestral form for both Neandertals and a.m. *H sapiens*. While some workers such as Wolpoff [1980b] have noted archaic characters in the small cranial capacity and relatively strong supraorbital torus, others have detected in the form of the brows, face, base, or occipital bone characters directly related to modern humans rather than to Neandertals [eg, Leakey, 1972; Vandermeersch, 1978; Laitman et al, 1979]. Assessment of the fossil has not been helped by its previous inaccessibility for study and poor preservation. Nevertheless, two areas of the specimen are particularly important in discussion of its classification—the midfacial area, and the occipital bone. The midfacial area has been described as possessing two non-Neandertal characteristics: midfacial flatness (ie, high subspinale angle [Stringer, 1978]) and a canine fossa [Vandermeersch, 1978]. While both features certainly seem to be present, they may be partly produced by postmortem distortion of this region. The occipital bone is thin and rounded, without lambdoid flattening, and with a gracile morphology of the occipital torus. It is certainly morphologically and metrically distinct from those of both *H erectus* and the more robust European Middle Pleistocene specimens. This difference is seen by such workers as Vlček [1978] as evidence for the coexistence of distinct hominid groups in Europe. Alternatively, Wolpoff [1980b] has suggested that the differences could be accounted for by high sexual dimorphism, whereas Stringer [1981] has argued for chronological and evolutionary differences. The apparent absence of Neandertal apomorphies might be explicable if Steinheim predated the emergence of such characters, but nasal form and occipital torus morphology, with an apparent suprainiac fossa [Hublin, 1978a], can be viewed as synapomorphies with Neandertals. If not caused by damage, the apparent modern characters of the Steinheim face could still be explained as plesiomorphies, and a similar interpretation, viewing the posterior vault form as rather unspecialized, except for one or two probable Neandertal characters, can be applied to both the Steinheim and Swanscombe fossils.

The Weimar-Ehringsdorf hominid material, collected between 1908 and 1925, in adjacent travertine quarries, may represent as many as nine individuals. Some of the material recovered in 1916 and representing a child's partial skeleton, has not yet been fully described because it has required extensive preparation from its surrounding matrix. The cranial material shows significant variation in parietal morphology [Jelínek, 1969] and this is especially true now that the 1925 cranial vault (Ehringsdorf 9 or H) has been reconstructed by Vlček in an attempt to compensate for distortion [Feustel, 1978]. Previous reconstructions of Ehringsdorf 9 by E and K

Lindig, Weidenreich, and Kleinschmidt showed varying degrees of resemblance to Neandertal crania in the orientation of the frontal bone, degree of lambdoid flattening, and rounded or even flattened cranial contours in *norma occipitalis* [Steiner, 1979]. From the new reconstruction by Vlček, the Ehringsdorf specimen has been assigned to *H sapiens praesapiens*. Vlček's new reconstruction displays a moderately vertical frontal bone, parietal bones that approximate the modern form "en maison," but a rather angulated occipital bone.

There is little doubt that the Ehringsdorf parietal bones do indeed seem "progressive" compared with typical examples of *H erectus* or Neandertal crania [see Stringer and Trinkaus, 1981], but this parietal morphology may be apomorphous compared with *H erectus* and plesiomorphous with respect to Upper Pleistocene hominids [Hublin, 1982, 1983]. As in hominids such as Petralona and Broken Hill, the parietal form may more closely resemble the form "en maison" of modern humans because the latter have retained the Middle Pleistocene ancestral pattern to a greater extent than the Neandertals [Hublin, 1982, 1983]. On the Ehringsdorf 9 occipital bone, however, there is a wide suprainiac fossa [Hublin, 1978a] as in Neandertals.

The other material from the site, including mandibular and postcranial parts from adult and immature individuals, shows little approach to the modern human skeletal configuration. The adult mandible (6 or F) specifically resembles Neandertals in the position of the mental foramen and the presence of retromolar spaces. Similarly, Ehringsdorf mandible G (possibly representing two individuals [Legoux, 1966; Oakley et al, 1971; Skinner and Sperber, 1982]) displays a moderate degree of taurodontism in the molars, and the anterior teeth are large, both arguably Neandertal characteristics.

A recent dating of the Ehringsdorf material by Schwarcz [reported in Cook et al, 1982] from uranium series, places the lower travertine and associated hominids at more than 200 ky rather than the expected last interglacial age. While such an age would be quite consistent with the non-Neandertal aspects of Vlček's reconstruction of Ehringsdorf 9, there are still problems of integrating this new date with the extensive archeological, floral, and faunal data from the site.

Feldhofer (Neandertal) and Salzgitter-Lebenstedt

The Feldhofer calotte and partial skeleton were discovered in 1856–57 during quarrying operations in a cave located in the Neander Valley, near Düsseldorf. The specimen was not only named as the type of the first recognized archaic hominid species (ie, *H neanderthalensis* [King, 1864]), but was quickly drawn into disputes about the validity of Darwin's evolutionary ideas as applied to humans [see Spencer, this volume]. Unfortunately, the specimen is undated, since no associated materials were recovered. However, its close resemblance to the Spy 1 fossil suggests that it may date to the last glaciation. Although the relatively small cranial dimensions and capacity might indicate that this individual was female, the supraorbital and occipital tori are relatively well developed, which along with the pelvic remains suggest sexing as male. The overall cranial morphology of the specimen conforms to the typical Neandertal pattern, and there is a wide and simple suprainiac fossa on the occipital bone [Hublin, 1978a].

Excavations conducted during the 1950s at the open site of Salzgitter-Lebenstedt, near Hanover, produced a nearly complete occipital bone and associated parietal fragments, which were not in fact recognized as being hominid until 1963. The site has also produced extensive faunal and archeological evidence [reviewed by Butzer, 1971], suggesting a series of summer occupations in at least the early last glaciation (the associated radiocarbon date of c 55 ky must be regarded as a minimum figure). The hominid remains show a clear Neandertal morphology in their large dimensions, curvature of the occipital plane and possession of a typical occipital structure, including a suprainiac fossa [Hublin, 1978a, 1982].

Late Pleistocene Hominids

As well as the famous late Pleistocene Oberkassel (Magdalenian) skeletons from a basalt quarry near Bonn, there is now a considerable number of German fossil hominids claimed to represent earlier a.m. *H sapiens*. Some of the specimens are dated by archeological or faunal associations eg, Stetten 1 and 3 (Middle Aurignacian) and Dobritz 2 (late Aurignacian or early Magdalenian). Additional cranial specimens have also been collected in recent years from such sites as Paderborn (Sande), Binshof (Speyer), Altrup (Ludwigshafen), Gleidingen/Laatzen (Hanover), Kärlich (Mulheim), and Kelsterbach (Frankfurt-Main) [see Henke and Xirotiris, 1982]. Some of these specimens have yet to be described and many depend heavily or totally for their assignment to the Pleistocene on radiocarbon and amino acid dates provided by R. Protsch. These include the crania from Kelsterbach (c 31 ky [Protsch and Semmel, 1978]), Paderborn (c 27 ky [Henke and Protsch, 1978], and Binshof (c 21 ky [Henke, 1982]). In addition, the 1913 partial skeleton from Neuessing cave 2 (located near Regensburg), a burial associated with red ochre and situated close to a Mousterian horizon, has been dated by the same methods to c 18 ky by Protsch and Glowatzki [1974]. Gieseler [1977] was clearly sceptical about the accuracy of this and other dates. However, the partial cranial vault has never been fully described, and is interesting for its relatively low cranial height and short projecting occiput. None of the other recently described specimens appear to display characters that could be interpreted as "archaic," and because of their gracile morphology, many have been sexed as female [Henke and Xirotiris, 1982]. Statistical analysis relates specimens such as Paderborn and Binshof to Holocene crania as much as to those associated with the Upper Paleolithic [Henke and Protsch, 1978; Henke, 1982, 1983].

The Hahnöfersand frontal bone [Bräuer, 1980, 1981] is an interesting specimen, both from its morphology and claimed antiquity. It was discovered in 1973 without any associated materials on the shore of the River Elbe, by the island of Hahnöfersand, near Hamburg. Initial amino acid dating by Protsch gave the specimen an age of c 36 ky, and this was subsequently matched by a radiocarbon determination, also by Protsch, of 36.3 ± .6 ky [Bräuer, 1980]. Details of the dating procedures have not yet been published, but the nitrogen content of the bone is apparently quite high for a specimen from river sands, given the high fluorine and uranium levels it also displays [Bräuer, personal communication]. The bone is unusually thick at the coronal margins, even compared with Neandertal specimens, and is broad and fairly flat, both sagitally and transversely. While the frontal sinus is developed laterally (as in Neandertals and some a.m. H sapiens), it also extends superiorly (as in a.m. H sapiens). The supraorbital torus is completely modern in form, with a moderately developed central portion, rather similar in form to that of fossils such as Qafzeh 9. Multivariate analyses place the frontal bone closest to Neandertal specimens such as Amud 1 and La Ferrassie 1, rather than to European or southwest Asian early modern human specimens [Bräuer, 1980, 1981].

Although the Hahnöfersand fossil appears to be unusual amongst the European sample, its principal dimensions can be matched quite closely by some of the most robust early a.m. H sapiens specimens from eastern Europe, eg, Pavlov 1, Mladeč 5, and especially Podbaba [Vlček, 1956]. Unfortuantely, Bräuer was unable to include these specimens in his multivariate analyses. He has proposed a model of hybridization between Neandertals and immigrant modern humans to explain the unusual morphology of this specimen [Bräuer, 1981], and this is considered by us to be quite plausible, given the published age for the fossil of more than 30 ky. Alternative views would identify the specimen as a "transitional" fossil between Neandertal and a.m. H sapiens [eg, Smith, 1982], or as a modern but unusual hominid bearing a resemblance to eastern European early modern fossils [Stringer, 1982]. If the radiocarbon accelerator dating program develops as hoped [Stringer and Burleigh, 1981], we would like to see this method applied to the Hahnöfersand specimen in order to confirm its position as probably the oldest European specimen that shows an apomorphous a.m. brow ridge structure. At present we believe it is very difficult to choose between the three competing models to explain its unusual morphology, namely whether it is transitional, hybrid, or merely a robust a.m. *H sapiens*.

IBERIAN PENINSULA

The fossil hominid record from Portugal, Spain, and Gibraltar is somewhat devalued by its very poor chronological control. Material believed to be of Middle or early Upper Pleistocene age which cannot be discussed here are the Lezetxiki (near Bilbao) humerus, and undescribed specimens, including an M^1 from the last interglacial (or "Riss Interstadial") site of Pinilla del Valle, near Madrid [Alferez-Delgado and Molero, 1982]. There is also a cranial fragment claimed to be from the Lower Pleistocene discovered at Venta Micena [Orce, Granada: Gibert et al, 1983]. Other specimens (predominantly cranial) of claimed late Pleistocene antiquity for which we have insufficient data to comment include the material from the caves of Urtiaga (near San Sebastián), Nerja (near Málaga), Barranc Blanc and Parpalló (near Valencia), and Camargo and Castillo (near Santander). It appears that none of these specimens differ from the anatomically modern pattern, and only the probable Holocene calvaria from Cuartamentero, near Oviedo [see Garralda, 1982] is very robust, especially in the morphology of the supraorbital torus and mastoid region.

Portugal

Very few Pleistocene hominids have been reported from Portugal [Veiga Ferreira and Leitao, 1981]. The Salemas quarry site (near Lisbon) has produced faunal material, as well as Paleolithic and Neolithic archeological materials. Some of the skeletal material from the highest levels in the site is clearly Neolithic. However, a partial skeleton of a morphologically modern adolescent has been found which may be derived from the Upper Paleolithic (Solutrean) levels. A deciduous molar found in a lower level is perhaps associated with a Mousterian industry [Ferembach, 1962, 1964–65a]. The cave of Columbeira (Bombarral, near Lisbon) has produced a lower molar crown of moderate size which derives from Mousterian level [Ferembach, 1964–1965b].

Spain

A number of unstratified hominid specimens have been recovered from the site of Atapuerca, near Burgos, since 1976. They have been assigned to the Middle Pleistocene on the basis of their morphology and presumed faunal associations [Cook et al, 1982]. The most important of these specimens are the mandibular body and posterior dentition of a young adult (Atapuerca 1), and the anterior fragments of an adolescent mandible (Atapuerca 2). There are also some isolated teeth representing at least two further adults, and some parietal fragments [Aguirre et al, 1976; Aguirre and de Lumley, 1977; Saban, 1980]. The anterior teeth represented in this collection are robust, while the posterior teeth are relatively small, especially in Atapuerca 1. Although the mandibular bodies are very low and robust, with a receding symphysis and a well-developed alveolar plane, they are very Neandertal-like in the position of the mental foramina (under M_1) and in the presence of retromolar spaces in the more complete specimen. The mandibles appear more like those of Neandertals than does the Banolas specimen (see below), and Wolpoff [1980a] has noted their resemblance to some of the Upper Pleistocene specimens from Krapina, Yugoslavia.

The right parietal bone found in the Cova Negra cave, near Valencia, in 1933, is not well dated since its stratigraphic position was uncertain. The specimen has been tentatively assigned by various workers to the Middle or Upper Pleistocene. The bone is fairly thick, but appears to indicate a broad, rounded cranial contour like that of Neandertal fossils. Although de Lumley [1973] has linked the specimen with the Middle Pleistocene fossils from Arago, Swanscombe and Fontéchevade, rather than with Neandertals, Payá and Walker [1980] considered the archeological, faunal, and stratigraphic evidence to be strongly in favor of a last glaciation age, and this assessment does not seem unreasonable from a morphological standpoint.

The Banolas mandible, found in a travertine deposit in a quarry close to the shore of Lake Banolas, near Gerona, in 1887, is poorly dated. There were no direct faunal or archeological associations, and the only published absolute dates known to us are a dubious radiocarbon date of less than 18 ky on the travertine matrix, and a date of c 120 ky (apparently a uranium series determination) mentioned without further details by H. de Lumley [1978]. The mandible was studied in detail by de Lumley [1973], and is certainly archaic, without being clearly Neandertal-like. Although the dentition shows very heavy wear, particularly on the buccal surfaces, the teeth are nevertheless still visibly large. Compared with the Atapuerca 1 mandible, the Banolas teeth are much larger, yet the mandibular body is less robust, with a thinner and more vertical symphysial area. However, the ascending ramus of the Banolas specimen is large and especially high. Another contrast with the Atapuerca material is that in the Banolas specimen there are no retromolar spaces and the mental foramina are positioned anterior to M_1, suggesting less midfacial prognathism. The lingular area of the mandibular foramen is, unfortunately, damaged. It is a pity that the relative dating of the Atapuerca and Banolas specimens is so uncertain, because their morphological differences could be very instructive in evolutionary terms. From a strictly morphological point of view it seems premature to accept their proposed relative dating to the Middle and Upper Pleistocene, respectively.

A possible early Upper Pleistocene ectocast of a human cranium and cervical vertebrae has been reported from Alicante [Payá and Walker, 1980]. The specimen is formed from a biocalcarenite that has been stratigraphically dated to at least the early Upper Pleistocene. The find, however, was made from rubble in a commercial excavation, and the original site has not been preserved. Given the unusual, but decidedly modern morphology of this cranial ectocast, it seems much more probable that the specimen was formed during the late Pleistocene or early Holocene, either through ectocast formation from an intrusive burial, or perhaps less likely, given its detailed morphology, by direct human action (ie, sculpture). If it does represent a genuine early Upper Pleistocene fossil, its high rounded cranial shape, without a supraorbital torus, and its anatomically modern facial morphology are quite unprecedented in the European fossil record. However, without better data or the benefit of personal study, it is impossible to evaluate this find more clearly.

The hominid material from the Cariguela cave (Píñar, near Granada) was discovered in 1955. The material falls into two groups, and although both have been claimed to derive from Mousterian levels, only the older specimens, from levels 6–7, appear to be nonmodern in morphology. The frontal bone of a child aged about 6 years has a more developed browridge and appears relatively lower and narrower compared with the Devil's Tower frontal bone (see section on Gibraltar), which is of similar size. Two somewhat less ancient, small and uninformative parietal fragments probably also derive from deposits dating from the earlier part of the last glaciation. From level 2 at the site, three further hominid finds were made in a level supposedly containing a Mousterian industry with "Aurignacian influences." These consist of another uninformative parietal fragment, and a mandible and tibia that are morphologically modern in apperance. Absolute dates for these specimens would be very useful, given their supposed archeological associations.

Gibraltar (Fig. 3)

The 1848 cranium from Forbes' Quarry, Gibraltar, was found without a stratigraphic, faunal or archeological context, but it may have been derived

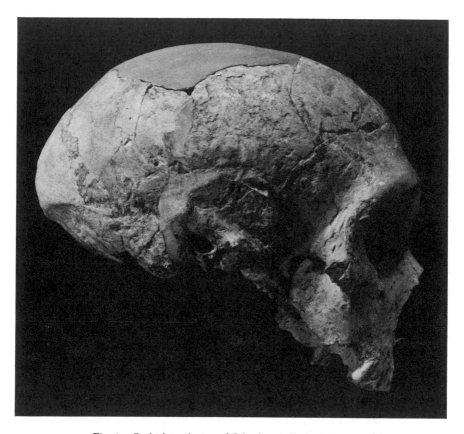

Fig. 3. Right lateral view of Gibraltar 1 (Forbes' Quarry).

from a brecciated cave talus. Payá and Walker [1980] argued that the fossil is probably of last interglacial (*sensu lato*) age, and therefore considerably older than the Devil's Tower material. While this assignment is reasonable given the clear morphological resemblances between the Forbes' Quarry cranium and the probable early upper Pleistocene Saccopastore 1 and Krapina C specimens, it cannot be demonstrated properly from the surviving evidence. All three of these southern European specimens are fairly gracile by Neandertal standards, leading to the reasonable assumption that they represent female individuals. While it is unfortunate that more of the Forbes' Quarry material could not have been recovered, there are enough obvious Neandertal characters in the specimen's mid and upper facial morphology, as well as occipital (projection and suprainiac fossa), and

mastoid area (well-developed occipitomastoid crest) to make the affinities of
the cranium clear. The supraorbital torus in this specimen, however, is
weakly developed, and basicranial flexion appears more marked than in the
specimens from La Ferrassie, La Chapelle, and Guattari (Circeo) [but compare Laitman et al, 1979]. Like Weidenreich [1943b], Condemi [1983] has
noted differences in the form of the sphenoid bone from typical Neandertals, and from the specimens from Saccopastore.

The remains of at least one Neandertal infant were excavated from
cemented blown sand in the mouth of the Devil's Tower Cave, Gibraltar,
in 1926. The age of these remains has been somewhat dubiously inferred to
be greater than 48 ky, based on correlations with a dated Mousterian level
in nearby Gorham's cave. Although generally assumed to represent the
remains of the cranium of a single infant, aged about 5 years, a recent study
by Tillier [1982] has suggested that two children are in fact represented.
Based on their size and developmental morphology, Tillier confirmed that
most of the fragments derived from an infant aged about 5 years (Gibraltar
2). The frontal bone is large and gracile compared with other Neandertal
immature specimens, though the parietal bone already displays a rounded
Neandertal form. According to Tillier [1982], the maxilla, like that of the
Gibraltar adult cranium, does not show signs of the typical Neandertal
morphology (the "extension" type of Sergi [1948]) found in the Guattari
(Circeo 1), La Ferrassie and La Chapelle adult crania, and already incipi-
ently developed in other Neandertal children. The Saccopastore crania also
lack this morphological character, and it is conjectured that this difference
may be due either to evolutionary factors, if the Gibraltar hominids are also
relatively ancient, or geographic factors. The mandible of Gibraltar 2 shows
a predominantly pleisomorphous rather than strictly Neandertal morphology, particularly with regard to its large deciduous teeth and poorly developed chin region. There are, however, two characteristics which are arguably
modern, namely the lack of a genioglossal fossa and the orientation of the
digastric impressions. As mentioned above, Tillier [1982] has proposed that
the right temporal bone (which cannot be articulated with other fragments)
represents another younger individual (Gibraltar 3). According to her, its
developmental morphology is indicative of a 3-year-old child. There are no
duplications of parts in the fragments, and our understanding of Neandertal
ontogeny is so imperfect as to leave room for some doubt about the separate
status of this specimen.

The Iberian Peninsula, with its wealth of Paleolithic sites, is potentially
very important in the documentation of Pleistocene hominid evolution,
though this potential is still nowhere near realization. Much of the hominid

material is poorly dated and without clear archeological associations. Absolute dating of the Neandertal and early modern material would provide a valuable comparison with the material from the French, Upper Paleolithic, and it is hoped that the dating of the Atapuerca and Bañolas material can be clarified by further work. Unfortunately, at present, the data from this region cannot contribute significantly to the problem of the origin of modern humans in western Europe. However, what evidence there is indicates that there are parallels between the archeological sequences of Spain and France, with an interface between Mousterian and Upper Paleolithic (Châtelperronian and Aurignacian) industries occurring between c 30 and 35 ky [Butzer, 1981; Moure Romanillo and Garciá-Soto, 1983].

ITALY
Middle Pleistocene Hominids

Ascenzi [1982] has briefly reviewed possible Middle Pleistocene hominid material from this area. The postcranial fragments from Sedia del Diavolo (near Rome) have been attributed to the "Riss" on the basis of faunal and archeological associations, while those from Cava Pompi (Pofi, Frosinone) are also now believed to date to the Middle Pleistocene based on recent excavations and absolute dates of c 400 ky. Claimed Middle Pleistocene cranial and femoral fragments have been recovered from Castel di Guido (near Rome), while Fontana Ranuccio (Anagni, Frosinone) has produced at least one tooth that may date from more than 400 ky. None of these finds are especially informative in comparison with other Middle Pleistocene material, although the Cava Pompi material does deserve restudy. From Mandrascava, Sicily, remains of a very modern-looking cranium are claimed to be associated with an Acheulian industry in a Middle Pleistocene context [Bianchini, 1982]. Further details of the site are needed before these claims can be assessed.

In 1968, part of the right innominate of an adult female was recovered from the Grotte du Prince, located in Grimaldi, Liguria. This material was studied in detail by de Lumley [1973], who identified various Neandertal characteristics, but noted the shallow acetabulum. The breccia which contained the fossil occurs immediately above the oldest deposit in the cave, a beach that has been interpreted as representing a marine transgression of "Mindel-Riss" or "Riss interstadial" age. The bone has been directly dated by gamma-ray spectrometry to c 230 ky. Barral and Simone [1982] have suggested that the find actually dates to an equivalent of oxygen isotope stage 8 (ie, c 240–280 ky). Also, it has been suggested by Sigmon [1982] that the Prince innominate shows resemblances to the earlier Arago specimen as well as to those of Neandertals.

Upper Pleistocene Hominids

Saccopastore (Fig. 4). The Saccopastore crania were found in a low terrace of the Aniene tributary of the Tiber, near Rome, in 1929 and 1935. Stratigraphic, floral, molluscan, and vertebrate evidence suggest assignment to the warmest stage of the last interglacial (oxygen isotope stage 5e, c 120 ky). Saccopastore 1, the more complete but smaller of the specimens, is generally regarded as a female, while Saccopastore 2 is usually sexed as a male. Saccopastore 1 is rather similar to the Gibraltar 1 and Krapina C fossils. The supraorbital torus was damaged, apparently during excavation, but was probably not strongly developed, while the frontal sinus was

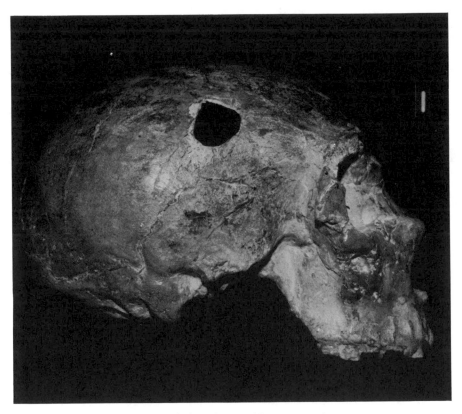

Fig. 4. Right lateral view of Saccopastore 1.

obviously complex and laterally extensive. The cranial vault is rather small, with an endocranial capacity of only about 1,250 cm^3. Although the cranial shape in *norma occipitalis* is "en bombe," in lateral view there is a rounded occiput, which is unlike that of Gibraltar 1 and the typical Neandertal pattern. A suprainiac fossa and occipito-mastoid crests are well developed, but there do not appear to be anterior mastoid tubercles. The face has a large, projecting nasal opening, and its proportions are like those of typical Neandertals [Stringer and Trinkaus, 1981]. Although the extent of midfacial projection is difficult to measure because of damage, it appears to be somewhat less than in specimens such as Guattari (Circeo) 1, and is certainly more marked than in anatomically modern crania or the Petralona and probably Arago 21, and Steinheim specimens. Like typical Neandertal crania, the specimen has a high bregma angle (relatively long nasion-basion dimension), but its low height is reflected in an exceptionally small nasion-bregma angle. Another archaic feature is the low prosthion angle, indicating a greater degree of total facial prognathism than in typical Neandertals or a.m. *H sapiens* [Stringer, 1978; Stringer and Trinkaus, 1981].

Saccopastore 2 is a less complete but visibly larger and more robust specimen. Facially it resembles a reduced version of the Petralona facial morphology more than that of Saccopastore 1 or typical Neandertals, as is indicated by facial proportions [Stringer and Trinkaus, 1981]. The degree of midfacial projection is again difficult to determine in this specimen, but is certainly more marked than in Petralona or a.m. hominid crania—though less marked than in typical Neandertal fossils [Stringer, 1978]. The occipito-mastoid crest is less developed than in Saccopastore 1 or typical Neandertals, and there does not appear to be an anterior tubercle on the surviving mastoid process. Both specimens show Neandertal features in the sphenoid [Condemi, 1983], but have relatively small teeth.

It has been claimed that the Saccopastore crania differ significantly from those of typical Neandertal specimens in a number of features. For example, they have a more concave zygomatic-maxillary structure [Sergi, 1948]. However, comparison of the Petralona and La Ferrassie specimens in this region indicates that this distinction between early and late crania is not always as clear as has sometimes been claimed. Differences in midfacial projection and overall form between the Saccopastore and late Neandertal crania are probably also less significant than has sometimes been claimed [eg, Stringer, 1978; Stringer et al, 1979]. Another claimed distinction is the greater (and more "modern") degree of basicranial flexion compared with the flatter cranial bases of typical Neandertals [Sergi, 1974; Howell, 1951, 1957; but compare Laitman et al, 1979]. There are certainly differences, which are also reflected in the short, deep infratemporal fossae of the Saccopastore (and Gibraltar and Petralona) crania compared with other Neandertals. However, this morphology may well be plesiomorphous, and is retained in modern humans as well as in the Shanidar 1 hominid.

Monte Circeo (including the Guattari Cave). A Neandertal cranium (Circeo or Guattari 1) was discovered isolated in a ring of stones in the Guattari Cave, Monte Circeo (near Rome) in 1939. At the same time, in a different part of the cave, a damaged adult mandible with one molar tooth was found. A more complete mandible was discovered in a breccia at the cave entrance in 1950 (specimen 3). The cave has been partially excavated [Piperno, 1976–1977; Taschini, 1979] and the dating of the hominids seems to be securely placed in the last glaciation, since the marine transgression of the last interglacial isolated Monte Circeo as an island, and the associated littoral deposits underlie the fossiliferous cave deposits.

Guattari 1 (Fig. 5) is a large cranium with many typical Neandertal characteristics in the face and cranial vault [eg, see Sergi, 1974; Hublin, 1978a; Wolpoff, 1980a; Stringer and Trinkaus, 1981]. Neandertal facial proportions and midfacial projection are well developed (although the nasio-frontal angle has a high and "modern" value), the face and supraorbital torus are highly pneumatized, and the cranium shows lambdoid flattening

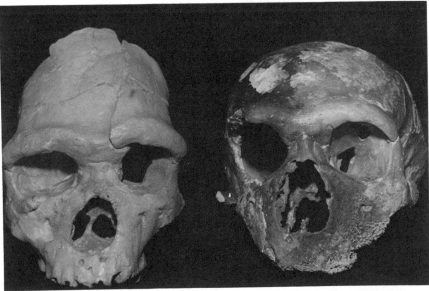

Fig. 5. Middle and Upper Pleistocene hominids form southern Europe. Cast of unreconstructed Arago 21 face (left) and Guattari (Circeo) 1 skull.

and occipital projection in lateral view, and the form "en bombe" in *norma occipitalis*. There is a suprainiac fossa, large occipito-mastoid crest, and an anterior mastoid tubercle (on the left side). In our opinion this cranium is morphologically and metrically very similar to French, Belgian and German Neandertal crania, and less like the Saccopastore specimens. However, an alternative view suggests there was a north-south cline in Europe, whereby the Italian fossils (as well as those from Spain and Yugoslavia) are similar to each other and different from the northern specimens, perhaps because of a lesser degree of cold adaptation [Wolpoff, 1980a]. We contend, however, that this view does not match the data and fails to take into account a number of important temporal differences. For example, the Italian crania derive from different climatic stages and may well differ in age by some 70 ky). Furthermore, the model does not explain how specimens much more distant from western Europe can still seem to show "classic" Neandertal features that have been interpreted as "cold-adapted" [Stringer and Trinkaus, 1981].

The Guattari 2 and 3 mandibles show obvious Neandertal features in the presence of retromolar spaces and the position of the mental foramina. However, Guattari 3 is certainly more "advanced" (in the sense of being more like anatomically modern humans) in its reduced robusticity, smaller teeth, incipient development of a chin, and an inferred lesser degree of midfacial projection (based on the relatively small retromolar space, and position of the mental foramina somewhat anterior to M_1). A more precise chronological placement for the Guattari specimens should be obtainable and would be very valuable. An additional mandibular fragment of a child c 10 years of age was discovered in a Mousterian level of the Fossellone Cave (Fossellone 1 or Circeo 4).

Archi (Reggio Calabria). The mandibular body from an infant aged c 5 years was found at this open site in 1970 [Ascenzi and Segre, 1971]. All the permanent teeth back to M_1 were developing but unerupted, with the deciduous molars and left canine present on the mandible. The specimen is robust with a slightly receding symphysis but some chin development, and overall it bears a marked resemblance to the Devil's Tower (Gibraltar 2) mandible, despite somewhat larger molars. The fossil is attributed to the Upper Pleistocene but the associated fauna lacks species, apart from the Great Auk, which might be diagnostic of a last glaciation age further North. The specimen was found in sands and gravels without reported archaeological material, overlying "grey sands" containing marine molluscs, interpreted as reworked from a Tyrrhenian beach. However there is evidence of a subsequent marine transgression as well, followed by over 90 meters of uplift, which greatly complicates the dating of the sequence. It might be possible to date the fossil by radiocarbon determinations on faunal material stratified with or above the fossil, and a minimum age might be obtained by potassium-argon dating of volcanic debris stratified immediately below the hominid level.

Fate (Liguria). Excavations in this cave during the last century produced a Pleistocene fauna and Mousterian industry, and three hominid specimens have recently been recognized amongst this material [Giacobini and de Lumley, 1982].

Fate 1, a frontal fragment (child aged c 9–10 years), shows a developed supraorbital torus, while Fate 2 is a left mandibular fragment of a child of about the same age, with C-M$_2$ erupted or developing. The specimen is said to be large and robust, with a retreating symphysis. Fate 3 is a right posterior mandibular fragment showing a retromolar space behind a single molar. This tooth is large, while the M$_1$ of Fate 2 is rather small by Neandertal standards. Both dentitions are said to show taurodontism. The mandibles are said to resemble other Mediterranean Neandertals, particularly those from Hortus, while Fate 1 is said to resemble the Cariguela 1 frontal bone.

Late Pleistocene Hominids

The Grimaldi Caves (Liguria). Excavations at the Grotte des Enfants site between 1874 and 1901 produced six hominid partial skeletons, a double inhumation of two children aged c 4–6 years (skeletons 1 and 2), an adult female (skeleton 3), an adult male (skeleton 4), another adult female (skeleton 5), and an adolescent male (skeleton 6)—see Figure 11. The first three skeletons were found considerably higher in the stratigraphic sequence than the better known specimens 4–6. The two children's skeletons were said to have "Aurignacian" associations, but unlike skeletons 4–6 they lacked the associated red ochre and rich ornaments, and may relate more closely in date to skeleton 3 which has been radiocarbon dated to c 12 ky using associated shells [Thommeret and Thommeret, 1973]. The children's skeletons are virtually complete, but have still not received proper study in over a century. The three best-known skeletons (4–6) may have been burials from an early Upper Paleolithic (Aurignacian or Gravettian) level into a sterile layer that may represent an interstadial period. This last level immediately succeeds the latest Mousterian horizon in the cave. The female (5) and the adolescent (6) may represent a double inhumation [but see Sauter, 1983], and these specimens have been termed "negroid." Some, but not all, of the relevant characteristics can be attributed to poor reconstruction. The strong alveolar prognathism and related poor prominence of the chin in the adolescent specimen can be partly attributed to a large dentition. The specimens are not otherwise robust, and possess short, gracile faces. The crania are narrow, with a parallel sided cranial vault in *norma occipitalis*. The post cranial remains are fairly robust, but like those of most Upper Paleolithic fossils, have quite different limb proportions to those of Neandertals (although the adult male has a low brachial index [Trinkaus, 1981]).

Excavations at the caves of Baousso da Torre, Barma del Caviglione, and Barma Grande between 1872 and 1894 also produced fragmentary and more complete skeletal remains, some of which have "Aurignacian" associations. These specimens are modern in morphology, having large endocranial volumes (inordinately so in the Barma Grande 2 reconstruction), broad but low faces and orbits, and little alveolar prognathism. Limb bones are generally very long, with typical Upper Paleolithic proportions, although the Baousso da Torre material (apparently now unlocated) may have had rather low brachial and crural indices [Trinkaus, 1981].

Other early Upper Paleolithic sites. One of the burials from Arene Candide (Liguria) probably dates to the Upper Gravettian or "epi-Gravettian" (c 19 ky [Sergi et al, 1974]). The specimen represents an adolescent individual, with robust postcrania and proportions like those of most Upper Paleolithic skeletons, except for a rather low crural index. The cranium is similar to the Grotte des Enfants specimens in its narrow form, with parallel-sided vault and gracile facial morphology.

In 1971 a partial skeleton, still not fully described, of a male individual aged c 12–13 years was excavated from the base of a Gravettian level in the Paglicci cave (Foggia [Mezzena and Palma di Cesnola, 1972]). Radiocarbon determinations on charcoal and burnt bone from the same level give ages of c 23–25 ky [di Cesnola, 1975]. The skeleton was associated with red ochre and ornaments, and is described as from a tall and long limbed individual. The cranium is not well preserved, but the dentition of this specimen, rather than the "epi-Gravettian" material from the site, was apparently measured by Frayer [1978], whose data indicate a dentition of typical Upper Paleolithic size, but with relatively large anterior teeth.

Excavations of the Cavallo cave (Uluzzo Bay, Puglia) produced three deciduous molars in 1964–1965 [di Cesnola and Messeri, 1967]. The first specimen (dM_2) was associated with a temperate fauna and Mousterian industry, said to contain denticulate and La Quina types, but the other specimens (dM^1 and dM^2) were found in a level succeeding the latest Mousterian at the site, with an early Upper Paleolithic industry termed "Uluzzian." This level is radiocarbon dated from charcoal to >31 ky [Alessio et al, 1970], while elsewhere, finite ages for Uluzzian levels of c 33 ky have been obtained [Gambassini, 1980]. The Uluzzian has been compared with the French Châtelperronian, and di Cesnola [1980] has speculated that the Italian early Upper Paleolithic ("Aurignacian") in fact had a dual origin. One source was the local Mousterian, via the Uluzzian, the other source was allochthonous, with greater parallels with the French Aurignacian proper. Taking the analogy with the French sequence even further, one might expect that the Uluzzian was associated with late Neandertals, as now seems to be the case for the Châtelperronian. The morphology and size of the Cavallo teeth does not contradict such a suggestion particularly in the case of the dM^2, but much more material would be needed to confirm such a model.

Lack of chronological control again hampers assessment of the hominid record from this area, which is all the more regrettable since many of the specimens could be dated radiometrically. However, comparing the early Neandertal crania (Saccopastore) with the only relatively complete later Neandertal specimen (Guattari) shows little evidence of change in the direction of modern humans. In fact, in several respects, the later specimen appears more distinct from modern humans, although the same cannot be said for the Guattari 3 mandible when it is compared with earlier European specimens. However, the available early modern skeletal material appears to show much less sign of archaic characters than the eastern European, or even the French early modern specimens. This may be explicable by the later date of the Italian specimens, but fossils such as Grotte des Enfants 4–6 are not well dated. Archeological analysis suggests a close similarity between the Pontinian Mousterian (as at Guattari) and the French Quina Mousterian [Bietti, 1982], and the available absolute dates suggest that the Mousterian-Upper Paleolithic interface occurred at about the same date in Italy as in France, and perhaps Germany [Hahn, 1981] and Spain, ie, between 30 and 35 ky. It may be that the first appearance of modern humans in Italy also occurred at this time.

FRANCE
Tautavel (Arago) (Figs. 2,5)

La Caune de l'Arago, near Tautavel, 19 km NW of Perpignan (Eastern Pyrenees), is a large cave which has been the subject of systematic excavations since 1964 by a team led by H. de Lumley. The entrance of this cavity opens at about 100 m overlooking a plain and is linked to the plateau above by an aven. The Quaternary fill consists of sediments 11 m thick in which more than 20 occupation floors have been recognized. The deposits have been subdivided into 4 units (ensembles) from bottom to top: I (beds L, K), II (beds J, I, H), III (beds G, F, E, D), IV (beds C, B, A). Stone artifacts are abundant throughout. The industry from almost all the levels has been reported as early Tayacian, but the uppermost level of unit III contains a Middle Acheulian assemblage. The fauna, also rich, provided the initial means of dating. Originally its age was considered to be "Riss I" [de Lumley, 1969–1971]. This assessment was maintained by Chaline [in press] who studied the fossils remains of rodents from the site. Nevertheless, Guérin [in press] later suggested a "pre-Rissian," even "Mindel" age based on the studies of fossil remains of rhinoceroses. Numerous absolute dates have been obtained at the site and the greater part of the results were presented at a CNRS International Colloquium: *Datations absolues et analyses isotopiques en préhistoire, méthodes et limites* [see de Lumley et al, 1984]. The methods applied include amino acid racemisation, uranium series (US), gamma ray analysis, thermoluminescence (TL), electron spin resonance (ESR), fission track, and paleomagnetism. The results from these various methods are contradictory and generally affected by considerable uncertainty. In their conclusions, de Lumley et al (in press) essentially retain the results yielded by US and ESR and propose the following ages:

Upper stalagmitic floor	> 90 ky
Unit V	> 115 ky
Decalcified pocket	> 130–400 ky
Unit IV	> 400 ky
Unit III	> 450 ky
Break in sedimentation	> 450–500 ky
Unit II	> 500 ky
Unit I	> 550 ky
Basal unit	> 550–700 ky
Lower stalagmitic floor	> 700 ky

A more critical synthesis of the various results may be found in Cook et al [1982], as well as Blackwell [1981]. Furthermore, new results have been produced by Aitken [in press] that suggest a maximum age of 450 ky for the lower stalagmitic floor using TL and ESR, and doubt has been raised about the reliability of some of the other determinations [Skinner, 1983; Debenham, 1983].

To date, the series of human fossil specimens discovered at the site amount to more than 50. The main pieces came from beds G and F. The former yielded a mandible (Arago 2), an associated face and parietal (Arago 21 and 47) and a left innominate (Arago 44), whilst bed F produced a half mandible (Arago 13). Numerous isolated teeth and fragmentary postcrania add to these fossils.

Arago 21 and 47 constitute a partial skull that is the most significant specimen from the site. It should be noted, however, that their association has been placed in doubt by Holloway [1982]. Arago 21 belongs to a young adult (the M^3 shows no signs of wear). The bones are considerably distorted and despite an attempt at reconstruction, many of their characters remain difficult to interpret. The face is broad and marked by a well developed supraorbital torus (Fig. 5). The *arcus superciliaris* is particularly distinct. The lower parts of the orbits are broad and rectangular in outline, and are separated by a wide interorbital space. This part of the face does not show the marked midfacial projection that characterizes the Neandertals [Stringer, 1978]. The nasio-frontal angle measured on the reconstructed skull is 154° 30′, well above that of the Neandertals and comparable to that of Skhūl 5 or Zuttiyeh. However, it is difficult to assess the role of postmortem deformation in this flattening [Hemmer, 1982]. The zygomatic bone is gracile relative to the fronto-sphenoidal process and it is positioned obliquely. The maxilla, lacking a canine fossa, is of the extension type. The midfacial region suggests an incipient Neandertal morphology and alveolar prognathism is marked. The palate is wide and deep and shows external tori. M^3 is reduced and M^2 is larger than M^1.

The forehead is flattened and elongated. The maximum breadth is generally lower than found among Neandertals. The parietal is large and shows a discontinuity in its curvature at the level of the temporal line. Seen from the back, the shape is approximately pentagonal, similar to that of Swanscombe. However, the parietal tubers remain less prominent than those of modern humans and the maximum breadth of the skull is situated fairly low. This form separates Tautavel man from the classic *Homo erectus* of Asia and Africa, but it is not incompatible with a Neandertal connection [Hublin, 1982]. The cranial capacity measured on the reconstruction by Holloway is about 1166 cm^3. According to Holloway [1982], the encephalization is a modern feature especially as it occurs in Arago 47. The two mandibles Arago 2 and 13 are quite different and this could be due to marked sexual dimorphism. Arago 13 (Fig. 2) is attributed to a young male and is characterized by considerable robusticity and primitive traits [de Lumley, 1976a]. The horizontal ramus is massive and shows a mandibular torus, and the mental foramen is situated inferiorly. The receding symphysis lacks a chin, but it shows an alveolar plane and genioglossal fossa bordered by two tori. The digastric impressions turn downwards. The ascending ramus is broad, and the sigmoid notch is not very deep. The teeth are comparable in size to those of H erectus with $M_2 > M_1 > M_3$. Arago 2 is more gracile and could have belonged to an old female. The teeth are much more worn. This mandible adds desired Neandertal characters to the primitive characters: the dental arcade tending to flatten in the anterior part with a bi-canine enlargement, posteriorly situated mental foramina and, above all, very marked retromolar spaces.

The complete description of Arago 44 (the pelvis) has not been published yet. According to de Lumley [in press], it is exceptionally robust and the curvature of the iliac crest exhibits a marked angle in its lower part. Such characteristics are found on African fossils as old as O. H. 28 and KMN-ER 3228, and Day [1982] and Sigmon [1982] have stressed the marked morphological stability of this bone in features such as an iliac pillar throughout the long period covered by these three fossils.

The hominid fossils from la Caune de l'Arago could be considered as primitive H sapiens. Some of the possible derived Neandertal characters, notably those of the Arago 2 mandible and less certainly those of the Arago face, may link them to Homo sapiens neanderthalensis of which they could represent a very primitive form, quite different from the classic Neandertals of the early last glaciation.

Montmaurin

The caves of Montmaurin are situated in the Haute-Garonne about 20 km north of Saint Gaudens. They occur on three terrace levels and exploitation of a quarry on the Middle Level cut into the cave of Coupe-Gorge and a vertical fissure called La Niche. In 1949 a mandible in very good state of preservation was found accidentally in the lower part of the fill of La Niche.

In the absence of stratigraphy and correlations between the deposits of La Niche and those of other cavities, the age of this fossil is difficult to establish, all the more so because quarrying had left little of the fill in place. At first the La Niche deposits were linked to those at the bottom of Coupe-Gorge, considered as interglacial, either "Riss-Würm" or "Mindel-Riss," on the basis of the fauna. After excavating the various cavities, Méroc [1963] subsequently worked out a correlation between La Niche and the cave of

La Terrasse situated in the upper system and dated to the "Mindel-Riss." More recently the research of Tavoso [1982] placed this sequence at the end of the "Riss" or the beginning of the "Riss-Würm." A late "Riss" age is supported by the palynological studies of Girard and Renault-Miskovksy [1983]. This later work tends to reduce the age of this fossil which might then be contemporary with some of the La Chaise hominids.

The mandible was recently the subject of a very detailed study by Billy and Vallois [1977]. It is very well preserved and retains its six molars. It belonged to a young adult, for its teeth are only slightly worn and the dentine was not exposed. Their small size have caused them to be attributed to a female but the overall massiveness and robusticity of the mandible compares with the archaic mandibles of Mauer and Arago or those of H erectus. The horizontal ramus, moderately high but very thick, has an index of robusticity of 58.6, close to that of Mauer. Its lower edge is very thick like that of H erectus, and the perimeter of the mandible thickens slightly toward the back. The external face in the region of the symphysis does not show a chin. It does not have an incurvatio mandibulae, but a slight projection in the lower half could correspond to a tuber symphyseos. The internal face exhibits from top to bottom a vast, very raised alveolar plane, an upper transverse torus, a spacious genioglossal fossa, and a lower transverse fossa. The digastric fossae on the lower edge are large. These characters, once again, recall those of H erectus and those of the Neandertals.

The morphology of the mandible body on the external as well as the internal surface evokes that of the "Atlanthropines" and the Mauer and Arago mandibles. But it also has numerous characters in common with the Neandertals: the muscle insertions, while showing a quite archaic arrangement, are not very developed, and the retromolar spaces are moderate.

The very vertical ascending rami have relatively slight dimensions, but their indices match those of H erectus and those of Neandertals. Both externally and internally their relief is marked, particularly the eminentia lateralis and the torus triangularis. The condyle has an articular surface similar to that of Mauer and, as with Neandertals, an important part of this structure is to be found on the outer surface of the ascending ramus. The sigmoid notch is quite deep and the coronoid process large. By its length and the weak divergence of its rami, the alveolar arcade is reminiscent of that of H erectus but its forward flattening recalls that of Neandertals. The teeth are elongated, have five cusps, and their crowns are convex. Radiographs show moderate taurodontism.

Thus a mixture of archaic characters reminiscent of H erectus and of the Mauer and Arago mandibles, and a certain number of traits tending towards Neandertals may be observed on this mandible. It most probably belongs to the pre-Neandertals but its archaic character does not agree well with the relatively recent age given to it in recent studies. If it really dates to the end of the "Riss" or the "Riss-Würm," it is then much more primitive than the mandibles contemporary with it. Thus it would indicate a very great morphological variation in this bone within the Neandertal line.

Lazaret

The cave of Lazaret opens on the western slopes of Mount Boron, east of the port of Nice. Excavated during the nineteenth century, this site was worked during the 1950s by Octobon in Locus VIII, a recess of the cave, also known as Lympia cave. During the 1960s, H. de Lumley excavated the entrance of the cave.

The main part of the deposits belong to the "Riss" [de Lumley, 1969] and the levels from which human remains were discovered, albeit with some uncertainty, to "Riss III" [de Lumley, 1961]. At Lazaret an Upper Acheulian industry is associated with the "Riss III" deposits at Lazaret.

A deciduous left upper incisor and lower right canine, both large, have been recovered from this site. However, the most interesting hominid fossil is a right parietal attributed to a child of about 8 or 9 years old [de Lumley, 1973]. This specimen exhibits a lesion near its sagittal edge, surrounded by a broad zone where the bone is very thin. On the endocranial surface numerous perforations are to be observed in the same area and the courses of the branches of the meningeal artery are modified. According to M.A. de Lumley this may indicate meningitis. It is a bone of elongated proportions and the parietal boss is poorly characterized. The maximum height of the parietal is comparable with or slightly less than that found among the Neandertals. Apart from it being a juvenile specimen, the fragmentary character and pathology make the interpretation of this fossil difficult. It therefore seems questionable to remove it from the pre-Neandertal line as de Lumley has suggested [1973, p 104].

Fontéchevade

The site of Fontéchevade takes its name from a village situated 25 km northeast of Angoulême. The cave forms a huge rectilinear cavern, the vault of which reaches a height of 12 m near the entrance. This site has been excavated since the turn of the century by Valade, Saint Perrier, David and G. Henri-Martin, who pursued research from 1937 to 1955. The latter [1957] proposed the following stratigraphy:

- B Middle Aurignacian
- Bs cemented breccia
- C1 Mousterian with points
- C2 Mousterian with bifaces
- D sterile rubble
- Eo, E1′, E″ upper Tayacian levels
- Bear layer
- E2′, E2″, E2‴ lower Tayacian levels
- Sterile clay
- Bedrock

However, the author points out that "the three subdivisions of E1 and E2 are artificial levels and have no stratigraphic significance." Apart from the Aurignacian remains, the site yielded three hominid fossils from the Tayacian levels (E1): a piece of frontal bone (Fontéchevade I ⟨4⟩), a calotte (Fontéchevade II ⟨5⟩) and a very worn parietal fragment. A fifth metatarsal reported by G. Henri-Martin from the Mousterian of Acheulian Tradition level should also be noted, but it could have been derived from overlying levels [Vallois, 1958].

The stratigraphy of the site has been the subject of lengthy discussions, particularly concerning the age of the Tayacian beds considered as "Riss-Würm" by G. Henri-Martin [1957, 1965]. This dating is based mainly on the presence of animals such as *Dama, Sus scrofa,* and *Testudo graeca* amongst the fauna. However, these "warm" elements in the macrofauna do not seem to have a very clear climatic significance [Debénath, 1974]. Today a consensus seems to be emerging for the attribution of these levels to a "Riss" age. This hypothesis finds corroboration as much from the study of the industry [Bordes, 1968; de Lumley, 1976] as from that of microfauna [Chaline, 1972] or, again, that of the pollen [Bastin, 1976].

The human fossils from Fontéchevade assume a particular importance from the extent that they have served as the foundation of the European presapiens theory along with the Piltdown remains and those from Swanscombe [Vallois, 1949, 1958; also see Spencer, this volume].

Fontéchevade I ⟨4⟩ is a frontal fragment comprising the glabellar region and a small part of the forehead. The absence of supraborbital torus was the main argument used by Vallois [1958] to place this fossil in the lineage of the European presapiens. He argued the same for the calotte, Fontéchevade II ⟨5⟩, which consists of the posterior part of the frontal, the almost complete left parietal, and part of the right parietal. Vallois attempted to reconstruct the supraorbital region of the frontal, relying partly on the position of a small cavity assumed to be a remnant of the sinus and he arrived at the conclusion that the supraorbital torus was absent. However, this interpretation of the material from Fontéchevade has aroused numerous controversies.

In the case of Fontéchevade I ⟨4⟩, Trinkaus [1973] believes that the various characteristics of this fossil, particularly the development of the frontal sinus, fit two hypotheses. It may be an adult individual completely lacking a supraorbital torus or an adolescent of about 12 years of Neandertal type in which the sinus has not yet completed its development and which could have acquired a fully developed supraorbital ridge at a more advanced age. For Tillier [in Vandermeersch et al, 1976] it is an adult or an adolescent (14 to 16 years old) of modern morphology. Moreover, certain doubts exist about the stratigraphic origin of this specimen [Howell, 1958] which was not discovered in situ but removed from a block of matrix brought back to the laboratory. Attention has been drawn to a number of stratigraphic uncertainties at this site [Vallois, 1958, p 157; Henri-Martin, 1957, p 31, note 6; Oakley and Hoskins, 1951, p 241]. However, it should be noted that the fluorine content [Oakley, 1980] agrees with the hypothesis of a Tayacian age for this fossil.

As for the Vallois' reconstruction of the frontal sinus, and the supraorbital relief not preserved on Fontéchevade II ⟨5⟩, it has been criticized by Trinkaus [1973], and Tillier [1977], who reckon that several completely different reconstructions may be derived from the surviving pieces. Tillier also notes that Vallois' reconstruction implies the existence of a sinus with an exceptional morphology. For some, the other characteristics of this poorly preserved calotte are primitive. The thickness is marked and the biasterionic breadth is very large. On the other hand, Neandertal or pre-Neandertal metrical characters are present [Sergi, 1953; Weiner and Campbell, 1964; Corruccini, 1975].

The Fontéchevade fossils seem too fragmentary to be capable of supporting a hypothesis as loaded with implications as that of the European presapiens, particularly if one considers that the two other key specimens in this argument, Piltdown and Swanscombe, have been reinterpreted in completely different ways! It should also be noted that the presapiens character of the Fontéchevade fossils remain unique in the presumed absence of the supraorbital torus, and it is now known that during the Upper Pleistocene at least some of the earliest people of modern type (Skhūl, Qafzeh), still had marked supraorbital relief.

Biache-Saint-Vaast (Fig. 1)

The site of Biache-Saint-Vaast occurs in the fluvial deposits of a terrace on the Scarpe, equidistant from Cambrai and Arras. Discovered accidentally in 1976 in the course of earth moving, it has been the subject of an extensive excavation directed by A. Tuffreau. The stratigraphy has revealed a sequence of human occupations corresponding with temperate climatic conditions, separated by phases of abandonment corresponding with returning cold periods. This sequence was retrieved from thick "Saalian" and "Weichselian" loess deposits. Sedimentological palynological and malacological results agree in placing this sequence "at least at the beginning of the last interglacial cycle of the Middle Pleistocene" [Tuffreau et al, 1982]. The industry, lacking bifaces, but with levallois material, is similar to the more recent classic Mousterian assemblages of northern France and, despite its antiquity, Tuffreau had reason to link it to the Mousterian.

The hominid remains consist of the back half of a cranial vault, part of a maxilla with six molars and five isolated teeth (4 premolars and an incisor). The condition of the sutures allow these remains to be attributed to a subadult individual. The generally gracile character of the material and the weak development of its surface relief have led to it being considered a female. The skull is small and a preliminary evaluation of the cranial capacity gave a result of slightly less than 1200 cm^3. The sagittal profile exhibits a low vault with prelambdoid flattening and a marked chignon. The occipital is not strongly angled but the occipital torus is very clear. Its profile is quite similar to those of classic Neandertals. The same similarity is to be found in the transverse profile that shows the "en bombe" form characteristic of the Neandertal skull (see Fig. 1).

Consideration of the various morphological characters allows numerous resemblances to Neandertals to be emphasized, particularly the following: 1) In the midsagittal sagittal region the occipital torus is divided in two by a suprainiac fossa. Its points of maximum projection are situated at the end of this fossa. The torus fades laterally and does not reach the occipito-mastoid suture. 2) The mastoid region is elongated and oblique towards the sagittal plane. The mastoid process projects slightly but less than the occipitomastoid crest. On the lower face the mastoid notch is closed anteriorly.

The metrical results provide complementary information. Numerous measurements place the Biache skull between those of Steinheim and Swanscombe and thereby allow the latter to be linked to the Neandertal line. The Biache skull belongs to this line more or less indisputably: It is a pre-Neandertal. It occupies a special position both because of its chronological and geographical position. On the one hand, it is the only human fossil discovered in northern France; on the other hand it may be placed chronologically between the classic Neanderthals of the last glaciation and Swanscombe (but see earlier discussion). As its morphology shows certain resemblances to the latter fossil, it provides supplementary arguments for placing Swanscombe in the Neandertal line. Further, it shows us that essentially Neandertal characteristics were established by the beginning of the penultimate glaciation at least as far as the occipital region is concerned [Hublin, 1978a,b, 1982].

La Chaise

Situated about 25 km east of Angoulême (Charente), the sites of La Chaise consist of a series of caves opening onto a Bajocian (Middle Jurassic) cliff that overhangs the valley of the Tardoire. The two most important La Chaise sites are the caves of Bourgeois-Delaunay and Suard. The stratigraphy of the cave of Bourgeois-Delaunay was reported by A. Debénath to show the following sequence from top to bottom: Aurignacian: Würm III; Mousterian: Würm II; Mousterian: Würm I (conventional usage of these terms).

The Würm II/III interstadial is probably represented by a sterile level, in which case the Würm I/II and the last interglacial are characterized by stalagmite floors. The fill rests on a rubble deposit of the late "Riss."

Suard cave shows a sequence of levels from "Riss" II, II/III, and III. The interglacial is marked by heavy brecciation that is overlain by beds of the early last glaciation.

Uranium series dates have given c 245 ky for a level attributed to "Riss" II/III in Suard and 151 ky for a "Riss" level at Bourgeois-Delaunay [but see further discussion in Cook et al, 1982].

The "Riss" beds indicate very severe depositional conditions. The interglacial is marked by high humidity with increasing warmth. A relatively slightly colder climate recurs in the beds at the beginning of the last glaciation and after the marked improvement of Würm I/II, very cold conditions return again.

The pre-Würm lithic industries have been attributed to the Acheulian by Debénath [1974] but they show many characteristics which lead him to recognize in situ evolution toward the Mousterian.

Numerous human remains have been discovered in these sites. Fragmentary and very dispersed, they can be summarized as follows:

Sites	Teeth	Cranial fragments	Mandibular/ maxillary fragments	Postcrania
Bourgeois-Delaunay	12	5	2	3
Suard	25	25	4	5

The majority of the bones of adults belong to young individuals, and children are well represented. As far as the skulls are concerned the most abundant bones are the temporals, parietals, and occipitals; frontals are rare and incomplete and there are only two facial bones, a zygomatic from Bourgeois-Delaunay and a maxilla from Suard. Certain bones found separately can be associated with each other.

The most important specimens are two almost complete occipitals (S), two pieces of calotte (S and B-D), a child's calotte (S), four children's mandibles (S), and two adult mandibles (S and B-D), of which one is almost complete (B-D).

Suard Cave

As far as the adults are concerned the most important specimen is a portion of calotte consisting of an incomplete parietal and frontal, belonging to a young individual. The metopic suture is preserved. The sutures are simple, and the bone thicker than in Neandertal specimens. The height of the parietal lies between that of H erectus and that of Neandertals. On the isolated occipital, the angle of the two planes and the positions of the cerebral and cerebellar fossae are reminiscent of Neandertals [Piveteau, 1976]. A metrical study of this bone [Krukoff, 1970] indicates that certain dimensions are similar to H erectus, while others lie between H erectus and the classic Neandertals. The corresponding cranial capacity has been estimated as 1,050 cm^3. The temporal also presents a mixture of Neandertal characters (a high position of the external auditory meatus, an anteriorly closed mastoid notch, and sagittal position of the stylomastoid foramen) and more archaic traits, such as the large development of the mastoid process.

The children are mainly represented by a frontal, a mandible, and some teeth. The frontal does not have a true supraorbital torus, though it does have a depression between the ridge and the rounded forehead. The mandible has a symphysis in which the external profile is vertical, but the chin has a triangular outline. The deciduous teeth have affinities with those of Neandertals as well as with those of a.m. H sapiens of southwest Asia [Tillier and Genet-Varcin, 1980].

Bourgeois-Delaunay Cave

A skull is represented by a frontal, lacking the supraorbital region, and by a large part of the parietals. Its morphology and dimensions are reminiscent of Neandertals. The morphology of the occipital lies between that of Swanscombe and those of the classic Neandertals, though it is closer to the

latter rather than to that of Suard. The temporal also has a clearly Neandertal appearance. There is an almost complete mandible of which only the
external surface of the symphysis is missing. Numerous characters such as
the presence of a large retromolar space connect it with Neandertals, and
the internal surface corresponds with that observed among Neandertals,
though there is no taurodontism.

The diaphysis of a femur is relatively important since it is a rarely
encountered piece of the postcranial skeleton of pre-Neandertal fossil hominids. It shows most of the traits of the classic form, that is marked
curvature, with a circular section, despite a well developed *linea aspera*. The
medullary cavity is reduced, a character that already existed in *H erectus*.

The La Chaise fossils are indisputably pre-Neandertal. They provide
important information about the evolution of Neandertal characters during
the course of the period that preceded the in situ development of the last
glaciation form, to which they are very close, particularly the Bourgeois-
Delaunay specimens.

CHRONOLOGICAL AND GEOGRAPHICAL DISTRIBUTION OF THE FRENCH NEANDERTALS

From the anthropological point of view, the Neandertal concept is not
well defined. The geographical distribution and chronological limits of this
population have been modified as a result of new discoveries and their
interpretation. We will only consider here the Neandertals of the first part
of the last glaciation, often called the classic or typical Neandertals. However
these chronological limits are arbitrary. As we now know in the light of
Saint-Césaire, the Neandertals survived into the beginning of the Upper
Paleolithic period.

With respect to the geographical distribution of the fossils, a region of
very high density may be recognized in France bordered by the southwest
area of the Massif Central. Having more or less the form of an arc, it
comprises southern Charente, the Dordogne, and the extreme west of
Corrèze. It is in this region that all the skeletons (La Quina, Le Moustier,
La Ferrassie, La Chapelle-aux Saints), crania, and numerous isolated bones
discussed in this text have been found. Beyond this key region, fossil
remains are always fragmentary and generally dispersed. They are distributed around the Massif Central with a zone of slightly higher density in the
south. To the west the département of Vienne has produced several very
fragmentary remains, while to the east, teeth and bones have come from
Saone-et-Loire and the Côte d'Or. The most northerly Neandertals are
those of Arcy-sur-Cure in Tonne. In addition to the Massif Central, the

north and northeast of France, as well as Britanny are devoid of fossil hominids. Neandertal remains therefore come mainly from the calcareous areas penetrated by caves and cut into by rockshelters.

The chronological distribution of these fossils is also very uneven. Those from the beginning of the "Würm" are very rare.[1] With the exception of Régourdou, they are always very fragmentary or just isolated teeth. The important discoveries, in particular the complete skeletons found in the graves at La Ferrassie, La Chapelle-aux-Saints and La Quina, date to Würm II, and therefore to the late Middle Paleolithic. This very unequal distribution seems to be linked to the relative importance of deposits of the various stages of the Würm. Those of Würm I are effectively much less than those of Würm II. Not that the caves and rockshelters were less frequented in Würm I, but it seems that the Würm I/II interstadial was marked in many cavities by a renewal of karstic activity that washed out Würm I deposits.

Thus study of the distribution of Neandertal remains in France shows the main concentration of fossils in Würm II sites of the southwest. The fossils become rarer with increasing distance from this center. But this picture is linked both to the conditions of preservation and the ease of access for researchers. Caves and rockshelters are most suitable for the preservation of archeological levels and they are generally more easily accessible than the Mousterian open air sites, which are less well known because they have been covered by recent deposits which are often very thick. Few have been subjected to systematic excavation in the same way as cave sites. As a result, the known distribution of fossil hominid remains probably does not correspond with the density of Neandertal occupation during different periods of the Würm.

La Chapelle-aux-Saints (Figs. 6, 7). Discovered in 1908, the Chapelle-aux-Saints hominid was by no means the first Neandertal scientifically excavated in Europe, but it remains one of the best preserved and the exhaustive study made by Boule [1911–1913] established it as a reference specimen. Its preeminent position was heightened further by the fact that, according to its discoverers, the skeleton had been intentionally buried [Bouyssonie et al, 1908], an observation that upset the concepts of the period.

The small cave (Bouffia Bonneval) that produced this fossil is situated near Corrèze, 40 km southeast of Brive. The specimen consists of an almost complete skeleton attributed to a male individual, advanced in years and

---

[1]Retaining here the conventional usage of the Alpine terms.

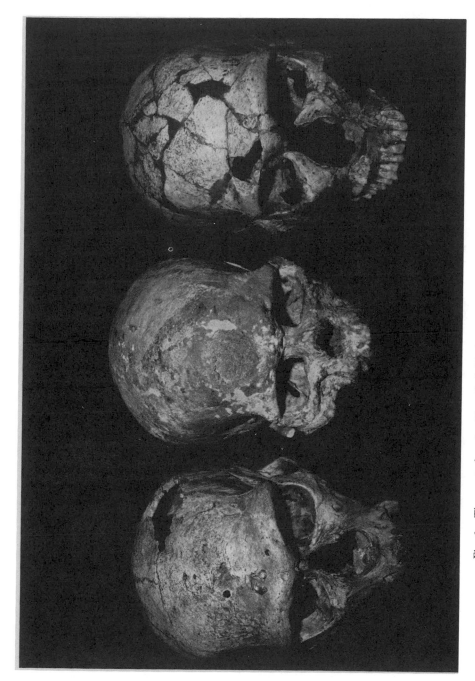

Fig. 6. The crania from La Chapelle (left), Cro-Magnon (center), and La Ferrassie (right).

Fig. 7. Occipital views of crania shown in Figure 6.

showing numerous pathological marks [Dastugue and de Lumley, 1976]: congenital deformation of the left acetabulum, *"patella partita,"* a crushed toe, developed arthritis of the cervical vertebrae and finally a rib broken not long before death.

The hominid type revealed by Boule's study is that of a male of medium height, but very strongly built. The large skull is elongated and flattened, and rounded in posterior view (Figs. 6, 7). The cranial bones are thicker than is typical of modern humans but do not reach the very considerable thickness of *H erectus*. The forehead is low. The brow ridge is marked and does not show a distinct *arcus superciliaris*. The orbits are large and rounded, the nasion is deep set, the nasal cavity is large, and the nasal bones are erect. The maxilla does not show a canine fossa, and the cheek region is gracile and receding. The whole midfacial region projects forward, explaining a series of anatomical details in this area. The temporal has a short and low squama, and the auditory meatus is situated at the level of the zygomatic process root. The mastoid process is small, overlapped below by the occipitomastoid crest. Its lateral face shows a marked anterior mastoid tubercle. The back of the skull is prominent. The broad, low occipital plane is markedly convex. It shows two suprainiac fossae positioned one above the other. The massive mandible does not show a definite chin but a receding anterior surface. However, it is almost completely toothless and it is not possible to observe the usual characteristics of the Neandertal dental arcade.

The postcranial skeleton exhibits a certain number of novel characteristics, some of which were wrongly interpreted by Boule as primitive characters. For example, the appendicular skeleton shows a shortening of the distal segments, which characterizes Neandertals and modern populations adapted to cold climates. The scapula is not represented and the pelvis is too poorly preserved to show the more typical Neandertal characteristics observed on other specimens.

The main derived Neandertal characters have been set out in the introduction to this chapter. The other traits are either primitive characters or derived characters common to the Neandertals and modern humans. The status of the Chapelle-aux-Saints hominid as a reference specimen is the cause of certain number of misconceptions or exaggerations concerning Neandertal characteristics [Hammond, 1982]. This group is homogenous but nevertheless shows notable variability. In relation to La Chapelle-aux-Saints, it should be noted particularly that its cranial capacity (1,625 cm^3) is one of the highest observed amongst European Neandertals and on the other hand, its height as estimated by Boule (154cm) is probably an underestimate [Endo and Kimura, 1970].

La Ferrassie. Of the French sites that have yielded Neandertals, La Ferrassie is the one that has produced the most important assemblage. Two adults represented by almost complete skeletons were, in effect, exhumed there along with six more or less well preserved children, ranging in age from the perinatal stage to about 12 years. The site is situated in the Dordogne, 40 km southeast of Périgueux. The important Mousterian and Upper Paleolithic sequence was mainly exploited at the turn of the century and the greater part of the discoveries were made in stages by Peyrony and Capitan between 1910 and 1921. However, one of the children was found in 1973 by Delporte during an excavation to ascertain the stratigraphy of the site. The assemblage of anthropological material has been studied by Heim [1976, 1982a].

Ferrassie I is considered to be a male adult of advanced years (Figs. 6–8). The skull exhibits the classic characteristics of the Neandertals, but the mandible is distinguished by an erect symphysis, with a slight protuberance of the chin. Although the teeth are very worn, the marked retromolar space and the alignment of the front teeth that characterize Neandertals are clearly represented. Unlike La Chapelle-aux-Saints, the scapulae of La Ferrassie I are well preserved and exhibit the form of lateral border encountered elsewhere among the Neandertals, with a broad dorso-lateral groove for the attachment of the *teres minor* muscle [Trinkaus, 1977]. In addition, La Ferrassie I is one of the rare European Neandertal fossils to show a well preserved pubis [Trinkaus, 1976a]. Like the pubic bones of Southwest Asian Neandertals, it shows a vertical flattening and medio-lateral expansion of the superior ramus (Fig. 8).

The children from La Ferrassie form the most complete series of juvenile specimens produced from a single site in France. Even if we do not endorse all the conclusions of the author, the publication of these specimens by Heim [1982b] is the first morphological and metrical study of the complete assemblage.

Less than 2 years after the discovery of the La Chapelle-aux-Saints skeleton, that of La Ferrassie I, and subsequently the other fossil hominids from this site, confirmed the existence of intentional burials in Mousterian levels. Their discoverers, Capitan and Peyrony [1912a,b, 1920, 1921] described the organization of these burials as a collective burial complex. After a critical examination of the results of these workers, Delporte [1976] has stressed the quality of their work for its time and endorses most of their conclusions.

La Quina. La Quina is the third large French site that yielded Neandertal remains during the great age of discoveries in the southwest,

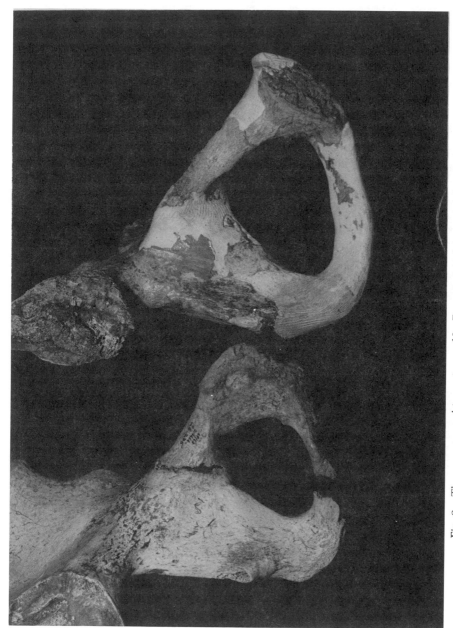

Fig. 8. The reconstructed innominate of La Ferrassie 1 (right) compared with the innominate from Pataud, to contrast the morphology of the superior pubic ramus.

just before the First World War. The site is situated in the departement of Charente, 25 km south of Angoulême. Discovered in 1892, the site, or rather the sites of La Quina that extend for several hundred meters, have been excavated since the end of the last century. However, it was only in 1905 that Dr. H. Martin began a long and exacting study that was subsequently taken over by his daughter, G. Henri-Martin. To date, at least 22 hominid fossils have been found in the Mousterian levels. Among them H5 and H18 are the most significant. H5 is is an incomplete skeleton that has a well-preserved and typically Neandertal skull. However, certain traits are once again special. The cranial capacity is moderate (1,367 cm^3) and the thickness of the wall is at the minimum recorded for Neandertals. These characters have been interpreted as feminine characters of a young adult subject since all the sutures are still open. The forehead is moderately high. Unlike the relatively gracile skull, the mandible is very robust. The postcranial skeleton shows a very marked assymetry of the humerus, possibly of pathological origin [Martin, 1923].

The child's skull (H18) is attributed to an individual about 8 years old. It already shows clear Neandertal characteristics [Martin, 1926]. Again it is notable that one of the mandibles from the site (H9) shows a prominent chin. Except for the numerous publications of Martin [cf Oakley et al, 1971], the La Quina remains have not been the subject of a wide range of publications [Guth, 1963; Vallois, 1969; Hublin, 1980a,b; Gambier, 1981].

Le Moustier. The site of Le Moustier (40 km southeast of Périgueux, Dordogne) was mainly excavated by the Swiss antiquary Otto Hauser, who found a skeleton of a young (perhaps 15-year-old) Neandertal there in August 1908. Removed in questionable circumstances [Virchow, 1939; Vallois, 1940], it was subsequently sold to a Berlin museum. Klaatsch made the first reconstruction of the skull [Klaatsch and Hauser, 1909; Klaatsch, 1909], which was very soon criticized. Others soon followed, including that of Weinert [1925]. In 1944 the skull, like that of Combe-Capelle, was separated from the postcranial skeleton that was virtually destroyed by the burning of the museum on Feburary 3, 1945. The few remaining pieces were collected from below the ruins by Kurth and Heberer in 1955 and have been published recently by Herrmann [1977]. Until then the skull was also thought to be lost, but it was subsequently handed back to authorities in East Berlin by the Soviet occupying force in 1958, and "rediscovered" in 1965 [Hesse and Ullrich, 1966]. The face however, has been largely destroyed. A new study of this fossil is required. Nevertheless its attribution to the Neandertal type is not contestable despite Klaatsch's creation of a new taxon *Homo mousteriensis hauseri*, the taxonomic significance of which did not deceive anybody [Klaatsch and Hauser, 1909, p 38 note].

The Le Moustier site seems jinxed since a child's skeleton discovered in a grave by Peyrony during his excavations [Peyrony, 1930] was also lost without being described.

Other Neandertals From Southwest France

The last major discovery to be considered from southwest France is that of Régourdou near Montignac (Dordogne), 300 m from the famous cave of Lascaux. In 1957, during excavation of a gallery to expose archeological levels, the owner of the site brought to light a Mousterian burial. The excavations in this ancient cave, the roof of which had collapsed onto the Mousterian fill, were then taken over by Bonifay and Vandermeersch. The human bones were found together beneath a sort of cairn of rocks. Part of a well-preserved postcranial skeleton as well as a mandible could be collected. The skull was not found but that it had originally been present at the site is clearly suggested by the presence and condition of the mandible and cervical vertebra. The mandible is one of the best Neandertal mandibles known (Fig. 2). It belongs to a young individual and the dentition is complete. It shows classic Neandertal characters although the teeth are rather small [Piveteau, 1963–65]. The sternum has been studied in detail [Vallois, 1965], as have the vertebrae [Piveteau, 1976]. The appendicular skeleton is at present being studied by Vandermeersch, and the astragalus has been described by Gambier [1982].

Several other Neandertals from the Southwest also deserve mention. In the Dordogne, three sites have yielded the remains of children: Pech de l'Azé, Roc de Marsal and Combe Grenal. At Pech de l'Azé it was once again, Peyrony who discovered the child's skull accompanied by its mandible in 1909. This fossil was the subject of a monograph by Patte [1957], then, after a new reconstruction, of a study by Ferembach et al [1970] who, despite the young age (two years) of the individual, suggested a certain number of "generally attenuated" Neandertal traits. The skeleton of a child from Roc de Marsal belongs to an individual of about three years. It is almost complete, but only the dentition and facial skeleton have been studied to date [Madre-Dupouy, 1976; Tillier, 1983a]. The latter author does not recognize any Neandertal character in the face except for the projection of the nose. Further miscellaneous Neandertal skeletal fragments have been found to date at Combe Grenal, which represent at least six individuals of various ages [Genet-Varcin, 1982]. A molar has also been described from Abri des Merveilles [Trinkaus, 1976b].

Several discoveries apart from the site of La Quina have been made in Charente. The site of Marillac has yielded a mandible and the back of a skull as well as teeth and several cranial fragments. The back of the skull is typically Neandertal [Vandermeersch, 1976a] and is remarkable for the apparent presence of cut marks in the area of attachment of the *galea aponeurotica* and *m. semispinalis capitis* muscles [Le Mort, 1981]. At Chateuneuf-sur-Charente, an isolated incisor and the remains of two children have been discovered. The latter consist of a mandible [Patte, 1957] and a badly damaged skull accompanied by its mandible and one vertebra. The teeth of a third individual have been described by Tillier [1979]. Two fragmentary mandibles, a maxilla, two isolated molars, and a fragment of the petrous part of a temporal have been discovered at Petit-Puymoyen. A Neandertal mandible was found by L. Dupont in 1974 at Montgaudier [Vandermeersch, 1976b]. Further fragmentary remains have come from the sites of Caminero (Castaigne), La Cave, and Le Placard. [Vandermeersch, 1976a]. In the Lot Valley, M. Lorblanchet and L. Genot found a cranial fragment and an I^1 at the site of Mas Viel in 1974. In Lot and Garonne, a series of fragmentary remains (mandible, maxilla, isolated teeth, and skull fragments) have come from Montsempon. They have been studied by Vallois [Coulanges et al, 1952].

The Neandertals of the Midi Region

The most southern areas of France have also produced Neandertal remains although they are not as complete as those recovered in the Southwest.

Two sites in Ariège, Soulabé and Le Portel, have produced isolated teeth, and from the latter, several cranial fragments. However, it is the Malarnaud mandible that particularly highlights this département in the list of French sites with Neandertal remains. Found in 1888, this fossil was the first undisputed Neandertal discovered in France. A reputedly Neandertal vertebra from Malarnaud is deposited in the museum at Bordeaux. It is also possible that other skeletal remains were discovered with the mandible but those have been lost [Pales, 1958]. The mandible belongs to a young individual because the M_3's had not erupted. It is quite gracile and notable for the absence of the I_2's. Its age remains uncertain ("Riss-Würm"?).

In the Mediterranean area of Midi, the richest site is l'Hortus situated about 20 km from Montpellier. This site has produced numerous but fragmentary human remains (between 20 and 36 individuals are represented). The condition of the material led M.A. de Lumley [1972] to resurrect the hypothesis of cannibalism. From a demographic point of view, the rarity of children under 15 should be noted. Morphologically these fossils show classic Neandertal traits (shovel-shaped incisors, taurodontism, alignment of the lower dentition, marked retromolar spaces...). However, as an assemblage they appear much smaller than other west European Neandertals. M.A. de Lumley [1973] distinguishes two assemblages that succeed each other stratigraphically and show increasing gracility of the dentition towards the end of the early last glaciation. According to this author, these features are a particular characteristic of Neandertal populations of the Mediterranean coast.

Some very fragmentary remains, mainly isolated teeth, have also been collected at the sites of La Crouzade (Aude), Bau de l'Aubesier (Vaucluse), Les Peyrards (Vaucluse), La Masque (Vaucluse), Rigabe (Var) and Pié Lombard (Alpes Maritimes).

The Neandertals of Central and Eastern France

The oldest known site in the east of France is that of the cave of Arcy-sur-Cure (Yonne) excavated during the middle of the last century. Discovered in 1859, 3 years after the original Neandertal (Feldhofer) remains, the Arcy mandible contributed to arguments at the time for the contemporaneity of man and extinct animals. However, its appearance is clearly modern. Among the other pieces brought to light at that time were an axis vertebra discovered there by l'abbé Parat and an atlas vertebra discovered by Franchet, which may be attributed to the Mousterian levels, but not with certainty [Leroi-Gourhan, 1958].

Since then, excavations undertaken in 1946 and onwards by A. Leroi-Gourhan's team have assembled a group of human fossils that have the merit of having been collected from known stratigraphic contexts. The Mousterian levels have produced a mandible, a maxilla, a metacarpal, the diaphysis of a fibula, parietal fragments, and isolated teeth. The main interest of the Arcy-sur-Cure fossils lies in the fact that the most recent Châtelperronian levels have yielded a series of isolated teeth that show perfect morphological continuity with those of the Mousterian levels [Leroi-Gourhan, 1958]. The publication by Leroi-Gourhan clearly poses the problem of the persistence of the Neandertals into the beginning of the Upper Paleolithic, a matter that could only be elucidated recently with the discovery of Saint-Césaire.

At Genay (Côte d'Or) cranial fragments and a series of 25 teeth from the same individual have been recovered from a Mousterian context [Joly, 1955]. The teeth of Neandertal morphology are the most robust in size of the sample of this group [de Lumley, 1976b]. Another important dental series comes from Vergisson (Soane-et-Loire). Potentially one of the largest Neandertal molars known [Legoux, 1972] was discovered in 1907 by A.

Laurent at the site of Rivaux, south of the Massif Central (Haute-Loire). Unfortunately its exact age is unknown. A further shovel-shaped upper left incisor, also large in size, has been discovered by Pradel in the Mousterian levels at Rousseau near Angles-sur-l'Anglin (Vienne) [Patte, 1960].

THE UPPER PALEOLITHIC
Saint-Césaire (Figs. 9, 10)

This deep rock shelter is situated in Charente-Maritime about 12 km from the town of Saintes on the banks of the Coron, a small tributary of the Charente.

At the top, the stratigraphy shows a group of Aurignacian beds comprising two levels of advanced Aurignacian, a level of Aurignacian I and a

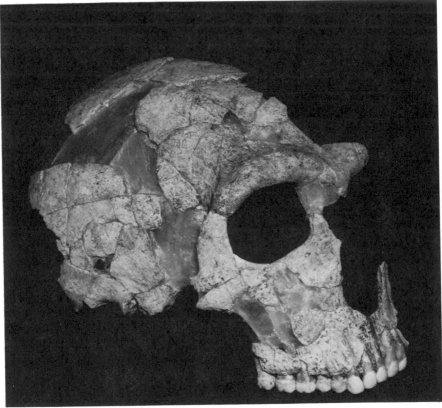

Fig. 9.   The Saint-Césaire skull as reconstructed by B. Vandermeersch.

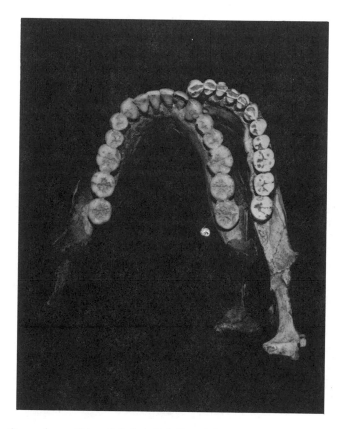

Fig. 10. Casts of mandibles of Qafzeh 9 (left) and Saint-Césaire. Note the great length, large retromolar space, and anterior placement of the (smaller) dentition in the French specimen.

proto-Aurignacian level. Beneath these beds, two Châtelperronian beds of yellow-orange color have been found, and below are the grey Mousterian beds.

The human remains found by Lévêque in 1979 came from the most recent Châtelperronian level. They consist of most of the right half of the skull, the right half of a mandible with the symphysial region, as well as numerous elements of the postcranial skeleton.

The vault of the skull is elongated and low with a supraorbital torus below a marked groove at the front. The forehead is receding. The parietal is elongated and only slightly convex front to back. The temporal squama is quite low. The face shows a large round orbit. The cheek region is sufficiently well preserved to show the absence of the canine fossa. The mandible is long, chinless, with a large retromolar space. The symphysial region is flattened and not evenly curved. This mandible is very similar to La Quina mandible No. 5.

These morphological characters suffice to show that this fossil belongs to the Neandertal population, and examination of the postcranial skeleton confirms this. For example: The radius shows a marked diaphysial curvature, in which the tuberosity is much farther from the proximal epiphysis than among modern *H sapiens* and the form of the distal epiphysis is of a Neandertal type. Similarly, the scapula does not show more than a single groove on the dorsal side as among the Neandertals.

The discovery of a Neandertal with an Upper Paleolithic industry has considerably modified our ideas about the extinction of Neandertals in western Europe and their replacement by a.m. hominids. The Neandertals did not completely disappear with the Mousterian at the end of the Middle Paleolithic. Certain groups persisted into the beginning of the Upper Paleolithic, and not just in backward or isolated areas either. Saint-Césaire is, in fact, situated in a region densely occupied during the Middle and Upper Paleolithic. If Neandertals were able to survive into the Upper Paleolithic, it was not because they were living in those regions ignorant of new arrivals, but more because they were able to adapt themselves as is shown by the technological progress evident in the transformation of the Mousterian into the Châtelperronian.

Nevertheless the Neandertals progressively gave way to a.m. *H sapiens* at the beginning of the Upper Paleolithic. This replacement was not instantaneous—it must have taken several thousand years.

Combe Capelle

In 1909, Otto Hauser, a dealer in antiquities, brought to light a skeleton under uncontrolled circumstances at the site of Combe Capelle (Dordogne). He then sold it to a museum in Berlin, where it was destroyed during the last war. According to Hauser the skeleton was found in a grave at the bottom of the sequence. The oldest level at this site contains a Châtelperronian industry. Given Hauser's personality and the uncertain circumstances of his discovery, the authenticity of this fossil has been argued about continually ever since it was found. Unfortunately, its destruction prevented it from being restudied, leaving its stratigraphic position and age uncertain.

Depite a small stature, less than 1.65 m, the Combe Capelle man, like the Cro-Magnons, shows an elongation of the lower arms and legs and a femur with a very marked *linea aspera*. But the skull is very different, particularly in its face. The skullcap is elongated and keeled, with an oblique forehead and marked orbital relief. The occipital region, evenly rounded, is not very prominent. The long, narrow face has large orbits which are more rounded than those of the Cro-Magnons. The nasal aperture is broad and bordered at the bottom by a prenasal fossa. On the mandible the external profile of the symphysial region is subvertical but there is nevertheless a well-marked, triangular chin. Overall, the skull gives the impression of robusticity. This robusticity and the allegedly archaic character of certain morphological traits (eg, obliquity of the frontal, marked supraorbital relief, vertical symphysis) combined with the supposed early date of the specimen have meant that the Combe Capelle man has often been considered as a rather archaic *H sapiens*, and sometimes even as a transitional form between Neandertals and a.m. *H sapiens*.

The discovery of the Saint-Césaire fossil hominid, indicating the persistence of Neandertals well into the beginning of the Upper Paleolithic and their link with the Châtelperronian, necessitates a reconsideration of the problem; all the more so as the Combe Capelle "type" is only represented by this single individual. While awaiting new discoveries, and bearing in mind the uncertainties that surround this skeleton, it is difficult to consider it as a representative of a Paleolithic population. However, this would then confirm that apart from the Saint-Césaire skeleton for the Châtelperronian, we know nothing of the population of the beginning of the French Upper Paleolithic, particularly those of the early Aurignacian.

Cro-Magnon (Figs. 6,7,11,12)

The most famous of Upper Paleolithic fossils were discovered in 1868 in a rock shelter called Cro-Magnon (Les Eyzies, Dordogne). Five numbered individuals and additional cranial and postcranial fragments are known (Figs. 6, 7, 11, 12). They have been extensively discussed, most recently in works such as Camps and Olivier [1970], de Lumley [1976], and Dastugue [1982], where paleopathological aspects of the material are examined. The specimens are associated with an evolved Aurignacian industry, and therefore cannot necessarily be said to be representative of the earliest a.m. hominids of France.

These fossils, and in particular Cro-Magnon 1 (the so-called "old man"), became the type specimens for a whole Upper Paleolithic "race." Its characters were often stated as including a long and voluminous skull with a "disharmonically" broad but short face, low orbits, long narrow nose, and tall stature. However, the Cro-Magnon specimens themselves show consid-

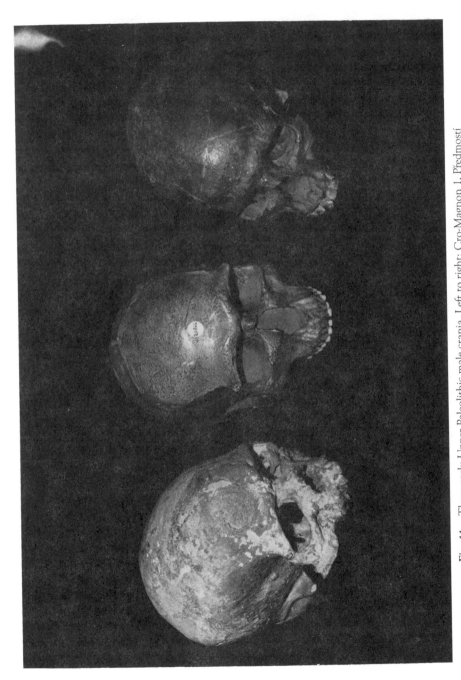

Fig. 11. Three early Upper Paleolithic male crania. Left to right: Cro-Magnon 1, Předmostí 3 (cast), and Grotte des Enfants 6 (cast).

erable variation in supraorbital and occipital morphology, with Cro-Magnon 4 showing the strongest brow ridge and a robust, but probably abnormal occipital morphology, unlike the pattern typical of Neandertals or Upper Paleolithic hominids. Cro-Magnon 3 has a very bulging occipital region, while Cro-Magnon 2 (probably the only female of the adult specimens) closely resembles crania such as Pataud 1 and Předmostí 4 in "modern" aspects of its morphology. The well-preserved face of Cro-Magnon 1 contrasts markedly with those of Neandertals in proportions and flatness (Fig. 6), but as Wolpoff [1980a] has noted, the face is positioned unusually far from the middle portions of the vault, resembling Neandertals in this respect.

Abri Pataud (Figs. 8,12)

The main excavations at this rock shelter at Les Eyzies (Dordogne) were conducted between 1958 and 1964 under the direction of Movius. Various aspects of the site, including the hominid material studied by Billy and Legoux, were described in the collection of papers edited by Movius [1975]. A large number of specimens representing mature and immature individuals were recovered across an area of 14 m^2 in level 2 (Proto-Magdalenian, c 22 ky) and more fragmentary material was recovered from the earlier Perigordian and Noaillian (level 4, c 27 ky) levels.

The most important specimens from the Proto-Magdalenian level are the skull and mandible of a late (female?) adolescent (2), a left ilium of a female (3), and two separate complexes of skeletal material from at least one adult and child (complex 22 and the material from trench VII) (note the different numbering of individuals in earlier publications, including Oakley et al [1971]). Overall, the adolescent skull closely resembles those of presumed females from Cro-Magnon and Předmostí, but the well-marked muscle insertions, great robusticity of the mandible, and dental peculiarities, including supernumerary teeth and remarkable M3 morphology are noteworthy. The robusticity of the jaws and teeth certainly appear archaic, but in fact contrast markedly with the pattern found in late Neandertals (eg, Saint-Césaire). The postcranial material shows a relatively low brachial index, closer to the means of Neandertal and recent European samples than to the Upper Paleolithic sample generally [Trinkaus, 1981]. In common with the rest of the Upper Paleolithic sample, the right scapula fragment shows the bisulcate pattern of the axillary border rather than the patterns typical of either Neandertal or recent samples. The morphology of the superior pubic ramus, however, is modern (see Fig. 8).

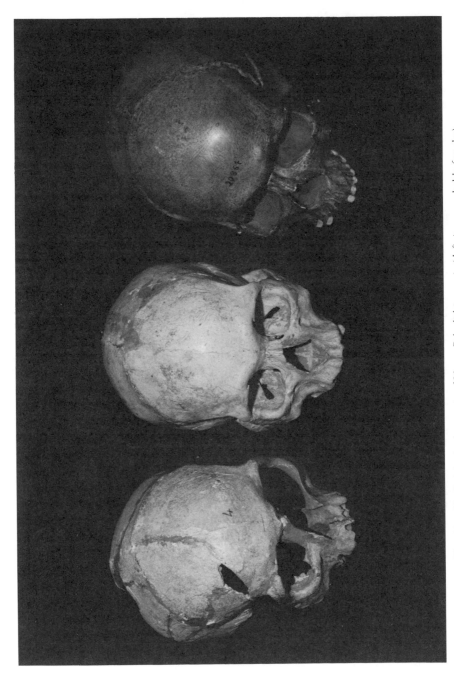

Fig. 12. Similarities in the crania of Upper Paleolithic crania (definite or probable females). Left to right: Cro-magnon 2, Pataud, and Předmostí 4 (cast).

Chancelade

The Raymonden rock shelter grave (Chancelade, Dordogne) was discovered by Hardy and Feaux in 1888. According to de Sonneville Bordes, it dates to the end of Magdalenian III or the beginning of Magdalenian IV. The skeleton was studied by the anatomist Testut who proposed that it should be made the type specimen of the "Chancelade race."

Chancelade man was small, 1.60 m, with a large, thickset skeleton. The skull is dolichocephalic with a raised vault, and markedly keeled. The frontal region rises vertically while the occipital projects only slightly. Even the supraorbital relief is very slightly marked. The face is very long and broad with the orbits square at the bottom and turning outwards towards the front. The nasal aperture is narrow and the facial profile shows perfect orthognathism. The robust and massive mandible has a markedly projecting chin.

Testut compared this morphology to that of Eskimos and envisaged a phyletic relationship between the latter and Chancelade, a hypothesis maintained by Sollas. This relationship seemed to find confirmation in the resemblances between the Magdalenian culture and that of Eskimos. This interpretation has now been abandoned because those resemblances between the Chancelade specimen and Eskimos correspond to secondary characters without evolutionary significance. The undoubted similarities that exist between the two cultures result from the simple phenomenon of convergence.

The Chancelade specimen is unquestionably different from the classic Cro-Magnons and the affinities which some workers have believed proven are not convincing. The same is true of the comparison with the other Magdalenian skeletons. So this skeleton seems to be well isolated in the French Upper Paleolithic and we do not know whether it corresponds to another population from the evolved Cro-Magnons or if it indicates the very great variability of the latter.

Belgium

Engis. Toward the end of 1829 or at the beginning of 1830, Paul Schmerling brought to light two human skulls as well as skeletal fragments of a third individual in the cave of Engis, near Liège. Historically, it was the first discovery of fossil human remains to come about during the course of an archeological excavation. A brilliant man before his time, Schmerling, who had recognized the artificial nature of flaked stone tools, published the main results as early as 1833. At that time, however, the importance of his work was not recognized by the scientific community. Like the discovery of the Gibraltar skull several years later, the Engis remains went nearly unnoticed.

This cave, now totally destroyed, contained Upper Paleolithic and Mousterian levels. After analyzing the results of the various excavations and an anatomical study of the human skulls, Charles Fraipont [1936] concluded that the Engis 2 cranium (sometimes numbered Engis 1) was that of a "*Homo neanderthalensis*" and had been discovered at the base of the Mousterian levels. Apart from the monograph by Fraipont [1936] this skull has been the subject of a recent re-examination by Tillier [1983b].

The Engis 2 fossil is attributed to a child between 5 and 6 years old [Tillier, 1983b]. Despite its fairly young age, it presents indisputable Neandertal characters. In her detailed study, Tillier separates the latter from the primitive and juvenile characters and here we reiterate some of these observations. The skull is broad by comparison with modern children (frontal, biporic, biasterionic, and maximum width) and the interorbital breadth is large. Nevertheless certain metrical attributes (the bregmatic and frontal angles of Schwalbe, height/width index of the temporal squama) distinguish Engis 2 from adult Neandertal characters. The orbits are large and rounded. The glenoid notch is low in relation to the external auditory meatus, and the occipito-mastoid crest is well developed. In posterior view the skull shows the "en bombe" form. In the occipital region a very distinct suprainiac fossa is visible, bordered by the two projections of the occipital [Hublin, 1980b], a Neandertal feature known elsewhere among other children. The preserved parts of the midfacial area seem to indicate that this region was swept back. The teeth are, however, of modest size compared with Middle Paleolithic samples.

In conclusion, the attribution made by Fraipont that this skull is of Neandertal type can hardly be questioned, even if on account of its age it does not show all the characters well (eg, marked supraorbital torus, pneumatization of the frontal, anterior mastoid tubercle, and so on).

The other hominid material found by Schmerling, and by Dupont in 1872, derives from the Upper Paleolithic level or levels. Although often described as Aurignacian, Otte [1979] considered the adult calvaria (Engis 1, sometimes numbered 2) to date from an Upper Perigordian level, with perhaps a succeeding late Paleolithic level as well. Engis 1 is entirely modern in morphology but more fragmentary material includes a maxillary fragment with highly worn molars and a large incisor. The hominid material was highly dispersed, perhaps by perturbation or by human activity [Otte, 1979].

La Naulette. The "Trou de Naulette" (Naulette Hole) is situated in the Lesse valley, near Dinant. This huge cavern (40 m deep, 10 m wide), was excavated in 1866 by the geologist Dupont, one of the founding fathers of

Belgian prehistoric archeology. He excavated a mandible, a fragmentary left ulna, and a third left metacarpal associated with a varied fossil fauna: reindeer, deer, chamois, mammoth, woolly rhinoceros, and wolf [Dupont, 1866]. However, no lithic industry was recovered along with these remains. At that time, the Naulette mandible aroused great interest. It was a human jaw undoubtedly dating back to the "age of the mammoth" [see Spencer, this volume]. Its main features were later studied by Topinard [1886]. Zeuner [1940] attributed an "early Würm" date to the levels that yielded this specimen, which can be regarded as the first Neandertal mandible ever discovered.

At the time of the first studies, this specimen provoked surprise over its great robusticity. However, even if the thickness of the horizontal ramus is high, it is nonetheless comparable to that of other Neandertals. In fact amongst this group, the Naulette mandible appears on the whole rather gracile, with a rather low horizontal ramus and the bony relief faintly marked on the external surface. It has no teeth, but the condition of the roots indicates that the M_3 had not completed its eruption. This specimen can therefore be attributed to an adolescent. The symphyseal region also provoked surprise. The front surface drops vertically, whereas the back surface shows "simian" characteristics: an alveolar plane, and superior and inferior transverse tori surrounding a genioglossal fossa. The mylo-hyoid line is well marked. The teeth were probably large in size and the molar alveoli increase in series toward the back. The retromolar space seems reduced but the eruption of M_3 is incomplete and the ascending ramus is broken. The mental foramen is situated below the P_4 alveolus in rather a low position. Overall, and certainly in part due to its incomplete condition, the Naulette mandible shows primitive rather than Neandertal characters. Although the edges of the canine alveoli are damaged, we nonetheless note that the alveolar arcade shows an alignment of the front teeth.

Spy. The discovery of Onoz-Spy (Namur province) took place in June 1886, and certainly was the most important excavation of Neandertal remains to occur in Belgium. The excavation of this cave, begun in 1885 by Lohest and de Puydt, led to the subsequent exhumation of three individuals: two adult skeletons, and tibia fragments and two teeth of a child. These fossils have played a considerable role in the history of human paleontology. Much more complete than the Naulette find, they put an end to the debate that had initially followed the discovery of the Feldhofer (Neandertal) remains in 1856 [see Spencer, this volume]. The uncovering of the human remains took place under the most exacting conditions that could be expected for that time. The stratigraphy of the deposits studied by the excavators showed without any possible doubt the association of the hominids, the Mousterian industry and an extinct fauna. Further, and this was the decisive factor, the monograph dedicated to these hominids by Julian Fraipont and Max Lohest [1887] established their resemblance to the Feldhofer fossil. Speculations about the pathological nature of the latter (as well as other arguments) had to be abandoned: The Neandertal type really existed.

The fauna has been attributed to a cold phase of the early Würm [Oakley, 1964]. The industry is of Quina Mousterian type [Bordes, 1959]. The two adult skeletons belong to an individual of about 35 years (Spy 1) and another (Spy 2) of about 25 years. Thoma [1975] estimated the cranial capacity of Spy 1 as 1,300 cm^3 and that of Spy 2, 1,504 cm^3. In common with Hrdlička [1930], Genovés [1954] and Twiesselmann [in Oakley et al, 1971], Thoma reattributed the most robust bones (notably the femur and tibia) to Spy 2, which led him to consider Spy 1 to be a female and Spy 2 as a male. Fraipont and Lohest applied themselves to demonstrating the similarities between the Neandertal remains, that of Naulette and those of Spy. The latter were considered as indisputable Neandertals that show in the preserved parts derived characters which we have cited at the beginning of this chapter. Nevertheless they have been regarded by certain researchers, namely following Hrdlička [1930], as possibly intermediate between Neandertal and modern man. Hrdlička's thesis is based on six anatomical arguments that have been rediscussed by Thoma [1975]. Thoma emphasizes in this criticism that the general form of the cranial vault of Spy is not "modern," that the size of the teeth of Spy 1 and 2 are within the variation of European Neandertals, and that the Spy 1 mandible, like other Neandertal mandibles, shows certain components of the bony chin but not the combination of characteristics of the modern chin. Further, the glabella region of Spy 2 is not preserved and the remaining parts of the supraorbital region show a Neandertal morphology. The occipital plane of Spy is slightly convex but in relation to its other dimensions it is on the whole a primitive character and the arc/chord ratio of the lambia-inion arc is very similar in Spy 1, Feldhofer, and Spy 2 [Hublin, 1982]. Finally, it remains true that the Spy 2 frontal is the most convex of the Neandertal frontals. Of the Neandertals, the Spy hominids remain among the more complete of the reference specimens. Apart from those already cited they have been the subject of several special studies [Fraipont, 1912, 1927; Trinkaus, 1978] and have been discussed in all the general literature on Neandertals as well.

Fond de Forêt. A fourth Belgian site has yielded human remains attributed to Neandertals, namely the cave of Fond de Forêt in the Vesdre valley near Liège. This discovery was made in 1895 by Tihon, an archeologist who found an isolated upper left molar and an important femur fragment [Tihon, 1898]. Subsequent excavations have confirmed that these remains came from the Mousterian levels [Hamal-Nandrin et al, 1934]. The femur was the subject of an important study by Tweisselmann [1961]. Four fifths of the lower part of the bone is preserved. Although his measurements and various indices failed to distinguish it from the total modern variation it was placed in an equi-probability curve representing the modern reference population with other Neandertals and it formed an homogenous group with them. This group is characterized by a marked robusticity, a low platymeric index, a femur head of large diameter in relation to total length, a very large superior length in relation to total length, a neck of notable diameter, and a very deep patellar surface. The medial section of the shaft is rounded and does not show the marked *linea aspera* that generally gives a triangular section to the modern femur.

DISCUSSION AND CONCLUSIONS

In our view, western Europe has no good evidence for the actual origin of a.m. *H sapiens*. Considering the probable Middle or earlier Upper Pleistocene hominids, one group of fossils shows no clear modern or Neandertal apomorphies (eg, Mauer, Arago 13, and Bilzingsleben). A second group has more debatable affinities (eg, Steinheim, Montmaurin, Arago 21, Atapuerca, Bañolas, and Saccopastore 2), while the third and largest group shows clear Neandertal affinities (eg, Biache, La Chaise, Salzgitter-Lebenstedt, Saccopastore 1, and probably Swanscombe and Arago 2). The only hominids that might still be used to support a classical presapiens scheme (eg, Fontéchevade 1 (4), Ehringsdorf 9, Alicante, and Mandrascava) have dating or interpretative problems. Taking the evidence overall, and bearing in mind the serious dating difficulties [Cook et al, 1982], the western European fossil record is most reasonably interpreted as showing the replacement of plesiomorphies by Neandertal apomorphies during the later Middle Pleistocene, followed by the establishment of the full array of Neandertal characters in the early Upper Pleistocene. While recognizing that the western European Neandertals show a significant degree of variation, some of it through time, less certainly through space, given the problems of absolute and relative dating even in the Upper Pleistocene, we do not recognize clear evolutionary trends toward a modern human morphology. This would require the ordered replacement of Neandertal apomorphies by modern apomorphies, as well as the redevelopment of characters otherwise interpreted as plesiomorphies (eg, flatter mid-face, presence of canine fossa, lower frequency of H-O mandibular foramen, modifications of temporal and occipital morphology). Furthermore, within the known dating framework, these changes throughout the skeleton would have had to occur in a fraction of the time over which Neandertal apomorphies were established.

In addition, we do not recognize any fossil specimens with an intermediate morphology between Neandertal and modern types, although the unusual form of the Hahnöfersand frontal bone could certainly be interpreted in this way. However, it shows a clear modern apomorphy in supraorbital form and resembles fossils such as Pavlov 1, Mladeč 5, and Podbaba as much as it resembles true Neandertals. Its cranial thickness and transverse flattening are unusual, and if its early date can be confirmed by further analyses, the hybridization model proposed to explain its morphology [Braüer, 1981] would be further supported. From Saint-Césaire there is good evidence of a late but quite typical Neandertal morphology surviving alongside modern hominids [Lévêque and Vandermeersch, 1981; and see earlier discussion]. However, it would be valuable to confirm the presumed totally modern nature of the broadly contemporaneous manufacturers of the earliest Aurignacian in the area, because it is in this sample that evidence of any gene flow from Neandertals should be most apparent. The suggestion of an appearance of the Aurignacian in France from the South, and a simultaneous northward movement of the makers of the Châtelperronian may give a clue to population movements during a replacement stage [Leroyer and Leroi-Gourhan, 1983].

COMPARISON WITH OTHER AREAS
East Europe

The Vindija and Šala fossils seem to display a less "specialized" or more "modern" morphology compared with western European Neandertals [Smith, 1982; and this volume]. Some of the moderate robusticity and midfacial prognathism of the specimens may be due to small body size, perhaps evolved in parallel with that of modern humans, but local evolutionary change or gene flow may have contributed progressive aspects to their morphology. On the available faunal evidence Šala is probably an early Neandertal, while the G_1 Vindija hominids probably predate the earliest local Upper Paleolithic, and a.m. hominids, dated at c 34 ky from Velika Pećina [Stringer, 1982].

Eastern European early a.m. hominid fossils display a greater degree of archaic characters (ie, those resembling nonmodern hominids) compared with their western European counterparts (Fig. 11). However, only one or two can be interpreted as relating particularly to Neandertals, and the lambdoid flattening and occipital protrusion that some of the most robust fossils display is not an exclusive Neandertal character, nor is it in fact typical of the known preceding Neandertals of the area. Fossils such as Mladeč 5 and Pavlov 1, which in some respects appear more archaic than their counterparts in western Europe or even some of those in southwest Asia, resemble African fossils such as the Irhoud remains where they are archaic, more than they resemble the European Neandertals. An anterior placement of nasion and subspinale in some of the European early modern fossils, however, does match more specifically the Neandertal pattern. On the other hand, the postcranial characters of the eastern European early modern specimens show no approximation to the Neandertals in either morphology or limb proportions.

The eastern European fossil record does provide better support than that of western Europe for some degree of local continuity at the time of appearance of a.m. humans. This continuity, however, may well be explained by the geographical position of the area closer to a source of "modern" characters. While some workers have proposed a central European origin, the early appearance of an apparently intrusive Aurignacian-like industry at Bacho Kiro in Bulgaria prior to 43 ky may be evidence of an exotic origin for the industry and its manufacturers. Unfortunately, the remains from the relevant layer 11 are too fragmentary for clear identification as early "modern" humans. Equally, there is no justification for identifying them as Neandertals [Kozłowski, 1982]. While the evolution of modern humans seems to be unconnected with the origin of the Aurignacian from its presumed Mousterian antecedent, the distribution of the earliest modern humans across Europe may still be correlated with the spread of this industry.

Southwest Asia

Two main scenarios have been proposed for Upper Pleistocene hominid evolution in this area. The first has local morphological and cultural continuity from Neandertal fossils such as those from Shanidar and Tabūn to Neandertaloid or early modern human fossils such as those from Skhūl and Qafzeh. Progressive features are claimed for the late Neandertals, and Neandertal features are claimed for the Skhūl and Qafzeh hominids [Wolpoff, 1980a]. A related view suggests that while there are clear differences between the Neandertal and modern samples, an in situ evolutionary transition probably did occur [see Trinkaus, this volume]. The second scenario parallels the replacement model of western Europe, except that the southwest Asian replacement occurs against a purely Mousterian background, rather than against a Mousterian/Upper Paleolithic interface or even a purely Upper Paleolithic background. Apparent cultural continuity provides something of a problem for the replacement model in this area, unless it is assumed that a widespread industry of general Middle Paleolithic affinities was made by both Neandertals and the precursors of modern humans. One suggestion identifies the Zuttiyeh fossil as representing a specifically modern precursor population temporarily replaced in the area by a Neandertal penetration from the North [Vandermeersch, 1981]. The specimen is incomplete and as it may even date from prior to 150 ky it is possible that some of the facial characters that separate it from Neandertal specimens are, in part at least, plesiomorphies rather than modern apomorphies. However, some Neandertal facial apomorphies were arguably established in Europe at this time, so their absence in the Zuttiyeh fossil may be significant.

The actual sequence of events in Upper Pleistocene hominid evolution in this area is difficult to reconstruct due to an inadequate dating framework (eg, see discussion in Ronen [1982]). While the youngest Neandertal specimens appear to date from c 50 ky (eg, Tabūn 1, see Jelínek [1982]), the Qafzeh hominids may be of comparable age. From the relative and absolute dating of transitional or early Upper Paleolithic industries in this area alone, the Mousterian levels of Qafzeh and Skhūl should date from at least 40 ky, indicating that any interface between Neandertals and modern humans in the area preceded that of western Europe by, at the most conservative estimate, more than 5 ky.

The morphological gap between Neandertal and early modern groups in this area is large, although not as great as that in western Europe because of a greater percentage of what we would interpret as plesiomorphous features that they have in common. One Neandertal specimen (Shanidar 1) does display modern parietal proportions and orthognathism, but has clear Neandertal apomorphies elsewhere in the face, vault, mandible, and postcranial skeleton. In a scheme that denies local continuity with modern humans, the parietal form of Shanidar 1 would have to be considered as a result of parallelism, gene flow or deformation [Thoma, 1965; Trinkaus, this volume]. In our view no unequivocal directional trends towards modern humans are evident in the Neandertal sample from this area as a whole, and we do not recognize any transitional fossils. However, if a rapid morphological change were possible, the Neandertals of southwest Asia would represent more plausible ancestors for early a.m. *H sapiens* than their western European counterparts [Stringer and Trinkaus, 1981], given their somewhat less specialized morphology.

Africa

Africa has recently been claimed as the source of the earliest a.m. *H sapiens* [see Rightmire; and Braüer, respectively, this volume] and if we accept the evidence, it seems well established that the earliest clear examples of modern humans derive from opposite ends of the African continent at a conservative estimate of more than 50 ky, and perhaps as far back as 120 ky.

At Omo-Kibish in Ethiopia, member 1 at the site contains evidence of a robust modern morphology (Omo 1) and a much more archaic hominid type (Omo 2) that nevertheless appears to demonstrate one modern apomorphy in the form of the supraorbital torus. These specimens have been assumed to be approximately contemporaneous in the late Middle or early Upper Pleistocene, but this is still not certain. However, unless they are intrusive specimens, they are both stratified well below levels containing molluscs dated by radiocarbon at greater than 37 ky. The relationship between the two morphological types is unclear, but there can be little doubt that Omo 1 demonstrates a modern pattern in the extant cranial and postcranial parts [Day and Stringer, 1982].

The Singa cranium from Sudan may be of comparable age, but does not appear to display a comparably modern morphology. Assessment of the specimen is hampered by its abnormal, probably pathological, form [Stringer, 1979; and research in progress].

In southern Africa, at Klasies Cave and Border Cave, examples of fragmentary but apparently modern hominids seem well dated to more than 70 ky and more than 50 ky, respectively [see Rightmire; and Braüer, respectively, this volume for further details and discussion; but compare Klein, 1983]. It has been claimed that these South African hominids closely resemble extant aboriginal populations of the same area, but caution must be used when assessing such incomplete specimens. At Florisbad, a fossil hominid of rather more archaic aspect may also date from the early Upper Pleistocene, and perhaps provides evidence of the origins of fully a.m. *H sapiens* forms from more archaic predecessors.

From eastern Africa there is, as yet, no firm evidence of a comparably early a.m. hominid presence, although this had been claimed on the basis of the fragmentary Kanjera hominids. Relevant material from West and East Turkana awaits detailed publication, including fossils that may elucidate the problems encountered in the wide variation of the Omo-Kibish sample farther to the north. Fragmentary, undescribed, but morphologically modern skeletal material from Ng'ira (Karungu) in Kenya may be asociated with Middle Stone Age artifacts and fauna suggestive of an early Upper Pleistocene age [Pickford, personal communication], but further research on the site is being conducted. The fossil hominid from Ngaloba (Laetoli hominid 18) probably dates from about 120 ky but does not, in our opinion, fall close to a modern cranial pattern. The flat mid-face, with a canine fossa, may represent a plesiomorphy present in other African Middle or early Upper Pleistocene fossils (although in this case the Bodo and Broken Hill 1 fossils display a more derived condition), and the cranial vault is morphologically nonmodern, even somewhat resembling Neandertal fossils in the shape of the rear of the skull and the large occipitomastoid crest. Probable earlier African hominids such as Bodo, Broken Hill, Saldanha, and Ndutu similarly do not show clear modern apomorphies in our opinion, although the poorly dated Eyasi occipital fragments approximate a modern pattern in the region of the occipital torus, with incipient development of a true external occipital protruberance, as also reported for Ndutu [Hublin, 1978a; Rightmire, this volume].

From North Africa there is evidence of hominids displaying some signs of local continuity from the early Middle Pleistocene (Ternifine) to the Upper Pleistocene (Djebel Irhoud and Dar-es-Soltan). The Middle Pleistocene Salé fossil appears to display a very modern-looking occipital profile, but there is some indication of pathology, as well as a flat morphology of the occipital torus. It is not until the Upper Pleistocene with the Irhoud and Dar-es-Soltan Aterian fossils that clear modern apomorphies appear. Irhoud 1 is a rather archaic specimen in the parietal and occipital regions, but the frontal bone is expanded, and the face is relatively short, broad and flat, possessing a moderate canine fossa like that of the Florisbad and Ngaloba fossils. Irhoud 2 has an even higher frontal bone with modern curvature and a laterally reduced supraorbital torus, while the Irhoud 3 child's mandible shows some development of a true chin. Modern apomorphies that are moderately developed in the Irhoud sample are much more fully expressed in the Dar-es-Soltan Aterian sample of which only the anterior cranium and mandible (specimen 5) have received preliminary description [Ferembach, 1976]. The specimen appears to show affinities to both the more archaic Irhoud and the "modern" Afalou samples from North Africa. The available dating framework cannot adequately place these hominids in relation to each other but the early Aterian industries probably predate 40 ky [Ferring, 1975]. North Africa, more than any other area, has fossils that bridge the gap between archaic and modern morphologies, and a secure chronological placement for the Irhoud and Dar-es-Soltan Aterian fossils would be invaluable in assessing the likely origin of modern human characters elsewhere around the Mediterranean. Neither sample appears to show Neandertal apomorphies. Instead they share some distinctive characters in facial proportions, such as a low nasal height in relation to upper facial breadth. The new skeleton from Wadi Kubbaniya, Egypt [Robert, 1982] may throw further light on the extent of modern or Neandertal characters in the cranium, mandible, and postcranial skeleton of these early Upper Pleistocene North African hominids. Overall, the African evidence suggests that a.m. hominids evolved during the timespan of the Middle Stone Age.

Far East and Australasia

It is difficult to determine the date of the earliest appearance of modern humans in this area [see Wolpoff et al, in this volume]. While clearly modern humans are known from Australasia at c 30 ky (Mungo), and perhaps from Borneo at c 40 ky (Niah), the earliest a.m. fossils from the Asian mainland (eg, Zhoukoudian Upper Cave, Liujang, and Ziyang) may all be younger than 20 ky [Zhou Mingzhen et al, 1982]. Although it is often stated that these fossils are closely related to their modern aboriginal counterparts, the extent of their similarities has rarely been carefully documented. Given the morphology and dating of the known material, it is still possible that the modern characters of the late Upper Pleistocene hominids of this area were ultimately derived from an exotic source such as Africa. Such a simplistic model, however, does not take account of possible modern characters in even earlier fossils such as the Da-li cranium from China (unless these are plesiomorphies or parallelism), does not explain the extremely robust nature of some of the late Pleistocene Australian material (unless these are local "reversions"), nor does it accord with a record of local cultural continuity and even apparent stasis. Overall, the available evidence is too poor for any convincing scenario, but a model involving varying amounts of gene flow from outside the area combined with local sources may explain the variation present in the fossil hominids. Identification of supposed "clade" characters that can be traced back from late Pleistocene Australian fossils to early Middle Pleistocene Indonesian specimens [see Thorne and Wolpoff, 1981; Wolpoff et al, this volume] must be viewed with caution until it has been determined whether or not such characters might be plesiomorphies found more widely in early a.m. *H sapiens*.

CONCLUDING REMARKS

The place of origin of the first hominids with a total morphological pattern matching that of recent humans is not identifiable from the present fossil record. Western Europe certainly does not seem to record in situ evolution of a.m. humans, but equally, even the African evidence provides scant indication of well-dated local evolutionary sequences from archaic to modern humans. North Africa provides the best approximation to such a sequence, but the relative dating of the fossils concerned is not good enough to exclude gene flow as a catalyst for the appearance of modern characters.

The essential problem with proposed local unilinear sequences that convert distinct geographical variations of archaic hominids into a.m. humans at opposite ends of the Old World, is that no mechanism (except gene flow, population movement, or less plausibly, orthogenesis) can be envisaged that could develop the anatomically modern pattern, in parallel in the whole skeleton, under diverse environmental and cultural conditions. If local evolutionary sequences were predominantly responsible for the evolution of modern humans, the model would predict Eurasian hominid relationships as in Figure 13. Without at least some gene flow, the local lineages would be expected to increasingly diverge.

We feel, however, that even the most committed proponents of local continuity will have great difficulty in demonstrating that on any reasonable assessment of the metrical and morphological evidence Figure 14 is not a more accurate representation of the true relationship involved.

Furthermore, the evolutionary changes involved in Figure 13 would be diachronous, and as already mentioned would occur against different environmental and cultural backgrounds (during the Mousterian of southwest Asia, but during the Mousterian/Upper Paleolithic interface in eastern Europe and even during the Upper Paleolithic of western Europe).

Because the fossil record presents only arguable examples of morphological sequences leading from archaic to modern humans, even in areas with

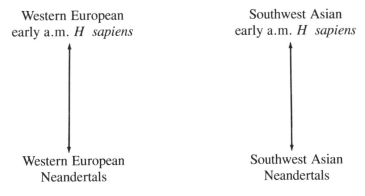

Fig. 13. Expected maximum similarities in skeletal characters following local continuity model.

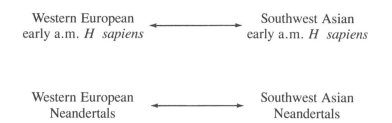

Fig. 14. Actual maximum similarities in characters, especially in synapomorphies.

the best hominid sequences, it remains quite possible that we do not yet have a record from the actual area of origin of the earliest modern humans, and that all other sequences are influenced by gene flow or population movement from that area. Alternatively, the evolution of modern humans may have been a punctuational event in terms of the Pleistocene time scale, so that few "intermediate" specimens have become fossilized. Such a model is consistent with some studies of genetic data, which indicate that there was "bottle-necking" in the evolution of recent human variation. Broad studies of genetic distances are generally consistent with derivation of all recent human variation within the last 200 ky [Jones, 1981; Nei and Roychoudury, 1982; but compare Cann, 1982]. Our view, then, favors a late Middle or early Upper Pleistocene development and radiation of the fully modern skeletal pattern from one area, on present evidence probably in Africa. However, evidence of local continuity suggests that, despite the western European record, replacement of more archaic morphologies may not always have been so rapid or total.

Because there is an absence of early modern skeletal material from most of Asia, and because the absolute ages of the Qafzeh and Skhūl material are not yet known, we must retain some caution before confirming Africa as the earliest source of modern skeletal characters, since such a claim is still dependent on negative evidence from elsewhere. However, if an Asian center of in situ evolution of modern humans were proposed, the parallel evolution of modern characters in geographically distant areas must again be explained, as well as the presence of clearly nonmodern hominids in western Asia for much of the Upper Pleistocene. If a model involving gene flow for the spread of modern characters from an original Asian center is also proposed, it must account for the appearance of modern features in southern Africa considerably earlier than they appear in Europe. This will be difficult, given the otherwise extensive floral and faunal connections between Asia and Europe, rather than between Asia and sub-Saharan Africa, during the later Pleistocene. On the other hand difficulties with a purely African center of origin for modern humans come in explaining why African cultural elements do not spread with the supposed African populations or genes, and in explaining the long nascence of modern characters in the area, and the reasons for their eventual spread from the continent. In addition, it is difficult to explain why some of the Eurasian (and also especially Australian) early modern hominid specimens are more robust or even more archaic than their supposed African ancestors, unless there was some subsequent acquisition of robusticity through local evolution or gene flow.

In our opinion there is not only a need for well-dated fossils to resolve the problem of the origin of modern humans, there is also a need for a different approach to the analysis of the evidence. We hope our chapter provides some indication of such an approach, even if we do not feel that our area of concern, western Europe, can contribute much positive evidence to the eventual solution of the problem.

ACKNOWLEDGMENTS

We would like to thank all the institutes that provided us with access to fossil material for study or photography, various grant-awarding bodies for funds for travel, particularly the Leakey Foundation (CBS), Mrs J. Cook for general assistance including translation work, and finally Drs. Frank Spencer and Fred H. Smith (editors) for their great forebearance in the face of our various difficulties.

LITERATURE CITED

Aguirre E, Lumley MA de (1977): Fossil men from Atapuerca, Spain: Their bearing on human evolution in the Middle Pleistocene. J Hum Evol 6:681–688.

Aguirre E, Basabe JM, Torres T (1976): Les fósiles humanos de Atapuerca (Burgos): Nota preliminar. Zephyrus 26–27:489–511.

Aitken MJ (In press): La Caune de l'Arago. Dating of stalagmitic floors by TL and ESR: Comment on the synthesis. In Lumley H de, Labayrie J (eds): "Datations Absolues et Analyses Isotopiques en Préhistoire. Méthodes et Limites." Tautavel: CNRS.

Alessio M, Bella F, Improta S, Belluomini G, Cortesi C, Turi B (1970): University of Rome carbon-14 dates VIII. Radiocarbon 12:599–616.

Alferez-Delgado F, Molero G (1982): Descubrimiento de un fosil humano (Riss-Würm) en Pinilla del Valle (Madrid). Abst 1er Congr. Int. Paléont Hum Nice, pp 103–104.

Angel JL, Coon CS (1954): La Cotte de St. Brelade II: Present Status. Man 76:53–55.

Ascenzi A (1982): Comparaison entre l'homme de Tautavel, les anténéandertaliens d'Italie et l'homme de Saccopastore. Prétirage 1er Congr Int Paléont Hum Nice, pp 918–933.

Ascenzi A, Segre AG (1971): A new Neandertal child mandible from an Upper Pleistocene site in southern Italy. Nature 233:280–283.

Barral L, Simone S (1982): Position chronologique de l'*Homo erectus* de la Grotte du Prince (Grimaldi, Ligurie Italienne). Abst 1er Congr Int Paléont Hum Nice, p 94.

Bastin B (1976): Etude palynologique des couches E_2, D et B^5 de la grotte de Fontechévade (Charente, France). Bull Soc R Belge Anthropol Prehist 87:15–27.

Bianchini G (1982): Scoperta, ricostruzione, posizione stratigrafica del cranio umano di Mandrascava e relativi dati sulle comparazioni morfologiche con i fossili di "Homo erectus" Europei. Abstr 1er Congr Int Paléont Hum Nice, pp 74–75.

Bietti A (1982): The Mousterian complexes in Italy: A search for cultural understanding of the industrial assemblages. In Jelínek J (ed): "Man and His Origins." Brno: Moravian Museum, pp 237–251.

Billy G, Vallois HV (1977): La mandibule pré-Rissienne de Montmaurin, Anthropologie 81:273–312, 411–458.

Blackwell B (1981): Absolute dating and isotopic analyses in prehistory; methods and limits. Geosci Can 8:174–175.

Bordes F (1959): Le contexte archéologique des hommes du Moustier et de Spy. Anthropologie 63:154–157.

Bordes F (1968): "Le Paléolithique Dans le Monde." Paris: Hachette.

Boule M (1911–1913): L'homme fossile de la Chapelle-aux-Saints. Ann Paleont 6:111–172, 7:3–192, 8:1–67.

Boule M (1923): "Les Hommes Fossiles, 2nd ed." Paris: Masson.

Boule M, Vallois HV (1957): "Les Hommes Fossiles, Fifth ed." Paris: Masson.

Bouyssonie A, Bouyssonie J, Bardon L (1908): Découverte d'un squelette humain moustérien à la bouffia de La Chapelle-aux-Saints (Corrèze). Anthropologie 19:513–518.

Brace CL (1967): "The Stages of Human Evolution." Englewood Cliffs: Prentice Hall.

Braüer G (1980): Die morphologischen Affinitaten des jungspleistozänen Stirnbeines aus dem Elbmündungsgebiet bei Hahnöfersand. Z Morphol Anthropol 71:1–42.

Braüer G (1981): New evidence on the transitional period between Neanderthal and modern man. J Hum Evol 10:467–474.

Breitinger E (1957): On the phyletic evolution of *Homo sapiens*. In Howells WW (ed): "Ideas on Human Evolution." Cambridge: Harvard University Press, pp 436–459.

Breitinger E (1964): Reconstruction of the Swanscombe skull. In Ovey CD (ed): "The Swanscombe Skull: A Survey of Research on a Pleistocene Site." London: Royal Anthropological Institute, pp 161–172.

Bridgland D (1980): A reappraisal of Pleistocene stratigraphy in North Kent and Eastern Essex and new evidence concerning former courses of the Thames and Medway. Q Newslett 32:15–24.

Burleigh R, Hewson A (1979): British Museum natural radiocarbon measurements XI. Radiocarbon 21:340.

Butzer KW (1971): "Environment and Archeology: An Ecological Approach to Prehistory, 2nd ed." Chicago: Aldine-Atherton.

Butzer KW (1981): Cave sediments, Upper Pleistocene stratigraphy and Mousterian facies in Cantabrian, Spain. J Archeol Sci 8:133–183.

Campbell BG (1964): Quantitative taxonomy and human evolution. In Washburn SL (ed): "Classification and Human Evolution." London: Methuen, pp 50–74.

Campbell JB (1977): "The Upper Palaeolithic of Britain: A Study of Man and Nature in the Late Ice Age." Oxford: Clarendon Press.

Camps G, Olivier G (1970): "L'homme de Cro-Magnon." Paris: Arts et Metier Graphiques.

Cann R (1982): Newslett Found Res Origin Man 4:1–2.

Capitan L, Peyrony D (1912a): Station préhistorique de La Ferrassie. Rev Anthrop 22:29–50, 76–99.

Capitan L, Peyrony D (1912b): Trois nouveaux squelettes humains fossiles. Rev Anthrop 22:439–442.

Capitan L, Peyrony D (1920): Nouvelles fouilles à La Ferrassie (Dordogne). CR Assoc Fran Av Sci 44:540–542.

Capitan L, Peyrony D (1921): Découverte d'un sixième squelette moustérien à La Ferrassie (Dordogne). Rev Anthropol 31:382–388.

Cesnola AP di (1975): Il Gravettiano della Grotta Paglicci nel Gargano. 1. L'industria litica e la cronologia assoluta. Riv Sci Prehist. 30:3–177.

Cesnola AP di (1980): L'Uluzzien et ses rapports avec le Protoaurignacien en Italie. In Bánesz L, Kozłowski JK (eds): "L'Aurignacien et le Gravettien (Périgordien) Dans Leur Cadre Écologique." Nitra: Archaeological Institute, pp 197–212.

Cesnola AP di, Messeri P (1967): Quatre dents humaines paléolithiques trouvées dans des cavernes d'Italie méridionale. Anthropologie 71:249–262.

Chaline J (1972): "Les Rongeurs du Pleistocene Moyen et Supérieur." Paris: CNRS.

Chaline J (in press): Les Rongeurs de la Caune de L'Arago à Tautavel et leur place dans la biostratigraphie européenne. In Lumley H de, Labayrie J (eds): "Datations Absolues et Analyses Isotopiques in Préhistoire. Méthodes et Limites." Tautavel: CNRS, pp 193–203.

Condemi S (1983): Comparaison des sphénoides des Hommes de Saccopastore I, II and de Gibraltar I. CR Acad Sci [D] (Paris) 296:389–392.

Conway BW, Waechter J d'A (1977): Barnfield Pit, Swanscombe. In: INQUA Congress Guide, South East England and the Thames Valley, Norwich: Geo Abstr pp 38–44.

Cook J, Stringer CB, Currant AP, Schwarcz HP, Wintle AG (1982): A review of the chronology of the European Middle Pleistocene hominid record. Yrbk Phys Anthropol 25:19–65.

Coon CS (1962): "The Origin of Races." New York: Knopf.

Corruccini RS (1974): Calvarial shape relationships between fossil hominids. Yrbk Phys Anthropol 18:89–109.

Corrucccini RS (1975): Metrical analysis of Fontéchevade II. Am J Phys Anthropol 42:95–98.

Coulonges L, Lansac A, Piveteau J, Vallois HV (1952): Le gisement préhistorique de Monsempron (Lot-et-Garonne). Ann Paléont 38:83–120.

Dastugue J (1982): Les maladies de nos ancêtres. Recherche 13:980–988.

Dastugue J, Lumley MA de (1976): Les Maladies des hommes préhistoriques du Paléolithique et du Mésolithique. In Lumley H de (ed): "La Préhistoire Francaise." Paris: CNRS, pp 612–622.

Day MH (1982): The Homo erectus pelvis: punctuation or gradualism? Prétirage 1er Congr Int Paléont Hum Nice, pp 411–421.

Day MH, Stringer CB (1982): A reconsideration of the Omo Kibish remains and the erectus-sapiens transition. Prétirage 1er Congr Int Paléont Hum Nice, pp 814–846.

Debenath A (1974): "Recherches sur les Terrains Quaternaires des Charentes et les Industries Qui Leur Sont Associées." Doctoral thesis: Université de Bordeaux.

Debenham NC (1983): Reliability of thermoluminescence dating of stalagmitic calcite. Nature 304:154–156.

Delporte H (1976): Les sépultures moustériennes de La Ferrassie. In Vandermeersch B (ed): "Les Sépultures Néandertaliennes." IXth Congr UISPP Prétirage. Nice: CNRS, pp 8–11.

Dupont E (1866): Etude sur les fouilles scientifiques exécutées pendant l'hiver de 1865 à 1866 dans les cavernes des bords de la Lesse. Bull Acad R Belg Cl Sci 22:31–54.

Endo B, Kimura T (1970): Postcranial skeleton of the Amud man. In Suzuki H, Takai F (eds): "The Amud Man and His Cave Site." Tokyo: Academic Press of Japan, pp 231–406.

Ferembach D (1962): La deuxième molaire déciduale inférieure de la grotte de Salemas (Portugal). Comun Servs Geol Port 46:177–187.

Ferembach D (1964–5a): Les ossements humains de Salemas (Portugal). Comun Servs Geol Port 48:165–185.

Ferembach D (1964–5b): La molaire humaine inférieure moustérienne de Bombarral (Portugal). Comun Servs Geol Port 48:185–190.

Ferembach D (1976): Les restes humains de la Grotte de Dar-es-Soltane 2 (Maroc) Campagne 1975. Bull Mem Soc Anthropol Paris 3:183–193.

Ferembach D, Legoux P, Fenart R, Empereur-Buisson R, and Vlček E (1970): L'enfant du Pech-de-l'Azé. Arch Inst Paleont Hum 33:1–180.

Ferring CR (1975): The Aterian in North African prehistory. In Wendorf F, Marks AE (eds): "Problems in Prehistory: North Africa and the Levant." Dallas: Southern Methodist University, pp 113–126.

Feustel R (1978): "Abstammungsgeschichte des Menschen." Jena: Gustav Fischer.

Fraipont C (1912): L'astragale chez l'homme moustérien de Spy. Ses affinités. Bull Soc Anthropol Brux 31:195–221.

Fraipont C (1927): Sur l'omoplate et le sacrum de l'homme de Spy. Rev Anthropol 37:189–195.

Fraipont C (1936): Les hommes fossiles d'Engis. Arch Inst Paleont Hum 16:1–52.

Fraipont J, Lohest M (1887): La race humaine de Néanderthal ou de Canstadt en Belgique. Recherches ethnographiques sur les ossements humains découverts dans les dépôts quaternaires d'une grotte à Spy et détermination de leur âge géologique. Arch Biol 7:587–757.

Frayer DW (1978): "Evolution of the Dentition in Upper Paleolithic and Mesolithic Europe." University of Kansas Publication in Anthropology, 10.

Gambassini P (1980): Le Paléolithique Supérieur ancien en Campanie. In Banesz L, Kozłowski JK (eds): "L'Aurignacien et le Gravettien (Périgordien) Dans Leur Cadre Écologique." Nitra: Archaeological Institute, pp 89–97.

Gambier D (1981): "Etude de l'Astragale Chez les Néandertaliens." Thesis, Université Paris VI.

Gambier D (1982): Etude ostéométrique des astragales néandertaliens du Régourdou (Montignac, Dordogne). CR Acad Sci [D] (Paris) 295:517–520.

Garralda MD (1982): The Cuartamentero skull (Llanes, Asturias, Spain). Abstr 1er Congr Int Paléont Hum Nice, p 125.

Genet-Varcin E (1982): Vestiges humains du Würmien inférieur de Combe-Grenal, commune de Domme (Dordogne). Ann Paléont 68:133–169.

Genovés S (1954): The problem of the sex of certain fossil hominids, with special reference to the neanderthal skeletons from Spy. J R Anthropol Inst 84:131–144.

Giacobini G, Lumley M-A de (1982): Restes humains de type néandertalien de la grotta della Fate (Finale, Ligurie Italienne). Abstr 1er Congr Int Paléont Hum Nice, pp 106–107.

Gibert J, Agustí J, Moyà-Solà S (1983): Presencia de Homo sp. en el yacimiento del Pleistoceno Inferior de Venta Micena (Orce, Granada). Paleont Evol (Special issue), Sabadell.

Gieseler W (1977): Das jungpaläolithische Skelett von Neuessing. Festschrift 75 Jahre Anthropol Staatssml Munich, pp 39–51.

Girard M, Renault-Miskowsky J (1983): Datation et paleoenvironment de la mandibule de Montmaurin (Montmaurin, Haut-Garonne); analyses polliniques dans la niche. CR Acad Sci [D] (Paris) 296:393–395.

Green HS, Stringer CB, Collcutt SN, Currant AP, Huxtable J, Schwarcz HP, Debenham N, Embleton C, Bull P, Bevins RE (1981): Pontnewydd Cave in Wales—a new Middle Pleistocene hominid site. Nature 294:707–713.

Guérin C (in press): Les Rhinoceros (Mammalia Perissodactyla) du gisement pleistocène moyen de la Caune de L'Arago à Tautavel. Signification stratigraphique. In Lumley H de, Labeyrie J (eds): "Datation Absolues et Analyses Isotopiques en Préhistoire. Méthodes et Limites." Tautavel: CNRS, pp 163–192.

Guth C (1963): Contribution à la connaissance du temporal des néandertaliens. CR Acad Sci [D] (Paris) 256:1329–1332.

Hahn J (1981): Abfolge und Umwelt der jüngeren Altsteinzeit in Südwest Deutschland. Fundber Baden-Württemberg 6:1–27.

Hamal-Nandrin J, Servais J, Louis M, Fourmarier P, Fraipont C (1934): Fouilles dans la terrasse des deux grottes de Fond-de-Forêt (Province de Liège) 1931–1933. Bull Soc Préhist Franç 31:484–505.

Hammond M (1982): The expulsion of the Neanderthals from human ancestry: Marcellin Boule and the social context of scientific research. Soc Stud Sci 12:1–36.

Heim JL (1976): Les hommes de La Ferrassie. I. Le gisement, les squelettes d'adultes: Crâne et squelette du tronc. Arch Inst Paléont Hum 35:1–368.

Heim JL (1982a): Les hommes fossiles de La Ferrassie II. Les squelettes d'adultes: Squelette des membres. Arch Inst Paléont Hum 38:1–272.

Heim JL (1982b): "Les Enfants Néandetaliens de La Ferrassie." Paris: Masson.

Hemmer H (1982): Major factors in the evolution of hominid skull morphology, biological correlates and the position of the ante-neandertals. Prétirage 1er Congr Int Paléont Hum Nice, pp 339–354.

Henke W (1982): Der Jungpaläolithiker von Binshof bei Speyer—eine vergleichend-biometrische studie. Mainzer Naturw Arch 20:147–175.

Henke W (1983): Faktorenanalytischer Versuch zur Typisierung der Jungpaläolithiker und Mesolithiker Europas. Z Morphol Anthropol 73:279–296.

Henke W, Protsch RRR (1978): Die Paderborner Calvaria—ein diluvialer Homo sapiens. Anthropol Anz 36:85–108.

Henke W, Xirotiris N (1982): New human Upper Paleolithic fossils of Middle Europe. In Jelínek J (ed): "Man and his Origins." Brno: Moravian Museum, pp 263–280.

Henri-Martin G (1957): La grotte de Fontéchevade, Part 1. Arch Inst Paléont Hum 28:1–288.

Henri-Martin G (1965): La grotte de Fontéchevade. Bull Assoc Franç Etude Q 3-4:211–216.

Herrmann B (1977): Über die Reste des postcranialen Skelettes des Neanderthalers von Le Moustier. Z Morphol Anthropol 68:129–149.

Hesse H, Ullrich H (1966): Schädel des "Homo mousteriensis Hauseri" wiedergefunden. Biol Rundschau 4:158–160.

Holloway RL (1982): Homo erectus brain endocasts: Volumetric and morphological observations with some comments on the cerebral asymmetries. Prétirage 1er Cong Int Paléont Hum Nice, pp 355–366.

Howell FC (1951): The place of Neanderthal man in human evolution. Am J Phys Anthropol 9:379–416.

Howell FC (1957): The evolutionary significance of variation and varieties of "Neanderthal" Man. Q Rev Biol 32:330–347.

Howell FC (1958): Upper Pleistocene men of the Southwest Asian Mousterian. In von Koenigswald GHR (ed): "Hundert Jahre Neanderthaler." Utrecht: Kemink en zoon, pp 185–198.

Howell FC (1960): European and northwest African Middle Pleistocene hominids. Curr Anthropol 1:195–232.

Howells WW (1973): Cranial variation in man: A Study by multivariate analysis of patterns of difference among recent human populations. Papers Peabody Mus 67:1–259.

Hrdlička A (1930): The skeletal remains of early man. Smithsonian Misc Coll 83.

Hublin JJ (1978a): "Le Torus Occipital Transverse et les Structures Associées: Evolution Dans le Genre Homo." Thesis, University Paris VI.

Hublin JJ (1978b): Quelques caractères apomorphes du crâne néandertalien et leur interprétation phylogénique. CR Acad Sci [D] (Paris) 287:923–926.

Hublin JJ (1980a): A propos de restes inédits du gisement de la Quina (Charente): Un trait méconnu des néandertaliens et des anténéandertaliens. Anthropologie 84:81–88.

Hublin JJ (1980b): La Chaise Suard, Engis 2 et La Quina H 18: Développement de la morphologie occipitale externe chez l'enfant prénéandertalien et néandertalien. CR Acad Sci [D] (Paris) 291:669–672.

Hublin JJ (1982): Les anténéandertaliens: presapiens ou prénéandertaliens. Géobios. Mém Special 6:345–357:

Hublin JJ (1983): Les origines de l'homme de type moderne en Europe. Pour La Science 64:62–71.

Jacobi RM (1980): The Upper Palaeolithic of Britain with special reference to Wales. In Taylor JA (ed): "Culture and Environment in Prehistoric Wales." Oxford: B.A.R. British Series 76, pp 15–100.

Jelinek A (1982): The Tabūn cave and Paleolithic man in the Levant. Science 216:1369–1375.

Jelínek J (1969): Neanderthal man and Homo sapiens in central and eastern Europe. Curr Anthropol 10:475–503.

Joly J (1955): Découverte de restes néandertaliens en Côte-d'Or. CR Acad Sci [D] (Paris) 240:2253–2255.

Jones JS (1981): How different are human races? Nature 293:188–190.

Kallay J (1963): A radiographic study of the Neanderthal teeth from Krapina, Croatia. In Brothwell DR (ed): "Dental Anthropology." Oxford: Pergamon Press, pp 75–86.

Keith A (1913): Problems relating to the teeth of the earlier forms of prehistoric man. Proc R Soc Exp Biol Med 6:103–104.

Keith A (1925): Neanderthal man in Malta. J R Anthropol Inst 54:251.

Keith A (1931): "New Discoveries Relating to the Antiquity of Man." London: Williams and Norgate.

Klaatsch H (1909): Die neuesten Ergebnisse der Paläontologie des Menschen und ihre Bedeutung für das Abstammungsproblem. Z Ethnol 41:537–584.

Klaatsch H, Hauser O (1909): Homo mousteriensis Hauseri. Arch Anthropol 7:287–297.

Klein R (1983): The Stone Age prehistory of southern Africa. Annu Rev Anthropol 12:25–48.

Kozłowski JK (ed) (1982): "Excavation in the Bacho Kiro Cave (Bulgaria): Final Report." Warsaw: Państwowe Wydawnictwo Naukowe.

Krukoff S (1970): L'occipital de La Chaise (Suard), caractères metriques distances de forme et de format. CR Acad Sci [D] (Paris) 270:42–45.

Laitman JT, Heimbuch RC, Crelin ES (1979): The basicranium of fossil hominids as an indicator of their upper respiratory systems. Am J Phys Anthropol 51:15–34.

Leakey LSB (1972): Homo sapiens in the Middle Pleistocene and the evidence of Homo sapiens evolution. In Bordes F (ed): "The Origin of Homo sapiens." Paris: Unesco, pp 25–29.

Le Gros Clark WE (1938a): General features of the Swanscombe skull bones. R Anthropol Inst 68:58–60.

Le Gros Clark WE (1938b): The endocranial cast of the Swanscombe bones. R Anthropol Inst 68:61–67.

Le Gros Clark WE (1955): "The Fossil Evidence of Human Evolution." Chicago: University of Chicago Press.

Legoux P (1966): "Détermination de l'Âge Dentaire de Fossiles de la Lignée Humaine." Paris: Maloine.

Legoux P (1972): La molaire néandertalienne des Rivaux. Anthropologie 76:665–684.

Le Mort F (1981): "Dégradations Artificielles sur des Os Humains du Paléolithique." Thesis, Université de Paris VI.

Leroi-Gourhan A (1958): Etude des restes humains fossiles provenant des grottes d'Arcy-sur-Cure. Ann Paléont 44:87–148.

Leroyer C, Leroi-Gourhan A (1983): Problèmes et chronologie: Le castelperronien et l'aurignacien. Bull Soc Prehist Franc 80:41–44.

Lévêque F, Vandermeersch B (1981): Le neándertalien de Saint-Césaire. Recherche 12:242–244.

Lumley H de (1961): La place du remplissage de la grotte du Lazaret (Alpes-Maritimes) dans la stratigraphie du Quaternaire de la région de Nice à Monaco. Bull Mus Anthropol Préhist Monaco 8:97–133.

Lumley H de (1969): Une cabane acheuléenne dans la grotte du Lazaret. Mem Soc Préhist Franc 7.

Lumley H de (1969–1971): "La Paléolithique Inférieur et Moyen du Midi Méditerranéen dans son Cadre Géologique. 1: Ligurie, Provence. 2: Bas-Languedoc, Roussillon Catalogne. Gallia-Préhistoire 5. Paris: CNRS.

Lumley H de (ed) (1976): "La préhistoire Francaise." Paris: CNRS.

Lumley H de (1978): Discussion. In Boné E et al: "Les Origines Humaines et les Époques de l'Intelligence." Paris: Masson, pp 176–177.

Lumley H de, Yokoyama Y, Park YC, Fournier A (in press): Bilan des datations du remplissage de la Caune de l'Arago à Tautavel. In Lumley H de, Labayrie J (eds): "Datations Absolues et Analyses Isotopiques en Préhistoire. Méthodes et Limites." Tautavel: CNRS.

Lumley M-A de (1972): Les nèandertaliens de la grotte de l'Hortus (Valflaunés, Herault). In Lumley H de (ed): "La Grotte Moustérienne de l'Hortus." Etud Q 1:375–385.

Lumley M-A de (1973): "Antenéandertaliens et Neándertaliens du Bassin Méditerranéen Occidental Européen." Marseille: Université de Provence.

Lumley M-A de (1976a): Les antenéandertaliens dans le sud. In Lumley H de (ed): "La Préhistoire Francaise." Paris: CNRS, pp 547–560.

Lumley M-A de (1976b): Les néandertaliens dans le nord et dans le centre. In Lumley H de (ed): "La Préhistoire Francaise." Paris: CNRS, pp 588–594.

Lumley M-A de (in press): L'homme de Tautavel. Critères morphologiques et stade évolutif. In Lumley H de, Labayrie J (eds): "Datations Absolues et Analyses Isotopiques en Préhistoire. Méthodes et Limites." Tautavel: CNRS, pp 259–264.

Madre-Dupouy M (1976): "Les Dents Temporaires de l'Enfant Néanderthalien du Roc de Marsal (Dordogne)." Thesis, Université Paris VI.

Martin H (1923): L'homme fossile de La Quina. Arch Morph Gen Exp 15:1–260.

Martin H (1926): "Recherches sur l'Évolution du Moustérien dans le gisement de La Quina, Charente; Vol 4. L'Enfant Fossile de La Quina." Angoulême: Impr. Ouvrière.

McBurney CBM (1960): Two soundings in the Badger Hole. Annu Rep Wells Nat Hist Archaeol Soc 71-72:19–27.

Méroc L (1963): Les elements de datation et la mandibule humaine de Moutmaurin (Haute-Garonne). Bull Soc Geol France 5:508–515.

Mezzena F, di Cesnola AP (1972): Scoperta di una sepoltura gravettiana nella Grotta Paglicci (Rignano Garganico). Riv Sci Preist 27:27–50.

Molleson TI (1976): Remains of Pleistocene man in Paviland and Pontnewydd Caves, Wales. Trans Br Cave Res Assoc 3:112–116.

Morant GM (1938): The form of the Swanscombe skull. J R Anthropol Inst 68:67–97.

Moure Romanillo JA, Garciá-Soto E (1983): Radiocarbon dating of the Mousterian a Cueva Millán (Hortigüela, Burgos, Spain). Curr Anthropol. 24:232–233.

Movius HL (ed) (1975): "Excavation of the Abri Patand, Les Eyzies (Dordogne)." Cambridge: Harvard University Press.

Nei M, Roychoudhury AK (1982): Genetic relationships and evolution of human races. Evol Biol 14:1–59.

Oakley KP (1964): "Frameworks for Dating Fossil Man." London: Weidenfeld and Nicolson.

Oakley KP (1980): Relative dating of the fossil hominids of Europe. Bull Br Mus Nat Hist (Geol) 34:1–63.

Oakley KP, Hoskins CR (1951): Application du test de la flourine aux crânes de Fontéchevade (Charente). Anthropologie 55:239–242.

Oakley KP, Campbell BG, Molleson TI (1971): "Catalogue of Fossil Hominids. Vol 2, Europe." London: British Museum (Natural History).

Otte M (1979): "La Paleolithique Supérieur Ancien du Belgique." Brussels: Musées Roy d'Art et d'Histoire.

Pales L (1958): Les Neándertaliens en France. In von Koenigswald GHR (ed): "Hundert Jahre Neanderthaler." Utrecht: Kemink en zoon, pp 32–37.

Patte E (1957): "L'Enfant Néandertalien du Pech de L'Azé. Paris: Masson.

Patte E (1960): Découverte d'un néandertalien dans la Vienne. Anthropologie 64:512–517.

Payá AC, Walker MJ (1980): A possible hominid fossil from Alicante, Spain? Curr Anthropol 21:795–800.

Peyrony D (1930): Le Moustier, ses gisements, ses industries, ses couches géologique. Rev Anthropol 40:48–76, 155–176.

Piperno M (1976–1977): Analyse du sol Moustérien de la Grotte Guattari au Mont Circé. Quaternaria 19:71–92.

Piveteau J (1963–1965): La grotte de Régourdou (Dordogne). Paléontologie humaine. Ann Paléont 49:285–304; 50:155–194; 52:163–194.

Piveteau J (1976): Les Anté-Néandertaliens du Sud-Ouest. In Lumley H de (ed): "La Préhistoire Française." Paris: CNRS, pp 561–566.

Protsch R, Glowatzki G (1974): Das absolute Alte des paläolithischen skeletts aus der Mittleren Klause bei Neuessing, Kreis Kelheim im Bayern. Anthropol Anz 34:140–144.

Protsch R, Semmel A (1978): Zur Chronologie des Kelsterbach—Hominiden. Eiszeitalter Gegenw. 28:200–210.

Robert C (1982): Découverte d'un homme fossile du paléolithique moyen (?) en Haute-Egypt. Anthropologie 85:332–333.

Ronen A (ed) (1982): The transition from the Lower to Middle Palaeolithic and the origin of modern man. Oxford: Br Archaeol Rep Int Ser 151.

Saban R (1980): Les empreintes vasculaires endocrâniennes (v.v. méningées moyennes) chez l'homme de l'Acheuléen en Europe en Afrique. Anthropologie (Brno) 18:133–152.

Saban R (1982): Les empreintes endocrâniennes des veines méningées moyennes et les étapes de l'évolution humaine. Ann Paléont 68:171–220.

Santa Luca AP (1978): A reexamination of presumed Neanderthal-like fossils. J Hum Evol 7:619–636.

Sauter M (1983): À propos de la presentation de la sépulture double de la grotte des Enfants aux Baoussi-Roussi (Grimaldi, Italie). Bull Soc Prehist Franc 80:139–140.

Schwalbe G (1904): "Die Vorgeschichte des Menschen." Braunschweig. Friedrich Viewug.

Schwarcz HP (1982): Uranium series dating of early man sites in Europe, Asia and North Africa. Abstr 1er Congr Int Paléont Hum Nice, p 95.

Sergi S (1948): Sulla morfologia della facies anterior corporis maxillae nei paleantropi di Saccopastore e del monte Circeo. Atti Accad Naz Lincei 4:387–394.

Sergi S (1953): Morphological position of the "Prophaneranthropi" (Swanscombe and Fontéchevade). Reprinted in Howells WW (ed): "Ideas on Human Evolution." Cambridge: Harvard University Press, 1962, pp 507–520.

Sergi S (1974): "Il Cranio Neandertaliano de Monte Circeo." Rome: Accademia Nazionale die Lincei.

Sergi S, Parenti R, Paoli G (1974): Il giovane paleolitico della caverna delle Arene Candide Mem Ist Ital Pal Umana 2:13–38.

Sigmon BA (1982): Comparative morphology of the locomotor skeleton of Homo erectus and the other fossil hominids, with special reference to the Tautavel innominate and femora. Prétirage 1er Congr Int Paléont Hum Nice, pp 422–446.

Skinner AF (1983): Overestimate of stalagmitic calcite ESR dates due to laboratory heating. Nature 304:152–154.

Skinner MF, Sperber GH (1982): "Atlas of Radiographs of Early Man." New York: Alan R. Liss.

Smith FH (1982): Upper Pleistocene hominid evolution in South-central Europe: A review of the evidence and analysis of trends. Curr Anthropol 23:667–703.

Sollas WJ (1913): Paviland Cave, an Aurignacian station in Wales. J R Anthropol Inst 43:337–364.

Spencer F, Smith FH (1981): The significance of Aleš Hrdlička's "Neanderthal Phase of Man": A historical and current assessment. Am J Phys Anthropol 56:435–459.

Steiner W (1979): "Der Travertin von Ehringsdorf und seine fossilen." Wittenberg: A Ziemsen.

Stewart TD (1964): A neglected primitive feature of the Swanscombe skull. In Ovey CD (ed): "The Swanscombe Skull: A Survey of Research on a Pleistocene Site." London: Royal Anthropological Institute, pp 151–159.

Stringer CB (1974): Population relationships of later Pleistocene hominids: A multivariate study of available crania. J Archaeol Sci 1:317–342.

Stringer CB (1978): Some problems in Middle and Upper Pleistocene hominid relationships. In Chivers DJ, Joysey K (eds): "Recent Advances in Primatology, Vol 3. Evolution." London: Academic Press, pp 395–418.

Stringer CB (1979): A re-evaluation of the fossil human calvaria from Singa, Sudan. Bull Br Mus Nat Hist (Geol) 32:77–83.

Stringer CB (1981): The dating of European Middle Pleistocene hominids and the existence of Homo erectus in Europe. Anthropologie (Brno) 19:3–14.

Stringer CB (1982): Towards a solution to the Neanderthal problem. J Hum Evol 11:431–438.

Stringer CB (In press (a)): The hominid remains. In Green HS (ed): "Pontnewydd Cave—a Lower Palaeolithic Hominid Site in Wales: An Interim Report." Cardiff: National Museum of Wales.

Stringer CB (In press (b)): Some further notes on the morphology and dating of the Petralona hominid.

Stringer CB, Burleigh R (1981): The Neanderthal problem and the prospects for direct dating of Neanderthal remains. Bull Br Mus Nat Hist (Geol) 35:225–241.

Stringer CB, Currant AP (In press): The human remains. In Callow P (ed): "La Cotte de St. Brelade: Excavations by CBM McBurney, 1961–1978." London: Academic Press.

Stringer CB, Trinkaus E (1981): The Shanidar Neanderthal crania. In Stringer CB (ed): "Aspects of Human Evolution." London: Taylor and Francis, pp 129–165.

Stringer CB, Howell FC, Melentis JK (1979): The significance of the fossil hominid skull from Petralona, Greece. J Archaeol Sci 6:235–253.

Stuart AJ (1982): "Pleistocene Vertebrates in the British Isles." London: Longman.

Taschini M (1979): L'industrie lithique de Grotta Guattari au Mont Circé, Latium: Definition culturelle, typologique et chronologique du Pontinien. Quaternaria 21:179–247.

Tavoso A (1982): Le cadre geochronologique de la mandibule de Montmaurin, examen des données disponibles. Abst 1er Congr Int Paléont Hum Nice, pp 96–97.

Thoma A (1965): La definition des néandertaliens et la position des hommes fossiles de Palestine. Anthropologie 69:519–534.

Thoma A (1975): Were the Spy fossils evolutionary intermediates between classic neandertal and modern man? J Hum Evol 4:387–410.

Thommeret J, Thommeret Y (1973): Monaco radiocarbon measurements IV. Radiocarbon 15:321–344.

Thorne AG, Wolpoff MH (1981): Regional continuity in Australasian Pleistocene hominid evolution. Am J Phys Anthropol 55:337–349.

Tihon J (1898): Les cavernes préhistoriques de la vallée de la Vesdre. Fouilles de Fond-de-Forêt (pt.2). Ann Soc Archéol Brux 12:165–173.

Tillier AM (1977): La pneumatisation du massif cranio-facial chez les hommes actuels et fossiles. Bull Mem Soc Anthropol Paris 4:177–189; 287–316.

Tillier AM (1979): La dentition de l'enfant moustérien Chateauneuf 2 découvert à l'abri de Hauteroche (Charente). Anthropologie 83:417–438.

Tillier AM (1982): Les enfants néanderthaliens de Devil's Tower (Gibraltar). Z Morphol Anthropol 73:125–148.

Tillier AM (1983a): L'enfant néanderthalien du Roc de Marsal (Campagne du Bugue, Dordogne). Le Squelette facial. Ann Paléont 69:137–149.

Tillier AM (1983b): Le crâne d'enfant d'Engis 2: Un example de distribution des caractères juvéniles, primitifs et néanderthaliens. Bull Soc R Belge Anthropol Préhist 94:51–75.

Tillier AM, Genet-Varcin E (1980): La plus ancienne mandibule d'enfant de counte en France dans le gisement de La Chaise Vouthoun (Abri Suard) en Charente. Z Morphol Anthropol 71:196–214.

Topinard P (1886): Les caractères simiens de la mandibule de La Naulette. Rev Anthropol 1:385–431.

Trinkaus E (1973): A reconsideration of the Fontéchevade fossils. Am J Phys Anthropol 39:25–36.

Trinkaus E (1976a): The morphology of the European and Southwest Asian neandertal pubic bones. Am J Phys Anthropol 44:95–104.

Trinkaus E (1976b): Note on the hominid molar from the Abri des Merveilles at Castel-Merle (Dordogne). J Hum Evol 5:203–205.

Trinkaus E (1977): A functional interpretation of the axillary border of the neandertal scapula. J Hum Evol 6:231–234.

Trinkaus E (1978): Les métatarsiens et les phalanges du pied des néandertaliens de Spy. Bull Inst R Sci Nat Belg 51:1–18.

Trinkaus E (1981): Upper Pleistocene hominid limb proportions and the problem of cold adaptation among the Neanderthals. In Stringer CB (ed): "Aspects of Human Evolution." London: Taylor and Francis, pp 187–224.

Trinkaus E, Howells WW (1979): The Neanderthals. Sci Am 241:118–133.

Twiesselmann F (1961): Le fémur néanderthalien de Fond-de-Forêt (Province de Liège). Mem Inst R Sci Nat Belg 148:1–164.

Vallois HV (1940): La découverte du squelette du Moustier. Anthropologie 49:776–778.

Vallois HV (1949): L'origine de l'*Homo sapiens*. CR Acad Sci [D] (Paris) 228:949–951.

Vallois HV (1954): Neandertals and presapiens. J R Anthropol Inst 84:111–130.

Vallois HV (1958): La grotte de Fontéchevade. 2nd part: Anthropologie. Arch Inst Paléont Hum 29:1–164.

Vallois HV (1965): Le Sternum néandertalien du Régourdou. Anthropol Anz 29:273–289.

Vallois HV (1969): Le temporal néandertalien H 27 de La Quina, étude anthropologique. Anthropologie 73:365–400, 525–544.

Vandermeersch B (1976a): Les néandertaliens en Charente. In Lumley H de (ed): "La Préhistorie Française." Paris: CNRS, pp 584–587.

Vandermeersch B (1976b): La mandibule moustérienne de Montgaudier (Montbron, Charente). CR Acad Sci [D] (Paris) 283:1161–1164.

Vandermeersch BV (1978): Discussions. In Boné E et al: "Les Origines Humaines et les Époques de l'Intelligence." Paris: Masson, pp 177–178.

Vandermeersch BV (1981): Les premiers *Homo sapiens* au Proche-Orient. In Ferembach D (ed): "Les Processus de l'Hominisation." Paris: CNRS, pp 97–100.

Vandermeersch B, Tillier AM, Krukoff S (1976): Position chronologique des restes humains de Fontéchevade. In Thoma A (ed): "Le Peuplement Anténéandertalien de l'Europe." IXth Cong UISPP Prétirage. Nice: CNRS, pp 19–26.

Veiga Ferreira O da, Leitão M (1981): "Portugal Pré-Histórico." Publicações Europa-América.

Virchow H (1939): Skelet von Le Moustier. Anat Anz 88:261–274.

Vlček E (1956): Kalva pleistocénního člověka z Podbaby (Praha XIX). Anthropozoikum 5:191–217.

Vlček E (1978): A new discovery of *Homo erectus* in Central Europe. J Hum Evol 7:239–251.

Weidenreich F (1943a): The "Neanderthal Man" and the ancestors of "*Homo sapiens.*" Am Anthropol 42:375–383.

Weidenreich F (1943b): The skull of *Sinanthropus pekinensis:* A comparative study on a primitive hominid skull. Palaeont Sinica D.10.

Weidenreich F (1947): Facts and speculations concerning the origin of *Homo sapiens*. Am Anthropol 49:187–203.

Weiner J, Campbell BG (1964): The taxonomic status of the Swanscombe skull. In Ovey CD (ed): "The Swanscombe Skull. A Survey of Research on a Pleistocene Site." London: Royal Anthropological Institute, pp 175–209.

Weinert H (1925): "Der Schädel der eiszeitlichen Menschen von Le Moustier in neuer Zusammensetzung." Berlin: J. Springer.

Wolpoff MH (1980a): "Paleoanthropology." New York: Knopf.

Wolpoff MH (1980b): Cranial remains of Middle Pleistocene European hominids. J Hum Evol 9:339–358.

Wolpoff MH (1982): The Arago dental sample in the context of hominid dental evolution. Prétirage 1er Congr Int Paléont Hum Nice, pp 389–410.

Zeuner FE (1940): The age of Neanderthal man with notes on the Cotte de St Brelade, Jersey, C.I. Occ. Papers Inst Archaeol Univ London 3:1–20.

Zhou Mingzhen, Li Yanxian, Wanglinghong (1982): Chronology of the Chinese fossil hominids. Prétirage 1er Congr Int Paléont Hum Nice, pp 593–604.

The Origins of Modern Humans: A World Survey of the Fossil Evidence, pages 137–209
© 1984 Alan R. Liss, Inc., 150 Fifth Avenue, New York, NY 10011

Fossil Hominids From the Upper Pleistocene of Central Europe and the Origin of Modern Europeans

Fred H. Smith

Department of Anthropology, University of Tennessee, Knoxville,
Tennessee 37916

> It is very difficult for French anthropologists to have a perfect knowledge of all that has been written on human remains found in central Europe and considered, rightly or wrongly, as Quaternary.

With these words, Marcellin Boule introduced an article by Hugo Obermaier entitled "Quaternary Human Remains in central Europe," published in L'Anthropologie in 1905. Obermaier's article, which despite the title had virtually nothing to do with the hominid remains themselves, was reprinted in English 2 years later [Obermaier, 1907]. Unfortunately, his efforts were apparently rather unsuccessful, because Upper Pleistocene fossil hominids from central Europe generally continued to be either excluded or given only cursory attention in discussion of Upper Pleistocene hominid evolution by most paleoanthropologists working west of the Rhine or across the Atlantic for the next 70 years. The reasons for this are complex and clearly involve sociopolitical and linguistic factors [Brace, 1964; Brace and Montagu, 1977], but much of the explanation for the emphasis on western European (especially French) fossil hominids in discussion of archaic/modern *Homo sapiens* relationships is related to the emergence of Boule as the leading and most influential expert on fossil hominids of his time. The various factors that contributed to this emergence are discussed in detail by Hammond [1980], but the primary one was unquestionably Boule's impressive, comprehensive study of the Neandertal skeleton from La Chapelle-aux-Saints [Boule, 1911/13; Boule and Anthony, 1911]. His analysis directly contradicted the unilineal interpretations of hominid evolution prevalent in the thinking of most central European anthropologists, most significantly Gustav Schwalbe [1906] and Karl Gorjanović-Kramberger [1906], and accommodated nicely the pre- and post-Piltdown evolutionary perspectives in England [Keith, 1911; Sollas,

1915; Hammond, 1979, 1980]. Boule's views quickly became almost universally accepted in French- and English-speaking anthropological circles.[1]
Furthermore, the La Chapelle study easily outdistanced the other detailed
monographical consideration of Neandertal morphology available, Gorjanović-Kramberger's [1906] descriptive analysis of the Krapina hominids, to
become the standard reference on Neandertals and archaic/modern *H
sapiens* relationships. These factors, plus the apparent capitulation of
Schwalbe [1914] under the influence of Boule's work on La Chapelle, and
the loss of prestige suffered by German science in general after World War I
[Brace, 1964], assured the supremacy of the French perspective on hominid
evolution in Europe, which in turn was based primarily on French specimens, for the next half century. I contend that, over the years, this led to
the assumption among most English- and French-speaking anthropologists
that patterns of Upper Pleistocene hominid evolution in other parts of
Europe simply reflected the patterns observed in France (western Europe)
and thus merited only peripheral consideration. While recent work [Jelínek,
1969; Frayer, 1978; Wolpoff, 1980a; Smith, 1982; Spencer and Smith, 1981]
has begun to refocus attention among non-central-European anthropologists on the importance of the central European material, the pro-western
Europe bias continues to represent the status quo. To take a very recent
example, Hublin [1983a], in a detailed presentation of his views on the
origins of modern *H sapiens* in Europe, bases his interpretation entirely on
western European and Near Eastern hominids and dismisses the entire
central European fossil hominid record in little more than a paragraph!

It is the purpose of this paper to discuss the Upper Pleistocene fossil
hominids from central Europe on an individual basis and to analyze the
pattern of evolution in this region separately from that of western Europe.
Following this, the pattern determined for central Europe will be compared
to those in other regions, including western Europe, in order to consider
the question of the origin of modern Europeans and of modern *H sapiens* in
general. I believe that a detailed analysis of the central European hominids
is necessary, first because they have been largely ignored for far too long,

---

[1]There were of course dissenters, the early ones including Verneau [1924] and Hrdlička
[1914, 1927, 1930]. It is interesting that Hrdlička received his early exposure to human
evolution at the Ecole D'Anthropologie in Paris [Spencer, 1979; and in this volume], where
unilineal views of cultural and biological evolution were fostered during the late 19th century.
The Ecole and its staff were scientific rivals of the Museum de Histoire Naturelle, and Boule
systematically attacked the scientific merit of the evolutionary perspectives prevalent there
[Hammond, 1980].

and second because the origin of modern H *sapiens* can only be understood if regional patterns of Upper Pleistocene hominid evolution are reasonably well-documented.

BACKGROUND TO THE STUDY OF UPPER PLEISTOCENE FOSSIL HOMINIDS IN CENTRAL EUROPE

Central Europe can be defined geographically in a number of ways, depending on the perspective of the definer. For purposes of this study, I have defined it as the Pannonian Basin and surrounding highlands (south-central Europe) and the region between the Rhine and Weichsel rivers to the north (north-central Europe). The former region is well-defined geographically, being enclosed by two crescent-shaped systems of mountains/ highlands. While clearly defined by these highlands, south-central Europe was probably never isolated from surrounding regions, although passage through the highlands was probably often restricted, especially during cold phases when snow lines were lower [Butzer, 1971]. Furthermore, numerous river valleys dissect the highland systems, many of which would have served as possible avenues of contact even under the harshest climatic conditions. North-central Europe is somewhat less distinctly defined from a geographic standpoint. It consists basically of the areas adjacent to south-central Europe to the north, extending from the Bavarian Plateau to the German/Polish Plain. It is separated from south-central Europe, though not completely, by the highlands of the Erzgebirge, Böhmerwald, and Bayerischerwald. Since two major river valley systems, the Danube and the Elbe, connect these two areas, and since an open corridor existed between them even during the maximum extent of Pleistocene glaciation [Kukla, 1978], there was most likely a considerable amount of interaction between their Upper Pleistocene hominid populations. In this respect, these two portions of central Europe can be considered natural extensions of each other. The differentiation of the German/Polish Plain as a geographic unit separate from adjacent regions to the east and west is, however, clearly somewhat arbitrary.

I have elected to use a slightly modified version of Musil and Valoch's [1966] chronostratigraphic division of the Upper Pleistocene for the Moravian Karst region of Czechoslovakia as a generalized chronological scheme for all of central Europe. In a region as large as this, there are certain to be differences in stratigraphy among subregions. It is obvious, for example, that chronostratigraphy in the Carpathian Basin [Gábori, 1976; Gábori-Csánk, 1983] and Germany [Mania and Preuse, 1975] is not identical to the Moravian karst. Thus the use of the Musil-Valoch scheme to provide a general chronological background for the entire region has been criticized [see comments to Valoch, 1968; Smith, 1982]. Nevertheless, when one compares the Upper Pleistocene chronostratigraphy in Musil-Valoch to that of Gábori [1976], for example, the general patterns are very similar. Both recognize a relatively long, complex Early Würm stadial, interrupted by warmer phases of varying lengths, and a more recent, colder Lower Würm stadial (Würm maximum in Gábori), followed by a Middle Würm intersta-dial (Podhraden-Hengelo in Musil and Valoch [1966]. Furthermore, the Musil-Valoch scheme works well as a *general* chronostratigraphic framework in the Hrvatsko Zagorje of Yugoslavia [Wolpoff et al, 1981; Smith, 1982] and agrees generally with chronological frameworks employed elsewhere in Europe [Butzer, 1971; Flint, 1971], including the well-documented sequences in the Netherlands [Zagwijn, 1974; Grootes, 1978] and northwestern France [Woillard, 1978; Woillard and Mook, 1982]. All things considered, I feel the use of the Musil-Valoch scheme to provide a chronological framework is justified so long as it is remembered that it can only provide a *general* picture of chronological relationships among sites from different subregions of central Europe. Unfortunately, though correlation between deep-sea cores and central European loess sequences are promising and may ultimately provide a much more precise chronostratigraphic scheme for this region [Kukla, 1975, 1978], it is not yet possible to utilize the deep sea core data to extensively aid chronological correlation of Upper Pleistocene hominid-bearing sites.

For purposes of this study, the following broad chronostratigraphic units (most recent listed first) will be utilized:

> Upper Würm Interstadial
> Middle Würm Stadial (28 to 32 ky BP)
> Middle Würm (Podhradem-Hengelo) Interstadial (32 to
>   38 ky BP)
> Lower Würm Stadial (38 to 45 ky BP)
> Early Würm period (45 to 70 ky BP)

The dates presented here are approximate and have been determined on the basis of the most up-to-date information. They do not necessarily correspond to those in Valoch [1968]. Furthermore, the Early Würm is a very complex period of climatic fluctuations, not a simple single unit. However, wrestling with the details of how it should be subdivided is not pertinent to the issues raised in this paper.

Finally, the archaeological framework for central Europe is discussed in several other places [eg, Valoch, 1968, 1977–1978; Bosinski, 1967; Gábori,

1976; Frayer, 1978; Hahn, 1977; Ivanova, 1979; Smith, 1982] and only very general considerations of archaeological matters will be included here. More detailed discussions of many of the south-central European sites and their background may be found in Smith [1982].

NEANDERTALS IN CENTRAL EUROPE

Archaic *H sapiens* remains have been recovered from a number of sites throughout central Europe (Table I), mostly associated with Middle Paleolithic industries. Unfortunately, most of the remains are fragmentary; and at many sites (Kůlna, Ochoz, Šala, Gánovce), only relatively small portions of one or two individuals are represented. In cases where several individuals are present (Krapina, Vindija), the fragmentary condition of the material makes the reconstruction of relatively complete crania or the determination of association between elements (crania to mandibles, crania to postcrania, etc) that might represent single individuals virtually impossible. Despite these problems, the collective sample of Neandertals from central Europe contributes significantly to our knowledge of archaic *H sapien* because *relatively* large samples of several critical anatomical regions (brow ridges, mandibles, maxillae, teeth, etc) are preserved. This provides an excellent perspective on the variability of European archaic *H sapiens* and, since there is a temporal difference between groups of specimens, the opportunity to search for morphological trends from earlier to later groups.

Stratigraphically, Middle Paleolithic industries exist from the Eemian or Riss-Würm interglacial, perhaps even earlier if sites like Ehringsdorf are found to be earlier than the Eemian [Cook et al, 1982], certainly to the end of the Lower Würm stadial and possibly into the Podhradem [Valoch, 1968; Gábori, 1976]. The earliest chronometric dates are uranium series dates for Tata (Hungary) and Ehringsdorf (Germany), both certainly in excess of 100 ky BP [Schwarcz and Skoflek, 1982; Cook et al, 1982]. Several radiocarbon dates for the central European Mousterian are known, which range from 55 to 40 ky BP [see Smith, 1982, p 670]; and the most recent, unequivocal Mousterian-associated radiocarbon date—from level 7a at Kůlna—is 38.6 + 0.95–0.8 ky BP (GrN-6024) [Valoch, 1977–78]. Several more recent dates on Hungarian Mousterian are discounted because they run contrary to well-established geological and paleontological sequences [Gábori-Csánk, 1970; Gábori, 1976].

South-Central Europe: The Hrvatsko Zagorje

The most well-known site from the mountainous karstlike Hrvatsko Zagorje in northwestern Croatia (Yugoslavia) is the Krapina rockshelter.

TABLE I. Upper Pleistocene Fossil Hominids From Central Europe

Site (country)	Year(s) of discovery	Human remains	Archeological association	Date
I. Neandertals—south-central Europe				
1. Šipka (ČSSR)[b]	1880	Mandibular symphysis of a child	Mousterian	Podhradem[c]
2. Krapina (Yugoslavia)	1899–1905	Cranial, mandibular, postcranial, and dental remains from individuals of all ages	Mousterian	Riss-Würm through Lower Würm[c]
3. Ochoz (ČSSR)	1905,1964	Adult mandible; adult molar; postcranial fragments	Mousterian	Early Würm[c]
4. Gánovce (ČSSR)	1926,1955	Natural endocast; cranial fragments; natural moulds of postcrania	Mousterian	Riss-Würm[c]
5. Subalyuk[a] (Hungary)	1932	Adult mandible and postcrania; child's cranium	Mousterian	Early Würm[c]
6. Šala (ČSSR)	1961	Adult frontal	None	Upper Pleistocene[d]
7. Kůlna (ČSSR)	1965	Adult right maxilla; adult parietal fragment, teeth	Mousterian	38.6 ± 0.92–0.8 ky BP[e] 45.6 ± 2.85–2.2 ky BP[e]
8. Vindija G_3 (Yugoslavia)	1974–1981	Cranial, mandibular, and postcranial remains from at least 8 individuals	Mousterian	Lower Würm[d,g]
II. Possible Neandertals—south-central Europe				
9. Džeravá Skála[a] (ČSSR)	1913	Lower M2	Szeletian/Aurignacian	Mid-Würm (?)[c]
10. Ohaba-Ponor[a] (Romania)	1923	Adult pedal phalanx	Mousterian with Upper Paleolithic Elements	Early Würm[c]
11. Vindija G_1-F (Yugoslavia)[c]	1974–1978	Adult mandibular ramus; parietal fragment; teeth	Aurignacian[g]	Podhradem[c,f]
III. Neandertals—north-central Europe				
12. Neandertal (West Germany)	1856	Calotte and postcrania of an adult	None	None[h]
13. Taubach (East Germany)	1887,1892	Lower M1; lower dm1	Middle Paleolithic	Riss-Würm[c]

continued

14. Ehringsdorf (East Germany)	1908–1925	Portions of crania, mandibles, and postcrania from as many as 9 individuals	Mousterian	Riss-Würm[c,j]
15. Wildscheuer[b] (West Germany)	1953	Cranial fragments	Mousterian	Early Würm[c]
16. Salzgitter-Lebenstedt[b] (West Germany)	1956	Adult occipital and parietal	Mousterian	55.6 ± 0.9 ky BP[e]
IV. Early modern *Homo sapiens*—south-central Europe				
17. Mladeč (ČSSR)	1881–1904	Cranial and postcranial remains from at least 12 individuals of various ages	Aurignacian[g]	Podhradem[c,j]
18. Podbaba[b] (ČSSR)	1883	Adult calvarium	Aurignacian[g]	Middle Würm[c,k]
19. Svatý Prokop[a] (ČSSR)	1887	Adult cranial and postcranial fragments	Undiagnostic	Middle Würm[c,k]
20. Brno (ČSSR)	1891	Adult calvarium and postcrania (Brno 2)	Eastern Gravettian	Middle Würm[j]
21. Předmostí[b] (ČSSR)	1894–1895 1927	Skeletons of from 27 to 29 individuals of all ages	Eastern Gravettian[q]	26.32 + 0.24 ky BP[e]
22. Willendorf (Austria)	1908	Adult mandibulary symphysis; femoral fragment	Eastern Gravettian	Middle Würm[c]
23. Miesslingstal (Austria)	1914	Child's mandibular corpus	Aurignacian	Middle Würm[c]
24. Dolní Věstonice (ČSSR)	1925–1951	Adult skeletal remains, crania, and teeth of 5 to 10 individuals, mostly adults	Eastern Gravettian	25.82 ± 0.17 ky BP[e]
25. Cioclinova[a] (Romania)	1942	Adult calotte	Aurignacian	Würm Interstadial[c]
26. Šilická Brezová[a] (ČSSR)	1948	Permanent lower molar	None	Early Würm[c]
27. Istállóskö[a] (Hungary)	1950	Germ of permanent lower molar	Aurignacian	30.9 ± 0.6 ky BP[e]
28. Zlatý Kůň (ČSSR)	1950–1952	Adult cranium and postcranial fragments	Aurignacian[g]	Middle Würm 25.02 ± 0.15 ky BP[e]
29. Pavlov (CSSR)	1954–1957	Adult skeleton; mandibular and maxillary fragments	Eastern Gravettian	26.62 ± 0.23 ky BP
30. Velika Pećina (Yugoslavia)	1961	Adult frontal	Aurignacian	> 33.85 ± 0.52 ky BP[e]

TABLE I. Upper Pleistocene Fossil Hominids From Central Europe (continued)

Site (country)	Year(s) of discovery	Human remains	Archeological association	Date
31. Bačo Kiro[a] (Bulgaria)	?	Very fragmentary portions of several bones	Aurignacian (?)[l]	> 43.0 ky BP[e,l]
V. Early modern *Homo sapiens*—north-central Europe				
32. Stretten (West Germany)	1931	Adult calvarium, mandible and postcranial fragments; right humerus[m], adult calvarium[n]	Aurignacian	Middle Würm[c]
33. Kelsterbach (West Germany)	1952	Adult calvarium	None	32.0 ky BP[o] 31.2 ± 0.6 ky BP[p]
34. Hahnöfersand (West Germany)	1973	Adult frontal	None	36.0 ky BP[o] 36.3 ± 0.6 ky BP[p]
35. Sande/Paderborn (West Germany)	1976	Adult calvarium	None	26.0 ky BP[o] 27.4 ± 0.6 ky BP[p]

[a]Indicates specimens I have not studied.

[b]Indicates study of casts only.

[c]Date geologically determined. Faunal association usually important.

[d]Date determined completely by faunal association.

[e]Radiocarbon (C14) date of associated stratum. Laboratory numbers are given in the text or in Smith [1982, p 670].

[f]Other portions of sedimentary sequence anchored by radiocarbon dates.

[g]Association between hominid remains and the cultural remains not certain.

[h]Considered to be Early Würm on the basis of morphological similarity to other Neandertals.

[i]Recently it has been suggested [Cook et al, 1982], partly on the basis of uranium series dating, that the Ehringsdorf hominids *might* predate the Riss-Würm.

[j]Cultural associations important in date determination.

[k]Exact stratigraphic location of hominid finds uncertain.

[l]Hominids undescribed and relationship between cultural material; date and hominids not clear.

[m]Humerus is reported to be from the base of the "Middle Aurignacian" level and may be a Neandertal (see Note 2).

[n]Second calvarium may not be contemporary with the Aurignacian.

[o]Dated by amino acid racemization.

[p]Radiocarbon date of residual bone collagen in the specimen. Laboratory numbers given in the text.

[q]Skeletons possibly, but not likely, associated with the earlier Aurignacian level at the site.

Excavation of this site was conducted by D. Gorjanović-Kramberger from 1899 to 1905 and produced some 800 hominid skeletal fragments, as well as extensive faunal and archaeological remains. Gorjanović-Kramberger spent virtually the rest of his life studying and publishing on the morphology of the fossil remains, the nature of the associated cultural remains, and their significance [Gorjanović-Kramberger, 1906, 1913; see also bibliography in Smith, 1976b]. In recent years, a number of newer studies have provided additional information on the reinterpretation of the Krapina hominid remains [eg, Smith, 1976b, 1978, 1982; Trinkaus, 1978; Brace, 1979a; Wolpoff, 1979, 1980a; Alexeev, 1979], site stratigraphy and chronology [Malez, 1970a, 1978a; Gábori, 1976], and faunal analysis [Malez, 1970b, 1978a].

The Krapina deposits are divisible into 13 stratigraphic units, the upper nine containing cultural material [Gorjanović-Kramberger, 1906; Malez, 1970a, 1978a]. The majority of the hominid skeletal remains are from cultural levels 3 and 4, designated the "hominid zone" [Gorjanović-Kramberger, 1906], which are correlated to the end of the Riss-Würm Interglacial by Gorjanović-Kramberger [1906, 1913] and Malez [1970a, 1978a]. Another possibility is that the hominid zone might represent an interstadial during the more recent, relatively mild Early Würm stadial, but on the basis of stratigraphy and geological factors, the hominid zone certainly predates the Lower Würm stadial [see Smith, 1982]. Even Gábori, who is quite critical of Malez's stratigraphic analysis at Krapina, would date the site prior to his equivalent of the Lower Würm [Gábori, 1976, pp 55–57]. A few specimens are known to derive from deposits more recent than the hominid zone at Krapina [Malez, 1978a; Smith, 1976b], some of which may be as recent as the end of the Lower Würm. All hominid specimens, including those stratigraphically younger than the hominid zone, are associated with the Mousterian; and although a few Upper-Paleolithic-like elements are found in the upper strata, all cultural strata are clearly representative of the Middle Paleolithic [Gorjanović-Kramberger, 1913; Malez, 1970c, 1978a; Gábori, 1976]. Two radiocarbon dates have been obtained on bone from Krapina [Vogel and Waterbolk, 1972]. One, taken on Pleistocene rhinoceros, possibly from level 1 or 2, is 3.2 + 0.78 ky BP (GrN-4938) and is obviously unrealistic. The other date, 30.7 + 0.78 ky BP (GrN-4299), is of unknown stratigraphic context and thus does not serve to anchor any part of the Krapina sequence.

The total morphological pattern of the Krapina hominids clearly aligns them with Neandertals, and although there is considerable variation, no feature excludes any Krapina specimen from this hominid group [Jelínek, 1969; Smith, 1976b, 1978; Trinkaus, 1978; Wolpoff, 1979]. Many specimens are from subadult individuals [Wolpoff, 1979; Smith, 1976b], so that some variation in the sample is unequivocally age related. However, even among adult specimens there is considerable variation. This is particularly true of anatomical regions represented by the largest samples—such as teeth, supraorbital tori, mastoid segments of temporals, mandibles, distal humeri, proximal radii and ulnae, and fibulae. Much of the non-age-related variation in cranial elements is probably explained by sexual dimorphism [Smith, 1976b, 1980; Wolpoff, 1980a; Zobeck, 1980; Smith and Ranyard, 1980], although (sexual) dimorphism is probably not as significant for explaining postcranial variation [Trinkaus, 1980].

Metrically, Krapina crania clearly fall in the Neandertal range (often toward the small end) in every measurable feature [Gorjanović-Kramberger, 1906; Smith, 1976b]; and no morphological feature qualitatively differentiates Krapina and western European Neandertal cranial form [Smith, 1976b, 1978; Wolpoff, 1980a; Alexeev, 1979; Jelínek, 1969]. Neurocranial specimens from Krapina exhibit long, broad, and relatively low vaults with receding foreheads and broad occiputs. The presence of lambdoidal flattening and occipital bunning in adults is difficult to identify, because no adult specimen preserves enough of the posterior neurocranium to allow an unequivocal judgment. The only specimen that preserves the posterior neurocranium sufficiently to allow an assessment of lambdoidal flattening and occipital bunning is the juvenile B skull, which clearly exhibits these features [Smith, 1976b]. This makes it quite likely that some adults exhibited them as well [Smith, 1982]. Temporal bones are characterized by robust mandibular fossae and mastoid processes, some of which project beyond the occipitomastoid crest of the cranial base. Faces are relatively large and robust with wide interorbital areas, pillarlike lateral orbital margins and well-developed supraorbital tori (see Fig. 1). Adult Krapina supraorbital tori exhibit considerable variability in thickness and projection, but all form continuous osseous bars over the orbits that are contiguous with the interorbital and lateral orbital regions in the characteristic Neandertal manner [Smith and Ranyard, 1980]. Frontal sinuses are large and restricted to the torus, a typically Neandertal feature [see Vlček, 1967]. Nasal root depressions and canine fossae are not present on any specimen, indicating that Krapina nasal areas project anteriorly at the midsagittal plane. This creates the "beaked face" morphology, noted by Coon [1962] for "classic" western European Neandertals. That the Krapina faces conform to this morphological pattern is clearly demonstrated by visual inspection of the C cranium (Fig. 1) and is supported by Alexeev's [1979] study of facial angles. However, metric assessment of midfacial prognathism [eg, Wolpoff et al, 1981] suggests

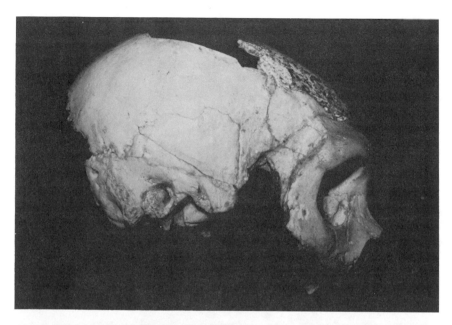

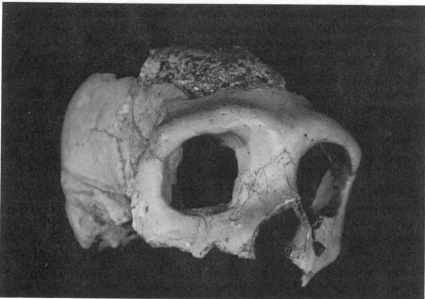

Fig. 1. Lateral and three-quarter views of the Krapina C cranium. Note the lack of nasal root depression and midfacial profile. These features indicate the presence of midfacial prognathism.

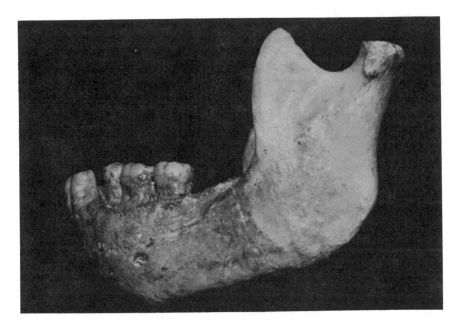

Fig. 2. The Krapina J mandible. Note the symphyseal contour and the retromolar space (obvious even though M₃ is missing).

that midfacial prognathism may not be as great in Krapina faces as in those of western European Neandertals. The two Krapina specimens for which nasal breadth is measurable (Maxilla C and E, museum numbers 47 and 49, respectively) yield values approximately 2 standard deviations below the western European Neandertal mean. While this might indicate that Krapina noses were narrower than western European Neandertal noses, it may also be simply a reflection of the subadult age [Wolpoff, 1979, p 112] of both specimens.

Prognathism of the lower face in the Krapina hominids is suggested by the presence of a retromolar space and angled alveolar plane in all the Krapina mandibles preserving the appropriate areas. Mandibles also lack both mental eminences and trigones, and their symphyses angle forward from the base to the alveolus (Fig. 2). Thus all symphyseal angles (measured from the basal margin) are greater than 90°, and the average is 99.6° (Table V). The lower face is also characterized by broad, but only moderately elongated, anterior maxillary alveoli (indicated by bicanine breadths and alveolar heights, respectively).

The anterior teeth in the Krapina sample are very large, both in crown and root dimensions. In fact, no other Pleistocene sample has anterior teeth as large as Krapina [Wolpoff, 1978, 1979; Smith, 1976b]. That the Krapina teeth are larger than western European Würm Neandertal anterior teeth (Table II) is not surprising since the majority of the Krapina teeth are probably earlier. Anterior dental size appears to reach its peak in Riss-Würm Europe and reduces steadily from this point through the later Pleistocene [Brace, 1979a; Wolpoff, 1980a; Frayer, 1978; Smith, 1976b]. Posterior teeth are also larger than in later European Upper Pleistocene hominids. Morphologically, Krapina molars are characterized by extensive taurodontism [Kallay, 1963] and by the tendency for well-developed anterior fovea on lower molars and by the presence of Carabelli-like features on many upper molars [Gorjanović-Kramberger, 1906]. Incisors, particularly maxillary incisors, exhibit well-developed marginal ridges and basal tuberales.

The postcranial anatomy of the Krapina specimens is comparable to that of other Neandertals (in that it is characterized by general postcranial robusticity), although the predominance of subadult material in the Krapina craniodental, and by implication postcranial, sample gives the impression of less rugosity than is common in western European Neadertals [Trinkaus, 1978]. Like other Neandertals, the Krapina people exhibit a complex of features in the pectoral girdle indicating a powerful, barrel-shaped upper thorax [Hrdlička, 1930; Smith, 1976b,c; Wolpoff, 1980a; Trinkaus, 1983b] and very muscular forearms [Trinkaus, 1983a,b] and hands [Musgrave, 1970, 1977]. One of the most distinctive features of the Neandertal postcranium is the presence of a dorsal groove on the axillary border of the scapula [Stewart, 1962, 1964]. This feature is rare in Upper Paleolithic hominids and virtually unknown in modern humans, where the groove is either ventral or there are smaller grooves on both the dorsal and ventral aspects (the latter is often referred to as the Chancelade pattern). A dorsal groove is present in 57.9% of Neandertals [Trinkaus, 1977]. At Krapina, 33.3% of adult scapulae exhibit this morphology, while 55.6% exhibit the Chancelade pattern [Smith, 1976b]. One Krapina scapula exhibits the only ventral sulcus known for a Neandertal. Trinkaus [1978] finds that the lower postcranial elements at Krapina exhibit the same morphological pattern as other Neandertals and differ from modern humans only by exhibiting an increased average degree of robustness and in the presence of an elongated, thinned superior pubic ramus [see also Trinkaus, 1976a; Smith, 1976b]. Increased robustness is best demonstrated by the hypertrophy of articular surfaces, increased cortical bone thickness and density in the shafts, and more pronounced areas of muscle attachment in Neandertals compared to more recent *H sapiens*.

TABLE II. Buccolingual (Breadth) Dental Dimensions: All Values in mm: If a Source Is Not Listed for a Specimen, the Values Given Represent Unpublished Measurements by the Author

Maxilla	I^1	I^2	C	P^3	P^4	M^1	M^2	M^3
Krapina[a] x̄ (σ)	8.9(5.2)	8.6(5.8)	9.8(.73)	11.4(5.7)	10.8(.37)	12.3(.41)	12.4(1.02)	11.5(1.39)
Range	8.1–9.7	7.7–9.5	8.1–11.2	8.5–11.9	10.3–11.4	11.0–13.2	10.5–14.0	8.7–12.5
N	19	15	16	11	14	11	17	6
Vindija G3[b] x̄ Range, N = 2							12.6 (11.8–13.0)	
Kůlna			9.7	9.8	9.8	11.0		
Mladeč 5457			10.6			13.6	13.8	
Mladeč 5458						12.2	12.0	11.0
Mladeč 1						11.7	11.8	
Vindija G1[b]	8.3		10.6					
Předmostí[c] x (σ)	7.5(.26)	6.9(.37)	8.6(.79)	9.5(.58)	9.7(.60)	12.2(.59)	12.3(.99)	12.2(1.13)
Range	7.1–7.9	6.4–7.3	7.8–9.8	8.7–10.6	8.9–10.8	11.4–13.0	11.1–13.7	10.8–13.1
N	8	6	6	8	7	9	8	4
Early upper x (σ)	7.5(.28)	6.8(.52)	9.0(.91)	9.6(.59)	9.6(.75)	12.3(.73)	12.3(.95)	11.4(1.16)
Paleolithic[d] Range	7.1–8.0	6.0–7.4	7.8–10.8	8.7–10.6	8.5–11.2	11.0–14.0	10.8–13.8	9.2–13.1
N	13	10	12	15	16	24	20	13

Mandible	I1	I2	C	P3	P4	M1	M2	M3
Krapina[a] x(σ)	7.7(.51)	8.3(.49)	9.4(.65)	9.1(.73)	9.7(.57)	11.4(.78)	11.7(.61)	10.8(.68)
Range	7.0–8.2	7.0–9.2	8.0–10.2	7.9–10.3	9.2–11.3	10.0–12.6	11.0–12.7	9.7–12.0
N	7	10	11	12	10	11	11	14
Ochoz[e]	7.7	8.0	9.7	9.6	9.3	11.1	11.6	11.7*
Subalyuk[e]	7.2	7.8	9.9	9.1*	9.0*	10.7	11.3	11.3
Ehringsdorf[e,f] (Range, N = 2)	7.7	7.9	9.0	9.0	9.2	11.0	11.1	9.4
Vindija G3[b] $\bar{x}$(σ)		7.7	7.9	9.0		11.1 (N = 3)	11.7 (N = 2)	11.8 (N = 2)
Range						10.5–11.4	11.2–12.3	11.7–11.8
Šipka[g]	7.0	7.0		8.0	8.0			
Předmostí[c] $\bar{x}$(σ)	6.2(.25)	6.7(.43)	8.8(.69)	8.4(.49)	8.5(.42)	10.9(.55)	10.8(.91)	10.7(1.03)
Range	5.9–6.5	6.0–7.0	8.0–9.8	7.8–9.0	8.0–9.1	10.0–11.8	10.0–12.0	9.9(11.9)
N	8	9	5	7	6	13	11	5
Early Upper x(σ)	6.4(.44)	7.0(.60)	9.0(6.7)	8.5(.54)	8.7(.58)	11.0(.61)	10.8(.80)	10.8(.96)
Paleolithic[d] Range	5.9–7.1	6.0–8.5	7.9–10.0	7.8–9.3	8.0–10.1	10.0–12.0	9.8–12.0	9.3–12.4
N	14	18	14	13	11	28	23	13

[a] After Smith [1976b].
[b] After Wolpoff et al [1981].
[c] Compiled from Matiegka [1934].
[d] After Frayer [1978].
[e] Value represents average of both sides of the individual, except those marked with an asterisk.
[f] Courtesy M. Wolpoff.
[g] After Vlček [1969].

The most important conclusion to draw from the morphology of the Krapina Neandertals is that, while a few features (occipital bunning, midfacial prognathism) may be less markedly expressed or present in lower frequencies in the Krapina hominids than in western European Neandertals, the former do not significantly differ in total morphological pattern from the latter. Certainly there is no basis for considering the Krapina Neandertals as more "progressive" vis-à-vis western European Neandertals. The one possible exception to these statements is the juvenile A cranium, which comes from level 8 and is the only cranial specimen known not to be from the hominid zone. Both Škerlj [1958] and Wolpoff [1980a] have suggested that this specimen exhibits a more modern H sapiens-like morphology than the other Krapina crania. Wolpoff bases his assessment on the fact that the vault is large for the degree of brow ridge development in the specimen, pointing out that vaults as large as the A cranium are found in older Neandertal specimens with more brow ridge development, while Neandertals with a similar level of brow ridge development exhibit smaller vault dimensions [Wolpoff, 1980a, pp 313–314]. Minugh's [1983] suggestion that brow ridge development occurs relatively late ontogenetically in early modern H sapiens but earlier in Neandertals seems also to support this suggestion, but other data do not clearly differentiate the specimen from other Neandertal children [Smith, 1976b, pp 30–36]. While the stratigraphic position of cranium A makes its interpretation as a "transitional" specimen between Neandertals and modern H sapiens attractive, so little is presently known about patterns of cranial growth in Neandertals that it is not possible to say whether the "discrepancy" between vault size/brow ridge development in cranium A is within the range of normal ontogenetic variability in Neandertals or indeed represents a transitional morphological pattern. Krapina A also represents the only example of this discrepancy, although the situation with Staroselje [Alexeyev, 1976] is perhaps similar. Thus the possible transitional status for Krapina A must presently be viewed very cautiously.

The second site in the Hrvatsko Zagorje yielding Neandertal remains is the site of Vindija, a large cave located only 50 km from Krapina. Since systematic excavation began in 1974, over 80 fragments of Pleistocene human skeletal material, an extensive collection of Pleistocene fauna, and archaeological remains have been recovered at Vindija [Malez, 1975, 1978b,c; Malez et al, 1980]. The stratified Pleistocene sediments provide an excellent sequence throughout much of the Upper Pleistocene [Malez and Rukavina, 1975, 1980; Wolpoff et al, 1981]. Mousterian lithic material characterizes the Riss-Würm through Lower Würm levels, with Aurignacian and Gravettian characterizing the Podhradem through the Upper Würm levels. A radiocarbon date of 27.0 + 0.6 ky BP (Z-551) is associated with the Aurignacian [Malez, 1978b]. Level G3 has been *tentatively* dated to 42.4 ± 4.3 ky BP (FRA-A-36) on the basis of amino acid racemization [Malez, personal communication].

The hominid remains from Vindija can be divided stratigraphically into three groups. The earliest and largest group comes from level G_3 and is associated with a late Mousterian assemblage. This stratum correlates to the Lower Würm Stadial based on stratigraphic, faunal, and cultural criteria [Malez et al, 1980; Wolpoff et al, 1981; Malez and Ullrich, 1982]. The middle group comes partly from level G_1, which probably represents the Podhradem Interstadial. The only diagnostic artifact from this level is a split-based bone point, which is from the top of the layer and suggests G_1 is an Aurignacian level. Also included with this group are three isolated teeth and a parietal fragment from the unquestionably Aurignacian levels Fd and Fd/d. The latest group comes from level D and is associated with a Gravettian industry. A list of the specimens from level D is given by Malez and Ullrich [1982].

The Vindija G_3 hominid sample consists of some 35 specimens. These remains are fragmentary, but fortunately various taxonomically relevant anatomical regions are well represented. An extensive descriptive and comparative study of these Vindija G_3 hominids [Wolpoff et al, 1981] has demonstrated that their total morphological pattern warrants inclusion in the taxon *Homo sapiens neanderthalensis*. This conclusion is based, in part, on the following criteria: the adult specimens preserving a supraorbital area (N = 5) exhibit supraorbital tori of Neandertal form [see also Smith and Ranyard, 1980]; frontal sinuses are large and are restricted to the torus proper (ie, they do not extend into the squama); the overall morphology and dimensions of the mandibles (N = 4), including the presence of retromolar spaces (although the length of this gap is less than in the Krapina mandibles); presence of a dorsal axillary groove on the single scapular fragment; the wide anterior alveolus of the maxilla (prosthion to postcanine distance); and various aspects of vault morphology (eg, presence of supra-iniac depressions and Breschet's sinuses, both characteristic Neandertal features).

Although unquestionably Neandertals, the Vindija G_3 hominids exhibit certain features in which they approach the morphology of early south-central European modern *Homo sapiens* more than the chronologically earlier Krapina Neandertals do. First, the supraorbital tori of the Vindija adult sample exhibits a distinct pattern of absolute and relative decrease in

TABLE III. Brow Ridge Dimensions in mm: All Measurements Taken by the Author

		Projection			Thickness		
		Lateral	Midorbit	Medial	Lateral	Orbit	Medial
Krapina	x̄ (σ)	24.3 (1.38)	23.9 (1.16)	20.3 (2.27)	12.5 (1.63)	10.7 (1.81)	17.6 (2.96)
	Range	23.0–27.0	23.0–26.0	17.5–23.0	10.3–16.0	7.0–14.3	15.8–22.0
	N	8	11	4	11	13	4
Western European Neandertals	x̄ (σ)	24.6 (2.1)	22.9 (2.3)	22.6 (1.9)	12.1 (0.5)	10.9 (0.5)	20.8 (1.2)
	Range	22.0–28.0	20.0–27.0	20.0–26.0	11.0–12.9	10.2–11.7	19.1–22.3
	N	7	7	7	8		8
Vindija	x̄ (σ)	22.1 (1.78)	19.0 (3.08)	—	10.6 (0.51)	8.6 (0.56)	—
	Range	19.5–24.5	16.0–23.0	—	10.0–11.3	8.0– 9.5	—
	N	5	5	—	5	5	—
Šala[a]		25.0	20.0	11.0	7.1	15.0	
Neandertal[a]		26.0	23.0	22.5	12.3	11.7	22.3
Ehringsdorf H		24.0	22.0	—	12.5	11.2	—
Mladeč 5[a]		23.0	19.0	17.5	10.1	7.7	23.7
Hahnöfersand		—	—	19.5	—	—	21.7
Zlatý Kůň[a]		22.0	19.0	16.0	8.4	5.8	16.7
Kelsterbach[a]		16.0	12.0	10.0	7.2	4.3	15.8
Brno 2[a]		21.0	16.0	12.0	9.3	6.3	20.3
Early Upper Paleolithic (south-central Europe)	x̄ (σ)	20.3 (2.58)	16.1 (3.37)	13.0 (3.01)	8.1 (1.39)	5.4 (1.72)	16.6 (3.33)
	Range	15.0–23.0	8.0–19.0	8.0–17.5	6.0–10.1	4.4– 7.7	11.5–23.7
	N	9	9	9	11	11	11

[a]Values are averages of right and left sides of the specimen.

both thickness and projection in comparison to the Krapina tori [Smith and Ranyard, 1980; see also Table III]. In addition, the relatively greater diminution in the midorbital region of the Vindija tori reflects an incipient tendency toward the division of the torus into medial and lateral positions (the *arcus superciliaris* and *trigonum supraorbitale*, respectively). In size and morphological form, the Vindija G_3 torus sample conforms to what one would expect as an intermediate between chronologically earlier Neandertals and early modern *Homo sapiens* (see Fig. 4). Secondly, the symphyseal angles of all four Vindija G_3 specimens preserving this area are practically vertical (see Table V); and there is a slight mental eminence on three of the four (Fig. 3). Thirdly, nasal breadths (which are estimated on the assumption of bilateral symmetry in nasal form) are very small for both Vindija maxillae. In fact, both yield dimensions more than three standard deviations *below* the western European Neandertal mean (see Table IV). Both also exhibit alveolar heights more than 2.5 standard deviations *below* the western European Neandertal mean. However, while certain dimensions of the lower face (like nasal breadth) exhibit significant degrees of reduction at Vindija compared to most other Neandertals, others (eg, prosthion to postcanine distance) exceed the Neandertal average. Fourthly, there is some indication of

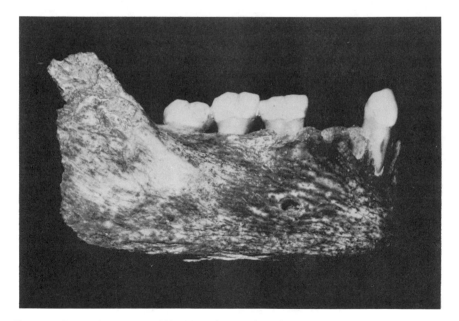

Fig. 3. The Vindija 206 mandible. Note the symphyseal contour and retromolar space.

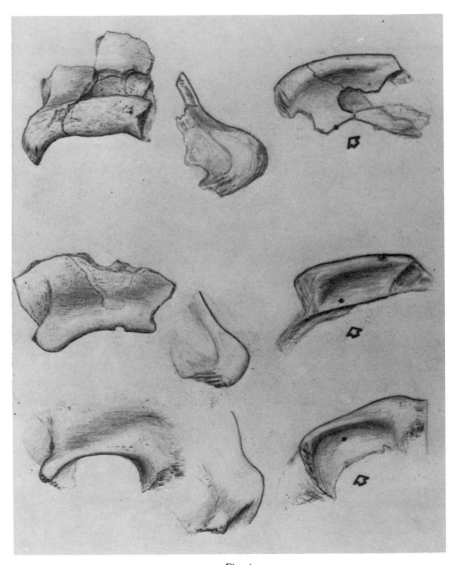

Fig. 4.

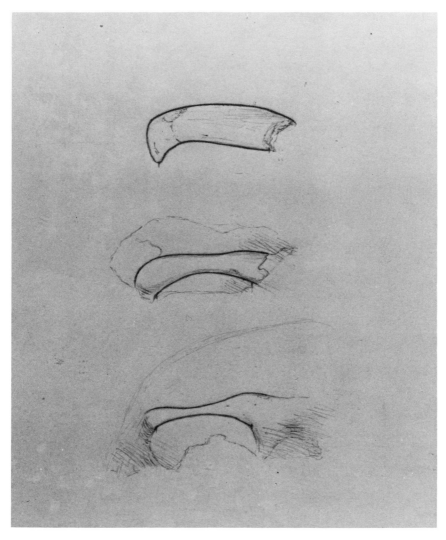

Fig. 4. Left: Anterior, lateral, and inferior views of Krapina 28 (top), Vindija 202 (center), and Velika Pećina (bottom) illustrating the changes in brow ridge form from an early central European Neandertal (Krapina), through a late central European Neandertal (Vindija), to early modern *H sapiens* (Velika Pećina). The dots (indicated by arrows) on the inferior views indicate the position of the internal wall of the frontal squama, thus giving an indication of brow ridge projection in the specimens. This page: The same three specimens oriented to emphasize the differences in thickness, particularly in the midorbital region. Note the intermediate morphology of the Vindija specimen in both drawings (drawing by Dr. M.O. Smith).

TABLE IV. Maxillary Measurements (in mm)

		Alveolar height	Nasal breadth	Prosthion to postcanine
Krapina E (#49)		22.5	30.5	25.2
F (#50)		24.2	—	—
Vindija 225		17.5	28.5*	27.3
259		16.3	26.2*	24.3
Kůlna		29.9	30*	25.1
Mladeč 1		18.8	25.1	22
5457		(20)	30.2	24
5458		—	28	—
Předmostí 3[a]		17.9	29	24
4[a]		17.6	27	22.2
Western European	$\bar{x}$ (σ)	26.1 (3.0)	33.3 (1.5)	24.9 (1.1)
Neandertal	Range	22.2–30.4	30.0–35.1	23.0–26.2
Early Upper	x (σ)	20.9 (2.3)	27.4 (1.7)	22.4 (1.9)
Paleolithic	Range	17.0–23.8	24.3–30.2	19.3–25.0

[a]Data taken from cast.
*Value determined by doubling the measurement for half the nasal aperture.

more vertical frontal squama in the Vindija sample than in most other Neandertals (Fig. 5). Finally, there is no indication of lambdoidal flattening or occipital bunning in the Vindija sample. It must be noted, however, that the pertinent morphological regions (occipitals and posterior parietals) are represented by only fragmentary specimens, making definite determination of the presence or absence of bunning and/or lambdoidal flattening rather tenuous.

While average body size in the Vindija G_3 sample was probably relatively small, it appears unlikely that these morphological reductions in the Vindija G_3 Neandertals are related to reductions in body size alone. For example, not all dimensions reduce. The prosthion-to-postcanine distance and posterior tooth breadths are essentially the same in the Vindija and Krapina Neandertal samples. Furthermore, the adult postcrania of earlier south-central European Neandertals (Krapina and Subalyuk) also indicate small individuals, without the transitional features noted for the Vindija cranial material.

From the stratigraphically more recent Vindija level G_1 comes a small sample of fragmentary hominid remains, which do not differ significantly from the G_3 Neandertals. An upper canine and central incisor from level G_1 fall metrically closer to the Krapina sample than to the early Upper Paleolithic sample in tooth breadths (Table II) but morphologically and

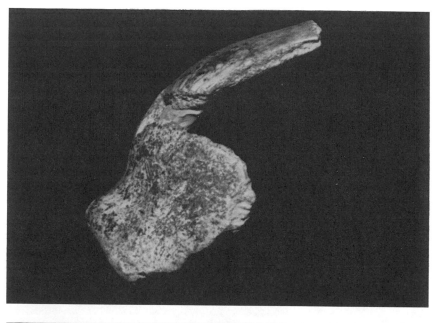

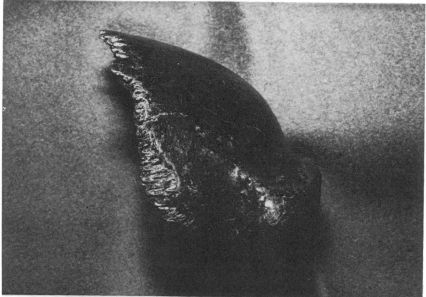

Fig. 5. Lateral views of late central European frontals: Vindija 261 (above) and Šala (below).

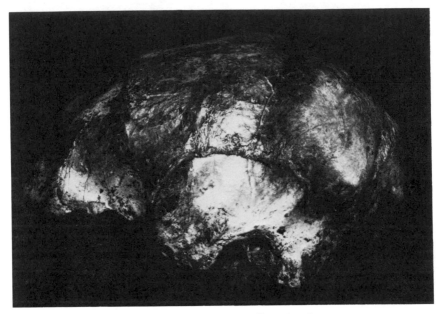

Fig. 6.   Lateral view of the Mladeč 5 male calvarium.

metrically could be accommodated in either an early Upper Paleolithic or Neandertal sample [Wolpoff et al, 1981]. A right mandibular ramus exhibits a retromolar space, but a low ramus height compared to other Neandertals [Wolpoff et al, 1981], and a left parietal portion is largely nondiagnostic (except for the presence of a moderately developed Breschet's sinus). There is insufficient information present to unequivocally classify the G_1 sample as either Neandertals or modern *H sapiens*.

The archaeological context of level G_1 is also somewhat unclear. The only diagnostic artifact is a split-based bone point, from the top of the stratum. This would suggest an Aurignacian classification, which would be reasonable given the fact that level G_1 is correlated on geological/stratigraphic grounds to the Podhradem [Malez et al, 1980; Wolpoff et al, 1981] and the earliest Aurignacian in south-central Europe is also encountered during this period [Smith, 1982]. Recently, Stringer [1982a,b] has questioned the association of the point with the G_1 hominids, noting that the action of cryoturbation at the site [Malez and Rukavina, 1975] could have resulted in the artificial mixture of the hominids and point. While this is certainly possible, cryoturbation is not noted in the portion of the cave where the hominid and cultural remains in question were recovered. Fur-

TABLE V. Data on the Mandibular Symphysis: All Measurements, Except Symphyseal Angles, in mm, and Made by the Author

		Symphyseal angle[a]	Symphyseal height	Symphyseal thickness (base)	Symphyseal thickness (alveolus)	Symphyseal index[b]	Alveolar index[c]
Neandertal[d]	x (σ)	98.5 (4.8)	35.3 (4.1)	14.9 (1.2)	11.1 (1.5)	42.7 (6.0)	73.3 (8.0)
	Range	87–106	30.7–42.4	13.6–17.4	8.6–14	33.1–51.6	58.9–81.5
	N	11	11	11	10	11	10
Krapina Neandertal	x (σ)	99.6 (1.6)	36.9 (4.8)	14.5 (1.1)	11.7 (0.7)	39.9 (5.4)	79.6 (2.4)
	Range	98–102	30.7–42.4	13.5–16.3	10.8–12.3	33.1–49.2	75.5–81.5
	N	5	5	5	4	5	4
Vindija Neandertal	x (σ)	87 (1.6)	30.8 (3.3)	15.2 (1.0)	10.3	49.5 (3.1)	64.9
	Range	85–89	26.2–33.5	13.9–16.4	10–10.5	45.4–53	64–65.
	N	3	3	3	2	3	2
Early Upper Paleolithic	x (σ)	76.8°(6.5)	32.3 (4.1)	15.1 (1.4)	9.8 (1.4)	49.4 (5.8)	65.5 (9.1)
	Range	66°–85°	26–40	14–18.2	8–12.5	38.8–57.3	54.9–83.3
	N	6	8	8	8	8	8

[a]Method after Olivier [1969].
[b]Symphyseal thickness (mental eminence)/symphyseal height × 100.
[c]Symphyseal thickness (alveolus)/symphyseal thickness (mental eminence) × 100.
[d]Western European Neandertals and Krapina Neandertals combined.

thermore, the morphology exhibited by the G_1 sample, such as it is, is not all that different from other Aurignacian-associated hominids in south-central Europe. The problem is really the paucity and fragmentary nature of the hominid remains from G_1, which renders any unequivocal interpretation of their significance impossible.

South-Central Europe: The Moravian Karst

Neandertal remains have been excavated from three sites from the Moravian karst region of Czechoslovakia (Table I). The first of these is Šipka, excavated by K. Maška from 1879 to 1883 [Maška, 1882; Vlček, 1969], which yielded a symphyseal fragment of a subadult mandible in 1880. This specimen comes from level 9 at the site, where it is associated with a faunal unit (Maška's horizon III) correlated to the Podhradem or Middle Würm interstadial by Musil [1965] and a Mousterian industry [Valoch, 1965, 1968]. However, both Kukla [1954] and Gábori [1976] suggest a correlation with the equivalent of either the Early or Lower Würm Stadial. If Maška's faunal unit IV is correlated to Würm 1 (Early Würm), as suggested by Kukla, then unit III would probably be no older than the Lower Würm stadial. Thus whichever age one accepts for the level containing the Šipka mandible, it is reasonably clear that it is a relatively late Mousterian specimen.

The morphology of the specimen is rather difficult to assess, because it consists only of the symphyseal portion of the mandible and also because the exernal alveolar surface is largely broken and distorted. Furthermore, the morphology exhibited by even an 8-to-10-year-old specimen may be altered before adulthood is reached, so certain "progressive" aspects of Šipka's morphology may be related to its young developmental age. In any case, the specimens could probably fit morphologically and metrically into either a late Neandertal or early modern *H sapiens* sample.

Thickness and height of the Šipka symphysis are within one standard deviation below both the Vindija and Krapina (adult) means (Table V). On the ventral (lingual) aspect, the alveolar plane is not strongly angled; and the superior transverse torus is weakly developed. The external aspect of the symphysis appears somewhat receding, but there does seem to be some development of a subalveolar concavity below the incisor roots. Both Jelínek [1965, 1969] and Vlček [1958, 1969] note that this contributes to the development of at least an incipient mental eminence, although they both see the alveolus projecting more anteriorly than the base of the mandible. Jelínek [1965] also notes evidence of an incipient mental trigone. The individual breadths of the incisors present are reduced compared to those of Krapina, Subalyuk, and Ochoz (Table II). The Šipka values lie approximately one standard deviation below the corresponding Krapina means. Several teeth are preserved in their crypts on the right side (C, P_3, P_4), and these also appear small. All the teeth, however, are within the size range of both Neandertals and early modern *Homo sapiens* [Jelínek, 1965; Smith, 1976b], although the incisors are above the Předmostí range [Matiegka, 1934] and slightly above the Early Upper Paleolithic means reported by Frayer [1978].

The second Neandertal specimen discovered in the Moravian karst is a mandible excavated from the Šedův stůl cave, near the village of Ochoz, in 1905 [Rzehak, 1905; Kříž, 1909; Bayer, 1925]. The mandible, along with two human cranial vault fragments and a human molar discovered in 1964 [Vaňura, 1965], is thought to correspond to level 2 of the "brown earth complex" [Klíma et al, 1962; Vlček, 1969], which contains the earlier Mousterian component at the site and is correlated to the Riss-Würm Interglacial. However, there is some possibility that the mandible is derived from the upper part of the complex and is thus more recent than Riss-Würm [Klíma et al, 1962].

The Ochoz mandible preserves the alveolar region and entire dentition, with the exception of the right M_3. The rami and corpus base are missing, probably broken from the remaining portions by geological processes after deposition. Prognathism is indicated by the angled alveolar plane and the certainty that a retromolar space was present. These and other features of the preserved portion of the mandible compare favorably with other Neandertals [Jelínek in Klíma et al, 1962; Jelínek, 1969; Vlček, 1969]. The dentition also conforms to the Neandertal pattern. The teeth are large, with all but one crown breadth (M_3) falling within a standard devation of the corresponding Krapina means (Table II). Only the values for C, P_3, and M_3 exceed the Krapina means. The anterior teeth, as in other Neandertals, are relatively expanded and quite heavily worn [Smith, 1976c]; and the molars are taurodont [Jelínek, 1969]. The 1964 tooth, a right M_3, morphologically and metrically indicates Neandertal affinities [Vlček, 1969]. The two cranial fragments, portions of a parietal and temporal squama, also found in 1964 are largely nondiagnostic [Vlček, 1969].

On the external face of the mandibular symphysis a gentle concavity begins just below the alveolar margin of the incisors and extends inferiorly to where the specimen is broken—a distance of some 25 mm. Rzehak [1905], Jelínek [1969; in Klíma et al, 1962], and Vlček [1969] all consider this to be incipient chin development, comparing it with Šipka and Tabūn II. However, since the base of the symphysis itself is missing, it is impossible to say whether an incipient mental eminence was present or not. Depressions

below the incisors are associated with incipient mental eminence development in the Vindija and Šipka specimens, but their presence does not
guarantee the presence of an incipient mental eminence. Both the Krapina
J and H mandibles, for example, have depressions comparable in development to Ochoz; and neither shows the slightest trace of a mental eminence.
In fact, the Ochoz remains would fit imperceptibly into the Krapina sample.

Kůlna Cave is the third site in the Moravian karst region to yield
Neandertal remains, but is perhaps even more significant because it preserves an excellent stratigraphic sequence throughout much of the Upper
Pleistocene as well as an extensive array of Middle Paleolithic cultural
components [Valoch, 1967, 1970, 1977–78]. In 1965, a right maxillary fragment of a Neandertal was excavated from the late Mousterian level 7a at
Kůlna [Jelínek, 1966, 1967]; and in 1970, a fragment of right parietal was
also found in the same layer [Jelínek, 1980, 1981]. Level 7a is correlated to
the Lower Würm stadial [Valoch, 1967; Jelínek, 1966, 1980] and is dated by
radiocarbon (GrN-6060) to between 45.66 + 2.85–2.2 ky BP and 38.6 +
0.92–0.8 ky BP (GrN-6024) [Valoch, 1977–78, 1981].

The maxillary fragment, which clearly is a Neandertal [Jelínek, 1966,
1967, 1969, 1981], is preserved from the midline to just behind M1 with the
canine, both premolars, and M1 present. Jelínek ages the specimen at 14,
which is a minimum estimate. The Kůlna maxilla has a narrow nose (based
on reconstruction of nasal breadth assuming bilateral symmetry) compared
to western European Neandertals, but its alveolar height and prosthion to
postcanine dimensions are at or above the western European Neandertal
mean (Table IV). Compared to the roughly contemporaneous Vindija
maxillae, it is much larger in all dimensions except prosthion-to-postcanine.
The posterior teeth are much smaller in breadth (each more than 2 standard
deviations) than the appropriate Krapina means, but the canine is slightly
larger than the Krapina mean (Table II). Also, there is a slight indication of
a shallow canine fossa on the maxilla. The relatively thick parietal exhibits
curvature similar to western European "classic" Neandertals and is thought
to be from a young individual, because the sagittal suture is completely
open [Jelínek, 1981]. Because of the subadult age of both specimens and
their location in the same stratum (though 20 m apart), Jelínek [1980, 1981]
suggests the parietal and maxilla are from the same individual.

South-Central Europe: Slovakia

In 1961, a human frontal bone was found in gravel deposits of a sandbar
in the Váh (Waag) River, near Šala in western Slovakia (ČSSR). Continued
searching in the sandbar yielded remains of Pleistocene fauna; and fluorine

tests support the contemporaneity of the fauna and hominid specimen, suggesting an Upper Pleistocene age for the Šala frontal [Vlček, 1968, 1969]. Presence of rhinoceros remains among the fauna excludes the cold phases of the Early and Lower Würm, but Šala could possibly correlate to an interstadial during these periods or to the Riss-Würm Interglacial [Vlček, 1969]. No cultural remains were recovered from the deposits.

The Šala frontal is complete, well preserved, and adult [Vlček, 1964a, 1967a, 1968, 1969; Jelínek, 1969]. The frontal squama is rounded in sagittal curvature and appears to be relatively higher than some western European and Krapina Neandertals. Vlček [1969, pp 161–163] gives an estimated value of 67° for the angle of frontal inclination for Šala. This is somewhat intermediate between the means of a Neandertal sample ($X = 57°$, $\sigma = 2.4°$, $N = 8$) and an early modern H sapiens sample ($X = 70°$, $\sigma = 6.5°$, $N = 9$). Šala's interorbital area is broad and projected anteriorly, indicating absence of a nasal root depression and presence of midfacial prognathism.

Šala's most diagnostic and salient feature is the presence of a distinct supraorbital torus, which projects markedly. Projection dimensions are close to the corresponding Krapina means and larger than the Vindija means (Table III).

An injury to the supraorbital torus has resulted in atrophy and thinning of the right lateral torus; but even excluding this, the thicknesses of Šala's torus are reduced and approach the Vindija means more closely than the Krapina means (Table III). There is relatively greater midorbit thinning of the torus (on the nonpathological, left side) than medially or laterally, and frontal sinuses are expansive and restricted to the torus.

Though the Šala frontal possesses some features that can be considered "transitional" or "progressive," its total morphological pattern is unquestionably Neandertal. Vlček [1969] believes it to be related to the "transitional Neandertals" of the Near East (especially Zuttiyeh and Skhul V), because of morphology and reduction of the supraorbital torus. The specimen is considered female [Vlček, 1968, 1969; Jelínek, 1969], which is reasonable because of its gracility and the gradual emergence of the torus from the squama in the glabellar area [see Smith, 1980].

In 1926, J. Petrbok collected an endocast from the Hrádok travertine deposits near Gánovce in northern Slovakia. The specimen was recognized as a hominid 11 years later, and subsequent excavations determined that the specimen was derived from the younger travertine at the site, which is correlated to the Riss-Würm interglacial [Vlček et al, 1958]. A Middle Paleolithic Taubachian industry is associated with the hominid specimen [Valoch, 1982].

From the few segments of the vault (portions of the left parietal, temporal, and occipital, which are inseparable from the endocast itself) and the general shape of the Ganovce endocast, an impression of the general shape of the cranium is possible [Vlček, 1953, 1955, 1969]. It is a long, quite low, and moderately broad specimen with its maximum breadth at the level of the mid-parietal area—a distinctive Neandertal feature [Wolpoff, 1980a]. In *Norma verticalis*, the occiput appears rather projecting; and there is definite indication of lambdoidal flattening. Very probably the cranium possessed an occipital bun. Furthermore, details of the morphology and measurements of the endocast itself are all within the Neandertal range [Vlček, 1953, 1955, 1969]. The low and flat frontal lobes, considered particularly primitive by Vlček, may be due to subsequent corrosion [Jelínek, 1969]. Cranial capacity is estimated at 1320 cc, which is only slightly below the Riss-Würm Neandertal average and above the average for female Neandertals [Smith, 1976b].

Excavations at the Hrádok travertines in 1955 also yielded natural molds of a partial left fibula and partial left radius [Vlček, 1955]. However, the bones were apparently not well preserved before they were molded, so that nothing significant can be determined about them [Vlček, 1969].

The site of Dzeravá Skála (Pálffy) in the Carpathian Mountains of eastern Slovakia was excavated by Hillebrand [1914] in the early part of this century. From a level contained Aurignacian and Szeletian components mixed by cryoturbation [Valoch, 1982], a germ of lower left M2 was recovered. Hillebrand [1914, p 15] describes this tooth as possessing a well-developed fovea anterior and being generally similar to the Krapina mandibular molars. No dimensions are given, nor is the specimen illustrated.

It is possible that the Dzeravá Skála tooth, like the Vindija G_1 hominids, is an example of the existence of Neandertals in association with early Upper Paleolithic in south-central Europe. However, as with Vindija G_1 material, the Dzeravá Skála molar cannot be conclusively classified as a Neandertal. In this case, one tooth just does not provide enough taxonomically relevant morphology to allow unequivocal classification. The morphology described by Hillebrand can also be found in early modern *H sapiens* samples.

South-Central Europe: Hungary and Romania

Subalyuk Cave is located on the southern slope of the Bükk Mountains overlooking the Hór Valley in Northern Hungary. In 1932, excavation of the Pleistocene deposits in the cave yielded the fragmentary remains of an adult and a child in association with Mousterian tools [Bartucz et al, 1940]. These specimens are said to come from the lower portion of level 14, which probably correlates to a period just prior to the Würm maximum (Lower Würm stadial) [Gábori, 1976, p 83].

The Subalyuk adult is represented by a mandible (preserving the left corpus, with P_4 through M_3, anterior portion of left ramus, symphysis with left C through right P_3, and the right molars) and a few postcranial elements, including a sacrum and manubrium. The manubrium is ventrally concave, indicating a typically Neandertal "barrel-shaped" upper thorax [Coon, 1962]. Both manubrium and sacrum indicate a very small individual. The mandible, which does not necessarily represent the same individual as the postcranial specimens, exhibits a retromolar space, receding symphyseal contour, no mental eminence, and taurodont molars [Bartucz et al, 1940]. The symphysis and left corpus are high but relatively thin compared to most Neandertals, and the alveolar plane is shorter and more vertically oriented than in the Krapina and Ochoz mandibles. Tooth breadths (buccolingual diameter) for the Subalyuk adult [Szabó in Bartucz et al, 1940] fall within one standard deviation of the Krapina means, with only the Subalyuk canine and third molars being larger (Table II).

The Subalyuk child is represented by a neurocranium, most of both maxillae, most of the deciduous dentition, and a nasal bone. Odontometric and morphological features show the Subalyuk child to be a Neandertal approximately 3 years of age [Thoma, 1963]. The maxillae exhibit no trace of canine fossae; the nose is broad (20 mm) for a young child; and the frontal process of the maxilla is broad [Thoma, 1963]. Furthermore, the curvature of the nasal bone [Bartucz et al, 1940: Plate VI; Jelínek, 1969] is vertical or slightly concave for some distance inferior to nasion and then becomes sharply convex. This suggests a projection of the nose along the midsagittal plane, a feature typical of Neandertal crania. A projecting supraorbital torus is not developed; but, like other Neandertal subadults [Smith and Ranyard, 1980], a distinct outline of a torus (in the form of a slight bulge) is observable [Vlček, 1970]. The neurocranium is long, relatively low, and broad compared to modern children of similar age [Thoma, 1963]. Biasterionic breadth is especially large in the Subalyuk child, and there is a distinct indication of lambdoidal flattening and an incipient occipital bun. As in most other Neandertal children [Vlček, 1970; Smith, 1976b], a metopic suture is present.

Bordul Mare Cave is located in central Romania, near Ohaba-Ponor. In 1923, a hominid second right proximal pedal phalynx was excavated in association with a Mousterian industry, containing Upper Paleolithic-like elements, and has been attributed to *H sapiens neanderthalensis* [Gaál, 1931; Necrasov and Cristescu, 1965]. The fauna is a cold fauna, indicating a correlation to the early part of the Würm [Nicolăescu-Plopşor, 1968], probably to the Lower Würm [Gábori, 1976].

Neandertal second proximal pedal phalanges differ from those of more modern *H sapiens* only in exhibiting greater robusticity [Trinkaus, 1975]. As Ohaba-Ponor has only been partially described [Gaál, 1931; Necrasov and Cristescu, 1965] and never systematically analyzed in comparison to other Neandertals [Trinkaus, 1975], little can be said about it in the context of the present inquiry.

North-Central Europe: Germany

North of the Pannonian Basin region in central Europe, five sites have yielded fossil remains described as Neandertals (Table I).[2] Three of these sites (Ehringsdorf, Taubach, and Neandertal) are among the earliest Neandertal localities discovered, with the original Neandertal (Feldhofer Kirche) skeleton being the first recognized Neandertal specimen.

The original Neandertal was accidentally discovered in 1856 in a cave known as Feldhofer Kirche or Grotto and initially reported and described by Fuhlrott and Schaaffhausen [1857; Fuhlrott, 1859; Schaaffhausen, 1858]. No archaeological or faunal material was directly associated with the specimen; and investigation of possible stratigraphic differentiation of the deposits in the cave was rendered impossible by the removal of the deposits by workers, during which the skeleton was discovered [see, for example, the account in Hrdlička, 1930, pp 149–150]. The "early Würm" date reported for the specimen by Gieseler [1971] is clearly an inference based on its morphological similarity to western European Würm Neandertals. Basically, the specimen is otherwise undated. The early controversy surrounding the interpretation and significance of the specimen is discussed by Brace [1964; Brace and Montagu, 1977].

The specimen consists of a well-preserved calotte (lacking both temporals and the nuchal plane) and several postcranial elements [listed in Schaefer, 1957, p 272; Gieseler, 1971, p 198] of an adult and has been described and analyzed by numerous individuals including R. Virchow [1872], Schwalbe [1901], Hrdlička [1930], Schaefer [1957], and Stewart [1962]. The cranial vault is long and relatively low, with a receding frontal squama (angle of inclination of the frontal = 53°) and marked lambdoidal flattening. The

---

[2]A partial right humerus from Stetten (Stetten 3) has also been suggested as a possible Neandertal [Gieseler, 1937, 1971]. Although it is quite robust, I find no basis for considering it Neandertal. Furthermore, Czarnetzki [1980] reports that its stratigraphic context (at the base of the Aurignacian at Stetten) is certain.

widest point on the cranium, viewed from the rear, is low on the parietals, the typical Neandertal condition [Wolpoff, 1980a]. The cranial capacity of the specimen is 1,233 cc [Schwalbe, 1901]. The frontal is broad with a very salient supraorbital torus, which is one of the thicker, more projecting of Neandertal tori (Table III). The total morphological pattern of the cranium aligns it clearly with the western European Würm and Krapina Neandertals and exhibits none of the "progressive" features noted for certain Neandertals from south-central Europe (eg, Vindija, Šala).

The postcranial remains indicate a very muscular, powerfully built individual. Articular surfaces are relatively large and areas of muscular attachment are generally quite robust. The right scapula exhibits a dorsal axillary groove [Stewart, 1962], and the right clavicle and rib fragments indicate a robust thorax. Parts of both upper limbs are present, with the left exhibiting evidence of an injury, probably a fracture of the proximal ulna early in life [Schaefer, 1957; Hrdlička, 1930]. This injury resulted in extensive remodeling of the humerus and ulna, some atrophy of the humerus, and considerable joint deterioration at the elbow. The right innominate exhibits a large iliac blade and a sciatic notch that is characteristically male in form, making the Feldhofer Kirche skeleton one of the few Neandertals whose sex can be postcranially established [Trinkaus, 1980; Smith, 1980]. The femora are especially well preserved and exhibit several features considered typical for Neandertal femora (antero-posterior bowing, well-developed trochanters and attachment of gluteus maximus, large head, and distal articular surfaces, high index of robusticity, lack of distinct pillaster, etc).

Probable contemporary remains described as Neandertal are known from two other sites in Germany. A complete occipital and partial right parietal of a hominid were excavated from the open site of Lebenstedt in the town of Salzgitter, central Germany. The specimen was found in association with a Mousterian industry [Tode, 1953] and correlated to the early Würm [Preul, 1953; Kleinschmidt, 1965]. A radiocarbon date of 55.6 ± 0.9 ky BP [Vogel and Zagwijn, 1967] is associated with the hominid remains. Two right parietal fragments (one adult, one subadult) from Wildscheuer Cave in Hessen were described by Knussmann [1965, 1967] as being possible Neandertals. Both specimens are from Mousterian levels in the cave, which are considered early Würm [Madera, 1954; Kutsch, 1954; Knussmann, 1967].

The Salzgitter-Lebenstedt hominid has yet to be described in detail, but a systematic study is in progress [Hublin, 1983b]. Typically Neandertal features exhibited by the specimen include a suprainiac fossa, lambdoidal flattening, and occipital bunning. The Wildscheuer specimens are small, nondiagnostic fragments and reveal no information pertinent for this study.[3]

Chronologically earlier hominid discoveries were made in several travertine quarries in the region around Weimar in eastern-central Germany [Soergel, 1927; Behm-Blancke, 1958, 1960]. In 1887 and 1892, hominid mandibular molars were discovered from supposedly contemporaneous levels at Taubach [Götze, 1892; Nehring, 1895], correlated to the last (Riss-Würm) interglacial [Soergel, 1927]. The permanent molar (a mandibular M1) is rather small, lying approximately 1.5 below the Krapina M_1 mean (Table II), but is morphologically very similar to Krapina and Ochoz lower molars, particularly in details of crown morphology and development of the fovea anterior [Nehring, 1895; Hrdlička, 1930; Vlček, 1969]. The deciduous molar (a mandibular dml) is also considerably smaller than the appropriate Krapina mean.

More extensive hominid finds were made at Kämpfe's and Fischer's quarries at Ehringsdorf between 1908 and 1925 [Behm-Blancke, 1960; Gieseler, 1971]. All of the hominid specimens were recovered from the lower travertine deposits, which have traditionally been considered Riss-Würm (Eemian) in age on the basis of geological and paleontological evidence [Soergel, 1927; Behm-Blancke, 1958, 1960; Guenther, 1964; Steiner, 1979]. This is basically supported by a number of uranium series dates [Rosholt and Antal, 1963; Cherdyntsev et al, 1975], although other determinations suggest these deposits may be somewhat older (despite the absence of classic pre-Upper Pleistocene faunal elements [Cook et al, 1982]). The complexity of the chronological situation at Ehringsdorf has recently been discussed by Feustel [1983], who notes that the lower travertine lithic industry is distinctly Mousterian in nature [see also Behm-Blancke, 1960], and thus not likely to be older than Upper Pleistocene [see Dennell, 1983].

The Ehringsdorf hominids have been analyzed several times over the years [Schwalbe, 1914; Virchow, 1920; Weidenreich, 1928; Hrdlička, 1930; Behm-Blancke, 1958, 1960; Feustel, 1976; Wolpoff, 1980a; Vlček, in press] and exhibit an interesting total morphological pattern. Adult (Ehringsdorf F) and subadult mandibles (Ehringsdorf G) and a rather complete, but difficult to reconstruct, neurocranium (Ehringsdorf H) are the best known of the remains. There are also three other parietals or parts thereof, an adult femoral diaphysis (Ehringsdorf E), and a partial subadult skeleton (Ehringsdorf G) partially still in matrix.

---

[3]From a study of casts of Wildscheuer A and B, I am of the opinion that they are not hominid fragments.

The cranium, discovered in 1925, lacks the base and has been reconstructed three times. The initial reconstruction by Weidenreich [1928] emphasized the "modern" aspect of the specimen by orienting the frontal so as to decrease projection of the supraorbital torus and increase the height of the frontal squama. The subsequent reconstruction by Kleinschmidt [Behm-Blancke, 1958: plates 42 and 43], gives the contour of the specimen a more typical Neandertal appearance. Occipital bunning is more clearly visible in the Kleinschmidt reconstruction; the frontal squama is less steep; and the maximum breadth is lower on the parietals. Vlček's [1978; in press] recent reconstruction is somewhat intermediate but tends to be more similar to Weidenreich's. Parietals B and D have higher parietal tubercles and exhibit a more acute angle between the side and top of the vault. This feature is considered by some as a "progressive" feature [Feustel, 1976] but is variably present also in earlier archaic *H sapiens* and *H erectus* [Wolpoff, 1980a; Bräuer, this volume]. Other aspects of cranial morphology reveal no "progressive" tendencies. For example, the H2 supraorbital torus is morphologically very similar to the Krapina tori and metrically is more easily accommodated in the earlier Krapina than in the "progressive" Vindija sample (Table III). Furthermore, a suprainiac fossa is present, and the degree of projection of the mastoid process fits in the Krapina range. The mastoid appears more projecting (and thus more "modern" in form), because the occipito-mastoid crest is missing.

The adult femur (Ehringsdorf E) exhibits considerable robustness, is bowed antero-posteriorly, and lacks a pilaster, thus aligning it with Neandertals.

Both mandibles have large anterior teeth. Posterior teeth are relatively somewhat smaller, but all teeth (except M3) are within a standard deviation of the Krapina means (Table II). The adult (Ehringsdorf F) mandible exhibits marked prognathism of the alveolus, resulting in a receding symphysis and retromolar space. The subadult (Ehringsdorf G) also has a receding symphysis, but due to its subadult age, does not exhibit as much prognathism as the adult. Both specimens exhibit concavities below the incisors, and several individuals have noted incipient mental eminences for one or both of them [Schwalbe, 1914; Hrdlička, 1930; Feustel, 1976; Wolpoff, 1980a]. In my opinion, incipient mental eminences like those ascribed for Vindija are not present [compare published photographs in Feustel, 1976, p 116; Wolpoff, 1980a, p 285].

Despite claims that the Ehringsdorf hominids may be modern *H sapiens*-like in some cranial vault features, I regard their total morphological pattern as aligning them with Neandertals. The femoral, mandibular, dental, and overall vault morphology at Ehringsdorf clearly fit the Neandertal pattern. The so-called progressive features seen in some parietals are due to increased parietal bossing and, in view of the variation in this feature at sites such as Spy, the range of expression of this feature at Ehringsdorf should not be considered phylogenetically significant.

EARLY MODERN *HOMO SAPIENS* IN CENTRAL EUROPE

Early modern *H sapiens* specimens (eg, those dated earlier than 25 ky BP) from this region are either associated with Upper Paleolithic (Aurignacian or eastern Gravettian) cultural components or have no cultural associations (Table I). None are associated a Middle Paleolithic component, and none are conclusively associated with the Szeletian or its variants in Central Europe.[4] The latter is particularly unfortunate because there is virtually complete agreement that, technologically, the Szeletian develops directly from central European Middle Paleolithic components [Valoch, 1968, 1972, 1976; Chmielewski, 1972; Gábori, 1976; Svoboda, 1980]. The Szeletian appears to date from as early as 43 ky BP to as late as 33 ky BP [Vogel and Waterbolk, 1972; Mook in Valoch, 1976] and overlaps the early Aurignacian during the last part of the Podhradem (Middle Würm) interstadial. The majority of radiocarbon dates would indicate a temporal range of 36 ky BP to 27 ky BP for the Aurignacian in south-central Europe [see Smith, 1982, pp 669–671] and perhaps slightly more recent in Germany [Hahn, 1977, p 168]. Earlier radiocarbon dates for the Aurignacian are known for Istállöskö in Hungary (44.3 ± 1.9 ky GrN-4659 and 39.7 ± 0.9 ky:GrN-4658) [Vogel and Waterbolk, 1972] but are discounted by Gábori-Csánk [1970] and Hahn [1977]. Hahn [1977, p 171] points out that the bone on which the dates were made may not be associated with the human occupation levels and that earlier radiocarbon tests dated the occupational levels to between 30 and 32 ky BP. Other early dates are known from two Bulgarian sites, Samuilica Cave (42,780 ± 1,270:GrN-5181) and Bačo Kiro level 11 (> 43,000) [Vogel and Waterbolk, 1972; Kozłowski, 1979]. The Aurignacian

---

[4]Possible hominid associations with Szletian are found at Dzeravá Skála (previously discussed), and Silická Brezová. Gábori [1976] also reports teeth associated with the early Szeletian (Jankovich), but I have located no analysis or description of the teeth themselves. The single tooth from Silická Brezová (a lower M1 or M2) is somewhat small (breadth measures 10 mm, one standard deviation below the early Upper Paleolithic mean—see Table II), and exhibits no taurodontism nor an anterior fovea [Vlček, 1957b]. It is not directly associated with any cultural material so it cannot be considered Szeletian.

status of neither site has been conclusively demonstrated, but the case seems quite strong for Bačo Kiro 11 [Delporte and Djindjian, 1979]. Furthermore, the origin of the central European Aurignacian is still debated, with some archaeologists suggesting that it represents an intrusive element from the Near East [Gábori, 1976; Kozłowski, 1979] while others [Valoch, 1972, 1976; Hahn, 1973, 1977; Svoboda, 1980] see its origin as an indigenous development from the central European Middle Paleolithic.

South-Central Europe: Aurignacian-Associated Modern *Homo sapiens*

Probably the most critical early *H sapiens* sample in central Europe comes from the Mladeč caves, next to the village of Mladeč (Lautsch) in the Moravian karst. A series of hominid remains were recovered from the main cave, Furst-Johanns-Höhle or Bačova díra, in 1881–1882 and 1903 and from a smaller, adjacent cave in 1904 [Szombathy, 1925; Smith, 1982]. In both caves, the human remains were excavated from distinctive reddish-brown clayey sediments, which constitute the fan of a talus cone formed by materials washed or thrown into the chimney of the cave, eventually closing the chimney in the main cave [Jelínek, 1976, 1978]. There is no indication of any habitation layer in the cave; but from apparently the same deposits as the hominid remains, a number of bone and lithic artifacts characteristic for the Aurignacian [Bayer, 1922; Szombathy, 1925; Hahn, 1977] were recovered. Because of the manner in which the deposits were accumulated, some doubt regarding contemporaneity of the hominid and cultural remains is inevitable [see Obermaier, 1905; Hahn, 1977]. However, there would appear to be little basis for serious doubt, because the excavators clearly note that the artifacts and hominids are from the same deposits; and there is no other cultural complex from those deposits [Szombathy, 1925].

Despite their importance, the Mladeč hominids elicited little more than passing mention after their initial description by Szombathy, until a series of recent publications reemphasized their significance [Jelínek, 1969, 1976, 1978; Frayer, 1978; Wolpoff, 1980a; Smith, 1982; Minugh, 1983]. The material excavated by Szombathy in 1881 and 1882 is preserved at the Natural History Museum in Vienna (Mladeč 1,2,3, and other fragments). The Mladeč 5 cranium miraculously survived the 1945 fire at Mikulov Castle and is housed in the Moravian Museum (Brno). A good cast of Mladeč 6 also exists, but all other specimens have been destroyed. From the postcranial sample preserved in Vienna and the descriptions provided by Szombathy of some of the fragments destroyed in 1945, a considerable variation in size and robusticity is evident. The two best-preserved femoral diaphyses in

Vienna are slender, but exhibit well-developed muscle markings and distinct pilasters. The two most complete humeral fragments are also slender and have more gracile muscle markings. Fragments of other arm bones, however, are obviously from larger and more robust individuals. Additionally, there are at least three innominate fragments from Mladeč in Vienna, one of which is clearly female. Unfortunately, no pubic bones are preserved. Likewise no axillary borders of scapulae are represented. This is disappointing as it would be interesting and potentially very significant to know what the morphology of these areas were like in this very early Upper Paleolithic hominid sample, particularly in light of the distinctive Neandertal pubic and axillary border morphologies. However, the postcranial elements that are present are clearly modern H sapiens in morphology and not specifically Neandertal-like in a single feature. For example, the articular surfaces are not relatively enlarged; and the femora exhibit well-developed pilasters and reduced relative shaft robusticity.

The Mladeč cranial sample also exhibits considerable variation, due largely to sexual dimorphism. The total morphological pattern of each cranial specimen is consistent with their classification in Homo sapiens sapiens, a conclusion supported by multivariate analyses. The latter place the Mladeč crania within an Upper Paleolithic group, differentiated clearly from Neandertals [eg, Morant, 1930; Stringer, 1974, 1978]. Despite this, the Mladeč crania exhibit features distinctly reminiscent of Neandertal cranial morphology.

The Mladeč 1 cranium is well preserved, although only four teeth remain (M1 and M2 on both sides). Breadth dimensions for these teeth are slightly below the corresponding early Upper Paleolithic means reported by Frayer [1978]. Upper facial height is moderate in comparison to other early Upper Paleolithic specimens, and facial breadth is small. Nasal breadth is moderate, and shallow canine fossae are present. There is a shallow nasal root depression; and the midface clearly does not project to the degree characteristic for European Neandertals [Wolpoff, 1980a, pp 312–313). Alveolar prognathism in Mladeč 1, however, is diminished to a lesser extent. The supraorbital region is divided into a superciliary arch and supraorbital trigone (ie, the modern European supraorbital morphology). Thicknesses of the supraorbitals consistently exceed the early south-central European Upper Paleolithic mean [Smith and Ranyard, 1980], but projection (though unmeasurable) appears less pronounced. On the basis of the relative size of the supraorbitals (compared to certain other Mladeč specimens) and lack of marked development of other cranial superstructures (also compared to other Mladeč specimens), Mladeč 1 appears to be female.

The vault contour [cf Szombathy, 1925; Morant, 1930] of Mladeč 1 conforms to a dolichocranic modern *H sapiens* pattern. In the lambdoidal area, however, there is distinct flattening associated with posterior projection of the occiput. The resulting structure is a type of occipital bun found rather commonly in Upper Paleolithic crania, particularly early central European specimens. Because it differs from the Neandertal occipital bun in a number of factors, this structure is referred to here as an occipital hemi-bun.

The Mladeč 2 calvarium has a rounder vault contour than Mladeč 1 and lacks an occipital hemi-bun. The preserved part of the supraorbital area is identical to Mladeč 1 but less robust. This, along with its small mastoid processes, indicates that Mladeč 2 is also female.

Mladeč crania 4, 5, and 6 are much more robust than Mladeč 1 and 2. Mladeč 4 consists basically of the anterior half of a cranial vault; Mladeč 5 is a virtually complete calvarium (Fig. 6), lacking most of the base (and slightly distorted by the 1945 fire); and Mladeč 6 preserves essentially the same areas as 5, except that the lateral supraorbital regions are missing. These three specimens are probably male and have long and relatively low vaults, lambdoidal flattening, occipital hemi-buns, expanded and robust nuchal areas, thick cranial vault bones, and massive supraorbitals. The supraorbital superstructures are basically modern (ie, somewhat divided into superciliary arches and superorbital trigones) but, especially in Mladeč 5, closely approach the condition of a Neandertal supraorbital torus (see Table III), particularly the form of late Neandertal tori in south-central Europe [Smith and Ranyard, 1980; Wolpoff et al, 1981; Smith, 1982]. Wolpoff [1980, p 311] notes that the cranial contour of Mladeč 5 is similar to that of La Chapelle-aux-Saints, except for a slightly higher forehead and less projecting occiput in the Mladeč specimen. Stringer [1978], however, finds the contours of Mladeč and other early Upper Paleolithic central European crania more angular than Neandertal contours. Both Mladeč 5 and 6 also have wide occipitals, and their maximum cranial widths are low on the parietals. Mladeč 6 exhibits a nasal root depression.

Two partial maxillae are also preserved in Vienna. Both have wide and shallow anterior palates, broad noses, and shallow canine fossae. One has anterior (C, I2) and posterior (M1, M2) tooth breadths slightly above Frayer's [1978] early Upper Paleolithic means, while the other's posterior breadths (M1, M2, M3) are slightly smaller than Frayer's means. All teeth exhibit extensive occlusal and interproximal attrition.

Finally, the partial cranium of an approximately 3-year-old child, Mladeč 3, has recently been studied by Minugh [1983]. It exhibits clear evidence of an incipient hemi-bun but no evidence of development of brow ridge. This would appear to constitute a difference in timing of certain aspects of cranial development from Neandertals, which exhibit relatively greater brow ridge development as well as occipital development at this age.

Like at Mladeč, the Pleistocene sediments in the Zlatý Kůň Cave in the Bohemian karst were deposited through the cave's chimney [Prošek et al, 1952] and do not represent hominid occupation of the cave itself. Hominid skeletal remains have been recovered from deposits correlated to the early Middle Würm stadial [Vlček, 1951, 1957a, personal communication], apparently associated with Aurignacian artifacts [Fridrich and Sklenář, 1976]. Again the nature of deposition leaves some doubt regarding the hominid/artifact association.

The Zlatý Kůň cranium is rather complete but lacks much of its base and face. Metrically, it resembles Mladeč 1, but it is intermediate in robustness between Mladeč 1 and 5. The Zlatý Kůň calvarium exhibits a well-developed hemi-bun and a supraorbital area divided into a superciliary arch and a supraorbital trigone. Thickness and projection values for this specimen are among the largest for south-central European early modern *H sapiens* [Smith and Ranyard, 1980].

The face is represented by both zygomatics and a right maxilla. The zygomatics lack the columnar, pillarlike frontal processes characteristic of Neandertals. The maxilla has a narrow nose and weakly developed canine fossa. The teeth (C through M2) are heavily worn. Their breadths fall consistently within one standard deviation below Frayer's [1978] early Upper Paleolithic means.

The Zlatý Kůň mandible is robust and clearly of modern *H sapiens* type. Its symphyseal angle is 81°, and a distinct mental eminence and mental trigone are present. Though both M3's were lost antemortem, there was clearly no retromolar space. The teeth (I2 through M2 on the left, C through M2 on the right) are heavily worn and smaller than the Early Upper Paleolithic means.

From the site of Velika Pećina in the Hrvatsko Zagorje of northwestern Croatia (Yugoslavia), a hominid frontal was excavated in association with artifacts which probably represent an early stage of the Aurignacian [Malez, 1967, 1974]. From the stratum immediately above the specimen, a radiocarbon date of 33.85 ± 0.52 ky BP (GrN-4979) is associated with a definite Aurignacian component [Malez, 1974, 1978b,c].

The Velika Pećina hominid specimen consists of the right half of an adult frontal bone [Smith, 1976a; Malez, 1978b,c]. Morphologically, it falls clearly within the early modern *H sapiens* group. This is particularly evident in the morphology of the supraorbital area [Smith, 1976a]. The dimensions of the supraorbital projections and the overall dimensions of the specimen are rather small [Smith and Ranyard, 1980] and suggest a female.

Three other finds of possible Aurignacian-associated cranial remains are of interest but, for different reasons, are poorly known and/or problematic. The first is the calvarium from Podbaba (now a part of Prague), Czechoslovakia, discovered in 1883 and accidentally destroyed in 1921 [Matiegka, 1924; Vlček, 1956]. The specimen preserves the frontal, most of the left parietal, and parts of the right parietal and left temporal.[5] The frontal squama is quite low, and the brow ridges are among the thickest and most projecting of the early modern H sapiens sample (Table III). In these features, Podbaba is very similar to the Hahnöfersand frontal, located further north in the same river valley system, which has been described by Bräuer [1980, 1981] as a hybrid between Neandertals and modern H sapiens. Interestingly, it was these two features which led Frič originally to describe Podbaba as a Neandertal. Later research [Matiegka, 1924; Vlček, 1956] demonstrated that the parietal vaulting and other features of the specimen were not Neandertal-like and that the total morphological pattern of Podbaba clearly aligned it with modern H sapiens. Unfortunately, while the specimen is likely to have been associated with the Aurignacian at the site, the exact stratigraphic position of the cranium is not certain [Obermaier, 1905; Matiegka, 1924]. The uncertainty is great enough that Vlček [1971] excludes Podbaba from fossil hominid sites in Czechoslovakia.

The second specimen is a calvarium from the site of Cioclinova in the Transalvanian Alps of Romania. Associated with three Aurignacian implements, this specimen has been described as a female between 30 and 40 years of age with typological affinities with the Předmostí hominids [Necrasov and Cristescu, 1965]. The specimen is clearly H s sapiens and exhibits the form of occipital bunning (slightly) and brow ridge development that characterizes early modern H sapiens from south-central Europe. The degree of development of the brow ridge, supramastoid region, and nuchal plane would appear to indicate a male rather than a female.

---

[5]This morphological assessment is based primarily on my study of a cast of this specimen located in the Laboratoire de Paléontologie Animale, Université de Liège, Belgium, undertaken through the courtesy of the director, Professor G. Ubaugh. My measurements on the cast are very similar to the appropriate values reported by Vlček [1956], which were taken on the original. Thus I am *reasonably* confident that the measurements made on the Liège cast, which were never made on the original (eg, the supraorbital dimensions) are as accurate as one can expect from a cast. Since the original no longer exists, they will have to do.

Third, very fragmentary hominid remains have been reported from level 11 at Bačo Kiro in Bulgaria [Kozłowski,1979] but are yet undescribed. Their fragmentary condition may preclude accurate anatomical assessment, but Kozłowski considers them primitive modern *H sapiens*. If these remains are clearly established as modern *H sapiens* and if they are indeed 43 ky old, they would indeed be exceedingly important for the understanding of the origin of modern Europeans.

Finally, dental and mandibular remains are associated with the Aurignacian at Istállöskö in Hungary [Malan, 1954], Miesslingstal in Lower Austria [Szombathy, 1950], and Vindija in northwestern Yugoslavia [Wolpoff et al, 1981]. The Istállöskö tooth is the fully formed crown of an unerupted mandibular permanent M2. Malan [1954] gives a breadth (buccolingual diameter) of 10.1 mm for the tooth, which is below the Vindija G_3 Neandertal range but within the Předmostí and Early Upper Paleolithic ranges (Table II). It exhibits no evidence of an anterior fovea. The Miesslingstal specimen is a partial juvenile mandible, clearly *H sapiens* in total morphological pattern [Szombathy, 1950], with permanent teeth also below the Vindija G_3 Neandertal range but within the Early Upper Paleolithic and Předmostí ranges. It is associated with Late Aurignacian at the site [Felgenhauer, 1950]. A right upper lateral incisor, left upper canine, and right lower lateral incisor (all permanent teeth) were excavated from the unquestionably Aurignacian level F_d at Vindija. Interestingly, all three of these teeth fall near the appropriate Krapina means and at the upper end or above the appropriate Early Upper Paleolithic means (Table II). Morphologically, they could be accommodated in either a Neandertal or early modern *H sapiens* sample.

North-Central Europe: Early Modern *H Sapiens* Specimens From Germany

Only one specimen, the Stetten 1 calvarium and mandible, is unquestionably associated with the Aurignacian in Germany [Gieseler, 1936]. This specimen was excavated from the Middle Aurignacian levels of the Vogelherd Cave near Stetten in the Lone Valley [Riek, 1932] and has been described by Gieseler [1936, 1937, 1941; see also Czarnetzki, 1980]. The calvarium is moderately robust in comparison to early modern *H sapiens* males from south-central Europe, but the rugose mastoid region and nuchal plane suggest the specimen is possibly male. The glabella region is broken away, revealing an extensive frontal sinus which extends into the squama. The lateral portions of the brow ridge are rather thick (Table III) and exhibit the morphological form typical of early modern *H sapiens* in south-central

Europe. Sagittal contours of the specimen reveal a relatively high forehead and occipital hemi-bun thus conforming to early modern H sapiens pattern. Maximum cranial breadth, like for the Mladeč males, is found at the supramastoid level. The mandible exhibits a distinct mental eminence but also has a small retromolar space and a rather vertical mandibular symphysis (85°), which overlaps the lower end of the Neandertal range (Table V).

A second calvarium (Stetten 2) was recovered from the top of the Upper Aurignacian level at Vogelherd but is possibly intrusive [Gieseler, 1936, p 157]. It is also robust, particularly in the lateral brow ridge region (glabellar area is missing), mastoid region, and nuchal plane. Maximum breadth is at the mid-parietal level, and there is more developing of frontal bossing than in Stetten 1. Morphologically, the specimen can certainly be accommodated in the early modern H sapiens group.

In addition to Stetten, three other German sites have yielded cranial remains which appear to be early modern H sapiens. These crania, which have been chronometrically dated between 26 and 36 ky BP, are: Kelsterlbach, near Frankfurt [Protsch and Semmel, 1978]; Sande/Paderborn, in Westphalia [Henke and Protsch, 1978]; and Hahnöfersand, from the Elbe Valley near Hamburg [Bräuer, 1980, 1981]. All were accidentally discovered, not recovered during controlled excavation; and none is associated with cultural material. The three specimens have each been dated by amino acid racemization and radiocarbon dating of residual bone collagen (see Table I) by R. Protsch and associates at the University of Frankfurt.

The chronologically oldest and in many respects most interesting of these specimens is the frontal bone from Hahnöfersand. It was recovered from a light-gray sand deposit, which obviously does not represent the primary depositional context of the specimen. It is not possible to determine how far down the Elbe the Hahnöfersand frontal has been transported, but the condition of the bone's surfaces do not suggest a long or high-energy transport. The specimen has been dated to 36 ky BP (Fra-a- 29) by amino acid racemization and to 36.3 ± 0.6 ky BP (Fra-24) by radiocarbon [Bräuer, 1980].

The frontal is largely complete, but unfortunately lacks the lateral and inferior portions of the brow ridge. The brow ridges appear to conform to the early modern H sapiens pattern, with well-developed superciliary arches that thin markedly laterally. The measurable dimensions of the arch (Table III) are among the largest of any early modern H sapiens. The supraorbital trigone areas are absent; but as Bräuer [1980] points out, a clear beginning of the division between the superciliary arch and supraorbital trigone is visible on the right side. The frontal sinus is extensive and extends slightly into the squama. The frontal squama appears relatively low. Bräuer [1980, p 7] estimates the angle of inclination at 60° ± 5°. The middle of this range is probably the most reasonable estimation. The squama is unusually thick but not pathological; the temporal lines are very markedly developed; and the minimum frontal breadth is relatively large (105.5 mm). The specimen also exhibits a distinctive, small midsagittal ridge, which extends from just above glabella some 30 mm outside the squama. This feature is also clearly seen in the Paderborn and Oberkassel male crania. Overall, Hahnöfersand is unquestionably a modern *H sapiens* but possesses some strongly Neandertal-reminescent features. Several of the bivariate and multivariate analyses of the specimen reported by Bräuer [1980, 1981] group it either with Neandertals or in an intermediate position between Neandertals and early modern *H sapiens*.

The Kelsterbach calvarium was recovered by workmen in the Willersinn gravel pit near Frankfurt in 1952 [Protsch and Semmel, 1978]. Its stratigraphic position was marked by one of the workmen. The age of the deposits that contained the calvarium clearly appears to be Upper Pleistocene on a number of grounds [Protsch and Semmel, 1978], but determination of their exact age is no simple matter. On geological grounds, a minimum age of 21 ky BP is supported; and mammoth remains, probably from a higher level than the hominid calvarium, yielded a radiocarbon date of 15.8 ky ± 0.41 ky BP [Protsch and Semmel, 1978]. Protsch and Semmel [1978] report a radiocarbon (collagen) date of 31.2 ± 0.6 ky BP (Fra-5) and an amino acid racemization date of 32 ky BP (no sample number given) for the cranium itself.

The hominid specimen is a calvarium, lacking much of the base, of a very gracile female. Brow ridges are among the thinnest and least projecting of the early modern *H sapiens* sample (Table III). The forehead is steep, and there is only a slight development of occipital bunning. Bräuer's [1980] bivariate analyses place the specimen clearly in the modern *H sapiens* group. Muscle attachments, especially on the nuchal plane, are very weakly marked. Some of this may be due to the extensive surface abrasion characterizing the specimen, but it is unlikely that this has a significant effect. Kelsterbach is morphologically the most gracile specimen in the entire sample, surpassing the supposedly more recent Gravettian-associated females like Dolní Věstonice 3 and Předmostí 4.

Finally, a complete hominid calvarium was accidentally discovered at the edge of a gravel pit in Sande near Paderborn and has yielded a radiocarbon date (made on residual collagen) of 27.4 ± 0.6 ky BP (Fra-15) and an amino acid racemization date of approximately 26 ky BP (no sample number given)

[Henke and Protsch, 1978]. The Paderborn calvarium exhibits well-developed brow ridges (Table III) of early modern *H sapiens* form. There is a slight occipital bunning and a very robust nuchal plane, inion, and mastoid/supramastoid region. These features indicate that the specimen is male. The contours of the specimen are clearly modern *H sapiens* in form, and both the multivariate and univariate analyses of Henke and Protsch [1978] and Bräuer [1980] place it unquestionably in the modern *H sapiens* group.

South-Central Europe: Hominids Associated With the Early Eastern Gravettian

Specimens from four sites in Moravia are associated with early eastern Gravettian (Pavlovian) components, dating in excess of about 25 ky BP. From Francouzská ulice in the city of Brno (Brünn), a grave of a male individual (Brno 2) covered with mammoth tusks and containing numerous grave goods was excavated in 1891 [Makowsky, 1892]. This mode of burial indicates an Eastern Gravettian cultural association for the specimen [Jelínek, 1957, 1976; Jelínek et al, 1959]. In 1894, a communal grave of at least 18 individuals covered with mammoth bones was discovered at Předmostí [Matiegka, 1934; Absolon and Klíma, 1978]. Further excavations yielded other individuals, a total of 29 in all. More recent work at the site indicates that the communal grave is contemporaneous with the main (Pavlovian) cultural level at the site [Absolon and Klíma, 1978], which is dated by radiocarbon to 26.32 ± 0.24 ky BP (GrN-6852). The Gravettian association is also supported by the style of burial. However, while the Gravettian association seems highly likely, it is impossible to totally exclude the possibility that the hominid remains are derived from the earlier Aurignacian component at the site. Finally, from the culturally impressive site of Dolní Věstonice [Klíma, 1962, 1963] and the nearby site of Pavlov [Klíma, 1955, 1959] come the burials of a female (Dolní Věstonice 3) and a male (Pavlov 1), respectively.[6] The Pavlov specimen is associated with radiocarbon dates of 25.02 ± 0.15 ky BP (GrN-1325) and 26.62 ± 0.23 ky BP (GrN-1272), while the Dolní Věstonice burial is associated with a date of 25.82 ± 0.17 ky (GrN-1286) [Vogel and Zagwijn, 1967; Klíma, 1963].

The male crania in this combined sample (Brno 2; Pavlov; Předmostí 1, 3, 9, 14, 18, 23) tend to be very robust but exhibit contours and overall morphological patterns that are clearly modern *H sapiens* [Morant, 1930;

[6]Other individuals were discovered at Dolní Věstonice, but were largely undescribed before their destruction in 1945 [see Smith, 1982]. Other fragments are also known from Pavlov.

Matiegka, 1934; Jelínek et al, 1959; Vlček, 1961; Jelínek, 1969, 1976; Smith, 1982]. Brow ridges are generally well developed and projecting (Table III), though of modern H sapiens form. Brno 2, for example, is surpassed only by Mladeč 5 and Hahnöfersand in the dimensions and archaic appearance of his supraorbitals. Faces are robust, but modern and are characterized by relatively narrow noses, canine fossae, moderate prognathism, and noncolumnar lateral orbital pillars. Several of the specimens (Předmostí 3 and 1, Brno 2) exhibit relatively lower cranial vaults/foreheads compared to more modern H sapiens; and all exhibit well-developed occipital hemi-buns, rugose nuchal planes, and robust mastoid/supramastoid regions. Předmostí 3, like Mladeč 6 and Stetten 1, has its maximum breadth relatively low (on the temporals). Mandibles are modern in form, exhibiting mental eminences and trigones, but several specimens maintain a small retromolar space (Brno 2, Předmostí 1 and 3).

Female crania (Dolní Věstonice 3, Předmostí 4 and 10) are considerably smaller and less robust than the males (Fig. 7) but still more robust than most more recent H sapiens [Matiegka, 1934; Jelínek, 1954, 1969, 1976]. Occipital hemi-buns are present but less developed and less robust. Supraorbital dimensions are markedly reduced compared to males, and mastoids are less marked. Dolní Věstonice 3 exhibits marked facial deformation on her left side, which has some interesting implications [see Smith, 1982, pp 681–682]. Interestingly, as was the case for Mladeč 3, several Předmostí juveniles (eg, 2 and 7) exhibit rather early and comparatively extensive development of hemi-buns but relatively little development of the supraorbitals.

Teeth in the sample as a whole tend to fall at or above the early Upper Paleolithic means for males and below the means for females [see Frayer, 1978]. All tend to be relatively heavily worn.

Postcranial morphology is variable but basically modern in the sample [Matiegka, 1938; Jelínek, 1954]. One Předmostí individual (14) exhibits the only dorsal axillary sulci of the scapula known among early modern Europeans; but pubic morphology, femoral form, and all other pertinent postcranial features conform to the total morphological pattern of modern H sapiens. For example, brachial and crural indices [Matiegka, 1938, p 70] show that the limbs of Předmostí hominids, like those of other Upper Paleolithic hominids, are relatively more elongated in their distal elements, unlike Neandertals [Trinkaus, 1981].

ANALYSIS OF MORPHOLOGICAL TRENDS IN CENTRAL EUROPEAN UPPER PLEISTOCENE HOMINIDS
Problems

Before discussing the pattern of morphological change in central Europe, I believe it is critical to discuss the two principal sources of uncertainty in

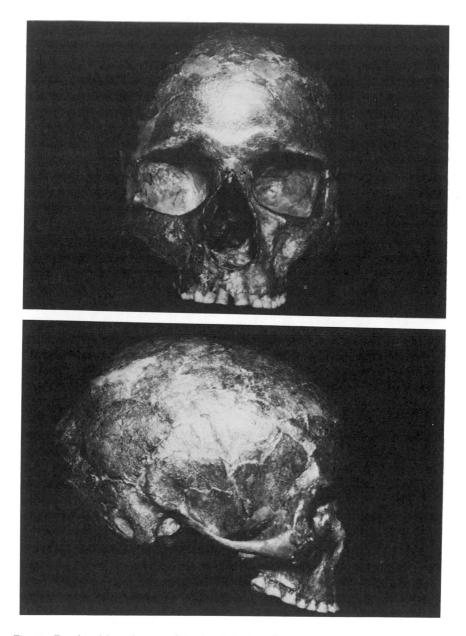

Fig. 7. Facial and lateral views of the female Dolní Věstonice cranium. Note the presence of a small, yet distinct, hemi-bun, form of the brow ridges, and the asymmetry of the face.

this analysis, and my rationale for the manner in which I have dealt with them. The first problem concerns inadequacies of dating and chronological placement of some of the critical specimens, a problem certainly not unique to the consideration of this sample. The second problem revolves around the fragmentary nature of much of the material and other factors that limit the scope of morphometric analysis.

It is obvious that any accurate assessment of evolutionary trends can be accomplished only within a reasonably precise temporal framework. For this analysis, it is necessary to establish that certain Neandertal specimens are chronologically older than others and that the early modern H sapiens sample is truly "early" (eg, in excess of 25 ky BP). Also, we need to determine if archaic and modern H sapiens lived as contemporaries in central Europe.

In south-central Europe, the existence of early (Riss-Würm and Early Würm) and late (Lower Würm) Neandertal groups is relatively easy to document. The pre-Lower Würm ages for Gánovce, Ochoz, and Krapina are well established and have been discussed in the text [see also Smith, 1982]. The Krapina hominid zone may not date to the Riss-Würm, but previous suggestions by Guenther [1964] that the hominids might be the equivalent of Podhradem in age were based (among other things) on the assumption that the Krapina deposits represented a relatively short temporal period. This is demonstrably not the case, and the Krapina hominid zone lies below strata that obviously represent a cold phase and that contain Mousterian artifacts [Malez, 1970a,c, 1978a]. Regardless of whether these strata represent the Lower Würm or an earlier stadial, the pre-Lower Würm age of the hominid zone is supported. Subalyuk also correlates to a period preceding the Hungarian equivalent of the Lower Würm [Gábori, 1976, p 87]. Thus, while not necessarily contemporaries in the strict sense of the word, I believe the designation of the hominids from these four sites as an early Neandertal (EN) sample is quite reasonable.

The later Neandertal (LN) group is represented by the specimens from Kůlna, Šipka, Vindija G₃, and perhaps Šala. The hominid remains from Kůlna, stratigraphically correlated to the Lower Würm, are bracketed between 45 and 38 ky BP on the basis of radiocarbon dates. The Vindija G₃ sample, though not as securely dated, is certainly more recent than the Krapina hominid zone sample and is stratigraphically directly below the Podhradem deposits at the site [Malez et al, 1980; Wolpoff et al, 1981]. Šipka is either early Podhradem or Lower Würm in age. That these three sites are relatively more recent than the early group appears certain, and the Kůlna radiocarbon dates help to demonstrate that this late group is only a few thousand years older than the earliest modern H sapiens material from central Europe.

The placement of the Šala frontal into this dichotomy is more difficult. There are only two clues to its age: its geological context/associations and its morphology. The former can only establish an Upper Pleistocene warm period age for the hominid specimen on the basis of its faunal association, particularly the presence of the steppe rhinoceros (*Dicerorhinus hemitoechus*). Since this species is present during warm periods well into the first half of the Würm in parts of Europe [Kurtén, 1968], its presence does not allow correlation of the hominid to any specific temporal period. I have argued elsewhere [Smith, 1977] that morphological "dating" is justified and reliable in Europe when used judiciously (eg, not as an argument against solid, independent dating by other means). Morphologically, as I have discussed elsewhere [Smith, 1982], Šala compares best to the later group, and this is where I am inclined to place it.

In north-central Europe, there is no evidence of late Neandertals. Ehringsdorf is at least Eemian in age [Cook et al, 1982], and Salzgitter-Lebenstedt clearly does not fall into the late Neandertal time range. Neandertal (Feldhofer Kirche) is undated but conforms morphologically to the early group.

The age range for early modern *H sapiens* in central Europe appears, on the basis of available chronometric dates, to be reasonably clear. The older end of the range is represented by ages of just over 34 ky BP for Velika Pećina and approximately 36 ky BP for Hahnöfersand. The possibility that the range may extend back to cover 40 ky BP (on the basis of Bačo Kiro) will be discussed, but judgment on this must await publication of the material. The other Aurignacian-associated remains can be reasonably dated between approximately 30 and 35 ky BP on the basis of the majority of radiocarbon dates available for Aurignacian in this region [Smith, 1982], excluding the questionable Istállöskö dates [Hahn, 1977; Gábori-Csánk, 1970; Gábori, 1976]. The early eastern Gravettian hominids included in the early modern sample are clearly at or older than the 25 ky BP limit, mostly on the basis of associated radiocarbon dates.

There are, however, some uncertainties associated with the early modern *H sapiens* sample. Primarily these deal with the stratigraphic or cultural associations of some critical specimens (specifically, those from Hahnöfersand, Kelsterbach, Paderborn, Mladeč, Zlatý Kůň, and Podbaba). In the case of the first three, radiometric dates place them within the temporal span appropriate for the early modern group; and although they have no cultural associations, I believe their inclusion is justified. I have discussed previously the situation at Mladeč. Though there can always be doubts expressed about the association of the tools and the human remains, I regard the possibility that the hominids are anything but Aurignacian as highly unlikely. The same agreement can be made, though perhaps with less certainty, for Zlatý Kůň. Consequently, I include the remains from both sites in the early modern sample. Similarly, though even greater doubts exist regarding the exact stratigraphic associations of Podbaba, I consider the possibility of its Aurignacian association and its distinct morphological similarity to the other hominids which make up the early modern sample as sufficient for its inclusion in that sample.

Finally, I do not believe the question of overlap between Neandertals and modern *H sapiens* in central Europe is answerable at this point. We are still uncertain whether the makers of the Szeletian were Neandertals, which appears to be the case for the makers of the Chatelperronian [Lévêque and Vandermeersch, 1980, 1981] or modern *H sapiens*, since only one human tooth is securely associated with the Szeletian. Neandertals of Podhradem age may be represented at Šipka and Vindija G1, and thus may overlap in time with Podhradem-age modern *H sapiens*. Such a situation is suggested by Jelínek [1969] for Šipka and Mladeč. However, Šipka may also date to a Würm stadial [Kukla, 1954; Gábori, 1976], and the specimens from Vindija G1 are too few and fragmentary to be certain of their taxonomic affinities. We will have to wait for more information, including the publication of the Bačo Kiro material, to shed more light on this issue.[7]

A glimpse of the second major limitation of the present analysis, the fragmentary nature of many of the remains, is provided in the previous paragraph, but the problem does not stop there. There are, for example, no central European Neandertal crania that preserve both a relatively complete brain case and a face, and only one example of a postcranial skeleton that can be directly associated with a specific cranium. Furthermore, most of the cranial specimens known from this region consist of individual bones (Kůlna, Šipka, Sala, Salzgitter-Lebenstedt) or relatively large series of fragments (Krapina, Ehringsdorf, Vindija). It is always possible that if more complete specimens were known, their total morphological pattern might be different than that indicated by individual fragments. This was pointed out by a number of commentors on a previous discussion of the south-central European hominids [Smith, 1982]. Furthermore, the lack of relatively complete Neandertal, especially late Neandertal, crania precludes the application of sophisticated multivariate analyses, like those of Stringer [1974, 1978] and Bräuer [1980], to the central European samples as a whole.

---

[7]The material may be too fragmentary for meaningful analysis.

Other frustrating inadequacies of the sample must also be noted. First, there are virtually no postcranial remains attributable to the late Neandertal group, making discussion of patterns of noncranial morphological change over time within the Neandertals impossible. Second, there are few anterior teeth in the Mladeč and late Neandertal samples. Third, the fragmentation of many specimens makes correct orientation for certain measurements (especially various frontal angles) very difficult. This problem is particularly apparent in regard to specimens like Šala, Podbaba, Hahnöfersand, and Vindija 261. Fourth, many of the original specimens have been destroyed, so that the collecting of additional data and the checking of reported measurements that seem unusual are impossible (except for casts). The major specimens that fall into this category are Šipka, the entire Předmostí series, Podbaba, Dolní Věstonice 1 and 2, Mladeč 6, and several other Mladeč specimens listed by Szombathy [1925].

While it would certainly be nice to have a more complete Upper Pleistocene fossil hominid record from central Europe (particularly more late Neandertal postcrania and more complete Neandertal and Mladeč crania), I believe it is useless to argue over what morphology *might* be like in anatomical elements which are not represented. It is only possible to deal with what information is present and capable of documentation. Luckily, there is reasonably good representation of critical anatomical regions (brow ridge, mandibular symphyses, nasal areas, maxillary alveoli, and teeth) in all three central European samples. Thus it is possible to investigate patterns of differences among the samples for a number of cranial characteristics and to determine whether any observed changes over time can be grouped into and explained in terms of morphological complexes relevant to the archaic/modern *H sapiens* transition.

Evolutionary Patterns

The EN from central Europe exhibit a total morphological pattern that differs only slightly from that of western European "classic" Neandertals. While in the past, remains of some early south-central European Neandertals (particularly Krapina) have been discussed as more "progressive" than western European Neandertals, there is no morphological basis for this assertion [Smith, 1976b, 1978]. The late Neandertals (LN) from central Europe, all of which are from the southern portion of the region, unquestionably qualify as archaic *H sapiens* on the basis of their total morphological pattern. However, there are a number of features in which the late Neandertals, as a group, consistently approaches the early modern *H sapiens* (EMH) condition to a greater degree than the EN sample does. Many of these features were pointed out some time ago by central European paleoan-thropologists on the basis of the rather sparse Czech LN material [Jelínek, 1965, 1967, 1969; Vlček, 1958, 1964b, 1969], but discovery of the larger Vindija sample was necessary to conclusively demonstrate the "progressive" nature of the central European LN sample. Furthermore, the central European EMH sample exhibits several features that suggest considerable mor-phological continuity with the LN sample and indicates an intermediate morphological position for it between Neandertals and later modern Euro-pean samples. Thus morphological continuity can be observed from the EN through the LN to the EMH samples, specifically in the five morphological complexes discussed below.

Brow ridge morphology. The EN sample is characterized by well-developed supraorbital tori, which tend to be quite thick (superio-inferior dimension) and projecting (anterio-posterior dimension) compared to mod-ern *H sapiens*. In comparison, the LN sample not only shows overall reduction of torus dimensions (Table III) but also a greater degree of relative midorbit thinning and reduction in projection [Smith and Ranyard, 1980, p 606]. These relative reductions result in an incipient division of the torus into supraorbital trigone and superciliary arch segments (Fig. 5), the char-acteristic modern *H sapiens* pattern [Cunningham, 1908]. Supraorbital su-perstructures in the EMH sample, though conforming to the modern pattern, are much more salient and robust than in later modern European populations; and the division between the superciliary arch and supraorbital trigone is often not as distinct. From a comparative perspective, brow ridge reduction is not found in the late Shanidar sample in the Near East [Trinkaus, 1983b] nor in the late western European Neandertal from Saint-Césaire.[8] Furthermore, the early modern *H sapiens* specimens from Skhul and Qafzeh do not exhibit the same pattern of reduction noted in central Europe [Smith and Jablonski, unpublished manuscript] nor is it evident in available African or Asian Upper Pleistocene hominids [see Howell, 1978; Wolpoff, 1980a]. Thus I believe the *pattern* of brow ridge reduction evident in central Europe is not only an excellent indicator of morphological continuity but also a distinctive European phenomenon.

Cranial vault form. Two frontals from the LN sample (Šala and Vindija 261) appear to exhibit less receding frontal squamae that are characteristic

---

[8]I am especially grateful to Professor B. Vandermeersch for his permission to study in detail the Saint-Césaire specimen. After thorough study, it is clear to me that my initial impression of considerable morphological similarity between Saint-Césaire and the central European sample [Spencer and Smith, 1981; Smith, 1982] was erroneous.

for Neandertals in general and in fact overlap the lower end of the EMH range. This overlap is clearly demonstrated by Bräuer [1980] in the case of Hahnöfersand. Moreover, every central European EMH specimen is characterized by some development of an occipital hemi-bun. These tend to be most well developed in males but are also present, though less saliently, in females (eg, Předmostí 4, Předmostí 10, Dolní Věstonice 3). Hemi-buns are positioned higher on the cranium and usually do not result in as much horizontal positioning of the nuchal plane as the occipital buns of Neandertals. As is the case for the pattern of brow ridge reduction, well-developed occipital bunning appears to be a typically European Upper Pleistocene hominid characteristic. It is, at best, only marginally developed in the Shanidar Neandertals [Trinkaus, 1983b]; absent in the Skhul-Qafzeh group [Vandermeersch, 1981; Wolpoff, 1980a]; and, so far as I am aware, not characteristic of Upper Pleistocene hominids in southern Africa [Howell, 1978]. Furthermore, regardless of whether one explains its development in terms of differences in timing of brain growth [Trinkaus and Lemay, 1982] or increasing efficiency of nuchal musculature [Brose and Wolpoff, 1971; Wolpoff, 1980a; Smith, 1983], it is certainly under some degree of genetic control. I would argue that the consistent presence of occipital bunning in both the European Neandertal and EMH sample argues convincingly for genetic continuity across the archaic/modern H sapiens transition in Europe.

Facial reduction. Their relatively long, prognathic faces are one of the major features that distinguish Neandertals, including the central European EN sample, from modern H sapiens [Smith, 1976b, 1983; Brace, 1979a,b; Trinkaus and Howells, 1979; Heim 1974, 1976a]. The LN sample from central Europe tends to be reduced in several critical facial dimensions, particularly nasal breadth and alveolar height, compared to Krapina and western European Neandertals and often overlaps the appropriate EMH ranges (Table IV).[9] Also, a shallow canine fossa is exhibited by Kůlna and perhaps also by the Vindija maxillae. Furthermore, it has been demonstrated that, on the basis of relative projection of nasion, both Vindija 224 and Šala exhibit less midfacial prognathism than characterizes other Neandertals [Wolpoff et al, 1981, p 525] and some early modern H sapiens (eg, Mladeč 5). Reduced alveolar prognathism for the LN sample is also suggested by the symphyseal angles (see below), although other indications of

---

[9]These reductions cannot be explained in terms of reduced body size in the LN sample. Postcranial dimensions (of adults) and posterior tooth breadths are virtually the same between the EN and LN samples. Plus, certain dimensions (for example, prosthion to postcanine, see Table IV) do not reduce.

alveolar prognathism (length of retromolar space and position of mandibular foramina) exhibit little difference between the EN and LN samples [Wolpoff et al, 1981]. Interestingly, on the basis of pertinent available studies [Endo, 1966; Hylander, 1977; Russell, 1983; Smith, 1983], one would predict that reduction of prognathism and increased height of the frontal squama should result in reduction of brow ridge. The central European LN sample appears to support this prediction.

Mandibular symphyseal morphology. Typically, the alveolus of Neandertal mandibles projects anteriorly more than the base, and the contour between these areas is essentially linear [Wolpoff, 1975; Weidenreich, 1936]. This results in the characteristic receding symphysis of Neandertals, which is found in all adult Krapina mandibles [Smith, 1976b]. All Vindija G₃ mandibles, however, exhibit relatively more projection of the base than the alveolus [Wolpoff et al, 1981] and consequently have symphyseal angles that are all less than 90° (Table V). Although there is only slight overlap with the EMH range, the Vindija values are clearly intermediate between the Krapina/western European Neandertals and the EMH sample. Furthermore, three of the four Vindija mandibles exhibit slight but distinct subincisor concavities, which result in the expression of an incipient mental eminence. A similar condition apparently also characterizes the Šipka mandible. Thus, while mental eminences and trigones are markedly more developed in modern *H sapiens* (even the EMH sample) than in the central European LN sample, a clear trend towards the EMH condition is presently in the LN group.

Dental reduction. Sample sizes for LN dentitions are unreliably small. Nonetheless, available evidence (Table II) suggests that LN anterior teeth tend to be slightly reduced in size compared to the EN sample. Posterior teeth do not show a clear pattern of reduction They are rather small in the Kůlna maxilla, but Vindija mandibular molars fall very close to the appropriate Krapina means. The central European Neandertals, especially the late ones, overlap the EMS sample to a considerable extent in anterior tooth size, and the EMS sample exhibits larger anterior teeth than later Upper Paleolithic specimens from Europe [Frayer, 1978]. Thus the LN sample fits appropriately into the sequence of anterior dental reduction documented in Europe from the Riss-Würm through the Neolithic [Frayer, 1978]. Anterior dental reduction is also indirectly suggested by the relative reduction in the thickness of the mandibular alveolus in the Vindija mandible compared to those of the EN sample. In fact, in both absolute and relative (alveolar index) symphyseal alveolar thickness, the LN sample falls closer to the EMH than to the EN samples (Table V). Reduced thickness of the alveolus indicates a reduction in size of incisor and canine roots, which is also supported by the decrease in symphyseal height of the LN sample compared to other Neandertals (Table V).

The data from these five complexes clearly document a pattern of change from early to late central European Neandertal morphology in the direction of EMH morphology and a high frequency of Neandertal-reminiscent or derived features in the EMH sample. I believe the evidence for this morphological continuum is all the more compelling because the patterns of change documented in these five complexes can be explained in the context of a single functional model. This model, the so-called teeth-as-tools hypothesis, would predict that reduction in the anterior dentition (brought about by reduced functional demands on it) would result in reduction of facial dimensions, prognathism, brow ridge development, occipital bunning, and other features [Brose and Wolpoff, 1971; Brace, 1979b; Wolpoff, 1980a; Smith, 1983]. Thus, the changes documented above are not unrelated morphological alterations, but can be interpreted as aspects of an integrated morphological transition brought about by shifts in functional demands due to behavioral factors [see Smith, 1983, for more details on the model].

Although the overall *pattern* of changes over time in these samples is clear, a detailed look at the data reveals the complexity of the evolutionary process. All characteristics do not change at the same rate nor do all specimens exhibit the same degree of salience or reduction in all elements. For example, the Kůlna maxilla exhibits very small posterior teeth for a Neandertal, but a rather large canine. The reverse is true for the Vindija 206 mandible (Table II). Hahnöfersand, Podbaba, and Mladeč 5 exhibit similar degrees of brow ridge development and are metrically very similar; but Mladeč 5 exhibits a higher frontal squama. Furthermore, one does not always see the *extent* of close interrelationship between features in single specimens one might expect. The Šala frontal conforms to expectations by combining reduced midfacial prognathism with a higher frontal squama but still preserves a supraorbital torus of average projection for a Neandertal (Table III). Glanville [1969], Coon [1962], and others have noted a close correlation between nasal and anterior palate breadth, yet both Vindija maxillae have very low nasal breadths associated with rather broad anterior palates (Table IV). I believe that all of these "inconsistencies" serve as an excellent example of the mosaic nature of morphological change and demonstrate once again the need for attention to the role of variability in the interpretation of fossil samples. A general pattern or trend simply cannot be based on a single specimen or the isolated presence of one "progressive" feature.

There are two aspects of the central European Upper Pleistocene hominid record for which morphological continuity is impossible to demonstrate: the postcranial skeleton and overall cranial vault form. Trinkaus [1976a,b, 1977, 1981, 1983a,b] and others [eg, Heim, 1974, 1976a; Wolpoff, 1980a; Vandermeersch, 1981; Smith, 1976b] have documented several differences bewtween the postcranium of Neandertals and modern H sapiens. The central European EN sample conforms to the pattern of Neandertal post-cranial anatomy in every detail; and the EMH sample, despite its rather high incidence of Neandertal-reminiscent cranial features, exhibits no *specifically* Neandertal-like postcranial features (with the exception of the dorsal axillary grooves of Předmostí 14). However, there is only one diagnostic LN postcranial fragment, a scapula from Vindija. It also possesses a dorsal axillary border (Neandertal pattern) and metrically fits into the Krapina range, but there is simply insufficient material to investigate the possible presence of change in postcranial morphology within the central European Neandertals.

It is interesting that Trinkaus [1982, 1983b] found no pattern of postcra-nial changes between the early and late Shanidar Neandertal samples, even though numerous differences were found in the cranial samples. Perhaps, the differences in the postcranium between Neandertals and modern H sapiens are brought about by developmental responses to different degrees of stress during the growth period [Wolpoff, 1980a]. With the emergence of the Upper Paleolithic in Europe and western Asia, many of the stressors would appear to have been reduced [Trinkaus, 1983a]. In this case a rather rapid change in postcranial morphology would have been possible. At present, however, existing postcranial remains yield little evidence of Nean-dertal/modern H sapiens continuity in Europe.

Multivariate analysis of overall cranial form [Stringer, 1974, 1978; see also Bräuer, 1980; Trinkaus and Howells, 1979] group the EMH sample from central Europe with modern H sapiens and not in an intermediate position between Neandertals and later modern H sapiens. However, such analyses cannot include any representative of the central European LN sample due to the fragmentary nature of the remains. If one assumes that overall vault morphology in this sample had evolved to the level suggested by the observable anatomical features, the degree of distinction between Neandertals and the EMH sample might not be in reality as marked as the available multivariate studies indicate.

THE ORIGIN OF MODERN *HOMO SAPIENS* WITH PARTICULAR REFERENCE TO CENTRAL EUROPE

In my opinion, morphological continuity between Neandertals and the EMH sample in central Europe is clearly documented by the available information, despite the lacunae in the data noted earlier. Further, I maintain that this morphological continuity also establishes the presence of some degree of genetic continuity across the archaic/modern H sapiens transition in this region. The demonstration that a morphological and genetic continuum exists does not, however, prove that the origin of modern Europeans lies completely in the European Neandertals; but it does establish that European Neandertals at the very least contributed significantly to the gene pool of modern Europeans. In order to establish whether external factors also played a significant role in the origin of modern Europeans, it is necessary to briefly consider Upper Pleistocene hominid evolution in other parts of the Old World and the major hypotheses which have been formulated to explain the origin of modern Homo sapiens.

One long-standing hypothesis that I believe can be dismissed is the so-called presapiens hypothesis. This hypothesis suggests the existence of a European hominid lineage, separate from that leading to the Neandertals, which ultimately gave rise to modern Europeans. The origin of this hypothesis may be traced to Boule [1911/13, 1923] and was further developed by Vallois [1954, 1958; see also Boule and Vallois, 1957], Piveteau [1957], and others [eg, de Lumley, 1973]. Boule's candidates for membership in this lineage have long since been discounted [see Brace, 1964; Spencer, this volume]. The fossil material suggested to be representatives of the presapiens lineage by Vallois, the Fontéchevade and Swanscombe cranial remains, have also subsequently been shown not to differ significantly from other contemporary hominids and thus yield no evidence of a separate, early modern H sapiens lineage [Sergi, 1962; Brace, 1964; Corruccini, 1975; Wolpoff, 1980a,b].[10] More recent suggestions of a European second lineage have focused on the form of the parietal. For example, de Lumley [1973, 1976] recognizes two Middle Pleistocene hominid lineages on the basis of differences in overall parietal shape. The presapiens lineage is supposedly represented by a square parietal form, while the pre-Neandertal lineage supposedly exhibits parietals elongated antero-posteriorly. However, no other characteristic supports such a division, and the danger of basing phylogenetic conclusions on single features is obvious. Furthermore, parietal shape is noticeably variable in all Upper Pleistocene hominid samples. Finally, it seems very unlikely that two lineages of hominids could have coexisted and remained completely separate within Europe throughout the entire Middle

---

[10]These statements apply only to Fontéchevade 2. The context of the Fontéchevade 1 glabellar fragment is unknown [Vandermeersch et al, 1976] and cannot definitely be considered to be Riss in age. Moreover, it may be a juvenile.

Pleistocene. In my opinion [see also Wolpoff, 1980a; Hublin, 1982; Bräuer, this volume], the presapiens hypothesis is simply not supportable on the basis of present paleoanthropological knowledge.

In the same vein, Hrdlička's hypothesis that modern H sapiens evolved *only* from European Neandertals and then spread to other parts of the world [see Spencer, 1979; Spencer and Smith, 1981] is equally unsupportable in light of the existing fossil record. It also does not seem reasonable to believe that modern humans evolved in several areas of the Old World *completely independently* of each other [see Howells, 1976].

Excluding these unlikely explanations, there would appear to be three hypotheses that may reasonably be presented to explain the origin of modern Europeans: 1) the replacement of Neandertals by an influx of completely modern H sapiens into Europe from some adjacent area, with some assimilation of Neandertal genes into the invader's gene pool; 2) a transition catalyzed by the flow of "progressive" genes into Europe but without a large-scale population influx; and 3) a basically localized transition within Europe that did not result primarily from outside influence.

In order to substantiate large-scale population replacement (hypothesis 1), I believe it is necessary to demonstrate that modern humans are of greater antiquity outside Europe than in Europe *and* to provide evidence that non-European modern H sapiens groups physically moved into Europe. The first of these points appears to be established. Modern humans are found at the sites of Skhul and Qafzeh in the Near East, which probably date slightly earlier than the EMH sample in central Europe.[11] By itself this difference in age is probably not of great significance, but modern H sapiens appear to have been present in southern Africa significantly earlier. The recent description of the Klasies River Mouth Cave 1 hominids [Singer and Wymer, 1982] strongly supports earlier claims for the existence of modern

---

[11]The debate over the exact age of the Skhūl and particularly the Qafzeh hominids continues [see Vandermeersch, 1981; Bar Yosef and Vandermeersch, 1980; Farrand, 1979; Trinkaus, 1983b, and in this volume; Jelínek, 1981, 1982]. Jelínek's lithic analysis indicates that the Qafzeh hominid-bearing levels are contemporary with the Skhūl hominid-bearing deposits, or perhaps even more recent [Jelínek, 1982]. Skhūl is considered to be more recent than Tabūn B, and Tabūn B is dated radiometrically to 40 ky BP. Thus I regard the argument for a relatively late (35–40 ky BP) date for both samples as the more convincing [see Smith, 1982].

H sapiens in this region prior to at least 65 ky BP [Rightmire, 1979, 1981; Beaumont, 1980; Beaumont et al, 1978].[12]

On the basis of the above information, it is tempting to suggest a clinal pattern of the spread of modern *H sapiens* from Africa, through the Near East, into Europe. However, the evidence for actual large-scale movements of modern *H sapiens* groups into Europe is nonexistent. In my opinion, if the EMH total morphological pattern were introduced into Europe by such a process, we should see a number of specimens in the central Europe EMH sample that exhibit non-European features. For example, they should have *no* occipital bunning and *no* indication of the characteristic European EMH brow ridge form, because these features are not found in the Near Eastern or African EMH samples. Furthermore, I would expect at least *some* evidence for contemporaneity of Neandertals and early modern *H sapiens*, since some Neandertal groups should have endured in the region for some period of time after the beginning of the supposed "influx." If these two points could be established, then the morphology of Mladeč, Brno 2, Hahnöfersand, and other central European EMH specimens could more reasonably be considered "hybrids" or the result of the assimilation of Neandertal genes into the invaders' gene pool.

The only EMH specimen that does not clearly "fit" the central European EMH total morphological pattern is the Kelsterbach calvarium. Even if its suggested 32 ky BP age is accepted, I regard one specimen in a reasonably large sample as inadequate to establish the presence of a non-European influx. The Bačo Kiro hominid remains may provide further indication of such a group, but this cannot be determined at present. Also, it is important to stress that much of eastern Yugoslavia, Bulgaria, and Turkey are poorly known from a paleoanthropological perspective and may ultimately yield evidence of early modern, non-European-like *H sapiens*. At present, despite suggestions of how and why modern *H sapiens* might have invaded Europe [Boaz et al, 1982], there is no convincing evidence for the presence of such a group in the Upper Pleistocene of central Europe.

Similarly, while the possibility cannot be completely ruled out, there is no unequivocal evidence for the contemporaneity of any Neandertal and

---

[12]It is important to note, however, that dating of these southern African sites is going through a "transition" period. Some specimens are being moved back in age, while others are being moved up (for example Fish Hoek [Rightmire, 1978b]; and there is considerable disagreement about the age of such critical specimens as Omo, Florisbad, and Broken Hill [cf Protsch, 1975; Rightmire, 1978a; Bräuer, this volume; Howell, 1978].

any EMH specimen in central Europe. Certainly no fossil remains of Neandertals and modern *H sapiens* have ever been found in association with each other or dated chronometrically to overlapping time spans in this region.[13] Furthermore, as far as I am aware, there is no site in the region in which Mousterian is stratigraphically more recent than the Upper Paleolithic (either Szeletian or Aurignacian). If the latter was brought into Europe by modern *H sapiens* (and in Europe the Aurignacian is always associated with modern *H sapiens*, except possibly at Vindija), I believe there should be at least *one* instance where a Mousterian/Neandertal occupation followed an Aurignacian/modern occupation at the same site. Thus, I believe an actual population replacement in Europe (hypothesis 1) can, at present, also be considered an unlikely explanation for the emergence of modern Europeans.

Unfortunately, the state of the art in paleoanthropology does not, in my opinion, allow a definite choice between the two other hypotheses because of the inadequacy of existing paleoanthropological models. It is simply not possible at present to determine what effect gene flow (without *major* population influx) might have had on the archaic/modern *H sapiens* transition in central Europe, because we cannot differentiate between a morphological continuum in which gene flow has played a significant role from one in which it has not [see Weiss and Maruyama, 1976]. In my estimation, the increasing evidence for the early appearance of modern *H sapiens* elsewhere strengthens the possibility that unidirectional gene flow into Europe, as defined in hypothesis 2, played a significant role in the emergence of modern Europeans. However, even if gene flow's role were significant, the nature of morphological continuity demonstrates that the Neandertal gene pool was also a major contributor to that of early modern Europeans.

In other regions of Europe, there is much less evidence of a morphological continuum than in central Europe. In western Europe, most Neandertals appear to date to Würm II [Heim, 1976b; de Lumley, 1976; Vandermeersch, 1976; Stringer and Burleigh, 1981] and thus are rather late. The Saint Césaire specimen [Lévêque and Vandermeersch, 1980, 1981; ApSimon, 1980] indicates that typical "Classic" Neandertals survived in western Europe until 31 to 33 ky BP (see Note 9). Though some isolated "progressive" features are found in some Würm II Neandertals in western Europe [see Wolpoff, 1980a; Smith, 1982], there is no indication of a late Neandertal group exhibiting a consistent "progressive" morphology. Furthermore, the

---

[13]This may be a result of the inadequacies of the dating techniques available at present [see Stringer and Burleigh, 1981]. Also, the Šipka mandible and the Mladeč sample have been claimed to be contemporaneous, but this is far from convincing (see text).

earliest modern *H sapiens* specimens from western Europe exhibit less Nean-dertal-reminiscent morphology (and are also more recent) than the central European EMH sample.

In eastern Europe (Russia), Upper Pleistocene hominid fossils are rather scarce and generally not well dated. However, no Mousterian-associated hominid, with the possible exception of Staroselje [Alexeyev, 1976; Wolpoff, 1980a], exhibits any morphology that could be called transitional [see Yakimov, 1980]. Thus central Europe appears to be the only region in Europe where a morphological continuum can be clearly demonstrated at the present time. This contributes further to the interpretation that a clinal pattern of the spread of "progressive" genes can be traded from Africa through the Near East to central Europe and then to other regions of the latter continent.

On the other hand, I believe that it is also possible that gene flow played the role more of a "modulator" rather than of a "catalyst" in the emergence of the central European EMH sample. The transition from archaic to modern *H sapiens* is essentially a process of gracilization in all areas where it occurs [Wolpoff, 1980a; Smith, 1982; Howell, 1978; Thorne and Wolpoff, 1981; and other papers this volume]. In the cranium, this gracilization involves reduction in anterior tooth size, facial size and prognathism, and other cranial features which support the masticatory apparatus [for details see Smith, 1983]. This reduction can be reasonably explained as the de-creased functional demands on the masticatory apparatus. Once cultural/behavioral parameters reduced the selective necessity for maintenance of the archaic *H sapiens* degree of expression in these features [Brace and Montagu, 1977; Smith, 1983], the complex would be expected to reduce to the modern *H sapiens* level relatively quickly. This would be due, in my opinion, to the biomechanical advantage of reduced prognathism and the selective advantage resulting from the fact that nutritional requirements would decline with decreased robusticity. A similar argument can be made for the postcranium [Trinkaus, 1983b].

The manner and form in which reduction in this complex could be expressed is severely limited by developmental/functional factors associated with the human cranium [see Enlow, 1975]. Therefore, it would have to follow basically the same pattern and result in a basically similar morphol-ogy even if it occurred in several areas essentially independently of each other. Furthermore, while the changes involved in this reduction certainly would have had a genetic background, they may not have resulted from exactly the same genetic changes in all groups. Because of the complexity of the mechanisms involved in craniofacial growth and development, similar patterns of reduction could probably be brought about by genetic changes affecting different steps in the developmental process. Thus the general similarity of early modern *H sapiens* in Africa, Europe, and western Asia may not necessarily be the result of identical genes that spread from one point of origin and transformed existing archaic *H sapiens* groups. Viewed in this perspective, a largely independent origin of modern *H sapiens* in several regions can be hypothesized without shades of orthogenesis. In this case, gene flow might serve only the function of reducing differences between populations, not as a causal factor in evolutionary change.

The Upper Pleistocene fossil hominid record in central Europe demonstrates the usefulness of somewhat restricted, regional analyses of the archaic/modern *H sapiens* transition. Specifically, it indicates that a considerable degree of continuity exists in the gene pool across this transition in Europe and, in my opinion, shows that Neandertals are reasonable candidates for the ancestral stock of modern Europeans. Certainly, the hypothesis that they are cannot be so easily dismissed as it might be if only western European Upper Pleistocene hominids are investigated. On the other hand, we are not in a position to suggest that a completely indigenous origin of modern Europeans is a proven fact. Such proof or falsification must, I believe, await a more complete understanding of the effect of gene flow in paleoanthropological samples and a broader knowledge of the behavioral and adaptive basis of the archaic/modern *H sapiens* transition.

ACKNOWLEDGMENTS

This paper was written, and much of the data and observations on which it is based were collected, during my tenure as an Alexander von-Humboldt Foundation Fellow at the University of Hamburg, Germany. I am grateful to the Foundation for its support and to my hosts in Hamburg (R. Knussmann, V. Chopra, and G. Bräuer) for numerous forms of assistance. I also thank Drs. M.O. Smith, T.S. Zobeck, and M.H. Wolpoff for help in preparing and critically reading the manuscript. The research on which this paper is based has also been supported by the Wenner-Gren Foundation, the National Academy of Sciences, and the Faculty Research Fund of the University of Tennessee at Knoxville. Finally, I am grateful to the following individuals for granting me access to the fossil remains under their care: I. Crnolatac and K. Sakač, Geological-Paleontological Museum, Zagreb; M. Malez, Geological-Paleontological Institute of the Jugoslav Academy of Sciences, Zagreb; J. Szilvássy, Natural History Museum, Vienna; E. Vlček, National Museum, Prague; J. Jelínek, Moravian Museum, Brno; M. Thurzo,

National Museum, Bratislava; G. Urbaugh, University of Liège; B. Vander-
meersch, University of Paris; J.-L. Heim, Museum of Man, Paris; R. Protsch,
University of Frankfurt; H.E. Joachim, Rhenish Staff Museum, Bonn; W.
Henke, University of Mainz; M. Schultz, University of Göttingen; G.
Bräuer, University of Hamburg; P. Schröter, State Anthropological Collec-
tion, Munich; and A. Czarnrtzki, University of Tübingen.

LITERATURE CITED

Absolon K, Klíma B (1978): "Předmostí: Ein Mammutjägerplatz in Mähren." Brno: Archeo-
logický Ústav ČSAV v Brne.

Alexeev VP (1979): Horizontal profile of the Neandertal crania from Krapina comparatively
considered. Coll Anthropol 3:7–13.

Alexeyev VP (1976): Position of the Staroselye find in the hominid system. J Hum Evol
5:413–421.

ApSimon A (1980): The last Neandertal in France? Nature 287:271–272.

Bartucz L, Danucza J, Hollendonner F, Kadić O, Mottl M, Pataki V, Pálosi E, Szabò J, Vendl
A (1940): Die Mussolini-Höhle (Subalyuk) bei Cserépfalu. Geol Hung Ser Paleontol 14.

Bar Yosef O, Vandermeersch B (1980): Notes concerning the possible age of the Mousterian
layers in Qafzeh cave. In: "Préhistoire du Levant." Paris: CNRS, pp 281–285.

Bayer J (1922): Das Aurignac-Atler der Artefakte und menschlichen Skelettreste aus der
Fürst-Johanns-Höhle bei Lautsch in Mähren. Mitt Anthropol Ges Wien 52.

Bayer J (1925): Das jungpaläolithische Alter des Ochozkiefers. Eiszeit 2:35–40.

Beaumont PB (1980): On the age of Border Cave hominids 1–5. Paleontol Afr 23:21–33.

Beaumont PB, De Villiers H, Vogel JC (1978): Modern man in sub-Saharan Africa prior to
49,000 years B.P.: A review and evaluation with particular reference to Border Cave. S
Afr J Sci 74:409–419.

Behm-Blancke G (1958): Umwelt, Kultur und Morphologie des eem-interglazialen Menschen
von Ehringsdorf bei Weimar. In von Koenigswald GHR (ed): "Hundert Jahre Neanderthaler." Köln: Bohlau, pp 141–150.

Behm-Blancke G (1960): Altsteinzeitliche Rastplätze in Travertingebiet von Taubach, Weimar, Ehringsdorf. Alt Thüringen 4:1–246.

Boaz NT, Ninkovich D, Rossignol-Strick M (1982): Paleoclimatic setting for *Homo sapiens
neanderthalensis*. Naturwissenschaften 69:29–33.

Bosinski G (1967): "Die Mittelpaläolithischen Funde im westlichen Mitteleuropa." Köln-
Graz: Fundamenta.

Boule M (1911/13): L'homme fossile de la Chapelle-aux-Saints. Am Paléontol 6:111–172;
7:21–56, 85–192; 8:1–70.

Boule M (1923): "Fossil Men." Edinburgh: Oliver and Boyd.

Boule M, Anthony R (1911): L'encéphale de l'homme fossile de la Chapelle-aux-Saints.
L'Anthropologie 22:129–196.

Boule M, Vallois HV (1957): "Fossil Men." New York: Dryden.

Brace CL (1962): Refocusing on the Neanderthal problem. Am Anthropol 64:729–741.

Brace CL (1964): The fate of the "classic" Neanderthals: A consideration of hominid
catastrophism. Curr Anthropol 5:3–43.

Brace CL (1979a): Krapina "classic" Neanderthals and the evolution of the European face. J
Hum Evol 8:527–550.

Brace CL (1979b): "The Stages of Human Evolution, 2nd ed." Englewood Cliffs: Prentice-Hall.

Brace CL, Montagu A (1977): "Human Evolution, 2nd ed." New York: Macmillan.

Bräuer G (1980): Die morphologischen Affinitäten des jungpleistozänen Stirnbeines aus dem Elbmündungsgebiet bei Hahnöfersand. Z Morphol Anthropol 71:1–42.

Bräuer G (1981): New evidence on the transitional period between Neanderthal and modern man. J Hum Evol 10:467–474.

Brose DS, Wolpoff MH (1971): Early Upper Paleolithic man and late Middle Paleolithic tools. Am Anthropol 73:1156–1194.

Butzer K (1971): "Environment and Archaeology, Revised ed." Chicago: Aldine.

Cherdyntsev VV, Semina N, Kuzmina EA (1975): Die Altersbestimmung der Travertin von Weimar-Ehringsdorf. (Über das Alter des Riss-Würm-Interglazials). Abhandl Zent Geol Inst (Berlin) 23:7–14.

Chmielewski W (1972): The continuity and discontinuity of the evolution of archaeological cultures in central and eastern Europe between the 55th and 25th millenaries B.C. In Bordes F (ed): "The Origin of Homon sapiens." Paris: UNESCO, pp 173–179.

Cook J, Stringer CB, Currant AP, Schwarcz HP, Wintle AG (1982): A review of the chronology of the European Middle Pleistocene hominid record. Yrbk Phys Anthropol 25:19–65.

Coon CS (1962): "The Origin of Races." New York: Knopf.

Corruccini R (1975): Metrical analysis of Fontéchevade II. Am J Phys Anthropol 42:95–98.

Cunningham DJ (1908): The evolution of the eyebrow region of the forehead, with special reference to the excessive supraorbital development in the Neanderthal race. Trans R Soc Edinburgh 46:283–311.

Czarnetzki A (1980): Pathological changes in the morphology of the young Paleolithic skeletal remains from Stetten (South-West Germany). J Hum Evol 9:15–17.

Delporte H, Djindjian F (1979): Note à propos de l'outillage aurignacien de la couche 11 de Bacho Kiro. In Kozłowski J (ed): "Middle and Upper Paleolithic in Balkans." Prace Archeol 28:101–103.

Dennell R (1983): A new chronology for the Mousterian. Nature 301:199–200.

de Lumley MA (1973): Anténéandertaliens et Néandertaliens du bassin méditerraneén occidental europeén. Etudes Quaternaires (Université de Provence), Mémoire 2.

de Lumley MA (1976): Les Néandertaliens dans le Midi méditerraneén. In de Lumley H (ed): "La Préhistoire Française, Vol 1." Paris: C.N.R.S.

Endo B (1966): Experimental studies on the mechanical significance of the form of the human facial skeleton. J Fac Sci Univ Tokyo Sec V (Anthropol) 3:1–106.

Enlow DH (1975): "Handbook of Facial Growth." Philadelphia: William Saunders.

Farrand WR (1979): Chronology and paleoenvironment of Levantine prehistoric sites as seen from sediment studies. J Archaeol Sci 6:369–392.

Felgenhauer VF (1950): Miesslingstal bei Spitz a. d. Donau, N-Ö, ein Fundplatz des oberen Paläolithikums. Archeol Austriaca 5:35–62.

Feustel R (1976): "Abstammungsgeschichte des Menschen." Jena: Fischer.

Feustel R (1983): Zur Zeitlichkeit und kulturellen Stellung des Paläolithikums von Weimar-Ehringsdorf. Alt Thüringen 19:16–42.

Flint RF (1971): "Glacial and Quaternary Geology." New York: Wiley.

Frayer DW (1978): Evolution of the dentition in Upper Paleolithic and Mesolithic Europe. Univ Kansas Publ Anthropol 10:1–201.

Fridrich J (1973): "Počátky mladopaleolitichého osídlení. Čech. Archeologické Rozhledy 25:392–442.

Fridrich J, Sklenář K (1976): "Die paläolithische und mesolithische Höhlenbesiedlung des Böhmischen Karstes." Prague: National Museum.

Fuhlrott C (1859): Menschliche Überreste aus einer Felsengrotte des Düsselthals. Verh Naturh 16:131-153.

Fuhlrott C, Schaaffhause H (1857): Correspondenzblatt des naturhistorischen Vereins der preussichen Rheinlande und Westphalens. Verh Naturh 14:50-52.

Gaál I (1931): A neandervölgyi ösember elsö erdély csontmaradványa. Pótfüzeteka Termószettudományi Közlönyhösz 63 [Suppl 181]:23-31, 61-71.

Gábori M (1976): "Les Civilisations du Paléolithique Moyen Entre les Alpes et l'Oural." Budapest: Akadémiai Kiadó.

Gábori-Csánk V (1970): C-14 dates of the Hungarian Paleolithic. Acta Archaeol Hung 22:1-11.

Gábori-Csánk V (1983): More on hominid evolution in South-Central Europe. Curr Anthropol, 27:235.

Gieseler W (1936): "Abstammungs-und Rassenkunde des Menschen I." Oehringen: Hohenlohesche Buchhandlung Ferdinand Rau.

Gieseler W (1937): Bericht über die jungpaläolithischen Skelettreste von Stetten ob Lontal bei Ulm. Verh Dtsch. Ges Phys Anthropol 8:41-48.

Gieseler W (1941): Die urgeschichtlichen Menschenfunde aus dem Lontal und ihre Bedeutung fur die deutsche Urgeschichte. Jahrbuch Akad Wissenschaften Tübingen 1:102-126.

Gieseler W (1971): Germany. In Oakley KP, Campbell BG, Molleson TI (eds): "Catalogue of Fossil Hominids. Part II: Europe." London: British Museum (Natural History), pp 189-215.

Glanville EV (1969): Nasal shape, prognathism, and adaptation in man. Am J Phys Anthropol 30:28-39.

Gorjanović-Kramberger K (1906): "Der diluviale Mensch von Krapina in Kroatien: Ein Beitrag zur Paläoanthropologie." Wiesbaden: Kriedel.

Gorjanović-Kramberger D (1913): Život i kultura diluvijalnoga čovjeka iz Krapine u Hrvatskoj. (German summary.) Djela Juglsl Akad Znan Umjet 23:1-54.

Götze A (1892): Die paläolithische Fundstelle von Taubach bei Weimar. Z Ethnol 24:366-377.

Grootes PM (1978): Carbon-14 time scale extended: Comparison of chronologies. Science 201:11-15.

Guenther EW (1959): Zur Altersdatierung der diluvialen Fundstelle von Krapina in Kroatien. Ber. 6. Tag. Dtsch Ges Anthropol, pp 202-209.

Guenther EW (1964): Zur Altersdatierung der "Homo"-Fundschicht von Ehringsdorf bei Weimar. Z Morphol Anthropol 56:23-32.

Hahn J (1973): Das Aurignacien in Mittel- und Osteuropa. Acta Praehist Archaeol 3:77-107.

Hahn J (1977): Aurignacien: Das altere Jungpaläolithikum in Mittel und Osteuropa. Fundamenta 9, series A, pp 1-355.

Hammond M (1979): A framework of plausibility for an anthropological forgery: The Piltdown case. Anthropology 3:47-58.

Hammond M (1980): The explusion of the Neanderthals from human ancestry: Marcellin Boule and the social context of scientific research. Soc Stud Sci 12:1-36.

Heim J-L (1974): Les hommes fossiles de La Ferrasie (Dordogne) et le problème de la définition des Néandertaliens classiques. L'Anthropologie 78:81-112, 321-377.

Heim J-L (1976a): Les hommes fossiles de La Ferrassie I. Arch Inst Paleontol Hum Mem 35:1-326.

Heim J-L (1976b): Les Néandertaliens en Périgord. In de Lumley H (ed): "La Préhistoire Francaise, Vol 1." Paris: C.N.R.S., pp 578-583.

Henke W, Protsch R (1978): Die Paderborner Calvaria: Ein diluvialer Homo sapiens. Anthropol Anz 36:85–108.

Hillebrand J (1914): Ergebnisse meiner Höhlenforschungen im Jahre 1913. Barlangkutatas 2:147–153.

Howell FC (1978): Hominidae. In VJ Maglio and HBS Cooke (eds.) Evolution of African Mammals. Cambridge: Harvard University Press, pp 154–248.

Howells WW (1976): Explaining modern man: Evolutionists versus migrationists. J Hum Evol 5:577–596.

Hrdlička A (1914): The most ancient skeletal remains of man. Annu Rep Smithsonian Inst 1913:491–552.

Hrdlička A (1927): The Neanderthal phase of man. J Anthropol Inst 57:249–274.

Hrdlička A (1930) The skeletal remains of early man. Smithsonian Misc Coll 83:1–379.

Hublin JJ (1982): Les Anténéandertaliens: Presapiens ou Prénéandertaliens? Geobios, Mem. Spéc. 6:345–357.

Hublin JJ (1983a): Les origines de l'homme de type moderne en Europe. Pour la Science 64:62–71.

Hublin JJ (1983b): "The Fossil Man From Salzgitter-Lebenstedt (FRG) and its Place in Human Evolution in Europe During the Pleistocene." Paper presented at the 18th Meeting of the Gesellschaft für Anthropologie und Humangenetik. Münster, West Germany.

Hylander WL (1977): The adaptive significance of Eskimo craniofacial morphology. In Dahlberg A, Graber T (eds): "Orofacial Growth and Development." The Hague: Mouton.

Ivanova S (1979): Cultural differentiation in the Middle Paleolithic on the Balkan peninsula. In Kozłowski J (ed): "Middle and Upper Paleolithic in Balkans." Prace Archeologiczne 28:13–33.

Jelinek A (1981): The Middle Paleolithic in the southern Levant from the perspective of the Tabun cave. In: "Préhistoire du Levant." Paris: CNRS, pp 265–280.

Jelínek A (1982): The Tabun Cave and Paleolithic man in the Levant. Science 216:1369–1375.

Jelínek J (1954): Nález fosilního člověka Dolní Věstonice III (English summary). Anthropozoikum 3:37–91.

Jelínek J (1957): La nouvelle datation de la découverte de l'homme fossile Brno II. Anthropol 61:513–515.

Jelínek J (1965): Srovnávací studium šipecke čelisti (German summary). Anthropos 17:135–179.

Jelínek J (1966): Jaw of an intermediate type of Neanderthal man from Czechoslovakia. Nature 212:701–702.

Jelínek J (1967): Der Fund eines neandertales Kiefers (Kůlna I) aus der Kůlna-Höhle in Mähren. Anthropologie (Brno) 5:3–19.

Jelínek J (1969): Neanderthal man and Homo sapiens in central and eastern Europe. Curr Anthropol 10:475–503.

Jelínek J (1976): The Homo sapiens neanderthalensis and Homo sapiens sapiens relationship in central Europe. Anthropologie (Brno) 14:79–81.

Jelínek J (1978): "Earliest Homo sapiens sapiens from Central Europe (Mladeč, Czechoslovakia)." Paper presented at the Xth Int Cong Anthropol Ethnol Sci, Delhi, India.

Jelínek J (1980): Neanderthals remains in Kůlna Cave, Czechoslovakia. In Schwidetzky I, Chiarelli B, Necrasov O (eds): "Physical Anthropology of European populations." The Hague: Mouton.

Jelínek, J (1981): Neanderthal parietal bone from Kůlna Cave, Czechoslovakia. Anthropologie (Brno) 19:195–196.

Jelínek J, Pelíšek J, Valoch K (1959): Der fossile Mensch Brno II. Anthropos 9:5–30.

Kallay J (1963): Radiographic study of the Neanderthal teeth from Krapina, Croatia. In Brothwell D (ed): "Dental Anthropology." Oxford: Pergamon Press, pp 75–86.

Keith A (1911): "Ancient Types of Man." London: Harper.

Kleinschmidt A (1965): Wichtigste Untersuchungsergebnisse der paläolithischen Grabung Salzgitter-Lebenstedt (Abstract). Eiszeitalter Gegenw 16:257.

Klíma B (1955): Přinos nové stanice v Pavlově k probematice nejstarších zěmědelských nástrojů. (German summary.) Památky Archeol 46:7–29.

Klíma B (1959): Objev paleolitického sidliště u Pavlova vo roce 1956. Archeol Rozhledy 11:305–316, 337–344.

Klíma B (1962): The first ground-plan of an Upper Paleolithic loess settlement in middle Europe and its meanings. In Braidwood RJ, Willey GR (eds): "Courses Toward Urban Life." Chicago: Aldine.

Klíma B (1963): "Dolní Věstonice: Výzkum Tábořiště Lovců Mamutů v Letech 1947–1952 (German summary)." Prague: Československá Akademie Věd.

Klíma B, Musil R, Pelísek J, Jelínek J (1962): Die Erforschung der Höhle Svéduv stůl 1953–1955. Anthropos 13, 1–297.

Knussmann R (1965): Die mittelpaläolithische menschliche Schädelfragmente von der Wildscheuer bei Steeden an der Lahn (Limburg). Ber 8 Tag Dtsch Ges Anthropol, pp 230–237.

Knussmann R (1967): Die mittelpaläolithischen menschlichen Knochenfragmente von der Wildscheuer bei Steeden (Oberlahnkreis). Nassau Ann 68:1–25.

Kozłowski J (1979): Le Bachokirien: La plus ancienne industrie du Paléolithique supérieur en Europa. In Kozłowski J (ed): "Middle and Upper Paleolithic in Balkans." Prace Archeologiczne 28:77–99.

Kříž M (1909): Die Schwedentischgrotte bei Ochoz in Mähren und Rzehaks Bericht über Homo primigenius wilseri. Verh Reichsanst 10:217–232.

Kukla J (1954): Složení pleistocénních sedimentů v kontrolním profilu v Šipce z roku 1950. Přirodovedecký Sbornik Ostraveského Kraje 15:105–124.

Kukla J (1975): Loess stratigraphy of Central Europe. In Butzer KW, Isaac GL (eds): "After the Australopithecines." The Hague: Mouton, pp 99–188.

Kukla J (1978): The classical European glacial stages: Correlation with deep-sea sediments. Trans Nebraska Acad Sci 6:57–93.

Kurtén B (1968): "Pleistocene Mammals of Europe." Chicago: Aldine.

Kutsch F (1954): Die Steedener Höhlen. 1. Übersicht über die Grabungen. 3. Die Funde der Nachuntersuchung. Nassau Ann 65:27–34, 42–45.

Lévêque F, Vandermeersch B (1980): Découverte de restes humains dans un niveau castelperronien à Saint-Césaire (CharenteMaritime). CR Acad Sci [D] (Paris), 281:187–189.

Lévêque F, Vandermeersch B (1981): Le néandertalien de Saint-Césaire. Recherche 12:242–244.

Madera H-E (1954): Die Steedener Höhlen. 2. Bericht über die Nachuntersuchung der Höhle "Wildscheuer" und ihres Vorplatzes 1953. Nassau Ann 65:35–42.

Makowsky A (1892): Der diluviale Mensch in Löss von Brünn. Mitt Anthropol Ges Wien 22:73–84.

Malan M (1954): Zahnkeim aus der zweiten Aurignacien Schicht der Höhle von Istállóskö. Acta Archeol Hung 5:145–148.

Malez M (1967): Paleolit Velike Pećine na Ravnoj gori u sjeverozapadnoj Hrvatskoj (German summary). Arheol Radovi Rasprave 4/5:7–68.

Malez M (1970a): Novi pogledi na stratigrafiju krapinskog nalazišta (German summary). In Malez M (ed): "Krapina 1899–1969." Zagreb: Jugoslavenska Akademija Znanosti i Umjetnosti, pp 13–44.

Malez M (1970b): Rezultati revizije pleistocenske faune iz Krapine. In Malez M (ed): "Krapina 1899–1969 (English and German summaries)." Zagreb: Jugoslavenska Akademija Znanosti i Umjetnosti, pp 45–56.

Malez M (1970c): Paleolitska kultura Krapine u svjetlu novijih istraživanja. In Malez M (ed): "Krapina 1899–1969 (English and German summaries)." Zagreb: Jugoslavenska Akademija Znanosti i Umjetnosti, pp 57–129.

Malez M (1974): Noviji rezultati istraživanja paleolitika u Velikoj Pećini, Veternici, i Šandalji (German summary). Arheol Radovi Rasprave 7:7–44.

Malez M (1975): Die Höhle Vindija: Eine neue Fundstelle fossiler Hominiden in Kroatien. Bull Sci (Yougoslavie) Section A 20:5–6.

Malez M (1978a): Stratigrafski, paleofaunski, i paleolitski odnosi krapinokog nalazišta. In Malez M (ed): "Krapinski Pračovjek i Evolucija Hominida (German summary). " Zagreb: Jugoslavenska Akademija Znanosti i Umjetnosti, pp 61–102.

Malez M (1978b): Populacije neandertalaca i neandertalcimi sličnih ljudi u Hrvatskoj. In Malez M (ed): "Krapinski Pračovjek i Evolucija Hominida (German summary)." Zagreb: Jugoslavenska Akademijja Znanosti i Umjetnosti, pp 331–371.

Malez M (1978c): Fossile Menschen aus Nordwestkoratien und ihre quartärgeologische, paläontologische und paläolithische Grundlage. Coll Anthropol 2:29–41.

Malez M, Rukavina D (1975): Krioturbacijske pojave u gornjopleistocenskim naslagama pećine Vindije kod donje Voće u sjeverozapadnoj Hrvatskoj (German summary). Rad Jugoslav. Akad Znanosti Umjetnosti 371:245–265.

Malez M, Rukavina D (1970): Položaj naslaga spilje Vindije u sustavu članjenja Kvartara šireg područja Alpa. Rad Jugoslav Akad Znanosti Umjetnosti 383:187–218.

Malez M, Smith F, Rukavina D, Radovčić J (1980): Upper Pleistocene fossil hominids from Vindija, Croatia, Yugoslavia. Curr Anthropol 21:365–367.

Malez M, Ullrich H (1982): Neuete Paläannthropologische untersüchungen am Material aus der Höhle Vindija (Kroatien, Jugoslawien). Palaeont Jugoslav 29:1–44.

Mania D, Preuss J (1975): "Zur Methoden and Problemen Ökologischer Untersuchungen in der Ur-und Frühgeschichte." Berlin: Symbolae Prehistoricae.

Martin Rudolf (1928): "Lehrbuch der Anthropologie, Vol 2. Kraniologie, Osteologie." Jena: Fischer.

Martin R, Saller K (1956): "Lehrbuch der Anthropologie." Stuttgart: Fischer.

Maška, K (1882): Über den diluvialen Menschen in Stramberg. Mitt Anthropol Ges Wien 12:32–38.

Matiegka J (1924): Lebka podbabská. Anthropologie (Prague) 2(1):1–16.

Matiegka J (1934): "Homo Předmostensis: Fosiliní Člověk z Předmostí na Moravě. 1. Lebky (French summary)." Prague: Česká Akademie Věd i Umění.

Matiegka J (1938): "Homo Předmostensis: Fosilní Člověk z Předmostí na Moravě. 2. Ostatní Cásti Kostrové (French summary)." Prague: Česká Akademie Věd i Umění.

Minugh NS (1983): The Mladeč 3 child: Aspects of cranial ontogeny in early anatomically modern Europeans (Abstract). Am J Phys Anthropol 60:228.

Morant GM (1930): Studies of Paleolithic man. 4. A biometric study of the Upper Paleolithic skulls of Europe and of their relationship to earlier and later types. Ann Eu 4:109–199.

Musgrave J (1970): How dextrous was Neandertal man? Nature 223:538–541.

Musgrave J (1977): The Neandertals from Krapina, northern Yugoslavia: An inventory of the handbones. Z Morphol Anthropol 68:150–171.

Musil R (1965): Zhodnoceni dřívějšick paleontologicých nálezů ze Šipky (German summary). Anthropos 17:127–134.

Musil R, Valoch K (1966): Beitrag zur Gliederung des Würms in Mitteleuropa. Eiszeitalter Gegenw 17:131–138.

Necrasov O, Cristescu M (1965): Données anthropologiques sur les populations de l'age de la pierre en Roumanie. Homo 16:129–161.

Nehring A (1895): Über einen menschlichen Molar aus dem Diluvium von Taubach bei Weimar. Z Ethnol 27:573–577.

Nicolăescu-Plopşor D (1968) Les hommes fossiles désouvertes in Roumanie. VIIe Cong Int Sci Anthropol Ethnol Moscou 3:381–386.

Obermaier H (1905): Les restes humains quarternaires dans l'Europe centrale. L'Anthropologie 16:385–410.

Obermaier H (1907): Quaternary human remains in Central Europe. Annu Rep Smithsonian Inst 1906:373–397.

Olivier G (1969): "Practical Anthropology." Springfield: Charles C Thomas.

Piveteau J (1957): "Traité de Paléontologie, Vol VII." Paris: Masson et Cie.

Preul F (1953): Die geologische Bearbeitung des paläolithischen Fundplatzes bei Salzgitter-Lebenstedt. Eiszeitalter Gegenw 3:149–154.

Prošek F, Stárka V, Hrdlička L, Hokr Z, Ložek V, Dohnal Z (1952): Výzkum jeskyně Zlatého Koně u Koněprus. Československý Kras 5:161–179.

Prótsch R (1975): The absolute dating of Upper Pleistocene sub-Saharan fossil hominids and their place in human evolution. J Hum Evol 4:297–322.

Protsch R, Semmel A (1978): Zur Chronologie des Kelsterbach-Hominiden. Eiszeitalter Gegenw 28:200–210.

Riek G (1932): Palaolithische Station mit Tierplastiken und menschlichen Skelettresten bei Stetten ob Lontal. Germania 16:1–8.

Rightmire GP (1978a): Florisbad and human population succession in southern Africa. Am J Phys Anthropol 48:475–486.

Rightmire GP (1978b): Human skeletal remains from the southern Cape Province and their bearing on the Stone Age prehistory of South Africa. Q Res 9:219–230.

Rightmire GP (1979): Implications of the Border Cave skeletal remains for later Pleistocene human evolution. Curr Anthropol 20:23–35.

Rightmire GP (1981): Later Pleistocene hominids of eastern and southern Africa. Anthropologie (Brno):19:15–26.

Rosholt J, Antal AP (1963): Evaluation of the Pa^{231}/Th^{230} method for dating of Pleistocene carbonate rocks. Pro Papers Geol Surv 450E:108–111.

Russell MD (1983): Browridge development as a function of bending stress in the supraorbital region (Abstract). Am J Phys Anthropol 60:248.

Rzehak A (1905): Der Unterkiefer von Ochoz: Ein Beitrag zur Kenntnis des altdiluvialen Menschen. Verh Naturf Ver Brünn 44:91–114.

Schaaffhausen H (1858): Zur Kenntnis der ältesten Rassenschädel. Arch Anat Physiol 25:453–478.

Schaefer U (1957): Homo neanderthalensis (King) I. Z Morphol Anthropol 48:269–297.

Schwalbe G (1901): Der Neanderthalschädel. Bonner Jahrbuch 106:1–72.

Schwalbe G (1906): "Studien zur Vorgeschichte des Menschen." Stuttgart: E. Scheizerbart.

Schwalbe G (1914): Kritische Besprechung von Boule's Werk: "L'Homme Fossile de la Chapelle-aux-Saints" mit einigen Untensuchungen. Z Morphol Anthropol 16:527–610.

Schwarcz H, Skoflek I (1982): New dates for the Tata, Hungary, archaeological site. Nature 295:590–591.

Sergi S (1962): Morphological position of the "Prophaneranthropi" (Swanscombe and Fon-téchevade). In Howells WW (ed): "Ideas on Human Evolution." Cambridge: Harvard University Press, pp 507–520.

Singer R, Wymer J (1982): "The Middle Stone Age at Klasies River Mouth in South Africa." Chicago: University of Chicago Press.

Škerlj B (1958): Were Neanderthalers the only inhabitants of Krapina? Bull Sci (Yougoslavie) 4:44.

Smith FH (1976a): A fossil hominid frontal from Velika Pećina (Croatia) and a consideration of Upper Pleistocene hominids from Yugoslavia. Am J Phys Anthropol 44:127–134.

Smith FH (1976b): The Neandertal remains from Krapina: A descriptive and comparative study. Univ Tennessee Dep Anthropol Rep Invest 15:1–359.

Smith FH (1976c): On anterior tooth wear at Krapina and Ochoz. Curr Anthropol 17:167–168.

Smith FH (1977): On the application of morphological "dating" to the hominid fossil record." J Anthropol Res 33:302–316.

Smith FH (1978): Some conclusions regarding the morphology and significance of the Krapina neandertal remains. In Malez M (ed): "Krapinski Pračovjek i Evolucija Hominida." Zagreb: Jugoslavenska Akademija Znanosti i Umjetnosti, pp 103–118.

Smith FH (1980): Sexual differences in European Neandertal crania with special reference to the Krapina remains. J Hum Evol 9:359–375.

Smith FH (1982): Upper Pleistocene hominid evolution in South-Central Europe: A review of the evidence and analysis of trends. Curr Anthropol 23:667–703.

Smith FH (1983): Behavioral interpretations of changes in craniofacial morphology across thhe archaic/modern Homo sapiens transition. In Trinkaus E (ed): "The Mousterian Legacy: Human Biocultural Change in the Upper Pleistocene." Br Archaeol Rep Int Ser 164:141–163.

Smith FH, Ranyard GC (1980): Evolution of the supraorbital region in Upper Pleistocene fossil hominids from South-Central Europe. Am J Phys Anthropol 53:589–609.

Soergel W (1927): Exkursion ins Travertingebiet von Ehringsdorf. Paläontol Z 8:7–33.

Sollas W (1915): "Ancient Hunters and their Modern Representatives." New York: Macmillan.

Spencer F (1979): "Aleš Hrdlička, M.D., 1869–1943: A Chronicle of the Life and Work of an American Physical Anthropologist." Ph.D. dissertation. University of Michigan, Ann Arbor.

Spencer F, Smith FH (1981): The significance of Aleš Hrdlička's "Neanderthal phase of man": A historical and current assessment. Am J Phys Anthropol 56:435–459.

Steiner W (1979): "Der Travertin von Ehringsdorf und seine Fossilen." Wittenberg: Ziemsen.

Stewart TD (1962): Neanderthal scapulae with special attention to the Shanidar Neanderthals from Iraq. Anthropos 57:779–800.

Stewart TD (1964): The scapula of the first recognized Neanderthal skeleton. Bonner Jahrbuch 164:1–14.

Stringer CB (1974): Population relationships of later Pleistocene hominids: A multivariate study of available crania. J Archaeol Sci 1:317–342.

Stringer CB (1978): Some problems in Middle and Upper Pleistocene hominid relationships. In Chivers D, Joysey K (eds): "Recent Advances in Primatology, Vol 3." London: Academic Press, pp 395–418.

Stringer CB (1982a): Towards a solution to the Neanderthal problem. J Hum Evol 11:431–438.

Stringer CB (1982b): Comment on Upper Pleistocene evolution in South-Central Europe: A review of the evidence and analysis of trends. Curr Anthropol 23:690–691.

Stringer CB, Burleigh R (1981): The Neanderthal problem and the prospects for direct dating of Neanderthal remains. Bull Br Mus Nat Hist (Geol) 35:225–241.

Svoboda J (1980): Křemencová industrie z Ondratic (English summary). Stud Archheol Ústavu Československé Akad Věd Brně 9(1):3–109.

Szombathy J (1925): Die diluvialen Menschenreste aus der Fürst-Johanns-Höhle bei Lautsch in Mähren. Eiszeit 2:1–34, 73–95.

Szombathy J (1950): Der menschliche Unterkiefer aus dem Miesslingstal bei Spitz, N-Ö. Archaeol Austriaca 5:1–5.

Thoma A (1963): The dentition of the Subalyuk Neandertal child. Z Morphol Anthropol 54:127–150.

Thorne A, Wolpoff M (1981): Regional continuity in australasian pleistocene hominid evolution. Am J Phys Anthropol 55:337–349.

Tode E (1953): Einige archäologische Erkenntnisse aus der paläolithischen Freiland Station von Salzgitter-Lebenstedt. Eiszeitalter Gegerw 3:192–215.

Trinkaus E (1975): "A Functional Analysis of the Neandertal Foot." Ph.D. dissertation. University of Pennsylvania, Philadelphia.

Trinkaus E (1976a): The morphology of European and Southwest Asian Neandertal pubic bones. Am J Phys Anthropol 44:95–104.

Trinkaus E (1976b): The evolution of the hominid femoral diaphysis during the Upper Pleistocene in Europe and the Near East. Z Morphol Anthropol 67:291–319.

Trinkaus E (1977): A functional interpretation of the axillary border of the Neandertal scapula. J Hum Evol 6:231–234.

Trinkaus E (1978): Functional implications of the Krapina neandertal lower limb remains. In Malez M (ed): "Krapinski Pračovjek i Evolucija Hominida." Zagreb: Jugoslavenska Academija Znanosti i Umjetnosti, pp 155–192.

Trinkaus E (1980): Sexual differences in Neanderthal limb bones. J Hum Evol 9:377–397.

Trinkaus E (1981): Neanderthal limb proportions and cold adaptation. In Stringer CB (ed): "Aspects of Human Evolution." London: Taylor and Francis, pp187–224.

Trinkaus E (1982): Evolutionary trends in the Shanidar Neandertal sample (Abstract). Am J Phys Anthrpol 57:237.

Trinkaus E (1983a): Neandertal postcrania and the adaptive shift to modern humans. In Trinkaus E (ed): "The Mousterian Legacy: Human Biocultural Change in the Upper Pleistocene." Br Archeol Rep Int Ser 164:165–200.

Trinkaus E (1983b): "The Shanidar Neandertals." New York: Academic Press.

Trinkaus E, Howells WW (1979): The Neanderthals. Sci Am 241:118,122–133.

Trinkaus E, LeMay M (1982): Occipital bunning among later Pleistocene hominids. Am J Phys Anthropol 57:27–35.

Vallois HV (1954): Neanderthals and presapiens. J Anthropol Inst 84:111–130.

Vallois HV (1958): La Grotte de Fontéchevade II. Anthropologie. Arch Inst Paléontol, Mém 29.

Valoch K (1965): Die Höhlen Šipka und Čertova dira bei Štramberk in Mähren. Anthropos (Brno) 17:5–125.

Valoch K (1967): Die Steinindustrie von der Fundstelle des menschlichen Skelettrestes I aus der Höhle Kulna bein Sloup (Mähren). Anthropologie (Brno) 5:21–32.

Valoch K (1968): Evolution of the Paleolithic in Central and Eastern Europe. Curr Anthropol 9:351–390.

Valoch K (1970): Early Middle Paleolithic (Stratum 14) in the Kůlna cave near Sloup in the Moravian karst (Czechoslovakia). World Archaeol 2:28–38.

Valoch K (1972): Rapports entre le Paléolithique moyen et al Paléolithique supérieur en Europe centrale. In Bordes F (ed): "The Origin of Homo Sapiens." Paris: UNESCO, pp 161-171.

Valoch K (1976): Die alsteinzeitliche Fundstelle in Brno-Bohunice. Stud Archeol Ústavu Československé Akad Věd Brně 4(1):3-120.

Valoch K (1977-78): Nové poznatky o paleolitu v Československu. Sbornik Praci Filoz Fak Brnenske Univ E 22-23:7-25.

Valoch K (1978): Die palaölithische Fundstelle Bořitov I (Bez. Blansko) in Mähren. Časopis Moraviského Mus 63:7-24.

Valoch K (1981): Einige Mittelpalaolithische Industrien aus der Kůlna Höhle im Mährischen Karst. Časopis Moravského Mus 64:47-67.

Valoch K (1982): Comment on Upper Pleistocene hominid evolution in South-Central Europe: A review of the evidence and analysis of trends. Curr Anthropol 23:602.

Valoch K, Pelíšek J, Musil R, Kovanda J, Opravil E (1970): Die Erforschung der Kůlna-Höhle bei Sloup im Mährischen Karst (Tschechoslowakei). Quartär 20:1-45.

Vandermeersch B (1976): Les Néandertaliens en Charente. In de Lumley H (ed): "La Préhistoire Francaise, Vol 1." Paris: C.N.R.S., pp 584-587.

Vandermeersch B (1981): "Les hommes fossiles de Qafzeh (Israël)." Paris: C.N.R.S.

Vandermeersch B, Tillier A-M, Krukoff S (1976): Position Chronologiques des restes de Fontechévade. In Le Peuplement Anténéandertalien de l'Europe. Colloque 9, Congres UISPP (Nice). pp 19-26.

Vanůra J (1965): Přjspěvek k poznání jeskyně Svéduv stůl v Moravském krasu. Československý Kras 15:59-63.

Verneau R (1924): La race de Néanderthal et la race de Grimaldi: Leur rôle dans l'humanité.

Virchow H (1920): "Die menschlichen Skeletrreste aus dem Kämpfeschen Bruch in Travertine von Ehringsdorf bei Weimar." Jena: Fischer.

Virchow R (1872): Untersuchung des Neanderthal-Schadëls. Z Ethnol 4:157-165.

Vlček E (1951): Pleistocenní člověk z jeskyně Sv. Prokopa (English summary). Anthropozoikum 1:213-226.

Vlček E (1953): Nález neandertálského člověka na Slovensku (English summary). Slovenská Archeol 1:5-132.

Vlček E (1955): The fossil man of Gánovce, Czechoslovakia. J Anthropol Inst 85:163-171.

Vlček E (1956): Kalva pleistocénního člověka z Podbaby (Praha XLX) (English summary). Anthropozoikum 5:191-217.

Vlček E (1957a): Pleistocénní člověk z jeskyně na Zlatém Koní u Koněprus (German summary). Anthropozoikum 6:283-311.

Vlček E (1957b): Lidský zub pleistocénního Staří ze Silické Brezové (German summary). Anthropozoikum 6:397-405.

Vlček E (1958): Die Reste des Neanderthalmenschen aus dem Gebiete der Tschechoslowakei. In von Koenigswand GHR (ed):"Hundert Jahre Neanderthaler." Utrecht: Kemink en Zoon, pp 107-120.

Vlček E (1961): Posůstatky mladopleistocénního člověk z Pavlova (German summary). Památky Archeol 52:46-56.

Vlček E (1964a): Neuer Fund eines Neandertalers in der Tschechoslowakei. Anthropol Anz 27:162-166.

Vlček E (1964b): Einige in der Ontogenese des modernen Menschen untersuchte Neandertalmerkmale. Z Morphol Anthropol 56:63-83.

Vlček E (1967): Die Sinus frontales bei europäischen Neandertalern. Anthropol Anz 30:166–189.

Vlček E (1968): Nález posůstatků neandertálce v Šali na Slovensku (English summary). Anthropozoikum 17:105–144.

Vlček E (1969): "Neandertaler der Tschechoslowakei." Prague: Academia.

Vlček E (1970): Étude comparative ontophylogénétique de l'enfant du Pech de L'Aze par rapport a d'austres enfants néandertaliens. In "L'enfant du Pech l'Aze." Arch Inst Paléontol Hum, Mém 33, pp 149–178.

Vlček E (1971): Czechoslovakia. In Oakley K, Campbell B, Molleson T (eds): "Catalogue of Fossil Hominids, Pt 2." London: British Museum (Natural History). pp 47–64.

Vlček E (1978): A new discovery of Homo erectus in Central Europe. J Hum Evol 7:239–259.

Vlček E (in press): Die altsteinzeitlichen Menschenreste von Weimar-Ehringsdorf Weimarer Monogr Ur-u, Frühgesch 6.

Vlček E, Prošek F, Wolfe J, Pelikan J, Knéblová, Fejfar O, Ložek V (1958): "Zusammenfassender Bericht über den Fundort Gánovce und die Reste des Neanderthalers in der Zips (ČSSR). Prague.

Vogel J, Waterbolk H (1972): Groningen radiocarbon dates X. Radiocarbon 14:6–10.

Vogel J, Zagwijn W (1967): Groningen radiocarbon dates VI. Radiocarbon 9:63–106.

Weidenreich F (ed) (1928): "Der Schädelfund von Weimar-Ehringsdorf." Jena: Fischer.

Weidenriech F (1936): The Mandibles of Sinanthropus Pekinensis: A Comparative Study. Paleontol Sinica Ser D 7:1–132.

Weiss KM, Maruyama T (1976): Archaeology, population genetics, and studies of human racial ancestry. Am J Phys Anthropol 44:31–49.

Woillard GM (1978): Grand Pile peat bog: A continuous pollen record for the last 140,000 years. Q Res 9:1–21.

Woillard GM, Mook WG (1982): Carbon-14 dates at Grand Pile: Corrleation of land and sea chronologies. Science 215:159–161.

Wolpoff MH (1975): Some aspects of human mandibular evolution. In McNamara JA, Jr (ed): "Determinants of Mandibular Form and Growth." Ann Arbor: Center for Human Growth and Development, pp 1–64.

Wolpoff MH (1978): The dental remains from Krapina. In Malez M (ed): "Krapinski Pračovjek i Evolucija Hominida." Zagreb: Jugoslavenska Akademija Znanosti i Umjetnosti, pp 119–152.

Wolpoff (1979): The Krapina dental remains. Am J Phys Anthropol 50:67–114.

Wolpoff (1980a): "Paleoanthropology." New York: Knopf.

Wolpoff (1980b): Cranial remains of Middle Pleistocene European hominids. J Hum Evol 9:339–358.

Wolpoff M, Smith F, Malez M, Radovčić J, Rukavina D (1981): Upper Pleistocene hominid remains from Vindija Cave, Croatia, Yugoslavia. Am J Phys Anthropol 54:499–545.

Yakimov VP (1980): New materials of skeletal remains of ancient peoples in the territory of the Soviet Union. In Könnigson L-K (ed): "Current Argument on Early Man." Oxford: Pergamon Press, pp 152–269.

Zagwijn WH (1974): Vegetation, climate, and radiocarbon datings in the late Pleistocene of the Netherlands. Part 2. Middle Weichselian. Mededelingen Rijks Geol Dienst 25:101–110.

Zobeck TS (1980): "An analysis of the Functional Significance and Sexual Dimorphism of the Neanderthal Mastoid Process. M.A. Thesis, University of Tennessee, Knoxville.

The Origins of Modern Humans: A World Survey of the Fossil Evidence, pages 211–250
© 1984 Alan R. Liss, Inc., 150 Fifth Avenue, New York, NY 10011

Biological and Cultural Change in the European Late Pleistocene and Early Holocene

David W. Frayer

Department of Anthropology, University of Kansas, Lawrence, Kansas 66045

This paper examines the nature of biological and cultural evolution in the Upper Paleolithic and Mesolithic human populations of Europe. These Late Pleistocene/Early Holocene groups are well known from their archaeological remains. Over the past 150 years, archaeologists have been involved in intensive analysis and interpretation of this sequence following the Mousterian and preceeding the Neolithic. Fruits of this work have made the archaeological and cultural aspects of Upper Paleolithic and Mesolithic society well-known among the profession and the public. It is common knowledge that the people of these periods developed complex tool and weaponry systems, spectacular cave-wall and mobilary art, and the first calendrical (notational) systems. Yet, the field is pervaded by the assumption that human morphology shows only slight or insignificant change from the Upper Paleolithic to recent populations. Statements such as "the appearance of fully modern *Homo sapiens*" underlie the general (but unfounded) conclusion that evolution stops with "Cro-Magnon" skeletons. There are at least four reasons for this: 1) Traditionally, human paleontologists have been enamoured with the discovery of the oldest human remains. There is a certain status one automatically acquires when "fortunate" enough to discover a Neandertal, *erectus*, or australopithecine. Consequently, there are considerably more paleoanthropologists involved in these pursuits than concerned with the evolution of fossil *Homo sapiens* of the Upper Paleolithic and Mesolithic. Numbers of active workers, in many respects, directly affect our state of knowledge of evolutionary patterns of the period. 2) Although archaeological assemblages and sequences of this period are well known, most of the interpretations have concentrated on establishing typological differences among the units, rather than emphasizing evolutionary developments through time. Following this approach, much of the research involving the human skeletal material has been a priori concerned with differentiating skulls into "racial" types. The typological approach of the archaeological research has directly influenced the little skeletal analysis that has been done, so that most conclusions drawn from the biological data emphasize the static nature of post-Mousterian groups. 3) Related to this, authors of textbooks and other reviews of the human fossil record have generally ignored the morphology of the humans that were responsible for the Upper Pleistocene and Mesolithic archaeological sequences. Partially due to the emphasis in paleoanthropology on "early man" and to the complexities and time depth of the fossil sequence, I suspect that most authors are relieved to announce the appearance of "modern" forms and conclude the work. This has led to the general assumption that nothing differentiates skeletons of 25 ky ago from modern inhabitants of Paris. 4) Upper Paleolithic and Mesolithic skeletons are scattered throughout western, central, and eastern Europe, and compiling a representative dataset demands a great deal of travel, effort, and financial support. Thus, most investigators have restricted metric and morphological analyses to circumscribed areas (France, northern Europe, Portugal, Italy) and, as stated above, concentrated on categorizing skulls in a synchronic perspective, rather than following trends over space and time. Thus, a variety of different factors have contributed to the consensus opinion that little biological change characterizes Upper Paleolithic and Mesolithic groups.

The approach and conclusions of this review deviate substantially from most previous work. Based on my review of much of the relevant material and my own a priori assumptions of the primacy of gradual evolutionary change, I present evidence for the significant modification of skeletal biology over time. After a review of techno-cultural evolution, data for dental, cranial, and postcranial change are presented. Since the size and nature of the datasets differ, the analysis of diachronic trends takes different forms. For the dentition, comparisons are made between earlier Neanderthal groups and the Upper Paleolithic, as well as trends over time within Upper Paleolithic and Mesolithic samples. Because of limitations in my comparative measurements for cranio-facial changes, only trends within the Upper Paleolithic and Mesolithic are reviewed. In the postcranial analysis, trends are reviewed over a longer period of time to provide a picture of body size changes (measured by stature) over the last 30 ky in Europe. Throughout this period reduction in size and in the degree of sexual dimorphism is stressed. Although much more work needs to be done, this review will hopefully show that the European Upper Paleolithic and Mesolithic is a dynamic phase in human evolution.

CULTURAL EVOLUTION IN THE LATE PLEISTOCENE AND EARLY HOLOCENE

Traditional reviews of the Upper Paleolithic and Mesolithic concentrate on the disjointed nature of the succession of industries, based primarily on the French Périgord sequence. The focus of much of the past archaeological interpretation has aimed at a typological assessment and classification of the diverse chronological and geographic industries into discrete units. Consequently, fundamental differences among industries have been stressed more than evolutionary developments in the techno-cultural system. In addition, the search for functional explanations in the classic industrial designations has been minimal, since more emphasis has been placed on separating them than relating them to specific adaptations. For example, for decades French Aurignacian and Upper Perigordian industries have been considered to be the product of different people, with little or no cross influence. Yet, recent analysis suggests that they

> can no longer be viewed as constituting successive chronological stages
> in unilinear sequences of development. Instead, there seems to exist a
> certain degree of temporal overlap, and even interstratification among
> certain industries . . . [Laville et al, 1980, p 286].

Along with regional analyses based on recent excavations [Freeman, 1981; Straus et al, 1980], there is a slowly emerging trend to view the Upper Paleolithic and Mesolithic sequence of Europe as one primarily influenced by changes in internal dynamics, modified by gradual evolutionary advances.

For this paper, intricacies of the sequence as they relate to classification are less important than changes in the adaptive system. Other than using the system to date the human remains, division of the Upper Paleolithic and Mesolithic into its various industrial periods is less critical than describing the overall patterns of change from the earliest to latest units. Detailed review of this sequence has been published elsewhere [Frayer, 1978]. Consequently, this review is primarily concerned with the chain of evolutionary developments in technology and culture that presumably had an effect on skeletal biology. For this synthesis, I have relied heavily on Bordes [1968], Jelínek [1972], Laville et al [1980], Mellars [1973], Odell [1980], and White [1982].

Probably the single most characteristic that is associated with the appearance of post-Mousterian groups is the wide usage of blade production techniques [Bordes, 1968]. Although blade tools are not absent in earlier periods, in the Upper Paleolithic and Mesolithic the prismatic core forms the nearly universal first step in stone tool production. Tools derived from these cores are a significant improvement over the earlier flake tools. Blade tools have extremely thin cross-sections and sharp edges, providing more effective cutting and scraping surfaces [Bordaz, 1970; Semenov, 1964]. In some cases, the nonworking edge was even dulled before usage, presumably so that the implement could be used without damaging the hand. The significant feature of blade tools is that they can be modified into a variety of forms, designed for specific purposes. This specificity of tool function is a cumulative trend within the Upper Paleolithic. For example:

> Before the Solutrean, the evolution of lithic assemblages seems to lead to the more efficient adaptation of different kinds of tools to different kinds of primary operations (slicers become better slicers, crushers better for crushing, and so on) regardless of the nature of the specific resource on which the operation was performed. From the Solutrean onward, we have increasing evidence for the special tailoring of specific tools to particular resources [Freeman, 1981, p 153].

Although this conclusion is based only on archaeological sequences in Cantabrian Spain, the generalization is probably applicable to overall lithic trends in the sequence of Upper Paleolithic and Mesolithic industries. Through time, the variety of tool types increases [Isaac, 1972] and more and more control over the intended form occurs. These innovations include thermal pretreatment that facilitates better-controlled knapping [Bordes, 1969], progressive "microlithization" of tools [Semenov, 1964], and, by the Mesolithic, regional tool types that are specifically designed for local ecosystems and subsistence/exploitive patterns [Rozoy, 1978].

In addition to producing sharper-edged tools, Upper Paleolithic groups also initiated the use of handles [Leroi-Gourhan, 1964], thereby increasing the leverage and power of the tools. Late in the Upper Paleolithic, blades are snapped into smaller units and inlaid in bone or antler, forming complex, composite tools. Magdalenian groups even produced "snap-on" tools and handles that were interchangeable [Bordaz, 1970]. These complex tools continue into the Mesolithic [Rozoy, 1978], diversifying in form and style.

Correlated with this mastery of lithic technology is a similar improvement in weaponry systems. Earliest Upper Paleolithic groups continued patterns of the past, in that game were killed with hand-held spears. However, tipping of spears with stone points in the earliest Upper Paleolithic undoubtedly improved the penetrability of the projectiles. By the Middle Upper Paleolithic, spearthrowers make their first appearance [Mellars, 1973] and in the Magdalenian there is evidence for the use of the bow [Rozoy, 1978; Rust, 1943]. The significance of this weaponry evolution is that human hunters, through technological inventions, increased the distance between themselves and the animals they killed. Spearthrowers increased the hunter's range of striking force, besides the ability to drive spears deeper into the quarry. The bow was, perhaps, a more significant advance in that it greatly separated the prey and predator and, thereby, reduced substantially the peril of the hunt. Studies by Pope [1962] on a variety of prehistoric and historic bows concluded that animals could be killed at distances up to nearly 100 yards. Coupled with the probable use of poison-tipped arrows, the addition of the bow to the hunting arsenal must have greatly reduced the likelihood and frequency of physical contact between hunters and their quarry. This technique also signals a change in hunting strategy, from directly killing an animal at close range, to wounding it, then trailing it until the animal is weakened by blood loss or the effect of a poison. In any case, the development of bow hunting marks a significant evolutionary advance in technological control over subsistence.

Besides innovations in stone tools, a greater variety of other materials appears to have been utilized in the Upper Paleolithic and Mesolithic when compared to earlier groups. Possibly related to the widespread production of burins, bone, ivory, antler, and, presumably, wood, became media for incorporation into the tool inventories. Eyed needles are first found in the Solutrean, and by the Magdalenian harpoons with detachable heads were commonplace [Bordes, 1968]. Mesolithic groups modeled a great variety of tools, mixing stone and organic materials for all kinds of food gathering uses [Clarke, 1976; Gabel, 1958]. From various sources, then, there is considerable evidence that over the Upper Paleolithic and Mesolithic human groups became increasingly sophisticated in their technological means of dealing with the environment.

The type of animals hunted also shows a change over the Upper Paleolithic and the Mesolithic. Varying with habitats and fluctuating environmental conditions, Upper Paleolithic groups can be classified in general terms as big-game hunters. Species providing the bulk of animal protein are dominated by medium-to-large animals, such as aurochs, bison, equids, mammoths, wooley rhinos, elk, and reindeer. Although Upper Paleolithic groups have often been characterized as specialized "reindeer hunters" [eg, Boule and Vallois, 1957], there is a great deal of variation in the reliance on reindeer. Some sites where reindeer make up nearly 100% of the recovered fauna are clearly seasonal kill sites [Bahn, 1977; Spiess, 1979], meaning that during other times of the year, a more diverse subsistence base was followed. With the gradual decline in glacial climate and the demise of big game, Late Magdalenian and Mesolithic groups exploited a broad spectrum of generally smaller, more solitary game [Meiklejohn, 1978; Straus et al, 1980]. In the Mesolithic there is an increased exploitation of coastal and riverine molluscs and an intensification of fishing, which signifies the development of a more broadly based economic system. Presumably, although there is little evidence to confirm it [Boone and Rinault-Miskovsky, 1976], gathering of vegetable food also changed over the Upper Paleolithic and between it and the Mesolithic. In general, one gets the impression that over time human groups became more proficient in their subsistence strategies. For the Iberian sequence, Freeman concludes that

> (by) the end of the Paleolithic,...peoples had evolved intricate and sophisticated interrelationships with a variety of environments. In some settings, human populations were apparently maintained at relative high levels by economic systems that relied on conscious or unconscious management and perhaps even deliberate manipulation of breeding reserves of those species which served, at least seasonally, as critical dietary resources. Such systems depend on attitudes and techniques which foreshadow those essential to food production...[Freeman, 1981, pp 163, 64].

There is also evidence for the development of more and more efficient cooking techniques over the Upper Paleolithic and Mesolithic. In addition to improvements in the implements for cutting up and preparing food, there are significant advances in the manner of cooking foods. Perlès [1975, 1976] has documented innovations in the construction of fire hearths and other cooking processes. By the Aurignacian and Gravettian, hearths were constructed with subterranean tunnels, which probably served as drafting systems. There are also above- and below-ground ovens. Stone-boiling practices (based on the large number of fire-cracked rocks) become commonplace. The significance of these cooking innovations is that humans adopted processes that reduced the amount of oral processing necessary to consume food. Development of pottery cooking vessels in the Neolithic is the culmination of a long series of improvements in "extra-oral" food preparation techniques [Brace and Mahler, 1971].

Archaeological information suggests that there are considerable changes during the Upper Paleolithic and Mesolithic and that the overall trend is for increasing sophistication and efficiency in dealing with survival. Progressively, techno-cultural innovations served to replace duties that were previously solved by human biology. Blades with their variety of shapes, specificity of function, and increased leverage through handles replaced manipulations that formerly required a biological solution. Weapon systems became more efficient in that new forms allowed a greater separation between the hunter and hunted. Cooking techniques more effectively reduced the need for prolonged chewing of food. These facets of the evolving cultural system are viewed as being largely responsible for the metric changes in Upper Paleolithic and Mesolithic groups.

THE SAMPLES—DENTAL, CRANIAL, AND POSTCRANIAL

Due to the timespan of the Upper Paleolithic and the various archaeological traditions that form the sequence, specimens attributed to this period can be subdivided into several different samples. Moreover, because dental, cranial, and postcranial measurements are derived from a variety of individuals in varying states of preservation, it is difficult to construct uniform sample divisions when comparing different morphological or metric systems. Consequently, because the dentition produces a reasonably large sample size, the dental sample for the Upper Paleolithic has been partitioned into three separate units and analyzed for change among discrete units of time. The smaller number of specimens with cranial and postcranial measurements necessitates forming samples that span greater periods of time.

For the dentition, the Upper Paleolithic has been divided into earliest, middle, and latest units. The EUP (Earliest Upper Paleolithic) sample is the smallest, consisting of data from nine different sites, all of which date to the very initial phase of the Upper Paleolithic (Table I). In this sample, an attempt has been made to distinguish the earliest Upper Paleolithic groups of western and central Europe from those later Aurignacian and Gravettian (Pavlovian) sites. For example, in earlier analyses [Frayer, 1977, 1978] I divided the Upper Paleolithic into two units. This was based on separating the Aurignacian and Szeletian industries from Gravettian (Pavlovian), Solutrean, and Magdalenian periods. Since some Aurignacian sites in western Europe (eg, Cro-Magnon) postdate some of the central European Pavlovian sites (eg, Předmost), this created some difficult decisions as to the proper placement of a few sites into either the early or late group. As Stringer [1982] has suggested, Předmost is probably incorrectly placed in my earlier analyses. Material from this site should be grouped with Dolní Věstonice, Pavlov, and other later Aurignacian and Gravettian materials from western Europe. One other specimen deserves special note concerning its placement in the present analysis. Dates and the stratigraphic position for Combe-Capelle have been debated for 50 years. Since it is most likely of an origin later than the Chatelperronian, it has not been placed in the earliest group.

Later Aurignacian, Gravettian (Pavlovian), and Solutrean sites comprise the MUP (Middle Upper Paleolithic). This sample has the best representation of individuals from western and central Europe, numbering 23 different sites. As with the EUP sample, specimens have been allocated to this period, following the discussion in Frayer [1978].

TABLE I. Upper Paleolithic Dental Sample Divided Into Three
Consecutive Groups (EUP, Earliest; MUP, Middle; LUP, Latest)

EUP	MUP	LUP
Arcy-sur-Cure	Abri Pataud	Arcy-sur-Cure
Brno II	Arene Candide	Aurensan
Les Cottés	Brno I and III	Aveline's Hole
Mladeč	Casterlmerle	Barma Grande
Le Rois	Combe-Capelle	Bédeilhac
Silická Brezová	Cro-Magnon	Brassempouy
Les Vachons	Dolni Věstonice	Bruniquel
Vindija	Fontéchevade	Cap Blanc
Zlatý Kůň	Fourneau du Diable	Cheddar
	Grotte des Enfants	Duruthy
	Isturits	Farincourt
	Kent's Cavern	Isturits
	Miesslingtal	Kent's Cavern
	Paglicci	Kostenki
	Pavlov	Lachaud
	Le Placard	Laugerie-Basse
	Předmost	Lussac-les-Châteaux
	La Quina	La Madeleine
	Les Roches	Le Morin
	La Rochette	Oberkassel
	Stetten	Oetrange
	Svitávka	Le Placard
	Les Vachons	Le Peyrat
		La Pique
		St. Germain
		St. Vincent
		Šandalja
		Veyrier
		Vindija

The LUP (Latest Upper Paleolithic) consists of specimens from 29 differ-ent sites dated to the Magdalenian or other contemporary assemblages. Except for the limited material from Kostenki, Šandalja, and Vindija, all data derive from the western sector of Europe and are associated with Magdalenian assemblages.

The Mesolithic sample includes material dated between the end of the Upper Paleolithic and beginning of the Neolithic. Because of the technolog-ical continuity across the Late Upper Paleolithic-Mesolithic transition and of fluctuations in environmental conditions at the Pleistocene/Holocene boundary, the division between Late Magdalenian and Early Mesolithic sites is somewhat arbitrary. Some, for example, would place Ofnet in the

LUP [Glowatzki and Protsch, 1973] based on the radiocarbon date associated with the skulls. However, I consider the site to be Mesolithic, due to the unique burial pattern and associated gravegoods [Saller, 1962]. The bulk of the Mesolithic sample is more easily placed, according to the occurrence of typical Mesolithic tools, temperate fauna, or radiocarbon dates.

Except for the inclusion of my own measurements of the Yugoslavian Iron Gates material from Vlasac [Nemeskéri and Szathmáry, 1978; Y'Edynak, 1978], the sample used here does not differ from my earlier analysis [Frayer, 1978; Table VII].

These four samples (EUP, MUP, LUP, and Meso) comprise the units of dental analysis. Each prepresents a time period of approximately 7 ky. The EUP dates from 33 to 26 ky BP; the MUP from 26 to 19 ky BP; the LUP from 19 to 12 ky BP; the Mesolithic from 12 to 5 ky BP.

Dental measurements consist of standard length and breadth dimensions of the crown. Measurement techniques and an error analysis are described in Frayer [1978]. Most of the measurements of Upper Paleolithic and Mesolithic teeth were taken by myself on original specimens. In some cases, when the originals were unavailable for study or have been lost or destroyed, I have relied upon published sources. Also, for a couple of specimens, I have included measurements from casts, but only after verifying the accuracy of the cast. Finally, I have used materials measured by M.H. Wolpoff (Michigan) and J. Verdène (Paris), after determining similar techniques in recording dental data.

The sample of cranio-facial measurements is comprised primarily of specimens I have measured. The major exceptions concern the Předmost data that were taken from Matiegka's invaluable monograph [1934] and a few specimens measured by M.H. Wolpoff. There are also some specimens that are represented by data culled from the literature. In all cases, measurements were taken following the standardized points developed by Martin [1928].

Since in this paper I have presented cranio-facial change along sex lines, it has been impossible to divide the Upper Paleolithic into smaller subsets. Unfortunately, available crania from this period are not numerous enough to permit dividing the sample into two or three successive series and still maintain adequate sample size for each. Consequently, analysis of craniofacial change is limited to a comparison of the total Upper Paleolithic to the total Mesolithic sample.

The Upper Paleolithic and Mesolithic postcranial dataset come primarily from the literature. Sources of these data and a discussion of their comparability are reviewed in Frayer [1981] and in a later section of this paper.

Tooth Size Changes

Changes in tooth size over the Upper Paleolithic and Mesolithic periods have been discussed in considerable detail elsewhere [Frayer, 1977, 1978]. The present analysis of dental evolution reviews only two segments of the numerous patterns of change characterizing the period. First, trends for reduction of tooth size between a late Mousterian sample (NEA) and the earlier Upper Paleolithic divisions are compared to trends occurring within the Upper Paleolithic period. The intentions are 1) to provide some basis for determining the likelihood of European Neanderthal ancestry to the Upper Paleolithic, 2) to illustrate the intermediate nature of the EUP and MUP samples (between NEA and LUP units), and 3) to demonstrate the substantial reduction in tooth size from the earliest to latest Upper Paleolithic samples. For the NEA sample, data have been compiled from sites listed in Table II. Allocation of those specimens to "late Mousterian" is based on information in Wolpoff [1980]. Most of the data derive from studies by Wolpoff (or myself) on the original materials.

The second topic of this section concerns changes in dental sexual dimorphism through time. This analysis provides a backdrop for the main focus of subsequent sections on cranio-facial and postcranial evolutionary trends.

Diachronic patterns—NEA to Meso. Changes in mandibular and maxillary anterior tooth size are reviewed in Figure 1 (complete statistics for the NEA and Upper Paleolithic samples are given in the appendixed tables). For this analysis only labial-lingual breadths are used, since any wear on the incisors critically reduces the "true" length dimension. In general, for both mandibular and maxillary anterior tooth breadths considerable reduction in size occurs from the NEA to the LUP sample.

In the maxillary incisors, the NEA sample has breadths that average 8.4% and 11.0% larger than the EUP, for I^1 and I^2, respectively. Maxillary central incisor breadth is essentially unchanged for the entire Upper Paleolithic, while I^2 breadth reduces 6.8% from the EUP to the MUP and 11.0% from the MUP to LUP, although sample size for the EUP is small. Maxillary canine breadths are better represented in all five samples and show continuous trends for reduction over the Upper Paleolithic. Maxillary canine breadths reduce 11.0% within the Upper Paleolithic and show virtually no difference between the NEA and EUP samples.

For the same three breadths in the mandibular teeth, NEA and EUP means cluster together differing by less than 2%. From the EUP to the LUP, mandibular I^1 reduces 17.6%; I^2 breadth 15.6%, and canine breadth 11.2%. Differences between the MUP and LUP samples are minimal, indicating

TABLE II. Late Mousterian Sites Grouped in NEA
Sample: Inclusion of Material in This Sample
Is Based on Wolpoff [1980]

Agut
Akhystyr'
Arcy-sur-Cure
Bau de l'Aubesier
Baume des Peyrards
Châteauneuf
Combe Grenal
Dzhruchula
Engis
Gibraltar
Hortus
Jersey
Krapina
Kůlna
La Chapelle-aux-Saints
La Ferrassie
La Quina
Le Moustier
Leuca
Marillac
Montgaudier
Mt Circeo
Pech de l'Azé
Petit Puymoyen
St. Brais
Sakajia
Šipka
Staroseljé
Teshik Tash
Vergisson
Vindija
Zaskalnaya

that the greatest reduction in mandibular anterior breadths occurs between the earliest and middle Upper Paleolithic samples. Thus, for the mandibular anterior breadths the EUP closely resembles the NEA sample. In the maxilla, the central and lateral incisor breadths are considerably larger in the late Mousterian sample.

Those favoring a "replacement" hypothesis for the NEA/EUP transition could cite these data as evidence for the distinctiveness of the NEA sample, although this only holds for the maxillary incisor breadths. I view the evidence as somewhat ambiguous, but lean more toward an interpretation

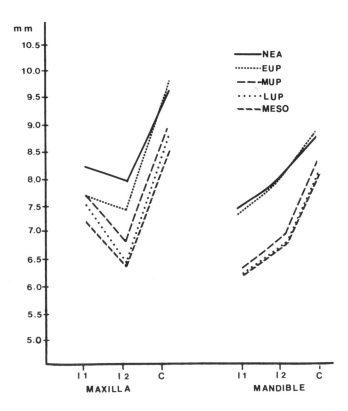

Fig. 1. Breadth profiles of anterior teeth (descriptive statistics in Appendix Tables I and II).

of continuity, based on the changes in maxillary canine breadth and all the mandibular breadths. However, since the mandibular and maxillary teeth *do* occlude with each other, it is somewhat disconcerting that the teeth from the two jaws show different patterns of reduction. This is more true for the incisors than the canines, and may be due to different sample compositions or simply to the small numbers of specimens for which there are measurements. Despite these inconsistencies, it is impossible to conclude that the Upper Paleolithic does not show substantial reduction in size for most anterior breadths.

Change from the LUP to Mesolithic is much less marked. In general, anterior tooth breadths are smaller in the Mesolithic, but the greatest reduction occurs earlier in the sequence.

Patterns of change in posterior tooth size also indicate a general trend for reduction through time. Here, to give an idea of changes in occlusal surface, only areas are plotted, although lengths, breadths, and areas are given in the appendixed tables. As with the anterior tooth breadths, sample sizes are small for the EUP, although in all but four tooth areas of the mandible and maxilla, the EUP means exceed those for any subsequent period.

In the maxilla (Fig. 2) the following trends are evident: 1) The EUP/NEA means are very similar to each other, except for P4 and M3. 2) Except for the considerably smaller canine size, the MUP sample is almost exactly intermediate between the NEA and LUP averages, and in most cases the EUP is intermediate between the NEA and MUP samples. 3) The LUP and Meso samples are similar in size and both groups differ considerably from the earlier Upper Paleolithic groups.

Mandibular posterior area profiles (Fig. 3) also show a trend for reduction over time with the earlier Upper Paleolithic groups larger than the subsequent LUP and Meso samples. 1) NEA and EUP means are similar to each other, except for M2 and M3. 2) The EUP is always larger than the MUP, again except for M2 and M3. 3) For most mandibular areas, the MUP sample is almost exactly intermediate between the means for the NEA and LUP samples. And finally, 4) the LUP and Meso means are very similar to each other, with both representing the smallest areas of all five samples.

In summary, for both the anterior tooth breadths and the maxillary and mandibular areas, there are distinct trends for reduction over time. The EUP sample is generally intermediate between the NEA and MUP samples, although small sample sizes for nearly every tooth do diminish the power of this conclusion. However, MUP averages also tend to occupy an intermediate position between NEA and LUP averages, which indicates the degree of reduction in overall tooth size within the Upper Paleolithic. Based on these data, there is no basis to argue that there is no strong trend for considerable tooth size change within the Upper Paleolithic [Stringer, 1982].

Sex differences in tooth size. In addition to the overall patterns of dental reduction from the earliest Upper Paleolithic to Mesolithic populations, there is also a trend for a decrease in sexual dimorphism in tooth size through time. This is primarily related to a greater magnitude of change in males, coupled with slower, less substantial change in female dental metrics. Although presumably this differential pattern of reduction between the sexes is a characteristic of the individual EUP, MUP, and LUP samples, this analysis is limited to only the total Upper Paleolithic and Mesolithic samples. Unfortunately, adequate sexable material is limited in the Upper Paleolithic, so that it is necessary to combine materials from the EUP, MUP, and LUP in one unit in order to produce reasonable sample sizes. Details concerning determination of sex for each specimen are discussed elsewhere

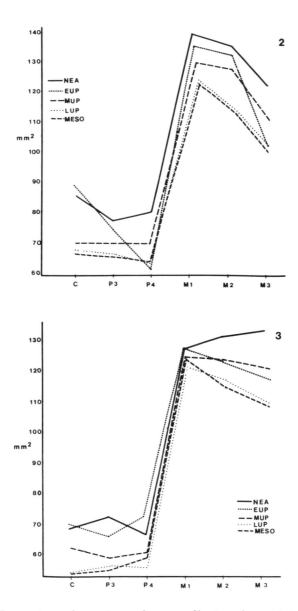

Fig. 2. Maxillary canine and posterior tooth area profiles (complete statistics in Appendix Tables I and II).

Fig. 3. Mandibular canine and posterior tooth area profiles (complete statistics in Appendix Tables I and II).

[Frayer, 1978, 1980], yet it is important to stress that sexing was based on cranial and, when available, postcranial material, never on the size of the teeth. Descriptive statistics for the male and female samples from the Upper Paleolithic and Mesolithic are presented in Appendix Table III.

Figure 4 summarizes data for the mandibular canine through M3 areas. Percent differences are used to compare levels of sexual dimorphism in each period and are calculated by the following formula:

$$\% \text{ dimorphism} = \frac{\text{male } \overline{X} - \text{female } \overline{X}}{\text{male } \overline{X}} \times 100$$

In addition to percent dimorphism, results of significance tests are given for each tooth, using student's t when variances are not significantly different or the Mann-Whitney U when F-tests show unequal variances between the two samples [Conover, 1971].

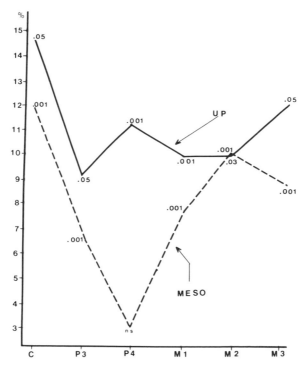

Fig. 4. Percent difference in sexual dimorphism for Upper Paleolithic and Mesolithic male and female tooth areas (mandible) (complete statistics in Appendix Table III).

For mandibular areas the Upper Paleolithic sample consistently shows a greater degree of sexual dimorphism than the Mesolithic. Canines prove to be the most dimorphic in both samples, averaging 14.6% larger in Upper Paleolithic males and 11.9% in the Mesolithic male/female index. Indices of sexual dimorphism for areas of all mandibular posterior teeth (except M2) are larger in the Upper Paleolithic and each is associated with a statistically significant difference. Sexual dimorphism in the Mesolithic is also associated with significant differences in male/female averages, although at a lower magnitude of percent difference.

Examination of maxillary canine and posterior tooth areas verifies the fact that Upper Paleolithic groups display more sexual dimorphism (Fig. 5).

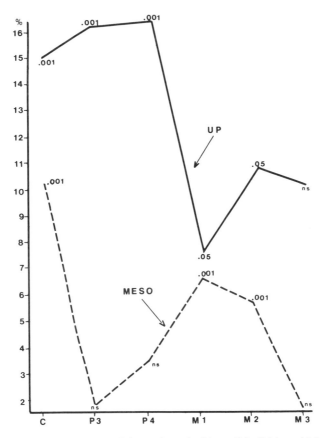

Fig. 5. Percent differences in sexual dimorphism for Upper Paleolithic and Mesolithic male and female tooth areas (maxilla) (complete statistics in Appendix Table III).

In this case, Upper Paleolithic and Mesolithic indices are completely distinct from each other. All Upper Paleolithic maxillary tooth areas exceed the dimorphism found in the Mesolithic. For example, Upper Paleolithic males have canines 15% larger than females, while the same tooth in the Mesolithic shows an index of dimorphism of only 10%. For the maxillary posterior tooth areas, Upper Paleolithic sexual differences range between 7.6 and 16.3%, with Mesolithic indices ranging from 1.7 to 6.4%. Furthermore, in the Upper Paleolithic all indices (but M3) are significantly different, while only the canine, M1, and M2 exhibit differences in the Mesolithic. Based on these data for the mandible and maxilla, there is a very substantial reduction in sexual dimorphism between the Upper Paleolithic and Mesolithic.

Using the same data presented in Appendix Table III, Figures 6 and 7 trace changes in tooth size within the same sex from the Upper Paleolithic to Mesolithic. These figures plot changes in male or female means between the two periods by the index

$$\% \text{ change} = \frac{\text{UP male (or female) } \overline{X} - \text{Meso male (or female) } \overline{X}}{\text{UP male (or female) } \overline{X}}$$

For the mandible, percent change in areas is greater in males for five of the six comparisons, and every case (where a statistically significant difference occurs) involves a reduction between 11.8% and 7.3%. Females show only two significant decreases in size through time (canine and M2) and consistently show lower indices of change. Only the mandibular M2 area exhibits an equivalent amount of change for males and females of both periods.

Not unexpectedly, maxillary area changes measured over the same time period show substantially more reduction in males (Fig. 7). Maxillary canine, P3, P4, and M2 areas reduce significantly in males, with only M2 area showing a significant reduction between Upper Paleolithic and Mesolithic females. Indices for maxillary area change in males range between 5.5% and 12.3%, while the greatest changes in females is 6.8%. Two areas (P3 and P4) have larger means in Mesolithic females, although neither are statistically significant.

Changes in the degree of sexual dimorphism between the Upper Paleolithic and Mesolithic samples, then, are the direct result of a greater reduction in male tooth size. Females over the same period exhibit less marked change. The combination of those two separate patterns of dental reduction accounts for the decrease in sexual dimorphism indices over time.

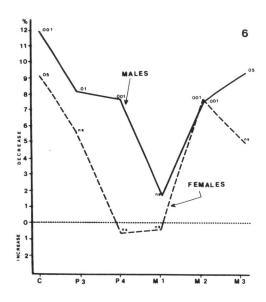

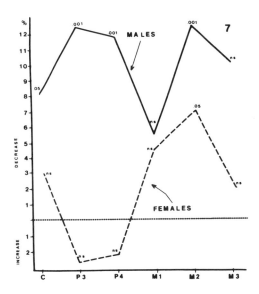

Fig. 6. Upper Paleolithic/Mesolithic male and Upper Paleolithic/Mesolithic female percent change (mandible) (complete statistics in Appendix Table III).

Fig. 7. Upper Paleolithic/Mesolithic male and Upper Paleolithic/Mesolithic female percent change (maxilla) (complete statistics in Appendix Table III).

In summary, considerable change in dental size occurs across the Upper Paleolithic-Mesolithic periods. When data are analyzed for change within the Upper Paleolithic (without regard for sex), anterior teeth show marked reduction in size from the earliest (EUP) to the latest (LUP) period. In general, the EUP sample either closely resembles the late Neanderthals (NEA) or is intermediate between them and the later Upper Paleolithic group. Reduction in areas of the mandibular and maxillary canine and posterior teeth also occurs within the Upper Paleolithic and the two earliest Upper Paleolithic samples prove to be intermediate in size between the NEA and LUP (or Mesolithic) samples. Although these data certainly do not *prove* that NEA groups were directly ancestral to the earliest Upper Paleolithic groups, they do not contradict the basis for this model either. Since the Mousterian-Upper Paleolithic transition (or break) is not the main focus of this paper, it is probably best to conclude that change in Upper Paleolithic/Mesolithic tooth size is part of a long pattern of dental reduction in the human lineage, begun well before the onset of the Late Pleistocene [Brace and Mahler, 1971; Frayer, 1977; Wolpoff, 1971].

Besides the trend for overall tooth size reduction, there is also a different rate of reduction between males and females. The consistent pattern of greater decrease in male tooth size from the Upper Paleolithic to the Mesolithic results in lower indices of sexual dimorphism in the Mesolithic. Selective mechanisms accounting for these changes, which are closely related to the evolution of technology and modifications in subsistence patterns, are discussed in the final section.

Cranial Variation and Change

Various analyses of cranio-facial change within the Upper Paleolithic and Mesolithic have followed both univariate and multivariate approaches and, in general, can be described as being relatively uninformative, or in the very least, confusing and contradictory. Univariate approaches suffer the drawback of being limited to comparisons of means with little capability to analyze the combined shape features of the face and vault. For example, in Morant's descriptive work [1928], the univariate section was summarized with the conclusion that Upper Paleolithic skulls were larger than more modern series. He speculated that this might be due to a sampling effect, where larger, more robust skulls would have a better chance of preservation [Morant, 1928, p 152]. More recent univariate studies have been concerned with the identification of racial types in the Upper Paleolithic and Mesolithic [Billy, 1976; Ferembach, 1969, 1978a; Tóth, 1972; Vlček, 1967]. In this regard it is interesting to note that although the race concept has received diminished emphasis in textbooks concerned with living people over the past few years [Littlefield et al, 1982], it certainly has not been dismissed by many authors who deal with human variation in the Late Pleistocene. Other studies have focused on gradualistic change in size [Riquet, 1968] or sexual dimorphism [Frayer, 1980]. Most of these reviews suffer from small, and in some cases, inaccurate samples; and, since they are almost always limited to comparisons of mean differences or two-dimension ratios, there has been little analysis of morphological or shape differences through time or between sexes.

The alternate multivariate approach has hardly been any more fruitful. Since there has been no comprehensive set of measurements collected for the Upper Paleolithic or Mesolithic and because there has been little theoretical justification for the various procedures used, multivariate studies based on original data are often limited in scope, and tend to conclude in typological generalizations [Bianchi et al, 1980; Ferembach, 1974, 1976]. On the other hand, in studies where data have been drawn exclusively from the literature, samples generally consist of many estimated data or are not representative of the Upper Paleolithic [Corruccini, 1976; Stringer, 1974]. Depending on the sample composition, the kind of multivariate test used, and the bias of the particular investigator, completely different results and conclusions have been generated. For example, Corruccini [1976] and Bilsborough [1972] argue for affinities between "Classic" Neandertals and the earliest Upper Paleolithic specimens, while Stringer [1974] finds justification for separating these two groups and contends that the mid-Würm Neandertals are not directly ancestral to earliest Upper Paleolithic populations. Similarly, Ferembach [1974] argues for the appearance of races in the Mesolithic (south of the France-Belgium border), while Constandse-Westermann [1974] (using the same statistical procedures) finds no evidence for racial differentiation in the northern European area and interprets cranial change in a clinal manner. Most recently, Henke [1981], using a large dataset I collected, concluded that 1) there is little evidence for the existence of racial types in the Upper Paleolithic or Mesolithic, 2) that there is a trend for a reduction in sexual dimorphism through time, and 3) that both males and females reduce in robusticity from the Upper Paleolithic to the Mesolithic. Henke's interpretations differ considerably from multivariate studies by others [for example, Bianchi et al, 1980]. Part of the reason for these contradictory results may relate to the different kinds of multivariate methods used and to the varying collections of specimens making up the samples. At present, however, the definitive work covering cranial form change has yet to be done. In many respects, until a detailed study of all the Upper

Paleolithic and Mesolithic crania is undertaken (which includes more than just the "traditional" set of cranio-facial metrics and concentrates on collecting measurements that relate to functional morphology), little can be said with certainty about patterns in or causes of Upper Paleolithic/Mesolithic cranial change.

With this state of affairs, it is perhaps "dangerous" to present any analysis of cranial variation in the Upper Paleolithic or Mesolithic without first restudying the entire collection. However, with respect to the pitfalls discussed above, the present analysis combines a univariate approach with an analysis of shape changes in an attempt to provide an indication of some of the morphological trends that characterize the transition from the Upper Paleolithic to the Mesolithic. Nearly all the measurements come from my own study of the specimens and in no instances are any data points estimated. Samples are divided by sex and analyzed separately, since previous work has demonstrated different trends between males and females in the late Pleistocene and Early Holocene [Frayer, 1980]. Reliability of sexing is always a problem, but these samples are sexed conservatively with an appreciation for the existence of robust females and gracile males in prehistoric populations. To my knowledge, this sample represents the most complete dataset available.

The technique followed here relies upon the reconstruction of lateral cranial contours that are generated by a series of measurements taken from common cranio-facial landmarks. The dataset consists of eight different measurements taken from the auricular point to various Martin points in the sagittal plane. These measurements, combined with chord measurements between the landmarks, result in geometric relations that have been plotted in lateral profile. The profiles are based on mean dimensions for Upper Paleolithic and Mesolithic males and females, with each of these four samples plotted separately. Comparisons between average lateral profiles are made by superimposing one profile on the other, with each oriented so that the auricular points are in an identical position, and so that the auricular point-bregma chords overlap exactly. Figures 8 and 9 show trends occurring between Upper Paleolithic and Mesolithic males or females while Figures 10 and 11 document differences in sexual dimorphism between Upper Paleolithic and Mesolithic samples. Unfortunately, due to the limitations of my dataset, it is impossible to reconstruct profiles in any other plane, so that this analysis is limited to differences in shape from the perspective of the lateral plane. Nevertheless, I feel this technique is superior to a simple comparison of means and it avoids some of the assumptions required of multivariate methods [Corruccini, 1978; Kowalski, 1972]. Comparisons with Mousterian groups have not been made since there are no comparable data.

TABLE III. Selected Upper Paleolithic and Mesolithic Cranial Measurements (X̄, n, and Standard Deviation; AP, Auricular Point)

	UP males	UP females	Meso males	Meso females
A P – Bregma	133.8(12)	125.3(12)	129.5(38)	122.8(35)
	7.9	6.6	5.6	4.4
A P – Glabella	122.5(11)	114.1(13)	116.7(30)	108.5(32)
	7.0	4.6	5.4	4.8
A P – Nasion	115.2(10)	108.6(13)	110.3(29)	103.7(31)
	5.8	4.5	4.6	3.9
A P – Nasospinale	116.9(8)	112.2(8)	113.3(18)	108.5(21)
	5.0	5.9	4.2	4.7
A P – Prosthion	121.7(7)	119.7(7)	121.3(15)	113.1(19)
	10.2	5.0	3.5	5.1
A P – Lambda	125.8(10)	119.9(12)	124.1(31)	120.7(31)
	4.9	7.5	6.1	6.0
A P – Inion	108.6(10)	104.5(9)	105.8(30)	101.2(28)
	6.6	4.6	4.4	5.1
A P – Opisthion	84.9(5)	77.4(8)	81.0(28)	76.3(26)
	6.4	4.1	4.4	3.9
Nasion – Bregma	116.3(24)	111.5(25)	111.6(52)	106.8(49)
	6.4	5.9	6.4	4.2
Nasion – Nasospinale	52.5(16)	48.1(15)	50.9(41)	48.3(36)
	4.2	3.8	3.6	3.7
Nasion – Prosthion	69.0(15)	63.5(15)	68.4(40)	63.3(37)
	5.5	3.7	4.2	5.5
Bregma – Lambda	121.9(23)	115.5(26)	116.8(48)	113.7(43)
	7.2	10.0	6.7	5.3
Lambda – Inion	61.2(13)	61.9(10)	68.8(33)	63.8(31)
	6.4	6.8	7.6	7.3
Lambda – Opisthion	98.3(16)	98.1(17)	100.6(41)	98.3(36)
	5.2	3.6	5.4	4.8
Lambda – Glabella	189.0(21)	179.1(19)	181.6(34)	173.6(35)
	6.5	6.2	8.0	6.9

From the data presented in Tables III and IV, it is clear that Upper Paleolithic males are considerably larger in most dimensions than Mesolithic males and that a similar pattern holds for the change between females from the Upper Paleolithic to the Mesolithic. Generally speaking, Upper Paleolithic skulls are longer and higher, with more projecting glabellar and nasal areas, longer frontal chords, and higher, more posteriorly placed nuchal areas. Figures 8 and 9 combine these data between the various means. Several differences can be discerned: 1) Common trends occur between the male and female comparisons, most notably the general trends for reduction in the amount of facial prognathism from the Upper Paleolithic to Meso-

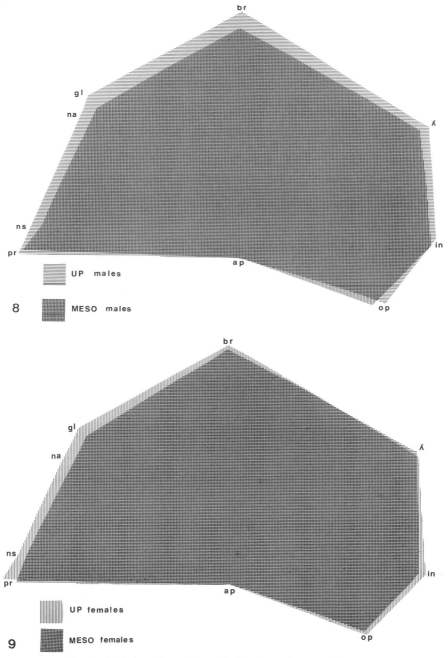

Fig. 8. Male lateral cranial profiles based on data in Table III.
Fig. 9. Female lateral cranial profiles based on data in Table III.

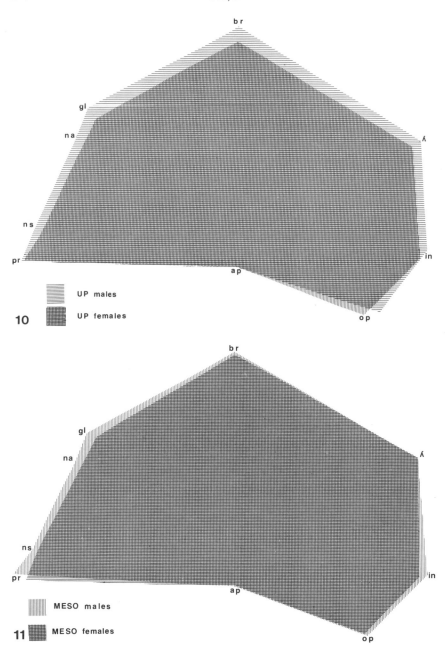

Fig. 10. Sexual dimorphism in lateral cranial profiles for the Upper Paleolithic (based on data in Table III).

Fig. 11. Sexual dimorphism in lateral cranial profiles for the Mesolithic (based on data in Table III).

TABLE IV. Percent Differences[a] and Significance Tests[b] for Means of
Selected Cranial Measurements

| | Upper Paleo-Meso by sex | | Sexual dimorphism | |
	Males	Females	UP	Meso
A P - Bregma	3.2/ns	2.0/ns	6.4/.003	5.2/.000
A P - Glabella	4.7/.026	4.9/.001	6.9/.003	7.0/.000
A P - Nasion	4.2/.033	4.4/.003	5.7/.008	6.0/.000
A P - Nasospinale	3.1/ns	3.3/ns	4.0/ns	4.2/.004
A P - Prosthion	.3/ns	5.5/.013	1.6/ns	6.8/.000
A P - Lambda	1.4/ns	+.7/ns	4.7/.037	2.7/ns
A P - Inion	2.6/ns	3.2/ns	3.8/ns	4.3/.001
A P - Opisthion	4.6/ns	1.4/ns	8.8/ns	5.8/.000
Nasion - Bregma	4.0/.005	4.2/.001	4.1/.018	4.3/.000
Nasion - Nasospinale	3.0/ns	+.4/ns	8.4/.013	5.1/.003
Nasion - Prosthion	.9/ns	.3/ns	8.0/.005	7.5/.000
Bregma - Lambda	4.2/.007	1.6/ns	5.3/.038	2.7/.023
Lambda - Inion	+12.4/.002	+3.1/ns	1.1/ns	7.3/.016
Lambda - Opisthion	+2.3/ns	+.2/ns	.2/ns	2.3/ns
Lambda - Glabella	3.9/.001	3.1/.005	5.2/.000	4.4/.000

[a]Percent differences are calculated by subtracting the earlier (or male) mean from the later
(or female) mean and dividing by the earlier (or male) mean.
[b]Significance measured by two-tailed Student's test; ns, not significant.

lithic. 2) Upper Paleolithic and Mesolithic males show considerable differences in overall size and morphology. Glabella, nasion, and nasospinale are
more forward-projecting in the Upper Paleolithic, while Mesolithic males
show a greater amount of alveolar prognathism. In addition, Upper Paleolithic males have a higher-placed bregma and lambda, with longer frontal
and parietal chords. At the back of the skull, lambda, inion, and opisthion
are more posteriorly oriented, and there is a higher and more perpendicular
nuchal plane. Thus, shape as well as size differences occur. 3) Upper Paleolithic and Mesolithic female contours are much more similar to each other,
particularly in features describing cranial shape. Upper Paleolithic females
have slightly higher skulls than females from the Mesolithic with only minor
differences in frontal and parietal chords. Also, their occipital regions are
virtually identical in shape and orientation. 4) Although not easily determined from the tabular data, the lateral profiles indicate that greater
reduction occurs in male cranio-facial features over the Upper Paleolithic
and Mesolithic.

Data for sexual dimorphism in the two periods are also reviewed in
Tables III and IV, and Figures 10 and 11. From Appendix/Table III, the
tabular data, it is clear that both samples are sexually dimorphic for a large

number of measurements. In the Upper Paleolithic, nine of the means are significantly different at the .05 level or better; in the Mesolithic, 13 reach significance. However, combining the measurements in lateral profile clearly shows that a greater degree of sexual dimorphism exists within the earlier period. Upper Paleolithic males have longer, higher skulls, with more facial and occipital projection and more vertically oriented nuchal planes. Besides their overall larger size, Upper Paleolithic males show a number of differences in shape, especially in the nuchal area. In the Mesolithic, males and females are more similar to each other in size for most dimensions of the vault. Height and length of the skull are very similar and the nuchal planes are oriented in the same manner. Besides the greater amount of sexual dimorphism in the Upper Paleolithic, the profiles also suggest that both sexes in the Upper Paleolithic are characterized by a greater amount of total facial prognathism, while Mesolithic males and females show a reduction in the upper face and more marked lower facial prognathism.

In sum, these lateral cranial profiles provide evidence to support the contention that males show a greater amount of reduction than females from the Upper Paleolithic to the Mesolithic. This results in a decrease in the degree of sexual dimorphism in Mesolithic populations and involves changes in both size and shape. Changes in cranio-facial contours are, then, consistent with the reported trends for reduction in dental sexual dimorphism between the two periods.

Stature

Estimates for stature, separated by sex, are given in Table V and Figure 12. Samples are divided into five periods (Upper Paleolithic, Mesolithic, Neolithic, Medieval, and Modern), and, in all but the modern case, stature was determined by using formulas presented by Trotter and Gleser [1952]

TABLE V. Stature Estimates for European Populations[a]

	UP	Meso	Neo	Med	Mod
Males					
X̄	1,743	1,678	1,664	1,685	1,735
n	20	41	62	41	20
s.d.	94	75	51	53	26
Females					
X̄	1,593	1,560	1,542	1,562	1,608
n	9	26	46	46	20
s.d.	41	68	65	65	29

[a]Statistics based on sum of population means. UP, Upper Paleolithic; Meso, Mesolithic; Neo, Neolithic; Med, Medieval; Mod, modern.

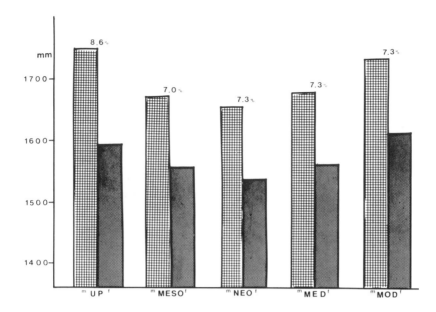

Fig. 12. Stature for various post-Mousterian samples (histograms based on data in Table V).

for white males and females. In some cases Steele's formulas [1970] for reconstructing maximum length from fragmentary long bones were used, with the resulting estimate substituted into the appropriate Trotter-Gleser formula. Because of additions to the Mesolithic and Neolithic samples, these data differ somewhat from my earlier analyses of body size [Frayer, 1980, 1981], although Upper Paleolithic groups still show the greatest average stature in all samples. In this analysis, Mesolithic skeletons from Vlasac [Nemeskéri and Szathmáry, 1978] have been added to the sample, based on my measurements of the material in Beograd. Since these individuals are considerably taller (in both sexes) than the previously published Mesolithic sample, stature estimates have been revised upward. The Neolithic sample in earlier publications consisted only of material from Vasterbjers (Sweden). Recently I have been able to add data from two Czech sites (Krškrany and Vedrovice) and a collection from the Ligurian Neolithic. The two Czech samples were measured in the Moravian Museum (Brno) and the Italian one comes from Parenti and Messeri [1962], where I calculated stature on their published maximum femur, tibia, or fibula lengths. Estimates for the

Middle Ages are from my measurements in Budapest of the postcranial remains from Zalavár (Hungary). Background data for this site can be found in Sós and Bökönyi [1963] and Howells [1973]. Whether these data are representative of stature in the European Middle Ages has not been determined. The modern sample has been assembled in a different manner. Here, data published by Eveleth and Tanner [1976] were combined for 20 different, recent European samples.

Trends in body size for both males and females show a clear reduction from the Upper Paleolithic to the Mesolithic and Neolithic, followed by an increase into the Middle Ages and further increase to modern European groups. However, with regard to the modern sample this increase is likely of very recent origin, related to rapid changes in European stature over the past 50 years [Olivier, 1980]. Over the period of reduction, males show a greater change than females; 4.5% for males from the Upper Paleolithic to Neolithic, 3.2% for females. From the Neolithic to modern samples, both males and females increase at approximately the same rate (4.3%). The resulting statures of modern groups are about the same as those for the Upper Paleolithic sample, although Upper Paleolithic males are still slightly taller than modern European males and modern females are larger than Upper Paleolithic females. Even though statures in these two groups are similar, from my unquantified comparisons, Upper Paleolithic populations are considerably more robust than the most recent European samples.

With respect to sexual dimorphism in stature, Upper Paleolithic groups show the most distinct differences between the sexes and, despite the general similarity between Upper Paleolithic and Modern male statures, the degree of sexual dimorphism in the modern sample is lower. It is also interesting that, notwithstanding an increase in stature from the Neolithic to the modern sample, the level of sexual dimorphism does not change, but remains at about 7.3%. The level of sexual dimorphism does not automatically rise with an increase in body size and is consequently not strictly related to body size changes.

These results are not substantially different from my earlier analyses [Frayer, 1980, 1981]. In the prehistoric and historic comparisons, Upper Paleolithic groups have a greater stature coupled with a greater degree of sexual dimorphism. Increasing the size and geographic spread of the Neolithic sample and inserting a Medieval sample shows that stature decreases to the Neolithic and increases afterward. Key [1980], using a sample drawn from the Mainz databank, arrives at the same conclusion, although his analysis is restricted to femur length. However, he also suggests an increase in sexual dimorphism from the Mesolithic to late Middle Ages for which I find no clear evidence. Differences in our interpretations relate primarily to sample composition that affects the sample means. Key's sample includes a much larger geographic area and is drawn from a number of different studies. In this analysis, I have attempted to restrict the samples to specimens I have either measured myself or measurements I can cross-check with photographs or other data. Consequently, I feel my samples are better controlled for sexing accuracy and more closely represent the actual trends occurring through time.

Yet, there are considerable problems in interpreting the post-Mesolithic dataset, particularly since these periods witnessed major population shifts and migrations of people back and forth across Europe, along with regionally different technological systems, subsistence patterns, and settlement types. Consequently, it is perhaps overly optimistic at the present state of data collection to have much faith in post-Mesolithic body size trends. They are included here, mainly for comparative purposes.

On the other hand, these data do indicate that, compared to all later groups, Upper Paleolithic populations are characterized by high stature and a high degree of sexual dimorphism. The available data also suggest that both stature and sexual dimorphism decrease considerably into the Mesolithic. In all respects, this is consistent with the patterns discussed above for the dentition and crania. With regard to changes in stature, these results are consistent with other studies by Ferembach [1979] and Key [1980] even though their results are based on geographically different and considerably smaller samples.

CULTURAL AND BIOLOGICAL CHANGES

Analysis of trends in tooth size, cranio-facial morphology, and stature from the Upper Paleolithic to Mesolithic indicates that this period of prehistory was a dynamic phase in the evolution of what are called "anatomically modern *Homo sapiens*." Rather than being static or characterized by random changes, there are substantial, directional trends in the size of teeth, the size and morphology of the face and cranial vault, and the dimensions of the body, measured by stature. In most respects, these three aspects of the biological system (teeth, skulls, body size) change in parallel ways, indicating that the whole morphological system responded over time to similar evolutionary forces. Moreover, since in all three systems, males show a greater magnitude of change (as compared to females), different kinds of intensities of selection were operating on the two sexes. Underlying the shared trend for size reduction, then, is another affecting the sexes differently.

The shared trends for reduction seem best related to evolving techno-cultural sophistication. Although this aspect of cultural evolution is one that characterizes the tempo of all human evolution, within the Upper Paleolithic there are a number of substantial improvements in cultural adaptation. Innovations such as blade tool production, greater incorporation of antler, bone, and ivory into tool inventories, increased use of handles and composite tools, and advances in preparing and cooking food mark the ability of humans to exert more cultural control over the environment. Patterns of tooth size and cranio-facial reduction clearly point to a link between cultural changes and human biological features.

When Upper Paleolithic groups are separated into three chronological units and analyzed irrespective of sex, dental trends consistently show size reduction through time. Earlier Upper Paleolithic groups (EUP, MUP) often more closely resemble late Neanderthal populations than they do the LUP or Mesolithic populations. At the very least, these dental trends indicate that EUP (and MUP) samples are intermediate between earlier and later populations. Thus, even though sample sizes are not large, there is considerable evidence for substantial dental change within the Upper Paleolithic, and a continued, but less marked trend for reduction into the Mesolithic. Cranio-facial size and morphology also decrease between the Upper Paleolithic and Mesolithic for both males and females. In general, Upper Paleolithic skulls are higher and longer and have more facial prognathism, from glabella to prosthion. In contrast, both male and female Mesolithic skulls have more gracile cranio-facial contours and reduced facial projection. Finally, there is a pattern of decreasing body size for both sexes over the Upper Paleolithic and Mesolithic periods. Each of these systems, considered apart or together, provide strong evidence for gracilization in size over time and affect both males and females.

Along with these overall patterns of reduction, there is also consistent evidence for changes in sexual dimorphism. Males from the Upper Paleolithic and Mesolithic show a greater rate of reduction in tooth size, cranio-facial dimensions, and stature when compared to females. The greater stability of females, coupled with more rapid reduction among males results in lower indices of sexual dimorphism in the Mesolithic. The persistence of these trends over time suggests that different evolutionary pressures affected male and female phenotypes, along with the shared tendency for overall reduction.

Demonstrating the precise evolutionary mechanism(s) for the general and specific trends proves more difficult than documenting the nature of the changes. However, it is possible to eliminate both genetic drift and gene flow, so that these trends are best allocated to changes in selective forces. Random change due to drift is an unlikely factor, since all current information suggests Upper Paleolithic groups were not isolated from each other. Reanalysis of older collections [Bahn, 1977] and interpretations drawn from recent excavations [Spiess, 1979] conclude that many sites are seasonal encampments and that Upper Paleolithic groups roamed over a wide geographic range during the yearly subsistence round. Occurrence of "foreign" lithics, Atlantic and Mediterranean shells, and other items in inland Upper Paleolithic sites attests to the movement through diverse areas [Bahn, 1977]. In addition, there are common art motifs and styles that criss-cross eastern to western Europe [Marshack, 1979] that argue against the likelihood of isolation. Thus, genetic drift is an unlikely cause, since no evidence points to group isolation for any significant period of time. If anything, Mesolithic groups are potentially more isolated from each other, yet even here there are a number of shared stylistic patterns that negate complete isolation [Rozoy, 1978]. Finally, considering the critical population size models formulated by Wobst [1976], a great deal of interaction across regions must have been the rule, since low population densities characterize both periods [Hassan, 1978]. Gamble [1980, 1982] suggests that Upper Paleolithic groups had wide interaction spheres, linked by commonalities in art styles that served as a mechanism for information exchange. Consequently, if these groups were sharing information, it also likely that they were building alliances through marriage exchanges and sharing genes. This would effectively eliminate the chance for drift to affect gene frequencies.

The importance of gene flow is restricted to the interchange of genetic information between areas as opposed to the mass migration of new people into the area. Several pieces of information support this contention: 1) Smith [1982] has shown the existence of continuous trends within populations of south-central Europe from the Mousterian to Middle Upper Paleolithic. As there is no abrupt break in this sequence, it is unlikely that a similar break occurs in more western parts of Europe, since the migrating group must have had to pass through south-central Europe. 2) Continued analysis of archaeological sequences suggest a continuity of assemblages through time instead of displacement [Laville et al, 1980; Straus, 1978]. 3) Even if some major displacements did occur (which is unlikely), with more "modern" or gracile groups prevailing, this does not provide a mechanism for reduction, since the displacing population must have at one time been descended from more robust ancestors.

Gene flow, then, is a factor only in that groups were probably organized in alliance systems with other groups for economic and marital reasons.

This produced a mechanism for the transfer of ideas and genes across space, so that, in general, all European Late Pleistocene/Early Holocene societies evolved in similar ways under broadly equivalent techno-cultural influences.

Thus, changes in selective forces operating on Upper Paleolithic and Mesolithic populations appear to be mainly responsible for the trends. Unfortunately, the critical factors responsible for the selective advantage of having smaller teeth, more gracile crania, and reduced stature over time and in the Mesolithic is exceedingly complicated and difficult to prove. In another place [Frayer, 1978], I have traced changes in absolute and relative variation across the Upper Paleolithic and Mesolithic. Several tests show that variation and reduction in tooth size are correlated; that is, as tooth size reduces there is a concomitant reduction in the variation. This pattern holds in analyses of the total sample, for trends among samples that appear to be drawn from true, biological populations, and for specific patterns of male/female reduction. Consequently, there appears to be little evidence for "relaxed" selection and the probable mutation effect [Brace, 1963, 1964]. Models involving directional selection would appear to be the most likely cause of the trends.

Although it is relatively straightforward to eliminate certain evolutionary models, it is much more difficult to identify specific factors responsible for the changes. This is due to (at least) two factors: 1) To date, we have no detailed knowledge of the underlying genetic system that accounts for such things as tooth size, cranial size or shape, or stature. Other than knowing that these systems are interrelated, of complex inheritance, and influenced greatly by nutrition, use, and other environmental factors, we have few direct clues about the way selection would work to reduce size. It is a sobering fact to realize that changes since 1945 in French stature [Olivier, 1980] are greater than those between the Upper Paleolithic and Mesolithic, and these are certainly not of genetic origin. 2) Even though reduction trends through time are marked, the rate of change necessary to produce the resulting trends is extremely small. For example, maxillary canine size reduces within the Upper Paleolithic from 89.4 mm^2 to 68.3 mm^2 [See Appendix]. This represents a reduction of 1.048 mm^2 per millenium, or 0.0201 mm^2 per generation, using a 21-ky span and a generation length of 20 years. Clearly, this kind of selection intensity, although effective over the long run, defies precise, causative identification.

Nevertheless, it is difficult to imagine that cultural changes did not have an effect on the biology of Upper Paleolithic and Mesolithic groups. For both sexes, technological improvements must have eased the necessity of direct biological solutions, thereby decreasing the selective advantage of larger size or greater muscularity. Since a greater body size demands a greater energy expenditure to maintain it, reduction in size (when technology reduces its need) would be an economical way for selection to proceed. However, the pace would be so slow that it would defy (statistical) proof, other than by inference.

Better evidence for the relationship between technological and biological change is the consistent trend for reduction in sexual dimorphism. Through virtually the whole Upper Paleolithic, hunters killed medium-large game with either hand-held spears or spearthrowers. Based on ethnographic analogy, this activity was the exclusive domain of males, with female subsistence activities more closely tied to gathering vegetable foods and running down smaller game [Murdock and Provost, 1973]. With the development of bow hunting in the LUP (but especially in the Mesolithic) and the change in the type of animals hunted to smaller, more docile species, I suspect that the biological requirements of typical Upper Paleolithic morphotypes were lessened. In the Mesolithic, a smaller male with a bow would be just as an effective hunter and would consume less energy due to reduced size. A form of energy conservation, then, may be responsible for the differential rates of change between males and females. This reduction assumes relative constancy in female gathering techniques and a more marked change in male economic activities from the Upper Paleolithic to Mesolithic. It also assumes that trends for smaller size in males is reflected by a suite of characters ranging from the dentition to stature, so that the shift concerns trends for gracilization in dimensions of interrelated units of the body. The result is reduced differences in male and female morphotypes through time.

With respect to stature, other theories have been proposed to account for both the decrease to the Mesolithic and the trend for reduced dimorphism. Trinkaus [1980], for example, argues for a relationship among body height, mating patterns, and sexual dimorphism. Although this model may describe the situation in many species of birds [Barash, 1977], mammals [Alexander et al, 1978], or even early hominids [Allen et al, 1982], I suspect it is of limited use in explaining sexual dimorphism in the recent evolution of human groups. For the ethnographic present, in societies where polygyny is practiced, aquisition of extra wives is generally based on a male's achieved or acquired social power and the ability (or need) to build alliances with other groups. Combative displays, vigorous territorial defense, and other "physical" measures to insure reproductive success have been replaced by social controls, so that it is not surprising that correlations between body size and marriage type in human groups are not significant [Alexander et al, 1978; Gray and Wolfe, 1980; Wolfe and Gray, 1982].

Other models for changes in dimorphism are based on the effect of poor nutrition on body size in human populations [Ferembach 1978b,c, 1979; Key, 1980]. For example, "la moindre différence sexuelle observée au Mésolithique tient très probablement à une alimentation plus pauvre en protéines animales" [Ferembach, 1979, p 185]. Yet, the basis for this conclusion rests on the traditional assumption that the Mesolithic represents a nutritionally and culturally impoverished period between the Upper Paleolithic and Neolithic. This position receives little support from recent surveys and detailed studies of the Mesolithic [Clarke, 1976; Mellars, 1976; Odell, 1980; Rozoy, 1978], and does not fit well with factors of human biology anyway. Although a possible force of short duration, it is inconceivable that human populations could withstand the pressure of inadequate nutrition for the entire Mesolithic. Consequently, trends for reduction in size (teeth, skulls, and stature) seem best related to modifications and improvements in techno-cultural systems.

ACKNOWLEDGMENTS

Initial support for data collection in 1973/74 was provided by NSF Grant #GS-38067. Two subsequent grants from the National Academy of Sciences (1981, 1983) allowed me the opportunity to collect additional information. Support from both agencies is greatly appreciated. Since the paper was completed while overseas, the office staff at the University of Kansas was instrumental in the completion of the manuscript. Without the help of G. Kirk and C. Riling, the paper would have never been finished. Even more importantly, J. Calcagno assisted in some of the statistical work and oversaw the organizational details and typing of the final version of the manuscript. The effort of these three friends is greatly appreciated. Finally, much of the dental analysis was accomplished through the cooperation of M.H. Wolpoff (Michigan). His constant, critical input has been essential to the genesis of the work and its moments of cohesion. However, the author is responsible for the conclusions in the paper.

LITERATURE CITED

Alexander RD, Hoogland JL, Howard RD, Noon KM, Sherman PW (1978): Sexual dimorphisms and breeding systems in pinnipeds, ungulates, primates and humans. In Chagnon NA, Irons W (eds): "Evolutionary Biology and Human Social Behavior: An Anthropological Perspective." North Scituate; MS: Duxberry Press, pp 402–435.

Allen LL, Bridges PS, Evon DL, Rosenberg KR, Russell MD, Schepartz LA, Vitzthum FJ, Wolpoff MH (1982): Demography and human origins. Am Anthropol 83:888–896.

Bahn PG (1977): Seasonal migration in South-West France during the late glacial period. J Arch Sci 4:245–277.

Barash DP (1977): Sociobiology and Behavior. Amsterdam: Elsevier.

Bianchi F, Borgognini-Tarli SM, Marchi M, Paoli G (1980): An attempt of application of multivariate statistics to the problem of the Italian Mesolithic samples. Homo 31:153–165.

Billy G (1976): Les hommes du Paléolithique supérieur. In de Lumley H (ed): "La Préhistoire Francaise." Paris:CNRS, pp 595–603.

Bilsborough A (1972): Cranial morphology of Neanderthal man. Nature 237:351–352.

Boone Y, Renault-Miskovsky J (1976): La cueillette. In de Lumley H (ed): "La Préhistoire Française." Paris:CNRS, pp 684–687.

Bordaz J (1970): "Tools of the Old and New Stone Age." New York:Natural History Press.

Bordes J (1968): "The Old Stone Age." New York:McGraw-Hill.

Bordes J (1969): Traitement thermique du silex au Solutréen. Bull Soc Préhist Fran 66:197.

Boule M, Vallois HV (1957): "Fossil Men." New York:Dryden Press.

Brace CL (1963): Structural reduction in evolution. Am Nat 97:39–49.

Brace CL (1964): The probable mutation effect. Am Nat 98:453–455.

Brace CL, Mahler PE (1971): Post-Pleistocene changes in the human dentition. Am J Phys Anthrop 34:191–203.

Clarke D (1976): Mesolithic Europe: the economic basis. In Sieveking GG, Longworth IH, Wilson KE (eds): "Problems in Economic and Social Archaeology." London:Duckworth, pp 449–482.

Conover WJ (1971): "Practical Nonparametric Statistics." New York:J. Wiley.

Constandse-Westermann TS (1974): L'homme Mésolithique du Nord-ouest de l'Europe, distances biologiques, considérations génétiques. Bull Mem Soc Anthropol Paris 1 (XIII):173–199.

Corruccini RS (1976): Calvarial shape relationships between fossil hominids. Yrbk Phys Anthropol 18:89–109.

Corruccini RS (1978): Morphometric analysis: uses and abuses. Yrbk Phys Anthropol 21:134–150.

Eveleth PB, Tanner JM (1976): "Worldwide Variation in Human Growth." Cambridge:Cambridge University Press.

Ferembach, D (1969): L'évolution des races au Mésolithique. In Nemeskéri J, Dezsö GY: (eds): "Evolutionary Trends in Recent and Fossil Hominids." Budapest: Akadémiai Kiadó, pp 83–86.

Ferembach D (1974): Les hommes de l'Epipaléolithique et du Mésolithique de la France et du Nord-ouest du bassin Méditerranéen. Bull Mem Soc Anthropol Paris 2(XIII):201–236.

Ferembach D (1976): Les hommes du Mésolithique. In deLumley H (ed): "La Préhistoire Francaise." Paris:CNRS, pp 604–611.

Ferembach D (1978a): Les Natoufiens et l'homme de Combe-Capelle. Bull Mem Soc Anthropol Paris 5(XIII):131–136.

Ferembach D (1978b): A propos des Magdeléniens et des Mésolithiques influence possible de modifications du milieu sur l'évolution morphologique. Bull Mem Soc Anthropol Paris 5(XIII):239–247.

Ferembach D (1978c): Nutrition et évolution morphologique: Application au passage Magdalénien-Mésolithique en France et á la différenciation de populations natoufiennes en Israël. Homo 29:1–6.

Ferembach D (1979): Filiation paléolithique des hommes du Mésolithique. In deSonneville-Bordes D (ed): "La Fin des Temps Glaciaires en Europe." Paris:CNRS, pp 179–188.

Frayer DW (1977): Metric dental change in the European Upper Paleolithic and Mesolithic. Am J Phys Anthropol 46:109–120.

Frayer DW (1978): "Evolution of the Dentition in Upper Paleolithic and Mesolithic Europe." Lawrence:University of Kansas Publications in Anthropology.

Frayer DW (1980): Sexual dimorphism and cultural evolution in the late Pleistocene and early Holocene of Europe. J Hum Evol 9:399–415.

Frayer DW (1981): Body size, weapon use, and natural selection in the European Upper Paleolithic and Mesolithic. Am Anthropol 83:57–73.

Freeman LG (1981): The fat of the land: notes on Paleolithic diet in Iberia. In Harding RSO, Teleki G (eds): "Omnivorous Primates." New York:Columbia University Press, pp 104–165.

Gabel WC (1958): The Mesolithic in Europe. Am Anthropol 60:658–667.

Gamble C (1980): Information exchange in the Paleolithic. Nature 283:522–523.

Gamble C (1982): Interaction and alliance in Paleolithic society. Man 17:92–107.

Glowatzki G, Protsch RRR (1973): Das absolut Alter der Kopfbestattungen in der Grossen Ofnet-Höhle bei Nordlingen in Bayern. Homo 24:1–6.

Gray JP, Wolfe LD (1980): Height and sexual dimorphism in stature among human societies. Am J Phys Anthropol 53:441–456.

Hassan F (1978): Demographic archaeology. In Schiffer MB (ed): "Advances in Archaeological Method and Theory, Vol I." New York:Academic Press, pp 49–103.

Henke W (1981): Entwicklungstrends und Variabilität bei Jungpaläolithikern und Mesolithikern Europas. Homo 32:177–196.

Howells WW (1973): "Cranial Variation in Man." Cambridge:Harvard University Press.

Isaac GL (1972): Chronology and the tempo of cultural change during the Pleistocene. In Bishop WW, Miller JA (eds): "Calibration of Hominid Evolution." Edinburgh:Scottish Academic Press, pp 381–430.

Jelínek J (1972): "Das Grosse Bilderlexikon des Mensch in der Vorzeit." Berlin:Bertelsmann Lexikon-Verlag.

Key P (1980): Evolutionary trends in femoral sexual dimorphism from the Mesolithic to the late Middle Ages in Europe. Am J Phys Anthropol 52:244.

Kowalski CJ (1972): A commentary on the use of multivariate statistical methods in anthropometric research. Am J Phys Anthropol 36:119–132.

Laville H, Rigaud JP, Sackett J (1980): "Rock shelters of the Périgord." New York:Academic Press.

Leroi-Gourhan A (1964): Chronologie des Grotte d'Arcy-sur-Cure (Yonne) Gallia-Préhistoire 8:1–64.

Littlefield A, Lieberman L, Reynolds LT (1982): Redefining race: The potential demise of a concept in physical anthropology. Curr Anthropol 23:641–656.

Marshack A (1979): Upper Paleolithic symbol systems of the Russian plain: Cognitive and comparative analysis. Curr Anthropol 20:271–311.

Martin R (1928): "Lehrbuch der Anthropologie." Jena:Fischer.

Matiegka J (1934): "Homo předmostensis: Fosiliní Člověk z Předmostí na Moravě." Prague: Česka Akademie Věd i Umení.

Meiklejohn C (1978): Ecological aspects of population size and growth in late glacial and early postglacial North-Western Europe. In Mellars P (ed): "The Early Postglacial Settlement of Northern Europe." London:Duckworth, pp 65–79.

Mellars P (1973): The character of Middle-Upper Paleolithic transition in Southwest France. In Renfrew C (ed): "The Explanation of Culture Change." London:Duckworth, pp 41–99.

Mellars P (1976): Settlement patterns and industrial variability in the British Mesolithic. In Sieveking GG, Longworth IH, Wilson KE (eds): "Problems in Economic and Social Archaeology." London:Duckworth, pp 375–400.

Morant GM (1928): Studies of Paleolithic man. Ann Eugen 4:109–214.

Murdock GP, Provost C (1973): Factors in the division of labor by sex: A cross-cultural analysis. Ethnology 12:203–225.

Nemeskéri J, Szathmáry L (1978): Individual data on the Vlasac anthropological series. In Srejović D, Letica Z (eds): "Vlasac: A Mesolithic Settlement in the Iron Gates." Beograd:Serbian Academy of Sciences and Arts, pp 285–426.

Odell GH (1980): Toward a more behavioral approach to archaeological lithic concentrations. Am Antiq 45:404–431.

Olivier G (1980): The increase of stature in France. J Hum Evol 9:645–649.

Parenti R, Messeri P (1962): I resti scheletrici umani del Neolitico Ligure. Palaeontol Ital:50.

Perlès C (1975): L'homme préhistorique et le feu. Recherche 60:829–839.

Perlès C (1976): Le feu. In deLumley H (ed): "La Préhistoire Francaise. Paris:CNRS, pp 679–683.

Pope ST (1962): "Bows and Arrows." Berkeley:University of California Press (re-issue of 1918 edition).

Riquet R (1968): La race de Cro-Magnon, abus de langage ou réalité objective? In Camps G, Olivier G, (eds): "L'Homme de Cro-Magnon." Paris:Arts et Métiers Graphiques, pp 59–72.

Rozoy JG (1978): Les Derniers Chasseurs. Bull Soc Arch Champenoise 2(special number), pp 1–1289.

Rust R (1943): "Die Alt-und Mittelsteinzeitlichen von Stellmoor." Neumünster:K. Wachholtz Verlag.

Saller K (1962): Die Ofnet-Höhle Funde in neuer Zusammensetzung. Zeit Morphol Anthropol 52:1–51.

Semenov SA (1964): "Prehistoric Technology." New York:Barnes and Noble.

Smith FH (1982): Upper Pleistocene hominid evolution in South-Central Europe: A review of the evidence and analysis of trends. Curr Anthropol 23:667–703.

Sós Á, Bökönyi S (1963): "Zalavár." Budapest:Akadémiai Kiadó.

Spiess AE (1979): "Reindeer and Caribou Hunters." New York:Academic Press.

Steele DG (1970): Estimation of stature from fragments of long limb bones, In Stewart TD (ed): "Personal Identification in Mass Disasters." Washington:Smithsonian Press.

Straus LG (1978): Of Neanderthal hillbillies, origin myths, and stone tools: Notes on Upper Paleolithic assemblage variability. Lithic Technol 7:36–39.

Straus LG, Clark GA, Altuna J, Ortea JA (1980): Ice-age subsistence in Northern Spain. Sci Am 242(6):142–152.

Stringer CB (1974): Population relationships of the later Pleistocene hominids: A multivariate study of available crania. J. Arch Sci 1:317–342.

Stringer CB (1982): Towards a solution to the Neanderthal problem. J Hum Evol 11:431–438.

Tóth T (1972): On the importance of the analysis of morphological modifications in paleoanthropology. In Törö I, Szabady E, Nemeskéri J, Eiben OG (eds): "Advances in the Biology of Human Populations." Budapest: Akadémiai Kiadó, pp 463–472.

Trinkaus E (1980): Sexual differences in Neanderthal limb bones. J Hum Evol 9:377–397.

Trotter M, Gleser GC (1952): Estimation of stature from long bones of American Whites and Negroes. Am J Phys Anthropol 10:463–514.

Vlček E (1967): Morphological relations of the fossil human types Brno and Cro-Magnon in the European Late Pleistocene. Folia Morphol 15:214–221.

White R (1982): Rethinking the Middle/Upper Paleolithic transition. Curr Anthropol 23:169–192.

Wobst HM (1976): Locational relationships in Paleolithic society. In Ward RH, Weiss KM (eds): "The Demographic Evolution of Human Populations." New York:Academic Press, pp 49–58.

Wolfe LD, Gray JP (1982): A cross-cultural investigation into sexual dimorphism of stature. In Hall RL (ed): "Sexual Dimorphism in *Homo sapiens.*" New York: Praeger, pp 197–230.

Wolpoff MH (1971): "Metric Trends in Hominid Dental Evolution." Cleveland:CWRU Press.

Wolpoff MH (1980): "Paleoanthropology." New York:Knopf.

Y'Edynak G (1978): Culture, diet, and dental reduction in Mesolithic foragers-fishers in Yugoslavia. Curr Anthropol 19:616–618.

APPENDIX

Appendix TABLE I. Late Mousterian and Upper Paleolithic Maxillary Dental Data

	NEA $\overline{X}$	NEA n	NEA s.d.	EUP $\overline{X}$	EUP n	EUP s.d.	MUP $\overline{X}$	MUP n	MUP s.d.	LUP $\overline{X}$	LUP n	LUP s.d.
I1 L	9.6	(15)	.8	9.7	(5)	.8	9.1	(13)	.5	8.7	(11)	.7
B	8.3	(17)	.5	7.6	(5)	.6	7.6	(18)	.5	7.5	(12)	.6
I2 L	8.0	(13)	.5	6.9	(2)	–	7.2	(12)	.7	6.8	(7)	.5
B	8.2	(16)	.5	7.3	(3)	–	6.8	(16)	.5	6.5	(11)	.6
C L	8.6	(12)	.7	9.0	(4)	.8	7.9	(18)	.6	7.9	(10)	.5
B	9.8	(13)	.5	9.9	(6)	1.1	8.9	(20)	.7	8.8	(11)	.6
P3 L	7.5	(13)	.6	7.3	(2)		7.2	(18)	.5	6.9	(13)	.8
B	10.3	(15)	.6	9.7	(2)		9.6	(19)	.7	9.6	(13)	.6
P4 L	7.5	(9)	.6	6.6	(4)	.4	7.1	(19)	.6	6.7	(11)	.7
B	10.4	(9)	.5	9.2	(5)	.5	9.8	(20)		9.5	(10)	.4
M1 L	11.5	(19)	.9	10.9	(9)	.5	10.7	(30)	.7	10.5	(18)	.5
B	12.0	(19)	.6	12.3	(9)	.8	12.1	(29)	.8	11.9	(18)	.6
M2 L	10.8	(17)	.6	10.5	(5)	1.1	10.3	(24)	1.1	9.7	(17)	.4
B	12.6	(17)	.8	12.4	(6)	1.1	12.4	(23)	1.1	11.9	(17)	.7
M3 L	9.5	(11)	.6	9.1	(5)	.4	9.3	(15)	1.1	8.8	(9)	1.0
B	12.9	(11)	.8	11.1	(5)	1.1	11.8	(14)	.9	11.5	(10)	.8
Areas												
C	85.0	(11)	10.6	89.4	(4)	16.4	69.8	(18)	10.3	68.3	(10)	10.0
P3	76.5	(13)	10.2	71.6	(2)		69.6	(18)	9.2	66.8	(13)	10.5
P4	78.1	(9)	9.3	61.2	(4)	6.0	70.3	(19)	11.0	62.5	(10)	8.1
M1	138.1	(19)	13.4	134.9	(8)	14.1	130.3	(29)	16.1	124.2	(18)	9.8
M2	134.7	(16)	12.3	131,6	(5)	21.9	128.4	(23)	20.3	114.8	(17)	10.5
M3	122.4	(11)	10.8	101.3	(5)	13.1	110.5	(14)	19.2	101.5	(9)	16.8

Appendix TABLE II. Late Mousterian and Upper Paleolithic Mandibular Dental Data

	NEA			EUP			MUP			LUP		
	$\overline{X}$	n	s.d.	$\overline{X}$	n	s.d.	$\overline{X}$	n	s.d.	$\overline{X}$	n	s.d.
I1 L	6.0	(6)	.3	6.3	(7)	.5	5.7	(15)	.6	5.2	(10)	.6
B	7.5	(11)	.4	7.4	(7)	.4	6.2	(17)	.4	6.1	(12)	.5
I2 L	6.6	(11)	.5	6.8	(6)	.4	6.3	(18)	.7	5.5	(12)	.5
B	7.8	(13)	.6	7.7	(6)	.4	6.7	(21)	.5	6.5	(16)	.6
C L	7.8	(16)	.5	7.5	(5)	.4	7.3	(16)	.6	6.7	(14)	.5
B	8.8	(19)	.8	8.9	(8)	.7	8.4	(17)	.7	7.9	(16)	.6
P3 L	7.9	(17)	.4	7.3	(5)	.7	7.1	(18)	.7	7.0	(16)	.6
B	9.0	(17)	.8	8.7	(6)	.6	8.3	(18)	.5	8.1	(18)	.8
P4 L	7.6	(15)	.6	7.9	(6)	.6	7.2	(17)	.5	6.8	(16)	.5
B	8.8	(13)	.8	8.9	(6)	.7	8.5	(17)	.4	8.3	(16)	.5
M1 L	11.6	(30)	.7	11.5	(6)	.8	11.5	(33)	.8	11.1	(29)	.7
B	11.0	(28)	.6	11.1	(7)	.6	10.9	(34)	.7	10.9	(32)	.6
M2 L	11.9	(19)	.6	11.1	(9)	.7	11.2	(26)	1.0	10.9	(29)	.7
B	11.1	(19)	.8	10.8	(10)	.7	10.8	(26)	.7	10.7	(29)	.7
M3 L	11.7	(17)	.6	11.0	(3)		11.2	(19)	1.3	10.5	(20)	.8
B	11.3	(19)	1.2	10.3	(3)		10.7	(18)	1.0	10.3	(19)	.8
Areas												
C	68.5	(15)	10.9	69.5	(5)	7.5	61.3	(16)	8.7	53.5	(14)	7.4
P 3	71.6	(17)	8.8	64.7	(5)	6.0	59.3	(18)	8.4	56.4	(16)	9.6
P 4	66.2	(13)	9.1	71.0	(5)	9.4	60.8	(17)	6.9	55.9	(15)	5.5
M 1	127.0	(28)	12.0	127.2	(6)	10.2	124.7	(32)	14.7	120.3	(29)	12.5
M 2	131.9	(19)	14.5	120.0	(9)	15.1	121.4	(25)	18.0	116.2	(29)	13.5
M 3	133.3	(17)	19.7	116.0	(3)		120.1	(18)	24.1	109.2	(19)	16.0

Appendix TABLE III. Male and Female Means for Mandibular and
Maxillary Canine and Posterior Tooth Areas

	Upper Paleolithic						Mesolithic					
	Males			Females			Males			Females		
Mandible	$\overline{X}$	(n)	s.d.	$\overline{X}$	(n)	s.d.	$\overline{X}$	(n)	s.d.	$\overline{X}$	(n)	s.d.
C	63.7	(13)	11.6	54.4	(13)	6.2	56.2	(47)	5.6	49.5	(36)	5.9
P3	60.2	(12)	8.4	55.0	(15)	5.4	55.7	(52)	4.9	52.0	(37)	5.3
P4	63.9	(13)	6.9	56.7	(14)	4.3	58.9	(49)	6.1	57.0	(38)	5.9
M1	130.4	(16)	11.8	117.8	(18)	11.5	128.1	(48)	11.2	118.0	(39)	9.8
M2	128.0	(17)	16.3	115.5	(18)	15.6	118.7	(52)	11.3	106.8	(41)	9.6
M3	122.9	(10)	18.0	107.4	(15)	19.2	111.9	(50)	12.6	102.2	(34)	10.3
Maxilla												
C	76.8	(12)	13.6	65.3	(14)	5.3	70.6	(40)	7.1	63.4	(30)	7.0
P3	74.2	(11)	9.1	62.2	(15)	6.5	65.1	(43)	7.0	63.8	(29)	6.5
P4	73.0	(10)	10.9	61.1	(15)	6.8	64.7	(44)	7.2	62.3	(33)	6.1
M1	133.5	(14)	11.1	123.3	(22)	12.6	126.1	(54)	9.5	118.0	(40)	11.1
M2	132.5	(15)	18.2	118.1	(20)	16.3	116.7	(56)	13.0	110.1	(44)	10.7
M3	110.1	(9)	21.7	98.8	(12)	13.3	98.5	(46)	13.0	96.8	(27)	13.0

The Origins of Modern Humans: A World Survey of the Fossil Evidence, pages 251–293
© 1984 Alan R. Liss, Inc., 150 Fifth Avenue, New York, NY 10011

Western Asia

Erik Trinkaus

Department of Anthropology, University of New Mexico, Albuquerque,
New Mexico 87131

INTRODUCTION

The past few decades have seen major increases in our knowledge of the human fossil record. For the phases of human evolution prior to the Upper Pleistocene, it has been primarily additions to the fossil record that have augmented our understanding of the course of human evolution. For the Upper Pleistocene, it has consisted more of a reevaluation of well known specimens in the light of new interpretive frameworks, supplemented by information and insights from new fossils. From this research, we are beginning to obtain a more sophisticated understanding of the later phases of human evolution than has been possible previously or is feasible for earlier periods of human evolution.

The primary focus of Upper Pleistocene human paleontology has been, and remains, the origins of anatomically modern humans. The vast majority of the analyses of human remains from the Upper Pleistocene, whether they bear directly or only indirectly on the question, contain references to the possible origins of modern humans, the evolutionary processes responsible for that evolutionary event, and the fate of the preceding archaic *Homo sapiens* (usually Neandertal) populations. Although this is not always fully justifiable, it is understandable given the importance of this evolutionary event to our perception of the adaptive significances of modern human morphology [Smith, 1983; Trinkaus, 1982a, 1983b].

For these reasons, I would like to review the evidence for the origins of anatomically modern humans in one region of the Old World, western Asia. This geographical area, which is here taken to include the Near East from the Mediterranean to the Hindu Kush, Soviet Central Asia, the Caucasus, and the eastern and northern shores of the Black Sea, has yielded sufficient human fossil remains from Upper Pleistocene contexts to provide an indication of human evolutionary processes during the past 125 ky. There are still many unanswered questions, ones that could be resolved by additional finds of prehistoric remains, but a sufficient amount is known to indicate the general patterns of later human evolution in this centrally located region of the Old World.

HISTORICAL BACKGROUND

The first major Paleolithic exploration of the Near East, with its associated discoveries of fossil hominids, began with the opening up of the Near East to Europeans after the fall of the Ottoman Empire and the establishment in 1920 of a French mandate in Lebanon and Syria and a British one in Jordan and Palestine. Various workers had previously collected prehistoric lithic implements eroding out of river beds and caves in the area, and a few fragmentary human remains had been discovered in 1893 in the Grotte d'Antélias, probably associated with an Aurignacian level of that cave [Zumoffen, 1893]. However, it was five individuals, D. A. E. Garrod, R. Neuville, A. Rust, M. Stekelis, and F. Turville-Petre, who, during the 1920s and 1930s, helped to inaugurate Paleolithic archeology in the Near East and discovered many of the human fossil skeletons that bear on the question of the origins of anatomically modern humans in this region.

In 1925 and 1926, Turville-Petre undertook the first of these Paleolithic expeditions and excavated the site of Mugharet el-Zuttiyeh in the Wadi Amud near Lake Kinneret [Turville-Petre, 1927]. In the lower levels of the cave he discovered an upper facial skeleton of the first archaic human to be known from western Asia [Keith, 1927]. Shortly afterwards, in 1928, Garrod excavated Shukbah Cave, near Jerusalem, and discovered a few human fragments, some of which were associated with Middle Paleolithic assemblages [Garrod and Bate, 1942; Keith, 1931]. This field work was followed by major excavations from 1929 to 1934 by Garrod in three caves in the Wadi el-Mughara in the Mount Carmel range south of Haifa: Mugharet es-Skhul, Mugharet et-Tabun, and Mugharet el-Wad [Garrod and Bate, 1937; Ronen, 1982]. The first two of these sites yielded some of the more important human skeletal remains known so far from western Asia, which were made all the more important since they were quickly described in detail by McCown and Keith [1939].

Additional discoveries of human remains were made by Turville-Petre in 1931 in the Aurignacian and Kebaran levels of the Mugharet el-Kebara, also in Mount Carmel [Garrod, 1954; McCown and Keith, 1939], and by Neuville and Stekelis at Jebel Qafzeh, near Nazareth, from 1933 to 1935 [Neuville, 1951]. These Levantine fossils have been added to recently by discoveries of archaic humans at the Amud Cave, also in the Wadi Amud

[Suzuki and Takai, 1970], the Mousterian levels of Mugharet el-Kebara [Smith and Arensburg, 1977], and Gesher Benot Ya'acov in the Jordan Valley [Geraads and Tchernov, in press]. In addition, there have been a number of early anatomically modern human partial skeletons discovered by B. Vandermeersch in the terrasse deposits of Jebel Qafzeh [Vandermeersch, 1981a, personal commun.], and an Aurignacian skeleton was unearthed at the site of Nahal 'En-Gev I on the eastern shore of Lake Kinneret [Arensburg, 1977].

During this same time, Soviet archeologists were beginning to explore the Paleolithic of the Crimea and regions to the east of it. The first significant human fossil find in this region was made at Kiik-Koba in the Crimea where, during excavations from 1924 to 1926, G. A. Bonč-Osmolovskij unearthed postcrania of an adult male and an infant [Bonč-Osmolovskij, 1940, 1941, 1954; Vlček, 1973]. Subsequently, in 1939, A. P. Okladnikov excavated a burial of an 8–10-year-old Neandertal in the cave of Teshik-Tash in southern Uzbekistan [Okladnikov, 1940; Gremyatskij and Nesturkh, 1949]. More recently there have been additional discoveries of Mousterian human remains in the Crimea; in 1953 and 1954 A. A. Formozov unearthed an infant's skeleton and fragments of an adult, both anatomically modern, from the site of Starosel'e [Formozov, 1954, 1957; Alexeyev, 1976], and in 1972 remains of an archaic human child were discovered at the site of Zaskalnaya VI [Kolosov et al, 1974, 1975]. In addition, in 1968 a partial human mandible was found associated with an Acheulian assemblage in Azykh Cave, in the Lesser Caucasus [Roginskij and Levin, 1978].

Relatively little was known of western Asian Paleolithic human remains outside of the Levant and Soviet Union until C. S. Coon discovered a human radial shaft in association with Middle Paleolithic tools in Bisitun Cave, western Iran in 1949 [Coon, 1951].[1] Shortly thereafter, R. S. Solecki began excavations in Shanidar Cave, in the Zagros Mountains of Iraqi Kurdistan. Between 1953 and 1960 Solecki unearthed the partial skeletons of nine Neandertals, providing the largest sample of Upper Pleistocene archaic humans from outside of Europe and one of the largest samples of fossil human partial skeletons known [Solecki, 1960, 1963, 1971; Stewart, 1977; Trinkaus, 1983a]. More recently, in 1966, L. Dupree excavated an anatomically modern human temporal bone from a probably late Mouster-

---

[1]The "human femur" listed by Coon [1951, 1975] from Tamtama Cave near Lake Urmia is non-hominid and probably *Cervus elaphus*, and the "human incisor" from Bisitun Cave is bovid.

ian level of Darra-i-Kur in northern Afghanistan [Angel, 1972; Dupree, 1972].

The fossil sample from western Asia also includes a number of isolated teeth and skeletal fragments [Howell and Fritz, 1975; Klein and Ivanova, 1971; Kolosov et al, 1975; Waechter, 1975], few of which provide significant paleontological data. Most of our paleontological information concerning the origin of anatomically modern humans in western Asia comes from nine sites: Amud Cave, Mugharet el-Kebara, Jebel Qafzeh, Mugharet es-Skhul, and Mugharet et-Tabun in Israel, Shanidar Cave in Iraq, Kiik-Koba and Starosel'e in the Crimea, and Teshik-Tash in Uzbekistan. Additional important information for late Middle and initial Upper Pleistocene hominids comes from Gesher Benot Ya'acov and Mugharet el-Zuttiyeh.

This sequence of discoveries of Upper Pleistocene human remains in western Asia, and the timing of these discoveries relative to finds elsewhere in the Old World, and especially in Europe, have influenced the ways in which the course of later human evolution in this region has been viewed.

When the Zuttiyeh face was discovered and described in the late 1920s, the available reference fossils consisted of abundant European specimens (mostly Neandertals and anatomically modern humans), the Broken Hill (Kabwe) remains from Zambia [Smith Woodward, 1921], and the Trinil *Homo erectus* remains from Java [Dubois, 1894]. With these reference samples available, it is not surprising that Keith [1927] found the greatest morphological affinities of the Zuttiyeh specimen to be with the European Neandertals, even though he recognized certain unique features of the specimen [Keith, 1931].

When the "Mount Carmel" remains from the sites of Mugharet es-Skhul and Mugharet et-Tabun were analyzed in the 1930s by McCown and Keith, they rejected the possibility that the skeletal remains represented two or more temporal samples and considered them to be one large and variable sample. Since the remains from Skhul were more abundant and complete, this had the effect of making their interpretations, and those of many subsequent workers, influenced primarily by the morphology of the Skhul specimens. Since the Skhul specimens are robust but anatomically modern, as was recognized by McCown and Keith [1939], they concluded that the "Mount Carmel" sample was intermediate between the Neandertals and early anatomically modern humans ("Cro-Magnons") but had its closest affinities to the latter group. It was not until the late 1950s that the Tabun and Skhul remains were separated for the purposes of analysis, and it became recognized that the Tabun remains were from an archaic (Neandertal-like) population and that the Skhul individuals were more closely related to other early anatomically modern humans [Howell, 1958, 1959].[2]

Using the faunal analyses of Bate [Garrod and Bate, 1937] as a guide, McCown and Keith [1939] dated the "Mount Carmel" remains to the end of the last interglacial; this had the effect of making them older than the European Neandertals of the early last glacial, and therefore providing a possible parent population for the early anatomically modern humans of the European Upper Paleolithic. Even though the dating of the Skhul Layer B remains to the last interglacial was rejected during the 1950s [Higgs, 1961; Howell, 1958], and it was shown that they derive from the middle of the last glacial, the idea that there were modern-appearing humans in western Asia at the same time as or before the Neandertals persists [Bar Yosef and Vandermeersch, 1981; Vandermeersch, 1981a,b].

It was about this time that the fossil humans excavated at Jebel Qafzeh by Neuville in Middle Paleolithic levels were recognized as fully anatomically modern and most similar to the specimens from Skhul [Howell, 1958, 1959]. The discoveries of additional specimens at Qafzeh and analyses of the entire sample by Vandermeersch [Vallois and Vandermeersch, 1972; Vandermeersch, 1981a] have more than documented this fact. However, it is still debated as to whether the Qafzeh hominids postdate the western Asian archaic humans or, as with the earlier interpretations of the Skhul material, were contemporary with archaic humans in this region [Bar Yosef and Vandermeersch, 1981; Jelínek, 1982a,b; see below].

The human paleontological research in the last few decades, both analyses of new specimens and reevaluations of previously known fossils, has served to document three things: first, the basically anatomically modern human morphologies of the Skhul, Qafzeh, and Starosel'e specimens [Alexeyev, 1976; Howells, 1970; Santa Luca, 1978; Stringer, 1978; Trinkaus, 1976a,b, 1983a; Vandermeersch, 1981a]; secondly, the close morphological affinities of the Amud, Shanidar, Tabun Layer C, and Teshik-Tash specimens to the European "Classic" Neandertals [Alexeyev, 1981; Santa Luca, 1978; Stewart, 1977; Suzuki and Takai, 1970; Trinkaus, 1983a]; and third, the distinctiveness of the Zuttiyeh facial skeleton from the Neandertals and its affinities to the Early Neandertals of Europe [Vandermeersch, 1981b;

---

[2]There is still uncertainty as to whether the Tabun C2 mandible is best included with the Skhul remains in an early modern sample or placed with the more archaic Tabun C1 and C3 remains [Vandermeersch, 1981a; Trinkaus, 1983a]. The implied midfacial prognathism resembles that of the Neandertals, whereas the symphyseal morphology is more closely aligned with that seen at Skhul and Qafzeh.

Trinkaus, 1982b, 1983a]. There are still a number of unresolved issues concerning these fossils and the course of Upper Pleistocene human evolution in western Asia; some will be solved by better geological dating, some require additional fossil evidence, and all need continuing analysis of the morphological patterns evident in the fossils.

THE FOSSIL SAMPLES

The Upper Pleistocene fossil human remains from western Asia fall largely into four samples, two of archaic humans and two of anatomically modern humans. The geologically oldest sample consists of remains that derive from late Middle Pleistocene and last interglacial (initial Upper Pleistocene) deposits; it consists of the specimens from Azykh, Gesher Benot Ya'acov, Tabun Layer E, and Zuttiyeh. The geologically youngest sample consists of modern-appearing humans from the Aurignacian of 'En-Gev I and Kebara. The two most important samples are those that abut the transition from archaic to anatomically modern humans. Both of these samples contain human remains from Middle Paleolithic contexts. The archaic group, referred to here as Neandertals because of their morphological resemblances to the European specimens of that group [Suzuki and Takai, 1970; Trinkaus, 1983a], contains the remains from Mousterian levels of Amud, Bisitun, Kebara, Kiik-Koba, Shanidar, Tabun Layers C and D, Teshik-Tash, and Zaskalnaya VI. The other sample, referred to here as early anatomically modern humans, consists of the large samples from Qafzeh and Skhul supplemented by the remains from Darra-i-Kur and Starosel'e.

These specimens are those that are diagnostic as to group; most provide significant paleontological data. There are a number of other specimens, such as those from Akhshtyr', Antélias, Dzhruchula, Ksar 'Akil, Masloukh, Ras el-Kelb, Rozhok, Shubbabiq, Shukbah, and Tabun Layer B [see references in Howell and Fritz, 1975; Klein and Ivanova, 1971; Waechter, 1975], that cannot be confidently assigned to one of these samples, provide little significant paleontological data, and/or have been very incompletely described and are not available for study. They will not be considered further.

The geological ages of most of these Upper Pleistocene hominid specimens are far from secure. However, it is possible to arrange most of the specimens in a reasonable chronological framework, one that permits the assessment of patterns and rates of change through time.

The earliest of these western Asian human specimens are those from Azykh, Gesher Benot Ya'acov, Zuttiyeh, and Tabun Layer E. The first is associated with an Acheulian industry and probably dates to the end of the

Middle Pleistocene or the very beginning of the last interglacial. The second are associated with a Middle Pleistocene fauna, probably from the late Middle Pleistocene [Geraads and Tchernov, in press]. The third almost certainly derived from the Yabrudian level in Zuttiyeh Cave [Gisis and Bar Yosef, 1974], which is generally known from the last interglacial [Jelínek, 1982a]; this date is supported by a thorium/uranium date of 148 ky BP for the travertine underlying the Yabrudian level at Zuttiyeh [Schwarcz, 1980]. The last derives from deposits that date to the end of the last interglacial, on the basis of archeological and geological analyses of the material from Tabun Cave [Farrand, 1982; Jelínek, 1982b].

The relative ages of the Neandertals are more difficult to determine. The specimens from Amud, Bisitun, Kiik-Koba, Teshik-Tash, and Zaskalnaya VI can be merely referred to the early last glacial, on the basis of their associations with Middle Paleolithic industries. The Shanidar Neandertals from the top of the Middle Paleolithic levels at Shanidar (Shanidar 1, 3, and 5) are associated with radiocarbon determinations that place their age at greater than 45 ky BP [Vogel and Waterbolk, 1963], but their actual ages could be considerably older given the difficulties in obtaining finite radiocarbon dates greater than 40 ky BP [Trinkaus, 1983a]. The earlier Shanidar specimens (Shanidar 2, 4, 6, 7, 8, and 9), from the middle of the Shanidar Middle Paleolithic levels, were originally assigned an age of 60 ky BP by Solecki [1963], but their actual ages may be older, possibly at the end of the last interglacial [Trinkaus, 1983a]. The Tabun C2 mandible, from Layer C of Tabun Cave, dates to between 50 and 60 ky BP [Jelínek, 1982a]. The Tabun C1 partial skeleton may derive from Layer C, but there is a good possibility that it comes from Layer D, which would make it 70 to 80 ky old [Jelínek, 1982a, personal commun.]. The Kebara 1 infant, on the basis of lithic comparisons with the Tabun sequence [Jelínek, 1982a], appears to be about 50 ky BP.

The early anatomically modern humans from Darra-i-Kur and Starosel'e appear to derive from the ends of the Mousterian sequences in Afghanistan and the Crimea, respectively [Dupree, 1972; Klein, 1965]. The Skhul Layer B remains likewise appear to be relatively late within the Middle Paleolithic, postdating Layer B at neighboring Tabun, and are probably in the vicinity of 40 ky BP [Higgs, 1961; Howell, 1959; Jelínek, 1982a; Jelínek et al, 1973].

Agreement on the age of the Qafzeh human remains has been more elusive. Comparisons of the lithics from the Qafzeh Mousterian levels to those from Skhul, Tabun, and other Levantine sites suggest that the Qafzeh levels are either similar in age to those of Skhul or slightly younger [Jelínek, 1982a,b]. Earlier assessments of the Qafzeh lithics [Neuville, 1951] and recent analyses of the Qafzeh microfauna, stratigraphy, and sedimentology have suggested that the Qafzeh human remains were contemporaneous with Layer D at Tabun [Bar Yosef and Vandermeersch, 1981; Farrand, 1979; Haas, in Jelínek et al, 1973; Vandermeersch, 1981a]. All of the techniques that have been used to date the Qafzeh remains have limitations, and each is insufficient to confirm a date. However, the younger date would alleviate the need to account for how two distinct groups of *Homo sapiens* could have coexisted for several tens of millennia in the small area of northern Israel, using the same cultural adaptive complex and yet remaining biologically distinct. For the purposes of this discussion, I will employ Occam's Razor, use the date for the Qafzeh remains that creates the fewest problems, and consider them to date to around 40 ky BP, about the same age as the Skhul Layer B remains.

Following these considerations, the western Asian Upper Pleistocene remains can be arranged chronologically as follows: a late Middle Pleistocene/initial Upper Pleistocene sample including Azykh, Gesher Benot Ya'acov, Tabun E, and Zuttiyeh; an initial last glacial sample with the early Shanidar sample and possibly Tabun C1; a later early last glacial sample with Kebara 1, the later Shanidar sample, Tabun C2, and probably the Neandertals from Amud, Bisitun, Kiik-Koba, Teshik-Tash, and Zaskalnaya VI; an early anatomically modern human sample from the middle of the last glacial including the specimens from Darra-i-Kur, Qafzeh, Skhul, and Starosel'e; and a later last glacial Upper Paleolithic sample. The arrangement of these specimens has been influenced by morphological considerations, but all of the proposed dates are well within the time limits indicated by their archeological and geological contexts.

ADAPTIVE CHANGES AT THE ORIGINS OF MODERN HUMANS: THE VIEW FROM THE FOSSILS

The evolutionary origins of modern humans have been investigated primarily from a phylogenetic point of view. The patterns of human anatomical change at the time of the archaic to anatomically modern human transition, and in the preceding and succeeding time periods, have therefore been investigated chiefly to determine whether early modern-appearing humans could have evolved locally out of preceding archaic humans or whether there had to have been significantly elevated levels of gene flow to explain the anatomical changes evident in the fossil record. The behavioral implications of these anatomical changes have been studied primarily to determine whether the observed changes could be accounted for by minor shifts in selective pressures or patterns of development.

These phylogenetically oriented investigations have accumulated considerable data on Upper Pleistocene human evolution, especially for Europe and the Near East, but they have brought us little closer to general agreement on this basic question (see Smith [1982] and Discussion therein). Once there is agreement that at least some of the Neandertals could have contributed to the gene pools of subsequent populations of early anatomically modern humans, it may not be possible to determine from the fossil record to what extent the Neandertals can be included in the ancestry of more recent humans.[3] It may therefore be more profitable to investigate this time period from a strictly functional point of view, leaving the phylogenetic question until there is a greater understanding of human adaptive evolution during the Upper Pleistocene.

For these reasons, I would like to review the human fossil evidence for changes in human behavior around the time of the transition from Neandertals to early modern-appearing humans in western Asia, including data from older and more recent specimens to the extent that they provide data on trends that help to elucidate changes at the transition.

Postcranial Robusticity

The most apparent difference between the postcrania of western Asian archaic humans and their early modern-appearing successors is in their levels of robusticity. The western Asian Neandertals, like their European contemporaries, were heavily built and the Qafzeh and Skhul postcrania appear gracile by comparison.

The western Asian Neandertal cervical vertebrae (those from Shanidar 1 and 2 [Trinkaus, 1983a]) have long, horizontal, and robust spines, that lie at the upper end of recent human ranges of variation. Those of Skhul 4 and 5, in contrast, are shorter (especially relative to body size) and thinner, although their C6 and C7 spines are relatively horizontal [McCown and Keith, 1939]. Similarly, the Shanidar and Tabun Neandertals, like the European ones, have ribs that are exceptionally thick in cross-section, whereas those from Qafzeh and Skhul are relatively thin, as are those of most recent humans [McCown and Keith, 1939; Trinkaus, 1983a; Vandermeersch, 1981a]; interestingly the Amud 1 Neandertal has relatively thin ribs similar to those from Skhul and Qafzeh [Endo and Kimura, 1970].

---

[3]The almost universal acceptance of the Neandertals and other archaic Upper Pleistocene humans into the species *Homo sapiens* indicates that human paleontologists accept the possibility of gene flow between these archaic humans and anatomically modern humans.

These suggest a reduction in the degree of hypertrophy of the nuchal musculature across the transition, associated with some decrease in the amount of stress being placed upon the trunk by muscles attaching to the ribs (eg, pectoralis major and minor, serratus anterior, and the erector spinae package).

In the upper limb there were a number of morphological shifts, all of which suggest a major decrease in muscle development across this transition. The most apparent are in the scapula and hand, but there are reflections as well in the arm bones.

Four western Asian Neandertals, Shanidar 1, 2, 3, and 4, provide scapular breadth ("morphological length") measurements (c 110.0 mm, c 116.0 mm, c 115 mm, and c 115.0 mm, respectively) that are well above those of Qafzeh 9 and Skhul 4 and 5 (93.0 mm, 95.0 mm, and 102.5 mm, respectively). The differences are more pronounced when these scapular breadths are normalized for body size using humeral maximum length (Shanidar 1, 3, and 4: c 32.8, c 36.1, c 37.7, respectively; Qafzeh 9: c 28.2; Skhul 4 and 5: 28.2 and c 27.0, respectively). Since the development of the supraspinatus and infraspinatus surfaces of the scapula is influenced by the degree of hypertrophy of the rotator cuff muscles [Doyle, 1977], these differences suggest a marked decrease in the sizes of these muscles in early anatomically modern human populations.

Similar changes are evident in the axillary borders of their scapulae. All of the Qafzeh and Skhul scapulae (N = 4) exhibit the bisulcate pattern on their axillary borders, whereas 83.3% (N = 6) of the western Asian Neandertal scapulae have the dorsal sulcus pattern (one, Shanidar 3, has the bisulcate pattern). This indicates a reduction in the average size of the teres minor muscle across the transition [Trinkaus, 1977], in association with the other reductions of the rotator cuff muscles.

In the forearm, there is a shift in the development of the muscles and their moment arms involved in pronation. The two pronators, pronator teres and quadratus, operate around an oblique axis of rotation through the radial and ulnar heads [Youm et al, 1979], and increases in the lateral bowing of the radius serve to augment the moment arms for these muscles. Western Asian and European early anatomically modern humans have radial curvature indices similar to those of recent humans (Table I), whereas Neandertals, especially those from Europe, tend to have highly bowed radii. Shanidar 4 and Tabun C1 have indices (5.48 and 5.81, respectively) similar to those of the European Neandertals, but Shanidar 1 and 6 have degrees of radial bowing similar to that of more recent humans (indices of 3.03 and 3.01, respectively). Associated with the average decrease in radial bowing

TABLE I. Comparisons of Upper Pleistocene Forearm Bones

		Radius shaft curvature index[a]	Ulna pronator quadratus index[b]
Western	$\overline{X}$	4.33	143.3
Asian	SD	1.52	11.7
Neandertals	N	4	4
Western Asian	$\overline{X}$	2.73	122.7
early anat.	SD	1.49	11.6
mod. humans	N	4	4
European	$\overline{X}$	5.86	148.0
Neandertals	SD	0.97	14.5
	N	7	3
European	$\overline{X}$	2.59	
early anat.	SD	0.52	
mod. humans	N	8	

[a](Subtense to the most lateral margin of the shaft from the lateral shaft chord/lateral shaft chord (M6)) × 100.
[b](Maximum shaft diameter across the pronator quadratus crest/minimum shaft diameter at same level) × 100.

across the transition, is a reduction in the development of the pronator quadratus crest on the distal ulnar diaphysis. Despite some overlap between the samples, there is a clear decrease in the relative size of the insertion crest for this important muscle for pronation between archaic and early modern western Asian human groups (Table I).

An identical pattern of decrease in muscular hypertrophy is evident in the hand remains of these western Asian Upper Pleistocene humans. This is apparent especially in the relative development of their palmar carpal tuberosities, thenar muscle insertions, pollical phalangeal proportions, and distal phalangeal tuberosities (Fig. 1). In all of these features the Neandertals (specimens from Amud, Kiik-Koba, Shanidar, Tabun, and Zaskalnaya VI) are considerably more developed than the early anatomically modern humans (specimens from Qafzeh and Skhul).

The palmar tuberosities of Neandertal carpals, especially their trapezial tuberosities, hamulae, and pisiform bones, are large, robust, and extend markedly from the bodies of their bones. This has the effect of creating large carpal tunnels for presumably large digital flexor tendons, large attachment areas for their flexor retinaculae and thenar and hypothenar muscles, enlarged moment arms for flexor carpi ulnaris, and enlarged moment arms for those muscles that act around the carpometacarpal articulations [Stoner,

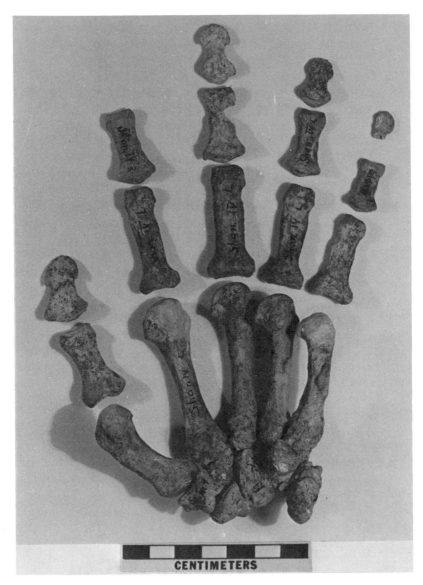

Fig. 1. Palmar view of the Shanidar 4 left hand. The carpals and metacarpals have been assembled by gluing their articular surfaces together, and the phalanges have been placed in their approximate anatomical positions. The heads of the second and fifth metacarpals have been reconstructed. Note the development of the trapezial tubercle, the hamulus, the muscular crests for the opponens pollicis and opponens digiti minimi muscles on the first and fifth metacarpals, respectively, the pits for the insertions of the long flexor tendons, and the expansions of the distal tuberosities, especially on digits three and four.

1981; Trinkaus, 1982c, 1983a]. The Qafzeh 8 and 9 and Skhul 4 and 5 carpals have, in contrast, palmar tuberosities similar in size to those of recent humans [McCown and Keith, 1939; Trinkaus, 1983a; Vander-meersch, 1981a].

The differences in thumbs are evident in several features. Amud 1, Kiik-Koba 1, and Shanidar 4 have exceptionally large opponens pollicis crests on their first metacarpal diaphyses, similar to those of European Neandertals [Kimura, 1976; Stoner, 1981; Trinkaus, 1983a; Vlček, 1975; Fig. 2], even though the opponens pollicis crest of Tabun C1 is rather modest [McCown and Keith, 1939]. Those of Qafzeh 8 and 9 and Skhul 4 and 5 are small, similar to those present on robust modern human hands. This large crest indicates hypertrophy of this important muscle for manipulation and an enlargement of its moment arm relative to the long axis of the first metacar-pal, around which the thumb rotates during opposition. The exceptional development of this muscle among the Neandertals appears early in their development, since the Kiik-Koba 2 infant (age 5–7 months) and the Zas-kalnaya VI child (age 10–12 years) show well-formed crests [Vlček, 1975], similar to those present on European Neandertal children (eg, La Ferrassie 3 and 6 [Heim, 1982; Trinkaus, personal observation]). All of the known western Asian Neandertal distal pollical phalanges (N = 5) have large and deep pits for the insertions of their flexor pollicis longus muscles, and they have distal tuberosities that have dimensions at the upper limits of more recent human ranges of variation [Trinkaus, 1983a]. Those from Qafzeh and Skhul are robust but closer in morphology to those more recent humans. These features all indicate exceptional strength in the thumbs of these Neandertals, and a reduction of that strength in the transition to the early anatomically modern human samples.

The muscular strength of these Neandertal thumbs was aided by their pollical phalangeal proportions.[4] Among recent and early anatomically modern humans, the distal pollical phalanx has a length about two thirds of that of the proximal pollical phalanx, whereas among the Neandertals they are subequal in length (index of phalangeal lengths: European Nean-dertals: 84.3–102.1 (N = 3); European Early Upper Paleolithic sample: 66.4 ± 5.6 (N = 7); Shanidar 4, 5, and 6: 90.5, 101.7, and 90.1, respectively; Qafzeh 9: 71.3; Skhul 4: 65.7). Phalangeal proportions such as those of the

[4]Contrary to early assessments [eg, Boule, pp 1911–13], Neandertals had first-to-third meta-carpal and summed-pollical-phalangeal-to-first-metacarpal length proportions indistinguish-able from those of early and recent anatomically modern humans [Musgrave, 1971; Stoner, 1981; Trinkaus, 1983a].

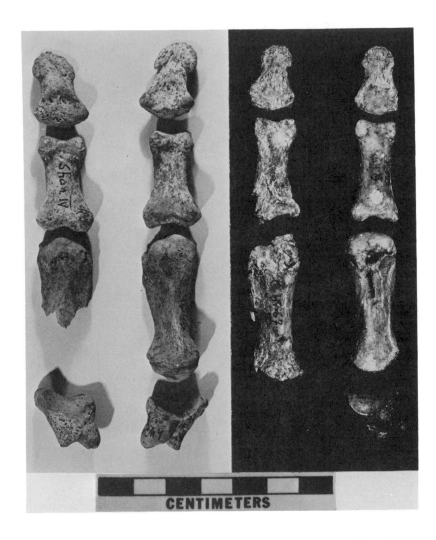

Fig. 2. Palmar views of the right and left thumb bones of Shanidar 4 (left) and Qafzeh 9 (right). The Qafzeh 9 left first metacarpal and proximal phalanx were partially crushed during fossilization. Note the differences in opponens pollicis crest development, phalangeal length proportions, proximal phalangeal breadth, flexor pollicis longus tendon insertion areas, and distal tuberosity size.

Neandertals would increase the load arm between the interphalangeal joint and the finger tip and decrease the load arm between the interphalangeal region and the metacarpophalangeal articulation. This would increase the mechanical advantage of the short thenar muscles when gripping objects across the interphalangeal region (the power grip) without changing the effectiveness of the flexor pollicis muscle. The change in phalangeal proportions with the appearance of anatomically modern humans produced a reduction in potential strength in the power grip, accompanied by an increase in the mechanical advantage of flexor pollicis longus when used in the precision grip. Strength was therefore maintained for the precision grip in the context of a generally weaker hand.

The last reflection of manual hypertrophy and its reduction in the transition to anatomically modern humans is in their distal phalangeal tuberosities. Those of the Neandertals, especially on digits 2 to 4, are large and round, as opposed to the narrower, hemiamydaloid shape seen among recent humans and the Skhul and Qafzeh specimens. Since the tuberosity supports the pulp and nail of the finger tip, its reduction implies a decrease in the habitual levels of stress passing through the fingers.

The reflections of robusticity differences between Neandertals and early anatomically modern humans in the lower limb are primarily in their femoral and tibial diaphyses, proximal femora, and patellae. These regions show a marked reduction in massiveness across this transition.

The robusticity indices of the western Asian femora and tibiae show a clear separation of the Neandertals and early modern-appearing humans (Table II). The western Asian Neandertals are close to their European relatives, whereas the Skhul and Qafzeh sample is close to a European Early Upper Paleolithic sample. There is an associated shift in the development of pilasters on their femora (indicated by their pilastric indices [Table II]) with the Neandertals lacking pilasters and having diaphyseal cross-sections that are either round or slightly ovoid (Fig. 3) and the early anatomically modern humans having clearly developed pilasters, some of them highly projecting (eg, Qafzeh 9 and Skhul 4 and 5) (Fig. 4). The geologically older femora from Gesher Benot Ya'acov 1 and 2 and Tabun E1 are similar to the Neandertals in this feature (pilastric indices of c 89.6, 96.4, and 92.9, respectively). This increase in the size of the pilaster appears to be an indirect response to the decrease in relative shaft breadth [Trinkaus, 1976a].

The above robusticity indices provide an indication of shifts in diaphyseal hypertrophy between these two samples, but they underestimate the degree of change, since it is the total amount of bone in the diaphyseal cross-section and its pattern of distribution in that cross-section that is important.

TABLE II. Comparisons of Upper Pleistocene Lower Limb Remains

		Femur shaft breadth index[a]	Femur pilastric index	Tibia robusticity index[b]	Patella thickness index[c]	Femur relative head diameter[d]	Femur neck angle
Western	$\overline{X}$	7.21	100.9	8.17	6.78	10.7	118.3°
Asian	SD	0.42	10.1	0.55	0.80	0.7	4.7°
Neandertals	N	6	6	6	5	4	3
Western Asian	$\overline{X}$	5.65	123.3	7.29	4.80	9.5	131.6°
early anat.	SD	0.42	9.8	0.66		0.2	6.6°
mod. humans	N	4	10	4	1	4	4
European	$\overline{X}$	6.92	99.5	8.02	6.31	12.0	121.2°
Neandertals	SD	0.16	7.7	0.35	0.60	0.6	5.2°
	N	6	8	4	3	5	7
European	$\overline{X}$	6.12	115.6	6.84	5.19	10.4	117.4°
early anat.	SD	0.41	14.3	0.70	0.54	0.7	7.7°
mod. humans	N	7	15	7	4	11	7

[a](Midshaft breadth/bicondylar length) × 100.
[b]((Midshaft AP diameter × midshaft ML diameter)$^{1/2}$/maximum length) × 100.
[c](Patella maximum thickness/tibia maximum length) × 100.
[d](Sagittal head diameter/bicondylar length) × 100.

Analysis of the Amud 1 and Shanidar 6 tibiae has shown that they are exceptionally strong, considerably more so than indices based on external shaft diameters indicate [Lovejoy and Trinkaus, 1980]. Similar analyses of their femora and of the Skhul and Qafzeh leg bones have not been done, but visual comparisons of the femoral midshaft cross-sections of two of these individuals (Shanidar 4 and Skhul 5 [Figs. 3, 4]) indicate that there was a marked decrease in both cortical thickness and diaphyseal diameters across the transition.[5]

The femoral heads of the western Asian Neandertals have relative sizes that are, on the average, above those of Qafzeh and Skhul specimens, even

[5]Lovejoy and Trinkaus [1980, p 467], quoting from Endo and Kimura [1970, p 354], stated that the Skhul 4 tibia was similar in robusticity to those of the Neandertals. The analysis by Endo and Kimura, however, was based on a drawing by McCown and Keith [1939: Fig. 20] that depicts only the periosteal contours accurately and is otherwise anatomically unusual. It is most likely that the Skhul 4 tibia had thin cortical walls, as do the Skhul 5 and 6 tibiae, and is therefore considerably less robust than those of the Neandertals. This is supported by comparisons of the Amud, Qafzeh and Skhul humeral diaphyses [Smith et al, 1983], which show the Amud 1 humerus to have been biomechanically markedly stronger than those from Skhul and Qafzeh.

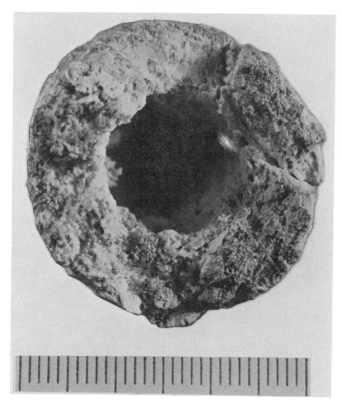

Fig. 3. Midshaft cross-section of the Shanidar 4 right femur, viewed in a distal-proximal direction. The linea aspera is at the bottom of the cross-section. Estimated bicondylar length of the specimen is 422.0 cm. Scale in millimeters.

though they are not especially high relative to those of European early anatomically modern humans (Table II). However, there is a marked increase in femoral neck angle across this transition in western Asia (Table II). Curiously a similar shift in femoral neck angles did not take place in Europe, since European Neandertal and Early Upper Paleolithic femora have angles that are indistinguishable from those of Amud 1, Shanidar 1, and Tabun C1; it is the western Asian early modern-appearing humans that appear unusual in this respect. Femoral neck angles are a good indication of activity level and biomechanical stress levels at the hip, since they decrease during development with accentuations in the level of habitual force at the hip [Houston and Zaleski, 1967].

Associated with these diaphyseal and proximal femoral changes is a decrease in the mechanical advantage of quadriceps femoris. Neandertals,

5 CMS

Fig. 4. Midshaft cross-section of the Skhul 5 fifth femur, viewed in a distal-proximal direction. The periosteal and endosteal contours were traced from a CT scan of the specimen, taken at Boston Brigham-Women's Hospital. Note the contrast in cortical bone thickness and pilaster development relative to the Shanidar 4 femur. Estimated bicondylar length of the specimen is 515.0 cm.

including Kiik-Koba 1, Shanidar 1, 5, and 6, and Tabun C1, have thick patellae relative to their tibial lengths (Table II). Since the patellar thickness makes up part of the power arm for quadriceps femoris and tibial length makes up most of its load arm, this implies greater effectiveness for this muscle. Skhul 4, like European Early Upper Paleolithic humans, has a relatively thinner patella.[6]

These data thus indicate that there was a marked decrease in human postcranial robusticity across the transition from Neandertals to their modern appearing successors in western Asia. The presence of a similar level of massiveness in the Tabun Layer E and Gesher Benot Ya'acov remains and a reduced amount of hypertrophy in the Nahal 'En-Gev and more recent Kebaran remains from Kebara [Arensburg, 1977; Trinkaus, personal observation] suggest that there was relatively little shift in these levels of robusticity before and after this Upper Pleistocene transition.

The pervasiveness of this robusticity among the Neandertals implies that it was an important part of their adaptation. It would have enabled them to generate and sustain more strength and higher levels of activity on an

---

[6]It may be argued that the differences in these indices are determined primarily by contrasts in crural indices between these samples [Trinkaus, 1981; see below]. However, the differences in relative patellar thickness remain if patellar thickness is compared to summed femoral and tibial lengths for those individuals that provide lengths for both bones [Trinkaus, 1983a].

habitual basis than could their early anatomically modern human successors. The development, maintenance, and operation of such massive bodies must have been energetically costly, a critical consideration for hunting and gathering populations that, like most natural populations, were frequently close to the limits of their energy reserves. This implies that their hypertrophy was necessary for survival; otherwise selection and developmental processes would have operated rapidly to reduce their level of robusticity.

The marked decrease in massiveness at the time of this Upper Pleistocene transition in western Asia therefore suggests that the early anatomically modern humans no longer needed the strength of the Neandertals for survival and successful reproduction. Since the postcranial anatomy would be stressed primarily during the food quest, these changes suggest that there were substantial improvements in technological and social aspects of food acquisition at this time in the Upper Pleistocene.

Limb Proportions

The European and western Asian Neandertals had relatively short distal limb segments and their early anatomically modern human successors had comparatively long ones [Trinkaus, 1981]. This is evident in comparisons of European Neandertal and Early Upper Paleolithic brachial indices (73.3 $\pm$ 2.5, N = 5 and 77.5 $\pm$ 2.5, N = 14, respectively) and especially crural indices (78.6 $\pm$ 1.8, N = 4 and 85.9 $\pm$ 1.8, N = 12, respectively). Western Asian Upper Pleistocene specimens are less complete, so that it is harder to determine the exact patterns of limb segments proportions between these respective samples. With respect to brachial indices, it is not possible to separate the western Asian Neandertals from their successors (Shanidar 4: c 78.0, Shanidar 6: c 74.2, Tabun C1: 77.4 Qafzeh 9: c 76.4, Skhul 4: c 81.3, Skhul 5: c 70.5), but their crural indices follow the European pattern (Amud 1: c 80.1, Shanidar 1: c 77.5, Shanidar 5: c 79.4, Shanidar 6: c 78.1, Tabun C1: c 77.8, Skhul 4: 88.6, Skhul 6: c 85.3). It should be emphasized that these differences, as with contrasts in brachial and crural indices among modern humans, exist within the context of the same intermembral proportions [Trinkaus, 1981].

Among recent humans, brachial and crural indices are correlated with mean annual temperature, since those populations that live in the coldest climates (Lapps and Eskimos) have the lowest indices and those that inhabit the hottest regions (Egyptians and equatorial Africans) have the highest indices, independent of stature [Trinkaus, 1981]. This association is supported by correlation coefficients of 0.856 (N = 12 samples) and 0.812 (N = 14 samples) between the mean annual temperature of the sites of origin and the mean brachial and crural indices respectively of the samples [Trinkaus, 1981]. It therefore appears that limb segment proportions are indicative of a population's pattern of thermal adaptation, as a special case of Allen's Rule [Allen, 1877; Trinkaus, 1981]. The shift in both brachial and crural indices across this Upper Pleistocene transition in Europe suggests that the succeeding populations of early anatomically modern humans were better able to deal with the thermal stress of the last glacial, both in terms of generating heat, through the metabolism of high-energy diets and carefully controlled fires, and at maintaining that heat around the body, with improved clothing and shelters. The situation is less clear in western Asia, since the observable shift exists solely in the lower limb. It may be that the sample sizes are too small, especially given the large overlaps between recent human samples with markedly different mean brachial and crural indices [Trinkaus, 1981]; if this is the case, the same implication of improved thermal adaptations would apply to the western Asian transition. If, however, the brachial and crural indices of these western Asian samples are representative, one would have to argue that the observed shift in crural indices was a result of a shift from an adaptation for power in the lower limb to one for speed at the expense of power. The relative elongation of the tibia would decrease the mechanical advantage of quadriceps femoris during acceleration while increasing the resultant excursion of the foot for a given contraction of quadriceps femoris.

Either interpretation of the shift in crural indices in western Asia at this time implies an increase in human abilities to deal with the pressures of the environment and subsistence. The thermal adaptation explanation implies differences in their abilities to generate and conserve heat, whereas the other provides one more example of biological change in response to increased cultural effectiveness (see robusticity discussion).

Pubic Bones and Gestation Length

These postcranial changes were associated with a marked change in pubic bone morphology [Stewart, 1960; Trinkaus, 1976b], a shift that was not accompanied by changes in overall iliac, ischial, or sacral morphology [Endo and Kimura, 1970; McCown and Keith, 1939; Trinkaus, 1983a; Vandermeersch, 1981a]. Three early anatomically modern humans from western Asia preserve pubic bones, Qafzeh 9 (F), Skhul 4 (M), and Skhul 9 (M?). All have pubic morphologies that are indistinguishable from those of European Early Upper Paleolithic and recent humans. The Neandertals that preserve portions of their pubic bones, Amud 1 (M), Shanidar 1 (M), 3 (M), and 4 (M), and Tabun C1 (F), have pubic regions that are elongated mediolaterally and superior pubic rami that are thin along their ventral margins.

The ventral superior pubic ramus thicknesses of the Neandertals (5.5 mm, 5.2 mm, 6.7 mm, and 7.8 mm for Amud 1 and Shanidar 1, 3, and 4, respectively, and 3.5 mm for Tabun C1) are separate from those of Qafzeh 9 (12.6 mm), Skhul 4 and 9 (13.0 mm and 9.5 mm), European Early Upper Paleolithic humans (14.5 mm and 17.5 mm), and recent males and females [Trinkaus, 1983a]. The significance of this feature is uncertain, and it probably is a secondary product of the mediolateral elongation of Neandertal pubic bones.

The acetabulosymphyseal lengths of the western Asian Neandertal pubic bones (Shanidar 1: c 93.0 mm, Shanidar 3: 80–90 mm, Tabun C1: 79.5 mm) are slightly below that of La Ferrassie 1 (98.0 mm) and above those of Qafzeh 9 (c 59.0 mm) and Skhul 4 (72.0 mm), but fall on either side of the exceptionally high value of c 83.0 mm for Skhul 9. The Skhul 9 measurement is above those of all known European Early Upper Paleolithic humans (58.0–73.0 mm, N = 6) and at the upper limits of two recent human samples (Amerindians: 55.8–82.7 mm, N = 100 [50/50, males/females]; Near Easterners: 59.5–83.0 mm, N = 34 [19/15, males/females]). When these pubic breadth measurements are standardized for body size using estimates of femoral bicondylar length, the resultant indices separate the samples even more (Shanidar 1: c 20.3, Shanidar 3: c 18.2–20.5, Tabun C1: c 19.4, Qafzeh 9: c 12.9, Skhul 4: 14.8). The western Asian Neandertals are close to that for La Ferrassie 1 (c 21.4) and Qafzeh 9 and Skhul 4 are close to those for European early anatomically modern humans (12.3–16.4, N = 5) and slightly above the means of two recent human samples [Trinkaus, 1976b, 1983a]. This pattern is further illustrated by comparisons of acetabulosymphyseal length to acetabular height (the latter being used to indicate body size). Tabun C1 has an index of c 165.6, which is close to that of La Ferrassie 1 (167.0), well above those of Qafzeh 9 (c 113.7), Skhul 4 (c 120.5), and European Early Upper Paleolithic humans (111.3–133.0, N = 7), and even above the high value for Skhul 9 (c 148.2).

This shift in pubic breadth had the effect of decreasing the horizontal diameters of the pelvis, both absolutely and relative to body size. One effect was a decrease in the interacetabular distance, which would have increased the efficiency of locomotion; an abbreviated interacetabular distance shortens the moment arm for the body's center of gravity relative to the hip articulation of the support leg during one-legged stance (about 95% of normal striding gait [Saunders et al, 1953]), which reduces the amount of force needed in the gluteal abductors to maintain balance during gait

[McLeish and Charnley, 1970]. This would reduce both the energetic expenditure in the gluteal abductor muscles and the joint reaction force at the hip. In addition, the decreased interacetabular distance would minimize displacement of the body's center gravity during gait and thus help to conserve kinetic energy [Saunders et al, 1953]. While these energetic efficiencies would have been selectively advantageous, it is doubtful whether the advantages they produced would have outweighed the increased difficulties in parturition that would have resulted from reduced pelvic diameters.

The dimensions of a fetus at any point in its development are determined by the size of its parents (especially maternal size), and its brain size is closely correlated with its expected adult cranial capacity [Leutenegger, 1982; Sacher and Staffeldt, 1974]. Western Asian Neandertals, like their European relatives, were on the average about 7% shorter than their early modern-appearing successors (Table III), but the differences in massiveness probably compensated for the shift in stature to provide them with similar mean body masses. The mean cranial capacities of the two samples are very similar and fall between those of European Neandertal and Early Upper Paleolithic samples (Table III). It is therefore likely that, at similar stages of development, the fetuses of the two groups were of similar size and had comparable head diameters. If gestation lengths remained the same across the transition, the early modern humans must have had considerably greater difficulties with parturition than their Neandertal predecessors. Despite cultural im-

TABLE III. Comparisons of Upper Pleistocene Statures and Cranial Capacities

		Male stature[a] (cm)	Female stature[a] (cm)	Cranial capacity (cc)
Western	$\overline{X}$	170.4	155.7, 158.5	1525.3
Asian	SD	4.9		198.0
Neandertals	N	6	2	4
Western Asian	$\overline{X}$	184.7	169.1	1545.0
early anat.	SD	3.7	7.7	26.8
mod. humans	N	4	4	5
European	$\overline{X}$	167.4	160.2	1509.8
Neandertals	SD	3.5	4.4	149.9
	N	6	3	6
European	$\overline{X}$	183.6	166.9	1576.8
early anat.	SD	5.1	1.6	135.5
mod. humans	N	12	4	11

[a]Stature estimates were computed using the formulae of Trotter and Gleser [1952] for Euroamerican males and females.

provements around this time, it is uncertain whether human obstetrical abilities improved as well.

Alternatively, it is possible that gestation length decreased across this Upper Pleistocene transition, with the Neandertals having had longer gestations than modern humans. Estimates from the available western Asian and European Neandertal pelvic remains suggest that a Neandertal pelvis, on the average, could have accommodated a sphere about 15 to 25% larger than one that could pass through an average modern human pelvis. They could thus have given birth to considerably larger-brained neonates. Brain growth data from modern human infants [Voigt and Pakkenberg, 1983] indicate that an additional 2 to 3 months postnatally is usually needed to achieve an increase of 15 to 25% in brain volume. This suggests that Neandertal gestation periods may have been 11 to 12 months, rather than the 9 months common for modern humans. Interestingly, correlations across mammals [Sacher and Staffeldt, 1974] predict gestation lengths of 12.5 to 13.2 months for mammals with adult brain sizes of 1300 cc to 1500 cc, and human postnatal growth trajectories [Martin, 1982] and patterns of human behavioral development [Deacon and Konner, unpublished manuscript] suggest that modern humans should have gestations considerably longer than the usual 9 months.

A decrease in gestation length with the advent of anatomically modern humans would require cultural compensations, since anatomically modern neonates would have less neuromuscular development and a less-developed immune system than Neandertal neonates. Both of these would require more maternal protection and support for survival, which would place greater restrictions on the mobility of their mothers, and hence on the abilities of those females with neonates to contribute to the subsistence of the social group. In addition, carrying an infant in one's arms would be energetically more expensive than carrying it in utero, and shifting an offspring from placental to lactational nourishment moves it one trophic level higher on the food chain, placing more energetic demands on the mother. Yet, a decrease in gestation length would expose the newborn to environmental stimuli earlier at a period of life when the brain is growing rapidly and is highly responsive to sensory stimuli [Conel, 1947–59; Yakovlev and Lecours, 1967]. The possible resultant headstart on neurological development could have conferred a selective advantage onto the infant, since it would have permitted the infant to respond adaptively to dangerous situations earlier in postnatal life. Furthermore, shorter gestations may have allowed for a reduction in birth spacing (if birth-to-conception intervals remained constant), thereby increasing the potential for population growth.

The reduction of gestation implied by the pubic bone alterations, despite the increased vulnerability of the neonate, therefore appears to have conferred a selective advantage to the early anatomically modern humans. This would have combined with energetic advantages during gait to promote these pelvic changes once the cutural system had evolved to the stage where altricial newborns could survive to reproductive age.

Dentitions and Anterior Tooth Use

Upper Pleistocene western Asian human remains retain a number of teeth, many of them in associated dentitions. Although not as abundant as those from Europe, they are sufficient to indicate patterns of change in size and occlusal wear through time. As in Europe, there appears to have been little change in cheek tooth size across the transition from archaic to modern-appearing humans in western Asia. This is evident in the dimensions of individual teeth [Frayer, 1978; McCown and Keith, 1939; Trinkaus, 1983a; Vandermeersch, 1981a] and in comparisons of their summed P3 to M3 "areas" (Table IV). There even appears to have been a slight, although statistically insignificant, increase in tooth size, a product of the modest dimensions of the Amud 1 and Tabun C1 teeth and the large sizes of some of the Qafzeh molars. A similar pattern of little or no change in size is evident in the available deciduous dentitions [Smith and Arensburg, 1977;

TABLE IV. Comparisons of Upper Pleistocene Dental Dimensions

		I^1-C^- summed breadths (mm)	I_1-C_- summed breadths (mm)	P^3-M^3 summed "areas" (mm^2)	P_3-M_3 summed "areas" (mm^2)
Western	$\overline{X}$	25.9	24.5	506.5	484.5
Asian	SD	0.8	1.2	33.8	30.2
Neandertals	N	4	5	5	5
Western Asian	$\overline{X}$	25.0	22.4	556.7	537.4
early anat.	SD	1.7	1.2	48.3	56.7
mod. humans	N	7	7	4	5
European	$\overline{X}$	27.4	24.7	572.6	554.6
Neandertals	SD	1.6	1.3	49.3	54.5
	N	11	12	6	12
European	$\overline{X}$	23.0	21.6	494.8	504.0
early anat.	SD	1.3	1.0	68.3	72.9
mod. humans	N	8	6	7	5

Trinkaus, 1983a], even though their occlusal surfaces became less complex across the transition.

In the anterior dentition, comparisons of summed I-1 to C breadths (Table IV) indicate that, on the average, there was a reduction in anterior tooth size across the transition, despite the rather large overlap between the two samples. This shift is similar to, but less pronounced than, one that occurred in Europe across the equivalent transition (Table IV). The reduction in anterior tooth size is more apparent when the summed anterior breadths are compared to the square roots of the summed posterior tooth "areas." The resultant maxillary indices are 114.2, 114.4, 117.7, and 118.7 for western Asian Neandertals and 106.0, 106.9, 107.1, and 115.2 for Skhul and Qafzeh individuals, and the western Asian mandibular indices are 107.3, 109.0, 111.7, and 111.8 for Neandertals and 91.9, 97.1, 99.2, and 100.0 for early modern-appearing humans. With the exception of the high maxillary value for Qafzeh 7 (115.2), there is a separation of the two samples, with the Neandertals having relatively larger anterior teeth. The same pattern, although more pronounced with no overlap, is present across the transition in Europe (Neandertals: maxilla: 112.3 ± 2.4, N = 4, mandible: 105.6 ± 4.6, N = 9; Early Upper Paleolithic: maxilla: 105.3 ± 6.0, N = 5, mandible: 94.0 ± 1.9, N = 4).

The frequency shift in anterior tooth dimensions suggests that there was a reduction in average levels of force and/or occlusal attrition placed on the anterior dentition across this Upper Pleistocene transition, since smaller teeth would both break more easily and, more importantly, wear at a faster rate. This interpretation is supported by the rates and patterns of anterior occlusal wear on these Upper Pleistocene fossils (Fig. 5). All of the western Asian Neandertal anterior dental remains show a pattern of wear in which the incisors and canines are relatively rapidly planed off. Among the early anatomically modern humans that preserve both anterior and posterior teeth, both regions of the dentition wear at about the same rate [McCown and Keith, 1939; Vandermeersch, 1981a]. The Neandertals with preservation of both anterior and posterior teeth, however, largely show a more rapid wear of the anterior teeth. This is apparent on the Shanidar 1 to 6 and Tabun C1 dentitions [McCown and Keith, 1939; Trinkaus, 1978, 1983a]; Amud 1 and Tabun C2 shows similar rates of wear on both dental regions [McCown and Keith, 1939; Sakura, 1970]. This suggests that these archaic humans were, on the average, subjecting their anterior dentitions to greater wear and stress than were their modern-appearing successors, implying a reduction in paramasticatory use of the dentition across the transition. This is supported by the presence on several of the western Asian

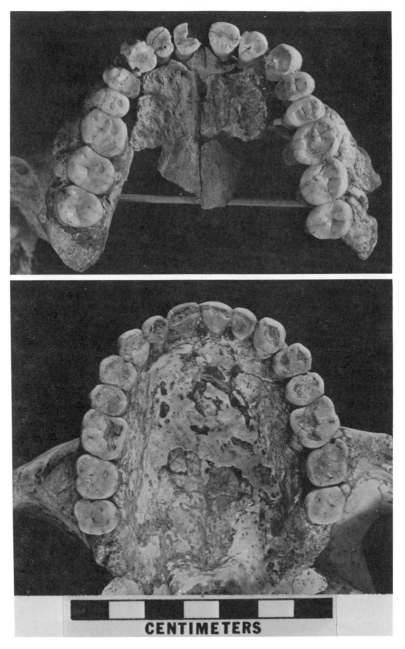

Fig. 5. Occlusal views of the Shanidar 2 (above) and Skhul 5 (below) palates. Their molars are of similar sizes and exhibit comparable amounts of occlusal wear. The anterior teeth of Shanidar 2, however, are larger and considerably more worn.

Neandertals (eg, Shanidar 1, 3, 4, and 5) of a labial rounding of their maxillary and mandibular anterior teeth that can only have been produced by pulling materials across their crowns linguolabially and/or holding objects between those teeth while placing anteriorly directed tension on them [Trinkaus, 1978, 1983a]. This interpretation is further supported by microscopic examinations of several of their occlusal surfaces [Ryan, 1980; Trinkaus, 1983a] that show labiolingually oriented striations on the occlusal surfaces and microchipping of the occlusal margins.

It therefore appears as though, in western Asia as in Europe, there was a reduction in paramasticatory use of the anterior dentition. This is evident primarily in the shift in patterns of occlusal wear, and is reflected more indirectly in the reduction of anterior dental dimensions relative to those of their premolars and molars.

The Facial Skeleton and Midfacial Prognathism

The configuration of the facial skeleton has long been considered to be one of the most important, and phylogenetically diagnostic, aspects of human skeletal morphology, and there have been numerous studies of the patterns of variation of human Upper Pleistocene facial shape. From these it has become apparent that the earliest anatomically modern humans have facial configurations that are similar to, although frequently more robust than, those of modern populations in the same geographical regions. Among archaic Upper Pleistocene humans, the European Neandertals appear to be distinct from their contemporaries in having pronounced midfacial prognathism: the anterior projection of the nasal region, alveoli, and associated regions of the maxillae and mandible relative to the zygomatic, orbital, and ramal regions. Several of the western Asian archaic humans share this facial configuration with the European Neandertals, but there is considerable variation among them [Keith, 1927; Stringer and Trinkaus, 1981; Trinkaus, 1983a; Fig. 6].

There are a number of reflections of midfacial prognathism. The most constant are low zygomaxillary angles, high zygomaxillare-molar alveolus radii differences, posteriorly placed anterior zygomatic roots (usually above M3 rather than M1-M2), large maxillary and mandibular retromolar spaces, anteriorly placed nasal apertures with relatively horizontal nasal bones, low nasiofrontal angles, flat anterior maxillae, and retreating zygomatic profiles (minimal angulation of the anteriolateral zygomatic surface). All of these measures are, or appear to be, correlated, since they are regional reflections of the same overall configuration [Trinkaus, 1982d, 1983a].

A number of the western Asian Upper Pleistocene crania provide accurate estimates of their zygomaxillary angles. The Qafzeh and Skhul speci-

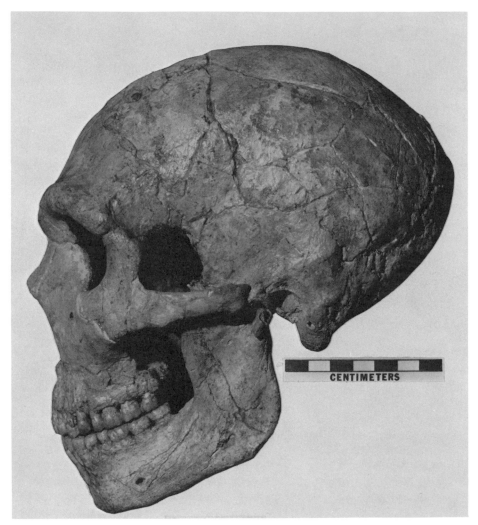

Fig. 6. The Amud 1 male Neandertal skull in *norma lateralis* left, as reconstructed by H. Suzuki.

mens have angles (c 124°, 131°, 133°) that are similar to those of European Early Upper Paleolithic humans (127.3° ± 8.8°, N = 6) and those of recent humans [Howells, 1973]. Four of the western Asian archaic humans (Amud 1, Shanidar 1 and 5, and Teshik-Tash 1) have angles (c 115°, 105°, c 115°, c 111°) that are similar to those of European Neandertals (105°, 109°, 114°,114°) and below those of almost all more recent humans.[7] However, two of the western Asian archaic humans, Shanidar 2 and 4, have angles (c 125° and c 123°) that are similar to those of anatomically modern humans, and it appears as though Tabun C1 had a relatively high angle (neither zygomatic bone is preserved on the Tabun C1 cranium). The archaic specimens that have low zygomaxillary angles also exhibit most of the other features associated with midfacial prognathism [Gremyatskij and Nesturkh, 1949; Suzuki and Takai, 1970; Trinkaus, 1983a]. Amud 1 has an exceptionally low nasiofrontal angle (129°), although the other individuals providing this measure of nasion projection have somewhat higher angles (Shanidar 5: 139°, Teshik-Tash 1: c 141°) The Qafzeh and Skhul specimens have relatively high nasiofrontal angles (139°, 147°, 152°), similar to those of European Early Upper Paleolithic humans (145.1° ± 7.9°, N = 15) and more recent humans [Howells, 1973]. The specimens with higher zygomaxillary angles tend to have less development of these features, since Shanidar 2 and Tabun C1 have small retromolar spaces and Shanidar 4 lacks one, their anterior zygomatic roots are located anterior to M3, Tabun C1 has a nasiofrontal angle of 142°, and at least Shanidar 2 and 4 have sharply angled zygomatic bones and concave anterior maxillae. Also, the earlier Zuttiyeh 1 specimen has a similarly angled zygomatic bone and lacks an anteriorly projecting nasion region (nasiofrontal angle: 150°).

Some of the early anatomically modern humans from western Asia have facial skeletons that appear archaic in comparison to modern humans. In particular, the males from Skhul (Skhul 4, 5, and 9) have large and robust faces and show an exceptional development of the supraorbital torus (Fig. 7). Yet, not all of these specimens show this facial massiveness; Skhul 2 and Qafzeh 3, 6, and 9 have more gracile faces and supraorbital regions that are well within the ranges of variation of more recent humans.

Despite the variation in facial shape among these archaic and early modern-appearing humans, there is relative constancy within each group as to the relative sizes of their faces, as indicated by their mandibular lengths

---

[7]All of the measurements provided for Teshik-Tash 1 are estimates of its "adult" size and proportions [Alexeyev, 1981].

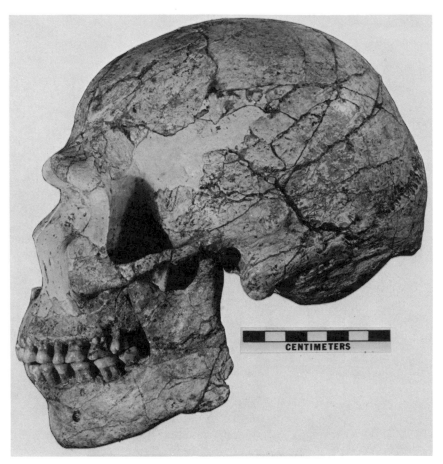

Fig. 7. The Skhul 5 male early anatomically modern human skull in *norma lateralis* left, as reconstructed by C. E. Snow.

(condyles to infradentale) relative to femoral and/or tibial length. In the mandible to femur length comparison, the western Asian Neandertals have indices (Amud 1: c 23.0, Shanidar 1, 4, and 5: c 25.1, c 26.3, and c 26.0, respectively, Tabun C1: 24.9) that are close to those for two European Neandertals (c 26.6 and c 26.7) and largely above those of early anatomically modern humans (Qafzeh 9: c 22.8, Skhul 4; 21.7, Skhul 5: c 20.7; European Early Upper Paleolithic: 20.2, 21.4, and 22.1). In the mandible-to-tibia length comparison, the differences are accentuated by the contrasts in crural indices (see above) (Amud 1: c 28.8, Shanidar 1: c 32.4, Shanidar 2: c 33.2,

Tabun C1: 32.0; Skhul 4: 24.5, Skhul 5: c 25.8, European Neandertals: c 33.0 and c 33.2; European Early Upper Paleolithic: c 23.1 and c 24.7). Among the western Asian specimens, only the Amud 1 specimen appears intermediate between the Neandertals and modern-appearing specimens with respect to relative facial length. These data thus indicate that there was a significant shift in total facial prognathism around the time of the transition to modern humans but little change prior to that, despite the differences in midfacial prognathism evident within the archaic group.

There thus appears to be considerable variation in facial form among these western Asian Upper Pleistocene humans, with Zuttiyeh 1, Shanidar 2 and 4, and perhaps Tabun C1 having archaic faces that in many ways resemble Early Neandertals from Europe, Amud 1, Shanidar 1 and 5, and Teshik-Tash 1 having facial configurations largely indistinguishable from those of European Neandertals, the Skhul males having robust but otherwise anatomically modern faces, and Skhul 2 and the Qafzeh specimens having fully anatomically modern human faces. The position of the Tabun C2 mandible among these specimens is uncertain, but it appears to fit best within the midfacially prognathic group [Trinkaus, 1983a].

It is possible to arrange these specimens into a morphological sequence, one that provides a relatively continuous pattern of change. This can be done without violating any of the indications available for their geological ages, if one assumes that the Qafzeh specimens are late, similar in age to the Skhul specimens and that Tabun C1 derives from Layer D of Tabun Cave. The rates of change in these features, however, appear to have fluctuated significantly during the Upper Pleistocene.

The earliest phase is represented by the Zuttiyeh 1 and Shanidar 2 and 4 specimens. These individuals all have relatively long and robust faces, but they lack the midfacial prognathism of the more recent specimens. This is reflected in high zygomaxillary angles, anterior positions of the anterior zygomatic roots, and small or absent retromolar spaces of Shanidar 2 and 4, absence of nasion projection in Zuttiyeh 1, and angled zygomatic bones of all three. The Tabun C1 specimen, on the basis of its high nasiofrontal angle, small retromolar spaces, and possibly concave anterior maxillae, may well belong to this group.

The second archaic phase is represented by the Amud 1, Shanidar 1 and 5, Teshik-Tash 1, and perhaps Tabun C2. They differ from the earlier group in having pronounced midfacial prognathism. Since their total facial prognathism relative to body size remains the same, this shift could not be due to an anterior migration of the dental arcades and nasal aperture. It must be the product of a posterior migration of the zygomatic region, including the lateral supraorbital torus and the associated anterior ramal margin. This interpretation is supported by comparisons of ramal breadth to mandible length for these individuals. The resultant indices for Shanidar 2 and 4 and Tabun C1 are c 39.2, c 41.0, and 37.3, respectively; the same indices are c 35.6, 35.7, and 32.8 for Amud 1, Shanidar 1, and Tabun C2, respectively. This posterior shift of the zygomatic and anterior ramal region would have the effect chiefly of moving the major masticatory muscles (temporalis and masseter) posteriorly relative to the dentition. It would thus decrease their moment arms relative to the instant centers of rotation of mastication. Since total prognathism did not change, there was no alteration in the moment arms for the bite force. The evolutionary change in overall facial proportions that led to the development of the Neandertal face from those of their predecessors thus involved a reduction in masticatory effectiveness.

The other changes evident in the shift from the earlier western Asian archaic group to the more recent one can be seen largely as secondary consequences of the posterior migration of the zygomatic region without any change in the position of the dentition. The large retromolar spaces were produced by moving the anterior margin of the ramus posteriorly relative to the M3, and the posterior position of the anterior zygomatic root was a direct consequence of the retraction of the zygomatic region. Furthermore, the posterior position of the later western Asian archaic human zygomatic bones relative to their dentitions reduced the effectiveness of the zygomatic region to absorb masticatory stress directed superoposteriorly through the facial skeleton. In compensation, they developed facial skeletons that were more anteriorly convex than those of their predecessors. Their anterior maxillae became largely flat and curved evenly onto their flat and retreating zygomatic bones. The lateral walls of their nasal apertures formed smooth trajectories up to the interorbital region, and their supraorbital tori projected anteriorly in the midline to form a relatively continuous arc from the nasal region to the frontal squamous. All of these changes would permit a more direct transmission of masticatory stress through the upper facial skeleton and minimize areas of stress concentration.

These alterations suggest a process of reduction in facial massiveness, in which different aspects of the face were reducing at different times during the Upper Pleistocene. This is not surprising, since some reduction of overall facial massiveness, as reflected in the rugosity of masticatory muscle attachments, is apparent between the earlier (Azykh, Shanidar 2 and 4, and Tabun C1) and later (Amud 1, Shanidar 1 and 5, and Tabun C2) archaic western Asian samples. This reduction was probably a continuation of previously existing trends evident in late *Homo erectus* and earlier *Homo sapiens* specimens, much of it undoubtedly a response to reduced loads on the masticatory apparatus once fire was domesticated in the Middle Pleistocene. It is uncertain why the Neandertals maintained their pronounced total facial prognathism, but it was probably in large part a response to their frequent use of their anterior teeth for nondietary purposes [Smith, 1983].

In light of these considerations, the shift in facial shape across the Neandertal to early anatomically modern human transition can be seen as a consequence of reduced total facial prognathism without any further change in the relative position of the zygomatic region, associated with further gracilization of the facial skeleton. Most of the associated changes in facial shape, such as maxillary concavity, zygomatic angulation, and loss of retromolar spaces, are spatial and/or biomechanical readjustments of the facial skeleton in response to the changed position of the alveolar process relative to the zygomatic/ramal region. The marked reduction and eventual loss of the supraorbital torus is probably a consequence of the contemporaneous increase in frontal squamous height (see below), which removed the structural need for a robust torus [Smith, 1983]. Subsequent changes in the human facial skeleton in western Asia during the Upper Paleolithic consisted entirely of gradual reductions in massiveness with little change in the overall configuration.

The reduction in total facial prognathism at the transition to modern-appearing humans would have the primary effects of decreasing the size of the bending moments that would be placed on the facial skeleton during mastication [Smith, 1983] and increasing the mechanical advantages of the masticatory muscles by reducing the moment arm for bite force. The faces of early anatomically modern humans therefore represent a more efficient system than those of their predecessors, both in terms of energy expenditure for its development and maintenance, given the reduced dimensions of the facial skeleton, and the muscle force required for normal levels of bite force. It is therefore not surprising that, once the demands of paramasticatory use of the dentition decreased, there was a relatively rapid reduction in total facial prognathism.

The changes in facial form during the Upper Pleistocene in western Asia thus followed a pattern of reduction of overall massiveness associated with a mosaic of changes in aspects of the facial skeleton. The initial shift consisted of a reduction in the size and mechanical advantage of the masticatory musculature, which was associated with a continued reliance upon nondietary use of the anterior dentition. The subsequent change, that associated with the change to anatomically modern humans, appears to have been primarily a response to a marked decrease in paramasticatory loading of the dentition.

Neurocranial Alterations

The transition from archaic to modern-appearing humans was also associated with a shift from relatively long and low neurocrania to ones that were higher and rounder (Figs. 6, 7). This has been demonstrated by a number of morphometric analyses [eg, Howells, 1970; Stringer, 1978] and is partially illustrated by their calotte height indices (relative to the glabella-inion line) (Amud 1:51.0, Tabun C1: c 46.6, Teshik-Tash 1: c 49.7; Qafzeh 6:53.1, Qafzeh 9:58.4, Skhul 5:52.5 (the artificially deformed Shanidar 1 cranium [Trinkaus, 1982e] has an index of 51.5). The Skhul and Qafzeh indices are similar to those of a European Early Upper Paleolithic sample (53.6 ± 2.8, N = 7), but the western Asian Neandertals have indices between those of these early anatomically modern human samples and European Neandertals (42.7 ± 3.0, N = 7). This shift in relative cranial vault height, however, was not associated with any change in brain size (Table III), so it represents merely an alteration of shape.

It is possible that this neurocranial shape change is a secondary consequence of the reduction in gestation length that took place across this transition. The human brain does not grow simply radially, but grows predominantly anterosuperiorly initially and more posteriorly later on [Trinkaus and LeMay, 1982]. If the offspring of early anatomically modern humans were experiencing a slightly earlier growth of the brain than those of their Neandertal predecessors, as a result of their shorter gestation periods, they may have had more anterosuperiorly direct brain growth during the early postnatal months when the cranial vault bones are incompletely formed. Since alterations in the curvatures of the cranial vault bones during the first year postnatally can have significant effects on adult vault shape, as indicated by the consequences of pressure applied to the vault for only a few months to produce artificial cranial deformation [Blackwood and Danby, 1955], a slight shift in brain growth timing may account for the observed shift in relative neurocranial height among adults at the time of this transition. It should be emphasized that this in no way implies that there were any differences in the neurological organizations of these two groups of Upper Pleistocene humans, especially given the general similarities of Neandertal and modern human endocranial casts [Holloway, 1981].

SUMMARY

The preceding considerations of human anatomical alterations during the Upper Pleistocene in western Asia and their behavioral implications indicate that there was a mosaic of adaptive changes during this time period. Some features, such as brain and posterior dental size, changed little if at all. Others, such as postcranial robusticity, pubic morphology, and anterior dental dimensions and wear, remained constant among archaic humans and then changed markedly at the time of the transition to early anatomically modern humans. Still others, such as facial robusticity, changed more gradually among both archaic and modern-appearing humans. This mosaic of changes suggests that those aspects of behavior most closely related to postcranial robusticity and nondietary anterior dental use, which would be primarily aspects of subsistence and technology, changed significantly primarily around the middle of the last glacial, whereas those behaviors that determine general levels of masticatory stress, such as food preparation, changed more gradually. It is likely that the shift in pubic morphology, and the associated changes in gestation length and neonatal care, occurred only when the cultural system had reached a sufficient level of elaborateness to permit the successful reproduction of those individuals with reduced pelvic aperture diameters. This appears to have occurred around the time of the shift from Neandertals to early anatomically modern humans, toward the end of the western Asian Middle Paleolithic.

PHYLOGENETIC CONSIDERATIONS

Is it possible, given these anatomical changes during the Upper Pleistocene of western Asia, to determine the phylogenetic relationships between these human samples? More precisely, can we reasonably assess whether the changes in morphology in this region were primarily due to in situ evolution with no elevation in the levels of gene flow or required a marked increase in gene flow? Did the observed changes occur within the context of local population continuity, or were there major movements of individuals between regions? The answers to these questions lie in assessments of the degree of gradual versus abrupt morphological change evident in the fossil record, the relative dating of the specimens, and the genetic versus environmental components in the observed morphology.

Within archaic and anatomically modern H sapiens groups, the evidence fits best with an interpretation of considerable local continuity of populations with little or no elevation in the levels of gene flow. From the earliest anatomically modern humans at the sites of Qafzeh, Skhul and Starosel'e to the more recent ones at Kebara, 'En-Gev, and subsequent Kebaran and proto-Neolithic sites, there is little morphological change. The evident alterations consist of a gradual gracilization of the entire skeleton, especially of the facial region, and some decrease in body mass. There are no changes of sufficient magnitude to suggest a significant influx of humans from neighboring regions.

The same applies to archaic *H sapiens* within western Asia. Regardless of how one arranges the various specimens, as long as Gesher Benot Ya'acov 1 and 2, Tabun E1, and Zuttiyeh 1 are the earliest, Shanidar 1, 3, and 5 and Tabun C2 are the most recent, Shanidar 2, 4, 6, and 8 are in between these early and late samples, and Amud 1, Kiik-Koba 1 and 2, Tabun C1 and C3, Teshik-Tash 1, and Zaskalnaya 1 are dated to the early last glacial, there are no abrupt changes in morphology. The postcranium, dentition, and neurocranium, despite some variation within temporal samples, exhibits considerable stasis. There is a decrease in facial robusticity during this time period, associated with a posterior retreat of the zygomatic region that produces the midfacial prognathism of the more recent specimens, but even this change exhibits a modest tempo during the more than 40 ky between the earlier specimens and the later ones. The existence of roughly contemporaneous facial changes in Europe [Trinkaus, 1982b] implies genetic contact between the regions, but the level of gene flow need not have been greater than it was prior to the Upper Pleistocene or once anatomically modern humans were established across the northwestern Old World.

In contrast, the morphological changes associated with the archaic to anatomically modern human transition suggest a more complex process, with significant elevations of interregional genetic exchange. There were at that time, as discussed at length above, significant alterations that affected all regions of the skeleton. The degree of alteration varied between and within anatomical regions, but the total amount of change was of a different order of magnitude than that which is evident within either the archaic or the anatomically modern *H sapiens* samples.

If one assumes that the chronological scheme employed here is accurate, with Neandertals present up to about 45–50 ky BP and anatomically modern humans appearing around 40 ky BP, there is an interval of about 5–10 ky in which the transition could have taken place. This could be adequate for the observed changes, assuming that there was a sufficiently large increase in selective pressure, the necessary alleles were in the parental population, and some of the changes were nongenetic developmental shifts. However, if the Qafzeh remains were in fact contemporaneous with the Neandertals in the Levant, as maintained by Bar Yosef and Vandermeersch [1981], the temporal overlap of populations would make it hard to argue for a strictly in situ transition.

The biological bases of the observed changes are extremely difficult, if not impossible, to determine. Many of the differences in robusticity, cranial and postcranial, can be produced by variation in levels of biomechanical stress during development; yet, the early appearance of at least some aspects of Neandertal postcranial robusticity (eg, opponens pollicis crest formation) and its stasis during most of the Upper Pleistocene suggest that there was a significant genetic component in their development. The pubic bone alterations can be produced by shifts in endocrine secretion levels and timing [Crelin, 1960; Riesenfeld, 1972], and the genetic basis of the change may have been simple; given the apparent selective advantages of the narrower pelves, once cultural adaptations could maintain highly altricial neonates, human pubic morphology could have changed within populations relatively rapidly at this time in the Upper Pleistocene. The facial changes are more complicated, and it is difficult to sort out the probable genetic versus developmental components in the alterations, especially since many of the changes appear to have been secondary to more primary shifts. Even if each one of these morphological alterations could have taken place within local lineages, it seems unlikely that all of them could have occurred at the same time without elevated levels of gene flow from neighboring regions.

Unfortunately most aspects of human anatomy are unknown from neighboring areas other than Europe for the Upper Pleistocene. Only North Africa furnishes data, and that relates primarily to the facial skeleton. The known North African Upper Pleistocene facial skeletons, both archaic (Haua Fteah 1 and 2 and Jebel Ighoud 1 and 3 [Ennouchi, 1962; Hublin and Tillier, 1981; Tobias, 1967]) and anatomically modern (Dar-es-Soltane 1 [Ferembach, 1976]) are relatively short and lack the pronounced midfacial prognathism of the western Asian archaic humans. They could therefore have provided a source for gene flow into western Asia, although this does not necessarily mean that they did.

It thus appears that the transition to modern humans in western Asia could have been primarily a regional phenomenon, with only general connections to neighboring geographical areas. However, the complexity of the changes and their similarities to those that were occurring about the same time in Europe suggest that there was a marked elevation in the level of gene flow across western Asia at this time in the Upper Pleistocene.

CONCLUSIONS

The origin of modern humans in western Asia appears to have been a complex process in which robust but anatomically modern populations evolved during the middle of the last glacial possibly in part from preceding populations of archaic humans (Neandertals). The various anatomical alterations, and their behavioral implications, suggest that this transition involved major improvements in human sociocultural abilities. Some of the changes were continuations of previously existing trends, but others suggest a major shift in human adaptive efficiency. The archeological record from western Asia is equivocal as to whether such a shift occurred at this time, but evidence is accumulating to support such an interpretation [Jelínek, 1982a,c; Hietala and Marks, 1981; Marks, 1983; Marks and Freidel, 1977]. It is expected that further analyses of the adaptive implications of both the human paleontological and archeological records from western Asia will help to clarify the nature of this, the last major cultural and biological transition in human evolution.

ACKNOWLEDGMENTS

Many individuals have made original fossils from western Asia and casts thereof available for analysis (especially Drs. Muayed Sa'id al'-Damirji and Isa Salman [Iraq Museum, Baghdad], Drs. Avi Eytan and Joseph Zias [Rockefeller Museum, Jerusalem], Prof. Bernard Vandermeersch [Université de Bordeaux I, Talence], and Dr. C. B. Stringer [British Museum (Natural History), London]), and others, too numerous to name, have provided access to European and African Upper Pleistocene human remains. Dr. M. LeMay arranged for the CT scan of the Skhul 5 femur illustrated in Figure 4, and the Iraq, Rockefeller, and Peabody Museums permitted publication of photographs of specimens in their collections. In addition, Drs. W. W. Howells, F. H. Smith, F. Spencer, and K. Maurer Trinkaus kindly provided helpful comments during the preparation of this paper. To all of them I am very grateful.

This research has been supported in part by Wenner-Gren grant #2979 and NSF grants #BNS76-14344 and #BNS-8004578.

LITERATURE CITED

Alexeyev VP (1976): Position of the Staroselye find in the hominid system. J Hum Evol 5:413–421.
Alexeyev VP (1981): Fossil man on the territory of the USSR and related problems. In Ferembach D (ed): "Les Processus de l'Hominisation." Paris: Éditions du C.N.R.S., pp 183–188.
Allen JA (1877): The influence of physical conditions in the genesis of species. Radical Rev 1:108–140.
Angel JL (1972): A Middle Palaeolithic temporal bone from Darra-i-Kur, Afghanistan. In Dupree L (ed): "Prehistoric Research in Afghanistan (1959–1966)." Trans Am Philos Soc 62:54–56.

Arensburg B (1977): New Upper Palaeolithic human remains from Israel. Eretz-Israel 13:208–215.

Bar Yosef O, Vandermeersch B (1981): Notes concerning the possible age of the Mousterian layers in Qafzeh Cave. In Sanlaville P, Cauvin J (eds): "Préhistoire du Levant." Paris: Éditions du C.N.R.S., pp 281–285.

Blackwood B, Danby PM (1955): A study of artificial cranial deformation in New Britain. J Roy Anthropol Inst 85:173–191.

Bonč-Osmolovskij GA (1940): The cave of Kiik-Koba (in Russian). Paleolit Kryma 1:1–185.

Bonč-Osmolovskij GA (1941): The hand of the fossil man from Kiik-Koba (in Russian). Paleolit Kryma 2:1–172.

Bonč-Osmolovskij GA (1954): The skeleton of the foot and leg of the fossil man from Kiik-Koba (in Russian). Paleolit Kryma 3:1–311.

Boule M (1911–13): L'homme fossile de La Chapelle-aux-Saints. Ann Paléontol 6:111–172; 7:21–56, 85–192; 8:1–70.

Conel JL (1947–59): "Postnatal Development of the Human Cerebral Cortex." Cambridge: Harvard Univ. Press.

Coon CS (1951): Cave explorations in Iran 1949. Mus Monogr Univ Mus, U. Penn.

Coon CS (1975): Iran. In Oakley KP, Campbell BG, Molleson TI (eds): "Catalogue of Fossil Hominids III." London: British Museum (Natural History), pp 117–120.

Crelin ES (1960): The development of bony pelvic sexual dimorphism in mice. Ann N Y Acad Sci 84:479–512.

Doyle WJ (1977): Functionally induced alteration of adult scapular morphology. Acta Anat 99:173–177.

Dubois E (1894): "Pithecanthropus erectus, Eine Menschenaehnliche Uebergangsform aus Java." Batavia: Landesdruckerei.

Dupree L (1972): Tentative conclusions and tentative chronological charts. In Dupree L (ed): "Prehistoric Research in Afghanistan (1959–1966)." Trans Am Philos Soc 62:74–84.

Endo B, Kimura T (1970): Postcranial skeleton of the Amud Man. In Suzuki H, Takai F (eds): "The Amud Man and his Cave Site." Tokyo: Academic Press of Japan, pp 231–406.

Ennouchi E (1962): Un Néandertalien: L'homme du Jebel Irhoud (Maroc). L'Anthropologie 66:279–299.

Farrand WR (1979): Chronology and palaeoenvironments of Levantine prehistoric sites as seen from sediment studies. J Archaeol Sci 6:369–392.

Farrand WR (1982): Environmental conditions during the Lower/Middle Palaeolithic transition in the Near East and the Balkans. In Ronen A (ed): "The Transition From Lower to Middle Palaeolithic and the Origin of Modern Man." Br Archaeol Rep Intl Ser 151:105–108.

Ferembach D (1976): Les restes humains de la Grotte de Dar-es-Soltane 2 (Maroc) Campagne 1975. Bull Mém Soc Anthropol Paris Sér 13 3:183–193.

Formozov AA (1954): A new discovery of Mousterian man in the USSR (in Russian). Sov Etnogr 1:11–22.

Formozov AA (1957): New data on Palaeolithic man from Starosel'e (in Russian). Sov Etnogr 2:124–130.

Frayer DW (1978): Evolution of the dentition in Upper Paleolithic and Mesolithic Europe. Univ Kansas Publ Anthropol 10:1–201.

Garrod DAE (1954): Excavations at the Mugharet Kebara, Mount Carmel, 1931: The Aurignacian industries. Proc Prehist Soc 20:155–192.

Garrod DAE, Bate DMA (1937): "The Stone Age of Mount Carmel I: Excavations at the Wady el-Mughara." Oxford: Clarendon Press.

Garrod DAE, Bate DMA (1942): Excavations at the cave of Shukbah, Palestine, 1928, with an appendix on the fossil mammals of Shukbah. Proc Prehist Soc 8:1–20.

Geraads D, Tchernov E (In press): Fémurs humains du Pleistocène moyen de Gesher Benot Ya'acov. L'Anthropologie.

Gisis I, Bar Yosef O (1974): New excavations in Zuttiyeh Cave, Wadi Amud, Israel. Paléorient 2:175–180.

Gremyatskij MA, Nesturkh MF (eds) (1949): "Teshik-Tash." Moscow: Moscow State Univ.

Heim JL (1982): "Les Enfants Néandertaliens de La Ferrassie." Paris: Masson.

Hietala H, Marks AE (1981): Changes in spatial organization at the Middle to Upper Palaeolithic transitional site of Boker Tachtit, Central Negev, Israel. In Sanlaville P, Cauvin J (eds): "Préhistoire du Levant." Paris: Éditions du C.N.R.S., pp 305–318.

Higgs ES (1961): Some Pleistocene faunas of the Mediterranean coastal areas. Proc Prehist Soc 27:144–154.

Holloway RL (1981): Volumetric and asymmetry determinations on recent hominid endocasts: Spy I and II, Djebel Irhoud I, and the Salé Homo erectus specimens with some notes on Neandertal brain size. Am J Phys Anthropol 55:385–393.

Houston CS, Zaleski WA (1967): The shape of vertebral bodies and femoral necks in relation to activity. Radiology 89:59–66.

Howell FC (1958): Upper Pleistocene man of the southwestern Asian Mousterian. In von Koenigswald GHR (ed): "Hundert Jahre Neanderthaler." Utrecht: Kemink en Zoon N. V., pp 185–198.

Howell FC (1959): Upper Pleistocene stratigraphy and early man in the Levant. Proc Am Philos Soc 103:1–65.

Howell FC, Fritz MC (1975): Israel. In Oakley KP, Campbell BG, Molleson TI (eds): "Catalogue of Fossil Hominids III." London: British Museum (Natural History), pp 125–154.

Howells WW (1970): Mount Carmel man: Morphological relationships. Proc VIIIth Int Cong Anthropol Ethnol Sci 1:269–272.

Howells WW (1973): Cranial variation in man. Peabody Mus Papers 67:1–259.

Hublin JJ, Tillier AM (1981): The Mousterian juvenile mandible from Irhoud (Morocco): A phylogenetic interpretation. In Stringer CB (ed): "Aspects of Human Evolution." London: Taylor and Francis, pp 167–185.

Jelínek AJ (1982a): The Tabun Cave and Paleolithic man in the Levant. Science 216:1369–1375.

Jelínek AJ (1982b): The Middle Palaeolithic in the southern Levant, with comments on the appearance of modern Homo sapiens. In Ronen A (ed): "The Transition From Lower to Middle Palaeolithic and the Origin of Modern Man." Br Archaeol Rep Int Ser 151:57–101.

Jelínek AJ (1982c): Concluding discussion. In Ronen A (ed): "The Transition From Lower to Middle Palaeolithic and the Origin of Modern Man." Br Archaeol Rep Int Ser 151:327–328.

Jelínek AJ, Farrand WR, Haas G, Horowitz A, Goldberg P (1973): New excavations at the Tabun Cave, Mount Carmel, Israel 1967–1972: A preliminary report. Paléorient 1:151–183.

Keith A (1927): A report on the Galilee skull. In: "Researches in Prehistoric Galilee." London: British School of Archaeology in Jerusalem, pp 53–106.

Keith A (1931): "New Discoveries Relating to the Antiquity of Man." New York: Norton and Co.

Kimura T (1976): Correction to the *Metacarpale I* of the Amud Man. A new description especially on the insertion area of the M. *opponens pollicis*. J Anthropol Soc Nippon 84:48–54.

Klein RG (1965): The Middle Paleolithic of the Crimea. Arctic Anthropol 3:34–68.

Klein RG, Ivanova IK (1971): U.S.S.R. In Oakley KP, Campbell BG, Molleson TI (eds): "Catalogue of Fossil Hominids II." London: British Museum (Natural History), pp 311–335.

Kolosov YG, Kharitonov VM, Yakimov YP (1974): Discovery of the skeletal remains of a paleoanthrope at the Zaskal'naia VI site in the Crimea (in Russian). Voprosy Antropol 46:79–88.

Kolosov YG, Kharitonov VM, Yakimov YP (1975): Palaeoanthropic specimens from the site of Zaskalnaya VI in the Crimea. In RH Tuttle (ed): "Paleoanthropology: Morphology and Paleoecology." The Hague: Mouton Publ., pp 419–428.

Leutenegger W (1982): Encephalization and obstetrics in Primates with particular reference to human evolution. In Armstrong E, Falk D (eds): "Primate Brain Evolution: Methods and Concepts." New York: Plenum Press, pp 85–95.

Lovejoy CO, Trinkaus E (1980): Strength and robusticity of the Neandertal tibia. Am J Phys Anthropol 53:465–470.

Marks AE (1983): The Middle to Upper Paleolithic transition in the Levant. In Wendorf F, Close A (eds): "Advances in World Archaeology." New York: Academic Press, Vol 2.

Marks AE, Freidel DA (1977): Prehistoric settlement patterns in the Avdat/Aqev area. In Marks AE (ed): "Prehistory and Paleoenvironments in the Central Negev II: The Avdat/Aqev Area, Part 2 and the Har Harif." Dallas: Department of Anthropology, Southern Methodist University, pp 131–158.

Martin RD (1982): Human brain evolution in an ecological context. 52nd James Arthur Lecture on the Evolution of the Human Brain. Am Mus Nat Hist.

McCown TD, Keith A (1939): "The Stone Age of Mount Carmel II: The Fossil Human Remains From the Levalloiso-Mousterian." Oxford: Clarendon Press.

McLeish RD, Charnley J (1970): Abduction forces in the one-legged stance. J Biomech 3:191–209.

Musgrave JH (1971): How dextrous was Neanderthal man? Nature 233:538–541.

Neuville R (1951): Paléolithique et Mésolithique du désert de Judée. Arch Inst Paléont Hum 24:1–270.

Okladnikov AP (1940): Neanderthal man and his culture in Central Asia. Asia 40:357–361, 427–429.

Riesenfeld A (1972): Functional and hormonal control of pelvic morphology in the rat. Acta Anat 82:231–253.

Roginskij YY, Levin MG (1978): "Antropologiya" (in Russian). Moscow: Vyshaya Shkola.

Ronen A (1982): Mt. Carmel caves—the first excavations. In Ronen A (ed): "The Transition From the Lower to Middle Palaeolithic and the Origin of Modern Man." Br Archaeol Rep Int Ser 151:7–28.

Ryan AS (1980): "Anterior Dental Microwear in Hominid Evolution: Comparisons With Human and Nonhuman Primates." Ph.D. Thesis, Univ. Of Michigan.

Sacher GA, Staffeldt EF (1974): Relation of gestation time to brain weight for placental mammals: Implications for the theory of vertebrate growth. Am Nat 108:593–615.

Sakura H (1970): Dentition of the Amud man. In Suzuki H, Takai F (eds): "The Amud Man and his Cave Site." Tokyo: Academic Press of Japan, pp 207–299.

Santa Luca AP (1978): A re-examination of presumed Neandertal fossils. J Hum Evol 7:619–636.

Saunders JBdeCM, Inman VT, Eberhart HD (1953): The major determinants in normal and pathological gait. J Bone Joint Surg 35A:543–558.

Schwarcz HP (1980): Absolute age determination of archaeological sites by uranium series dating of travertines. Archaeometry 22:3–24.

Smith FH (1982): Upper Pleistocene hominid evolution in south-central Europe: A review of the evidence and analysis of trends. Curr Anthropol 23:667–703.

Smith FH (1983): A behavioral interpretation of changes in craniofacial morphology across the archaic/modern *Homo sapiens* transition. In Trinkaus E (ed): "The Mousterian Legacy: Human Biocultural Change in the Upper Pleistocene." Br Archaeol Rep Int Ser 164:141–163.

Smith P, Arensburg B (1977): A Mousterian skeleton from Kebara Cave. Eretz Israel 13:164–176.

Smith P, Bloom RA, Berkowitz J (1983): Bone morphology and biomechanical efficiency in fossil hominids. Curr Anthropol 24:662–663.

Smith Woodward A (1921): A new cave man from Rhodesia, South Africa. Nature 108:371–372.

Solecki RS (1960): Three adult Neanderthal skeletons from Shanidar Cave, northern Iraq. Annu Rep Smithsonian Inst 1959:603–635.

Solecki RS (1963): Prehistory in Shanidar Valley, northern Iraq. Science 139:179–193.

Solecki RS (1971): "Shanidar, the First Flower People." New York: Knopf Publ.

Stewart TD (1960): Form of the pubic bone in Neanderthal man. Science 131:1437–1438.

Stewart TD (1977): The Neanderthal skeletal remains from Shanidar Cave, Iraq: A summary of findings to date. Proc Am Philos Soc 121:121–165.

Stoner BP (1981): "A Statistical Analysis of the Neanderthal Thumb: Functional Adaptations for the Transmission of Force." B.A. Thesis, Harvard Univ.

Stringer CB (1978): Some problems in Middle and Upper Pleistocene hominid relationships. In Chivers DJ, Joysey KA (eds): "Recent Advances in Primatology." London: Academic Press, Vol 3, pp 395–418.

Stringer CB, Trinkaus E (1981): The Shanidar Neanderthal crania. In Stringer CB (ed): "Aspects of Human Evolution." London: Taylor and Francis, pp 129–165.

Suzuki H, Takai F (eds) (1970): "The Amud Man and his Cave Site." Tokyo: Academic Press of Japan.

Tobias PV (1967): The hominid skeletal remains of Haua Fteah. In McBurney CBM (ed): "The Haua Fteah (Cyrenaica)." Cambridge: Cambridge Univ. Press, pp 338–352.

Trinkaus E (1976a): The evolution of the hominid femoral diaphysis during the Upper Pleistocene in Europe and the Near East. Z Morphol Anthropol 67:291–319.

Trinkaus E (1976b): The morphology of European and southwest Asian Neandertal pubic bones. Am J Phys Anthropol 44:95–104.

Trinkaus E (1977): A functional interpretation of the axillary border of the Neandertal scapula. J Hum Evol 6:231–234.

Trinkaus E (1978): Dental remains from the Shanidar adult Neanderthals. J Hum Evol 7:369–382.

Trinkaus E (1981): Neanderthal limb proportions and cold adaptation. In Stringer CB (ed): "Aspects of Human Evolution." London: Taylor and Francis, pp 187–224.

Trinkaus E (1982a): A history of *Homo erectus* and *Homo sapiens* paleontology in America. In Spencer F (ed): "A History of American Physical Anthropology, 1930–1980." New York: Academic Press, pp 261–280.

Trinkaus E (1982b): Evolutionary continuity among archaic *Homo sapiens*. In Ronen A (ed): "The Transition From Lower to Middle Palaeolithic and the Origin of Modern Man." Archaeol Rep Int Ser 151:301–314.

Trinkaus E (1982c): The Shanidar 3 Neandertal. Am J Phys Anthropol 57:37–60.

Trinkaus E (1982d): "The Evolution of the Neandertal Face." Paper presented at the 81st meeting of the Am Anthropol Assoc.

Trinkaus E (1982e): Artificial cranial deformation in the Shanidar 1 and 5 Neandertals. Curr Anthropol 23:198–199.

Trinkaus E (1983a): "The Shanidar Neandertals." New York: Academic Press.

Trinkaus E (1983b): Neandertal postcrania and the adaptive shift to modern humans. In Trinkaus E (ed): "The Mousterian Legacy: Human Biocultural Change in the Upper Pleistocene." Br Archaeol Rep Int Ser 164:165–200.

Trinkaus E, LeMay M (1982): Occipital bunning among later Pleistocene hominids. Am J Phys Anthropol 57:27–35.

Trotter M, Gleser GC (1952): Estimation of stature from long bones of American Whites and Negros. Am J Phys Anthropol 10:463–514.

Turville-Petre F (1927): Researches in prehistoric Galilee, 1925–1926. In: "Researches in Prehistoric Galilee." London: British School of Archaeology in Jerusalem, pp 1–52.

Vallois HV, Vandermeersch B (1972): Le crâne moustérien de Qafzeh (Homo VI). Étude anthropologique. L'Anthropologie 76:71–96.

Vandermeersch B (1981a): "Les Hommes Fossiles de Qafzeh (Israël)." Paris: Éditions du C.N.R.S.

Vandermeersch B (1981b): Les premiers Homo sapiens au Proche Orient. In Ferembach D (ed): "Les Processus de l'Hominisation." Paris: Éditions du C.N.R.S., pp 97–100.

Vlček E (1973): Postcranial skeleton of a Neandertal child from Kiik-Koba, U.S.S.R. J Hum Evol 2:537–544.

Vlček E (1975): Morphology of the first metacarpal of Neanderthal individuals from the Crimea. Bull Mém Soc Anthropol Paris Sér 13 2:257–276.

Vogel JC, Waterbolk HT (1963): Groningen radiocarbon dates IV. Radiocarbon 5:163–202.

Voigt J, Pakkenberg H (1983): Brain weight of Danish children. A forensic material. Acta anat 116:290–301.

Waechter JA (1975): Lebanon. In Oakley KP, Campbell BG, Molleson TI (eds): "Catalogue of Fossil Hominids III." London: British Museum (Natural History), pp 161–165.

Yakovlev PI, Lecours AR (1967): The myelogenetic cycles of regional maturation of the brain. In Minkowski A (ed): "Regional Development of the Brain in Early Life." Oxford: Blackwell, pp 3–70.

Youm Y, Dryer RF, Thambyrajah K, Flatt AE, Sprague BL (1979): Biomechanical analyses of forearm pronation-supination and elbow flexion-extension. J Biomech 12:245–255.

Zumoffen G (1893): "Note sur la Découverte de l'Homme Quaternaire de la Grotte d'Antélias au Liban." Beirut: Imp. Catholique.

The Origins of Modern Humans: A World Survey of the Fossil Evidence, pages 295–325
© 1984 Alan R. Liss, Inc., 150 Fifth Avenue, New York, NY 10011

Homo sapiens in Sub-Saharan Africa

G. Philip Rightmire

Department of Anthropology, State University of New York, Binghamton,
New York 13901

Africa's role in the story of human evolution has long been recognized as central. Australopithecines recovered at Taung in 1924 and later at the Transvaal cave sites have been immensely important to our understanding of the Plio-Pleistocene hominids. Fossils found since the late 1950s at Olduvai Gorge and at other localities in the East African Rift Valley have proved equally valuable (Fig. 1). These discoveries shed new light on both *Australopithecus* and early *Homo*. Studies conducted in sub-Saharan Africa and in the North also bear on human evolution in the Middle Pleistocene. Remains of *Homo erectus* are now known from several sites in Tanzania, Kenya and Ethiopia as well as from the Maghreb. Later in the Middle Pleistocene, hominids best described as archaic representatives of *Homo sapiens* appear in the record. The Broken Hill cranium found in Zambia in 1921 is one of the most complete specimens, and additional material now helps to document the emergence of our own species in more detail. Human populations of this time are frequently linked with Acheulian cultures, and field projects are continuing to turn up traces of this way of life. Evidence from these later Middle Pleistocene localities will be emphasized in this review. Other African fossils provide information about the appearance of more recent people. By the beginning of the Upper Pleistocene, hominids producing a wide variety of Middle Stone Age tools are known to have inhabited much of Africa. It is likely that at least some of these populations were fully modern in their morphology, if rather fragmentary skeletons from several South African sites have been interpreted correctly. This very early occurrence of modern *Homo sapiens* in Africa will have to be confirmed by better dates and more study of the skeletal remains. My comments here are limited to the evidence as presently available in the sub-Saharan region. More discussion of the spread of later Pleistocene people into North Africa or beyond is provided by Bräuer (this volume).

TRACES OF ACHEULIAN OCCUPATION IN THE MIDDLE PLEISTOCENE

Throughout the long reaches of later Lower Pleistocene and Middle Pleistocene time, hominid populations of sub-Saharan Africa practiced an

Fig. 1. Map of Africa, showing locations of archeological sites and hominid discoveries discussed in the text.

Acheulian way of life. Probably well before 1,000 ky ago, Acheulian people were present in the Rift Valley and the adjacent highlands of eastern Africa, and there is evidence from sites such as Gadeb [Williams et al, 1979] and Melka Kunturé [Chavaillon, 1982] to show that they were able to live on the high plateaus of Ethiopia as well. Many of their sites are located near streams or river channels, at lake margins, or near other water sources. Acheulian assemblages occur frequently in open settings, both in eastern and in southern Africa. Some instances of cave occupation are also on record, as at Montagu in the Cape and the Cave of Hearths in the South African Transvaal. Earlier sites contain little direct evidence for the use of fire, although at Gadeb in Ethiopia, paleomagnetic measurements suggest that samples of welded tuff may have been burned [Clark and Kurashina, 1979]. Whether fire was controlled even by later Acheulian people is uncertain, but probably this skill was well known by the time of the Cave of Hearths occupation.

It is often claimed that hunting was an important part of Acheulian life, and at some localities the remains of numerous animals are associated with stone artifacts. This is the case at Olorgesailie in Kenya, where vast quantities of handaxes and other tools have been recovered. The animal whose skeletal parts occur most frequently at one of the Olorgesailie localities is *Theropithecus*, the giant gelada baboon. Isaac [1977], who has carried out extensive field work at the site, suggests that the baboons were systematically butchered by Middle Pleistocene hunters. The association of bones with artifacts does not by itself prove that the animals were killed by humans, but close analysis of the *Theropithecus* remains does lend support to Isaac's argument. Shipman et al [1981] note that juveniles and younger adult animals are heavily represented in the assemblage, showing attritional mortality. The bones themselves exhibit breakage patterns that may be related to butchering. At most other Middle Pleistocene sites, the evidence for hunting is not so clear, however. Fossil remains of bovids, suids, equids, and elephants are present, but it is difficult to rule out carnivores or other nonhominid agents as responsible for the bone accumulations. Large suids, elephants, and giant baboons would make formidable prey, perhaps not easily taken by people armed with Acheulian technology. It is not likely that Acheulian hunters were more efficient than those of the succeeding Middle Stone Age, and Klein [1983] points out that the latter do not seem to have been very skilled at killing larger, more dangerous game.

Stone artifacts found commonly at earlier Acheulian sites include handaxes and cleavers, and flake scrapers, choppers, polyhedrons, and spheroids of the sort associated also with industries of the Developed Oldowan tradition. Acheulian tools from eastern and southern Africa are generally similar, and little change in the artifacts themselves or in the makeup of the assemblages is apparent over long periods of time. At later sites, the bifacial tools that continue to dominate the Acheulian may be more carefully manufactured, and a wider variety of artifact types may be recognized. But few innovations occur, and the behavioral flexibility of the toolmakers would seem to be limited [Clark, 1980]. Regional specializations in stone technology do not become obvious in the archaeological record until very late in the Middle Pleistocene or at the beginning of the Upper Pleistocene, at the onset of the Middle Stone Age.

HOMINIDS ASSOCIATED WITH ACHEULIAN INDUSTRIES

The humans who made earlier Acheulian tools were most probably representative of the species *Homo erectus*. This association is not entirely clear at any of the southern African prehistoric sites, although it is likely that some of the fossil remains recovered both at Sterkfontein and at the Swartkrans cave should be grouped with the genus *Homo* [Tobias, 1978]. While there is some uncertainty as to whether the *Homo* material from Member 1 at Swartkrans should be classified as *Homo habilis* or as *Homo erectus*, the mandible (SK 15) from Member 2 deposits does resemble jaws of *Homo erectus*. Excavations conducted recently by Brain have turned up additional hominid specimens in the older component of Member 2, and some (not all) of these are also attributable to *Homo* [Brain, 1982]. Very few stone artifacts occur in Member 1 at Swartkrans, but Acheulian tools are found in Member 2.

In eastern Africa, no hominid fossils have been discovered at such tool-rich localities as Olorgesailie and Gadeb, but *Homo erectus* does seem to be linked to signs of Acheulian occupation at Olduvai Gorge in Tanzania. Postcranial remains of this species have been uncovered with a number of handaxes and cleavers, as well as a variety of smaller tools, at site WK in Bed IV [Leakey, 1971, 1980]. Elsewhere in the Gorge, skull and dental parts of *Homo erectus* occur at several sites, but the association of bones with artifacts is evident in only a few instances. At FLK in the Masek Beds, part of a mandible was found with large, finely trimmed quartzite handaxes and other tools [Leakey, 1980]. Unfortunately, this mandibular fragment is quite incomplete, so it is difficult to be very firm about its assignment to *Homo erectus* rather than to another *Homo* species [Rightmire, 1980]. In Ethiopia, pieces of a parietal and frontal bone probably attributable to *Homo erectus* are known from the Acheulian locality of Gombore 2 at Melka Kunturé [Chavaillon et al, 1974; Chavaillon, 1982].

These ties between archaic humans and earlier Acheulian industries are strengthened by archaeological evidence from Ternifine and other sites in northwest Africa, and there can be little doubt that *Homo erectus* practiced Acheulian life ways for a long time. At later sites, however, there are strong suggestions that the tool makers had changed. The stone industry is not greatly different from that produced earlier, but the hominids of the later Middle Pleistocene exhibit morphological features that set them apart from

Homo erectus. Fossil skulls, teeth (and a few postcranial parts of poorly established provenience) from this time period have turned up at several sites in sub-Saharan Africa. The material is still scarce, and many of the specimens are rather badly damaged. But a few crania are well preserved, and similarities to our own species are apparent. Although these skulls are still primitive in some respects, with heavy brows, a low flattened frontal, and a relatively broad base, the overall vault form and other characters, particularly of the occiput and temporal bone, are more modern. Most authorities now agree that these fossils should be classified as early *Homo sapiens.*

Some of the most famous remains are known from Zambia, where a human cranium was recovered from ancient cavern deposits at Broken Hill (now Kabwe) in 1921. The cranium itself is quite complete and undistorted and is a most important find. Other human material was also discovered at Broken Hill, and several different individuals are represented by cranial and postcranial fossils. But because the cave was destroyed by mining operations, it has been very difficult to ascertain whether the skull was actually associated with any of the other human bones or with faunal material collected at the site. The information available has been reviewed recently by Partridge [1982], who feels that the cranium is probably older than either the faunal assemblage or the bulk of the stone artifacts. This may be the case, but early notes kept on the stratigraphy of the cave deposits are very incomplete. About the only means of dating any of the material is provided by the faunal collection. This contains more than 20 species of large mammals, several of which are extinct [Klein, 1973]. This suggests that at least some of the Broken Hill bones and artifacts are of later Middle Pleistocene or earlier Upper Pleistocene date. Such an age for the stone industry would not be surprising, as no handaxes were recovered. The assemblage is probably best described as typologically late Acheulian, and it is known that final Acheulian industries were giving way to the Middle Stone Age at the close of the Middle Pleistocene, well over 130 ky ago [see Klein, 1983 for a recent review].

In South Africa, hominid remains associated with implements of later Acheulian manufacture have been found at two localities. On the farm Elandsfontein, near Hopefield in the southwestern Cape, stone tools and animal bones are present on deflation surfaces ("bays") cleared by wind action between shifting dunes of sand. For some time, this area has been recognized as a source of prehistoric material, and part of a human braincase was picked up in 1953. A fragment of mandible was also recovered, but further surveys and several excavations at the site have failed to produce any additional remains. The hominids and many of the other fossil bones occur in a horizon characterized by nodular calcrete development. Rather small, carefully finished Acheulian handaxes are also found on this surface. Butzer [1973] and earlier workers have suggested that the fossil horizon may be related to a former land surface, but Partridge [1982] doubts this to be the case. He proposes that wind action may have concentrated the remains at the level of an ancient water table. If this interpretation is correct, then it is unlikely that many of the tools or bones are preserved on occupation floors. Direct association of the Elandsfontein cranium with the Acheulian industry cannot be established, but it is probable that the hominids and artifacts are coeval with the fossil fauna.

As at Broken Hill, the best estimate for the age of the Elandsfontein assemblage is provided by analysis of the animal remains. Some 50 species of larger mammals have been identified, and at least 19 of these are extinct. This evidence is strongly indicative of Middle Pleistocene antiquity [Hendey, 1974; Howell, 1978]. Vrba's [1982] comments on the Elandsfontein bovids suggest that the faunal assemblage may have accumulated over a lengthy period of time, but the bulk of the material seems to be of later Middle Pleistocene age.

Far to the northeast, in the Transvaal, more human fragments have been extracted from the hard, brecciated deposits of the Cave of Hearths. Here the association of a broken lower jaw with later Acheulian handaxes and other tools is firmly established. Unfortunately, dating of the lower part of the cave sequence is problematical, as few faunal remains are preserved. At least some extinct species are present, including a large alcelaphine antelope [Vrba, 1982]. Study of the available collection has led most workers to conclude that the Hearths assemblage may be about as old as that from Broken Hill, or slightly younger. The hominid consists only of the right half of the body of a mandible, with three teeth in place. This individual is juvenile and may represent a population similar to those sampled at Broken Hill and Elandsfontein [Tobias, 1971]. However, because the fossil is so fragmentary, its significance is limited.

Rather more informative specimens are known from sites in eastern Africa. Several human fossils have been discovered in Bed IV and in the Masek and Ndutu Beds at Olduvai in northern Tanzania, but much of this material is likely to represent Homo erectus, as noted previously. Bed IV and the Masek levels are of earlier Middle Pleistocene age, and probably the first populations of Homo sapiens did not appear in East Africa until somewhat later on. A cranium that may well help to document this period in human evolution has come to light in excavations conducted at Lake Ndutu, located at the western end of the Main Gorge at Olduvai. Details of these excavations have been published by Mturi [1976]. The human cranium was uncovered in sandy clay deposits, apparently laid down at or near the western shore of the lake. Quantities of animal bones, mostly not well preserved, and some stone artifacts, were also found during the course of the 1973 field season. A few of the tools were in situ in the deposits, and these seem to be drawn from an Acheulian industry containing an unu-sually high proportion of spheroids and polyhedrons.

The greenish sandy clay in which the remains occur is overlain by a hard, rootmarked tuff that may be correlative with the Norkilili Member as known from the upper part of the Masek Beds exposed at Olduvai. If this is the case, then the cranium and artifacts may approach 400 ky or so in age [Hay, 1976]. However, Lake Ndutu lies outside of the immediate Olduvai tectonic zone, and it is not possible without more fieldwork to be entirely certain of this correlation with the Olduvai sequence. According to Leakey and Hay [1982], the Ndutu tuff is mineralogically similar to the Norkilili Member but different from the tuffs of the Masek lower unit. This suggests that the hominid is no older than the top of the Masek Beds. At the same time, mineralogic analysis does not fully distinguish between the upper Masek tuffs and tuffs of the overlying Ndutu Beds at Olduvai. It is possible that the remains are younger than Masek age, but the Ndutu lower unit covers a long span of time, perhaps several hundred thousand years. Dating is thus very imprecise, but a later Middle Pleistocene age is likely.

The discovery at Lake Baringo, Kenya, of a nearly complete hominid mandible associated with postcranial fragments and Acheulian tools has been reported by Leakey et al [1969]. Deposits containing this assemblage have been designated as the Middle Silts and Gravels Member of the Kapthurin Formation, which is apparently of Middle Pleistocene age. The sediments show normal magnetic polarity [Dagley et al, 1978] and are overlain by a tuff tentatively dated at 0.23 or 0.24 million years [Tallon, 1978]. More dates for this formation should result from new field studies that are underway at the lake. This work should also yield fresh samples of the lithic assemblage, perhaps along with fossil remains of the toolmakers themselves.

The Kapthurin mandible is somewhat less robust than other jaws from East Africa that have been referred to as *Homo erectus*. Such differences emerge from comparisons with the material found at Olduvai. It is clear, for example, that the superior transverse torus, which buttresses the chin internally, is less prominent in the Baringo hominid than in the most complete mandible from Olduvai Beds III/IV. Although some surface bone has been lost from the chin region, and the tooth sockets are eroded, the Baringo jaw may exhibit a more vertical symphyseal profile as well [Rightmire, 1980]. What these differences signify and whether the mandible should be identified as *Homo erectus* or as archaic *Homo sapiens* has not been settled. There are few firm criteria for distinguishing the jaws of earlier *Homo* species, and it may be that mandibles are simply not useful in this sort of diagnostic work. As with the Cave of Hearths material, it is difficult to obtain much information from a damaged specimen found without associated cranial parts.

Some uncertainty also surrounds the classification of important hominids from Ethiopia, although in the case of Bodo, there is more material to study. A fairly complete facial skeleton and parts of a braincase were recovered at this site in the Middle Awash Valley during 1976 and 1978 [Kalb et al, 1980]. Work resumed in the area in 1981, and a parietal bone from a second individual has since been collected. These fossils exhibit a number of archaic features, but on the basis of descriptions provided so far [Conroy et al, 1978], they have been regarded by most workers as early *Homo sapiens*. This interpretation is subject to change, as new information becomes available. The cranium has recently been cleaned and reconstructed by T. White at Berkeley and can now be studied in more detail. There is also some promise of obtaining radiometric dates for the lower members of the Wehaietu Formation, where the skull(s) were found. The fauna from the Bodo Member includes many mammalian species, and some of the material is well preserved. Comparisons carried out by Kalb et al [1982] suggest a Middle Pleistocene age, but just where within this span of time the hominid site may lie has not yet been established. Acheulian artifacts are plentiful at Bodo, and there seems to be no doubt that the tools are associated with the fauna and with the *Homo* fragments.

At Melka Kunturé, in the Ethiopian highlands south of Addis Ababa, where probable *Homo erectus* remains have been uncovered with early Acheulian tools, other discoveries document a hominid presence at Garba 3. This locality falls later in the Melka Kunturé succession. Tools at the site are of final Acheulian manufacture and foreshadow the coming of the Middle Stone Age [Chavaillon, 1982]. Faunal material occurs in some abundance, and fragments of a human cranium have also been picked up. A description of the hominid, which is probably (substantially?) more recent than the Bodo finds, has not been published. If the remains are reasonably complete, this should be an important specimen. Field projects are continuing at Melka Kunturé and elswhere in Ethiopia, and there is every reason to hope for other new discoveries. In any case, research carried out systematically at these Middle Pleistocene localities should add greatly to our knowledge of Acheulian people and their way of life.

MORPHOLOGY OF THE BROKEN HILL, ELANDSFONTEIN, NDUTU, AND BODO CRANIA

Although these African hominids are a rather scattered and fragmentary lot, there are now enough fossils on record to enable us to piece together a rough description of the anatomy of later Middle Pleistocene populations. The damaged jaws from the Cave of Hearths, Lake Baringo, and other sites are not very useful, but several of the crania associated with Acheulian artifacts do provide important information. One of the best specimens, in fact the first to be discovered, is the skull from Broken Hill (Table I). This cranium is low in outline, with massive brows. The supraorbital torus is thicker than that of most other later Pleistocene hominids and is even a little heavier than the torus of Hominid 9 from Olduvai. Greatest breadth of the braincase falls at the level of the mastoid processes rather than higher on the parietals. The sides of the vault are flattened and not rounded or bossed as in more modern humans. Because of such features, Broken Hill has sometimes been referred to the species *Homo erectus*. But cranial capacity is expanded well beyond the *Homo erectus* average of just under 1,000 ml, and other apsects of cranial morphology also point in the direction of *Homo sapiens*. The occipital bone is less strongly curved than in *Homo erectus*, and its upper scale is approximately vertical above the transverse torus. In Broken Hill, some surface bone has been lost from the center of the occiput, and most of the right side of this bone is missing altogether. The part of the torus that remains is sharply defined and overhangs the nuchal area below. The nuchal surface itself is flattened so as to lie fully in the horizontal plane, and the mastoid processes are not so laterally prominent. Studies of the Broken Hill cranial base, which exhibits a short basioccipital and steep clivus, indicate that this part of the specimen is essentially modern [Laitman et al, 1979]. Details of the sphenotemporal articulation and mandibular fossa are like those of more recent humans. The postcranial remains that have been examined also appear to possess a few of the characteristics associated with *Homo erectus*. However, it should be remembered that there is no clear association of the limb and pelvic parts with the adult skull, and hominid remains from different sections of the cave deposits may be of quite different age.

Individuals from Elandsfontein and Lake Ndutu are less complete, but there is general agreement that the Elandsfontein braincase shows a strong

TABLE I. Measurements of Middle and Early Upper Pleistocene Fossil Crania From Sub-Saharan Africa[a]

Measurement	Broken Hill	Elandsfontein	Ndutu	Florisbad	Omo 1	Omo 2	L.H. 18
Cranial length	205	200[b]	183[b]	—	—	—	204
Torus thickness							
Central	23	21	—	15	—	—	14
Lateral	16	15	10.5	12	—	—	—
Basion-nasion	108	—	105[b]	—	—	—	—
Basibregmatic ht.	127	—	—	—	—	—	—
Max. cranial br.	145	—	144	—	—	147	140
Biauricular br.	140[c]	—	128	—	—	132	125
Max. frontal br.	118	114	112(?)	132[c]	—	121	115
Biasterionic br.	—	—	113	—	—	—	117
Biorbital chord	125	—	—	124[c]	—	—	—
Nasion subtense	26	—	—	21(?)	—	—	—
Midorbital chord	76	—	72[c]	—	—	—	74[c]
Naso-orbital subtense	20	—	—	—	—	—	—
Frontal sag. chord	120	116[b]	—	120	—	—	116
Frontal subtense	21	20[b]	—	23	—	—	19
Occipital sag. chord	87	—	87	—	101	106	—
Occipital subtense	—	—	30	—	30	40	—
Lambda-inion[d]	—	53(?)	61	—	73	59	60(?)
Inion-opisthion[d]	—	—	45	—	47	74	—
Mastoid length	27	—	27	—	—	30	21

L.H., Laetoli hominid.

[a]All measurements are taken in millimeters on original fossil material, except where noted.

[b]These measurements are taken from a cast and/or make use of reconstructed landmarks.

[c]Values obtained by doubling a reading to the midline.

[d]These measurements are influenced by the way in which inion is defined. If this landmark is considered to be on the linear tubercle (at the junction of the superior nuchal lines), then length of the occipital upper scale will generally be greater than when inion is located at the center of the transverse torus. In the case of Ndutu and L.H. 18, inion is taken at the linear tubercle.

resemblance to Broken Hill. Similarities are apparent in morphology of the supraorbital region and form of the frontal. The parietal bone, well preserved on one side, may exhibit slightly more bossing, but maximum breadth of the vault is still low, near the (missing) mastoid processes. The base of the skull is broken away, but the form of the occiput seems to be like that of Broken Hill, even if a transverse torus is not so well developed.

The Ndutu cranium (Fig. 2), badly fragmented at the time of its discovery, has been reconstructed by Clarke [1976]. Only parts of the facial skeleton are preserved, and the frontal bone is also poorly represented. Other portions of the specimen have been fitted back together, although some

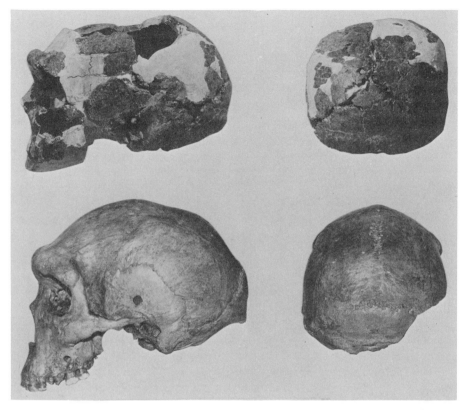

Fig. 2. Lateral and occipital views of the Lake Ndutu cranium (above) compared to corresponding views of the Broken Hill specimen (below).

damage to the parietals and the cranial base can not be repaired completely. In superior view, the Ndutu vault is somewhat more rounded than that of Broken Hill but is comparable in form to Elandsfontein. A general resemblance to the South African skull is again apparent when the two fossils are viewed from the rear. Both have walls that rise steeply from the supramastoid region, and the Ndutu parietals especially are more rounded than those of Broken Hill. Clarke [1976] is correct in stating that Ndutu shows more parietal bossing than would be expected in *Homo erectus*. Occipital morphology is broadly similar in all three individuals. The upper scale is close to vertical rather than forward sloping as in many *Homo erectus* specimens. In Ndutu, the lambda-inion chord is greater than the length of the nuchal plane. A relatively expanded upper scale is often said to be characteristic of

Homo sapiens, although this feature does not always distinguish archaic from more sapient humans.

In Ndutu, there is a "true" external occipital protuberance at the junction of the highest nuchal lines, and the transverse torus is moundlike rather than sharply outlined. In other aspects of occipital and temporal anatomy, this individual is close to Broken Hill. The nuchal plane is flattened, with some relief resulting from muscular insertions. A crest is well developed along the full length of the mastoid process, but there is no distinct tubercle (of the sort common in European Neandertals) behind the auditory opening. Near the junction of the occiput and temporal, there is a deep digastric incisure. There is some heaping up of bone on the medial side of this groove, to produce a juxtamastoid eminence. In Broken Hill, this is accentuated by some hollowing of the adjacent muscle attachment, but in neither case is an occipitomastoid crest in the sense of Weidenreich [1943] or Hublin [1978] developed. The eminence is not so well expressed as in archaic European crania and instead resembles the condition seen in many modern humans.

The glenoid cavity, somewhat deeper in Ndutu, is bounded anteriorly by an articular tubercle that is prominent enough to stand out against the preglenoid surface of the temporal bone. In both individuals, there is a strong postglenoid process. The medial wall of the cavity is broken in Ndutu, so it is not possible to tell whether a sphenoid spine is present in the entoglenoid region. Broken Hill is better preserved, and here it is clear that the entoglenoid process is primarily of temporal origin. However, a sphenoid spine is developed, in a manner not evident in *Homo erectus* specimens. This spine is not particularly prominent in Broken Hill but is oriented in about the same way as in recent crania. The posterior wall of the glenoid surface is made up of the tympanic plate. In both skulls, this plate is thickened locally to produce a sheath for the styloid process. The Broken Hill process is missing, although a circular opening marks its original position. On one of the Ndutu temporal bones, the root of the styloid process is still in place, and the adjacent inferior border of the tympanic plate is thin as in modern humans. A "spine of the crista petrosa" as described by Weidenreich [1943] for Chinese *Homo erectus* is not apparent.

One additional find from Ethiopia provides more information about the Middle Pleistocene populations of eastern Africa. The Bodo skull, reconstructed from many fragments, is comparable to Broken Hill in some respects, including frontal form. Anteriorly, the frontal surface is flattened, and a supratoral sulcus is not well developed. Glabella is prominent, and the torus itself is very thick. There is substantial postorbital constriction, although neither Bodo nor Broken Hill exhibits as much postorbital nar-rowing as does Hominid 9 from Olduvai. Both individuals are absolutely broad across the brows and upper face. The Bodo facial skeleton is heavily constructed, with a flattened nasal root, widely separated orbits, and a low broad nasal aperture. The maxillary walls and zygomatic bones are espe-cially massive, as noted by Conroy et al [1978]. Cheek height is much greater than in Broken Hill, and the surface of the zygomatic arch is more prominent (almost bulbous), relative to the orbital margin. The Bodo face is thus flattened in appearance, and differs from that of the Zambian hominid principally in the breadth and build of its middle parts. Whether these features suggest that Bodo is better grouped with Homo erectus or with Homo sapiens is presently not certain. More detailed study of the specimen is needed, now that cleaning and fresh reconstructive efforts have been carried out. If examination of the cranial base confirms ties to Broken Hill and Ndutu, then an assignment to archaic Homo sapiens will rest on firmer ground.

CLASSIFICATION OF THE LATER MIDDLE PLEISTOCENE POPULATIONS

Even if the position of Bodo is not entirely clear, these comparisons demonstrate that the African crania share a number of features linking them to Homo sapiens. Characters of the parietal and occipital bones, the occipitomastoid region, the tympanic plate, and the glenoid cavity all point in this direction. Of course there is considerable variation, extending to aspects of vault shape as well as size. The Ndutu cranium is smaller than Broken Hill, with a capacity estimated as only about 1,100 cm^3. It is possible that this individual is female, although the extent of sex dimorphism to be expected in Middle Pleistocene populations is difficult to assess. Only a few postcranial remains have been found at African localities, and it is difficult to assign sex to skulls alone. The Elandsfontein skullcap, for example, is so incomplete that no accurate sex determination can be made. In the case of Bodo, the massive build of the facial skeleton suggests that this Middle Awash specimen may be a male.

While most workers accept the fossils as early representatives of Homo sapiens, there is less agreement about the evolutionary significance of this species designation. Some paleontologists who recognize both Homo erectus and Homo sapiens emphasize the arbitrary nature of such distinctions. Gingerich [1979], for example, defines species as "arbitrarily divided seg-ments of an evolving lineage that differ morphologically from other species in the same or different lineages." For him, the boundaries between *Homo erectus* and *Homo sapiens* are not clearly demarcated. Other authorities support this view and add that the pace of evolutionary change within the *Homo* lineage appears to have been slow and steady. If this is correct, fossils displaying "intermediate" morphology should be common in the prehistoric record. Cronin et al [1981] have recently surveyed much of the available Plio-Pleistocene hominid material and have concluded that such transitional individuals do occur, especially in the later Pleistocene of Europe. Remains from Petralona and other localities are used to advance a claim for gradual change in populations linking late *Homo erectus* with early *Homo sapiens*. Some of the African evidence can also be interpreted in this way, and Clarke [1976] notes that the Ndutu cranium exhibits both archaic features and resemblances to more modern humans. A few anthropologists have carried this thinking a step further, by arguing that there is no need to define separate species in the Middle Pleistocene. In Jelínek's [1980a,b] opinion, all of the hominids from Europe and North Africa should be regarded as *Homo sapiens*, although evolution within this species is said to be a complex process, proceeding differently in areas such as the Maghreb and the Middle East.

A more explicit hypothesis of gradualism with regional continuity has been proposed by Thorne and Wolpoff [1981; see also Wolpoff et al, this volume]. These workers may recognize two *Homo* taxa in Southeast Asia and Australia, but these chronospecies are no more than segments of a morphological continuum. *Homo erectus* from Indonesian localities is described as part of a "regional clade," which also includes populations of later *Homo sapiens*. A similar case has been made for Europe by Wolpoff [1980, 1982]. Here the division between species is again said to be arbitrary, and Wolpoff seems to prefer a chronological criterion (the "end of the Mindel glaciation") to mark the boundary. A few of the European fossils may be *Homo erectus* by this reasoning, but all are taken as members of a single sexually dimorphic lineage, ancestral to later Neanderthals and recent humans.

General acceptance of gradualism and regional continuity in human evolution would provide an uncomplicated framework within which to describe relationships of the African populations. Finds from Broken Hill, Ndutu, and Bodo could presumably be assigned to a sort of pan-African lineage or clade, along with specimens from the Kapthurin Formation, Melka Kunturé, and Olduvai. The question of whether to refer (some of) the Olduvai and other early material to *Homo erectus* would become trivial if divisions between *Homo* species were considered to be arbitrary or based on dating rather than morphology. The crania from Broken Hill and other later Middle Pleistocene localities might be viewed as similar in "grade" to archaic *Homo* specimens from Europe and the Far East, but evolutionary continuity in Africa could be assumed. That is, populations sampled in southern and in eastern Africa would be seen as ancestral to later, more modern populations inhabiting these same regions.

Several aspects of this scenario can be questioned, however, and there are alternate ways to interpret the fossil evidence from the sub-Sahara. Firstly, it should be pointed out that the concept of paleontological species as arbitrarily defined segments of a lineage has not been embraced by all authorities. Some have doubted the prevailing view that *Homo* species are subject to gradual, progressive change. In particular, the assumption that *Homo erectus* populations all across the Old World merged imperceptibly with early *Homo sapiens* has been challenged [Eldredge and Tattersall, 1982; Gould and Eldredge, 1977; Rightmire, 1981a, 1982; see also a comprehensive review by Howells, 1980]. Delson et al [1977] have taken a rather extreme position on this issue and question whether there can be any continuity between the two species. Other workers contend that at least some groups of *Homo erectus* are likely to be the direct ancestors of later humans but suggest that the transition from archaic to more modern forms may have taken place just once, in a restricted geographic province. Various regions, including Africa and eastern Asia, have been identified as such evolutionary centers. Recently Delson [1981] has noted that evidence for speciation may be found in Europe. He hypothesizes that populations of *Homo sapiens* may have emerged there first, as a consequence of isolation due to glacial conditions. At present, this issue cannot be settled. What should be emphasized here is not the timing or geographic location of speciation, but rather the status of *Homo erectus* as a "real" species, distinct in important ways from *Homo sapiens*. Bonde [1981] has summed up much of the current thinking about species in paleontology, and he makes a good case for treating such groups as coherent entities. In this view, species are evolutionary units that cannot be subdivided arbitrarily.

From this perspective, African hominids such as Ndutu seem less clearly to be "intermediate" in their morphology. Such crania do show a few characters traditionally associated with *Homo erectus* (eg, a relatively small cranial capacity, thick vault bones, a low frontal profile, and a broad base), but these features may have little diagnostic value. A relatively flat frontal and a broad skull base are found not only in *Homo erectus* but also in later Pleistocene populations widely recognized to be more modern. Other aspects, particularly of the Nduto occiput and temporal bone, may reflect a new (or derived) state shared (only) with Homo sapiens, as noted here and elsewhere [Rightmire, 1983]. Skulls from mid-Pleistocene Europe also possess at least some of these characters [Stringer et al, 1979]. But more detailed studies of the distribution of such traits among Middle Pleistocene assemblages are needed before we can answer the question of where Homo sapiens arose first.

Just how the later Acheulian people of Africa are related to contemporary populations of Europe is another issue, too large to be explored here. Simply stating that Broken Hill, Ndutu, and perhaps Bodo belong to the same broad evolutionary grade as the Petralona and Arago remains does not convey much information. Perhaps it will prove more useful to treat the sub-Saharan fossils as representative of an African subspecies of Homo sapiens, distinct from other groups. This approach has been advocated before, although there may be problems in trying to apply too many formal subspecific designations [Howells, 1980]. Whether the Middle Pleistocene material can be linked to later African populations is also an important question. This cannot be addressed without commenting on additional fossils, found in South Africa and as far north as the Sudan.

FLORISBAD, LAETOLI, SINGA, AND THE OMO REMAINS

More evidence concerning early Homo sapiens in Africa has come from several localities. Some of the finds were made early in the century, when excavations were not conducted as carefully as is the rule today. In cases where the material is of known stratigraphic provenience, dates have proved difficult to ascertain. Artifactual associations are poorly documented or lacking altogether. Problems of this sort coupled with the fragmentary state of some of the fossils have resulted in a fair amount of controversy. The significance of these discoveries is not entirely clear, although all workers agree that they are crucial to our understanding of human evolution in the later Pleistocene.

One of the sites is Florisbad in South Africa. The Florisbad deposits are made up of organic clays or peats, layered between beds of coarse sand. A number of spring vents are present, and groups of these vents or "eyes" appear to have been active at different times. Excavations conducted at the site show that some of the old springs are associated with columns of relatively pure sand, which extend to varying levels within the stratigraphic succession. These sandy "caps" occur over basal mounds of cemented debris, consisting of material that entered the spring or was washed from overlying deposits during periods of activity. Bones and artifacts are found in these basal accumulations, and it was in such a deposit that part of a human cranium was recovered in 1932. Given the rather complicated situation at the site, it is not surprising that the provenience of the find was not specified too clearly. Excavations carried out later did not fully resolve the question of whether the human remains were in the lowest (basal) peat level or higher in peat I. Most accounts indicate that the spring eye had penetrated the peat I layer before being capped by deposits of hard greenish sand. This would suggest that the spring contents are roughly contemporary with peat I and certainly older than peat II, which lies above the sand. Partridge [1982] questions this, but new evidence bearing on the location of the original discovery has not yet been presented. R.J. Clarke is currently working at the site, and his findings should be useful.

Radiocarbon dates obtained for the Florisbad peat deposits more than 20 years ago have been treated with skepticism, and it has been assumed that at least the earlier peat levels are too old to be dated by this method. Samples obtained more recently confirm this, and peat II is now known to have an age in excess of 42 ky. Peat I may be substantially earlier. Several extinct species of mammals occur in the Florisbad fauna, but these forms have been found in Upper Pleistocene contexts elsewhere in southern Africa [Klein, 1980]. The faunal evidence is therefore consistent with an early Upper Pleistocene date for the hominid remains but does not provide much more information. An age as great as that of the Elandsfontein (or Broken Hill) specimens is unlikely. Stone tools distributed through the spring deposits belong to the Middle Stone Age rather than to an Acheulian industry, but association of these artifacts with the fossils has still to be established.

The Florisbad human remains consist of a rather well preserved frontal bone, parts of both parietals, and the incomplete right side of a face (Fig. 3). Following their discovery, these pieces were set in a plaster reconstruction by Dreyer, who noted the "primitive" appearance of the specimen. Dreyer also emphasized features of the face that he considered to be more modern. Several later workers expressed similar views, and links between Florisbad and recent San (Bushman) populations were thought to be most likely. Other anatomists doubted this and preferred instead to draw parallels with more archaic finds from southern Africa, including the skulls from Elandsfontein and Broken Hill. This latter approach to the fossils is probably more reasonable, even if the Florisbad spring deposits are not of confirmed Middle Pleistocene antiquity.

Comparisons between Florisbad and Broken Hill show that, while frontal chord length is the same for both individuals, the Florisbad forehead is higher and substantially broader. Postorbital constriction is less noticeable,

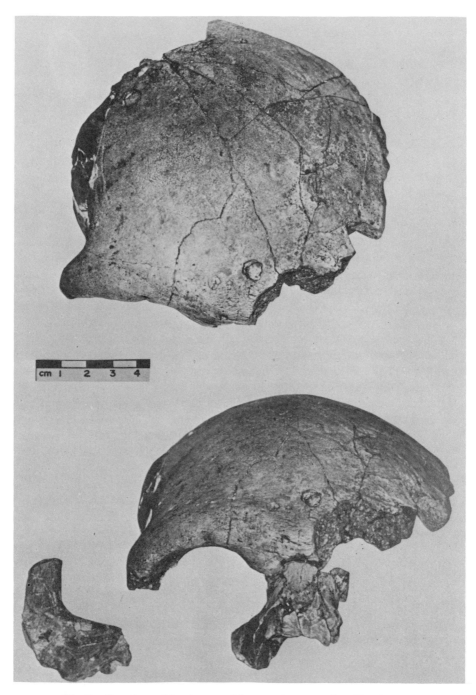

Fig. 3. Superior and facial views of the partial cranium from Florisbad.

and the brow ridge is not so thickened. At the same time, there are general resemblances to the more archaic hominid, particularly in the supraorbital region. In breadth across the orbits, the Florisbad face must be close to Broken Hill, and it is too bad that facial height cannot also be measured in both specimens. Enough of the Florisbad orbit is preserved to show that this is a little lower and wider than that of Broken Hill, but differences are not marked. More contrast is provided by the form of the zygomatic bone, which suggests some hollowing of the wall below the orbit. This depression of the infraorbital region was surely exaggerated in Dreyer's original reconstruction, as was prognathism below the nose. But even after appropriate modifications are made, there is some departure from the morphology of Broken Hill [Rightmire, 1978]. Shape of the cheek, along with the more domed frontal, combine to give Florisbad a more modern look, although the cranium is so fragmentary that comparisons with other fossils must be very limited.

Remains closely resembling Florisbad are not known from other South African localities, but several better preserved skeletons have come from the lower Omo Basin in southwestern Ethiopia. One individual, numbered Omo 1, was found in situ in deltaic beds assigned to Member I of the Kibish Formation. A Th/U date of 130 ky has been reported for shells from these deposits [Butzer et al, 1969], and Butzer has continued to express confidence in this result [Butzer and Isaac, 1975]. Unfortunately, faunal material including bovids and elephants collected at the site are not particularly helpful and do not independently confirm a later Middle Pleistocene age. A few stone artifacts recovered with Omo 1 have not been classified. Another specimen was picked up on the surface, several kilometers away. Geological investigations at this second site suggest that Omo 2 must also be derived from Member I sediments, although some uncertainty as to the stratigraphic provenience of this cranium remains. A few fragments representing a third Omo individual may have weathered out of deposits higher in the Kibish sequence.

The more complete cranium, lacking all of its face, is Omo 2 (Fig. 4). The supraorbital region is damaged, but the rest of the broad, flattened frontal is intact, and the braincase is relatively long and low. The back of the vault is strongly curved, and the nuchal area of the occiput is large. These features have prompted some comparisons to *Homo erectus*, but the endocranial volume of Omo 2, estimated as greater than 1,400 cm^3, is surely higher than would be expected in any member of that taxon. Other characters of the frontal and occiput (and probably the morphology of the temporal bones) are also in keeping with identification as *Homo sapiens*, and Omo 2 is broadly

Fig. 4. Lateral and occipital views of the Omo 2 braincase.

similar to Broken Hill [Rightmire, 1976]. Only the very wide frontal bone, carrying what was apparently a moderately developed supraorbital torus, is appreciably different, and here there is some resemblance to Florisbad.

Omo 1 is made up mainly of the rear of a braincase, including parts of the occipital and parietal bones. Some of the frontal and a few facial fragments are also available, and all of this material has been used by Day and Stringer [1982] in a recent reconstruction. There is little doubt about assignment to *Homo sapiens*, as at least the back of the skull displays modern aspects. The upper scale of the occiput is longer, relative to the reduced nuchal plane, than in Omo 2. The occipital bone as a whole is less strongly curved or angled. Viewed posteriorly, the vault seems higher, and more expansion (bossing) of the parietals is apparent. Several frontal pieces cannot be joined directly to the parietals but suggest a long flattened forehead not unlike that of Omo 2. Segments of the supraorbital rim that can be compared in the two specimens are also similar, and the torus is not massively developed.

Despite some resemblances to Omo 2, Omo 1 is more modern anatomically. Day's and Stringer's [1982] reconstruction certainly indicates this, even if the frontal and most of the face are too incomplete to provide much corroborative evidence. A major question then is whether the Omo skulls should be treated together or regarded as representatives of separate populations. This has been debated for some time. Stringer [1978] and Day and Stringer [1982] argue mainly from analysis of measurements that Omo 1 and Omo 2 must be drawn respectively from anatomically modern and from more archaic groups; Howell [1978] prefers to lump the Omo skulls together. He is confident that both individuals belong to a subspecies of *Homo sapiens* that is distinct morphologically from any to which Broken Hill and the East African Acheulian people might be assigned, and he expects that the Omo group is probably more recent. This interpretation is reasonable in that Omo 1 at least is likely to be younger than fossils such as Ndutu or Bodo. If Howell's proposal is accepted, then the Omo people, perhaps along with Florisbad [Beaumont et al, 1978], can be viewed as the ancestors of later *Homo sapiens sapiens*, known to have inhabited Africa and the Middle East in the Upper Pleistocene.

However, this reading of the evidence takes no account of the archaic features of Omo 2, which are brought out clearly by measurement and by anatomical comparisons. If the geological setting of the fossils provides no basis for separating Omo 2 from Omo 1 in time, then perhaps the relation-ship between the Kibish people and populations represented by Ndutu or Broken Hill is not so distinct. Given so many uncertainties, it is probably premature to point either to the Omo basin or to Florisbad as the source of populations directly ancestral to modern humans. The Florisbad cranium shows no special ties with San (Bushmen) or other living Africans in discriminant analysis [Rightmire, 1978]. Certainly the same can be said for the Omo remains, and even fossils such as Singa from the Sudan no longer appear to be closely linked to recent people. The famous Singa cranium was, like Florisbad, viewed initially as Bushman-like, although nearly all of the facial skeleton is missing. Current study of the braincase suggests that this interpretation is incorrect, and Stringer [1979] has emphasized that the specimen is best aligned with archaic rather than with fully modern humans.

Another discovery tht seems to bear directly on this question has been made in Tanzania. In 1976, a rather well-preserved cranium was recovered from the Ngaloba Beds at Laetoli, not far from Olduvai Gorge. Tentative correlation of a tuff occurring in these beds with a similar tuff in the Olduvai Ndutu Beds suggests an age for the hominid of about 120 ky [Leakey and Hay, 1982]. A preliminary description of the skull has been published by Day et al [1980], and it is clearly *Homo sapiens*. Archaic features are said to include a low thickened vault, a flattened frontal contour, and a well developed supraorbital torus. Other characters show more resemblance to the Omo skulls than to Ndutu or Broken Hill.

The Laetoli hominid is important, and in my own opinion it provides stronger evidence than does the Omo assemblage for a modern presence in East Africa. The frontal bone is only very slightly keeled, and there is little postorbital constriction. The brows are moderately thickened, but on each side the central portion of the torus is set off from the lateral supraorbital margin by a shallow depression. Glabella is inflated but does not project much beyond the nasal root. Although parts of the skull are thick, the parietal bones are large, and length of the bregma-lambda chord greatly exceeds that of either Ndutu or Broken Hill. In rear view, L.H. (Laetoli hominid) 18 displays a rounded profile, and the parietal vault is broader than the base. Neither the supramastoid crests nor the mastoid processes are as laterally projecting as in more archaic *Homo sapiens*, and there is little inturning of the mastoid tips.

The upper scale of the Laetoli occiput is high, and although length of the lower scale cannot be measured accurately, there is no doubt that the nuchal plane is relatively short. The rather prominent, moundlike occipital torus has no clear upper margin. Instead, swelling associated with the torus extends over much of the upper scale, to give the rear of the vault a bulging appearance. On the right, a few centimeters of the occipitomastoid junction are preserved, and the digastric incisure is deep and wide. On the medial side of this groove, there is some heaping up of bone to form an eminence. Because of damage, it is difficult to ascertain how far this ridge may have extended or how it is related to muscle scars that have been lost. This structure may best be described as a juxtamastoid eminence of indeterminate proportions. Both glenoid cavities are present and seem to be fully modern in their anatomy. An articular tubercle is clearly defined, and there is a strong sphenoid spine. The inferior aspect of the tympanic plate is thin.

Despite a generally robust appearance and some frontal flattening, the Laetoli cranium resembles recent humans in a number of respects, extending to details of parietal, occipital, and temporal anatomy as well as to overall size and shape of the braincase. While the face is very incomplete, there are indications here also that the specimen is modern. The zygomatic processes of the maxillae appear to be less heavily built than those of Broken Hill, and there is definite hollowing of the maxillary wall to form a canine fossa. Much bone below the nasal opening is missing, but there is some indication of alveolar prognathism. Further comparative study of L.H. 18 is needed, particularly to clarify similarities of this individual to material from Ethiopia and (perhaps) to other later hominids from eastern Africa. Work of this sort may produce solid evidence for continuity of later Middle Pleistocene populations with more recent groups. Information presently available does not satisfactorily settle the question of whether *Homo sapiens sapiens* evolved gradually, in several parts of Africa, or whether anatomically modern people arose from a single source, somewhere in the sub-Saharan region or outside of Africa entirely. However, it is increasingly likely that modern humans were established in southern Africa not long after the onset of the Upper Pleistocene, as discussed in the next section.

MIDDLE STONE AGE POPULATIONS

By the onset of the Upper Pleistocene, Acheulian industries had been largely replaced by Middle Stone Age toolkits. Middle Stone Age associations are not clear at Florisbad or at the Omo sites, although tools found in the Ngaloba Beds at Laetoli are said to be of Middle Stone Age manufacture. At other sub-Saharan African localities, traces of Middle Stone Age occupation are much more plentiful. It is now possible to reconstruct the lifeways of earlier Upper Pleistocene people in some detail, and there are signs of change from the preceeding Acheulian.

Middle Stone Age sites occur both along the African coastline and in a variety of inland settings. It is clear that people were equipped to occupy the more humid, wooded regions as well as open grasslands. Caves and rockshelters were utilized where these were available, and it is possible that some bands returned to the same caves on a seasonal basis. Artifact scatters in the open suggest more transitory hunting camps. There is clear evidence for the use of fire, and hearths have been located in many of the excavated sites.

Lithic assemblages of this period no longer include the bifacial handaxes and cleavers common in Acheulian times. Instead, projectile points, knives, and scraping tools predominate. In some of the more humid parts of Africa, heavy duty flaked core "axes" or adzes and long bifacially worked points also occur. Clark [1980] has suggested that some of these heavy tools were used for woodworking, although the precise function of most implements cannot be known with any certainty. In general, technological advances relative to the latest Acheulian do not appear to be striking. However, there is some evidence for the hafting of several sorts of stone artifacts, and Middle Stone Age people had probably learned to produce composite tools [Volman, 1984].

Faunal remains excavated in caves in southern Africa have been studied extensively by R.G. Klein, and his work has shed much light on Middle Stone Age subsistence. Mortality profiles constructed for some of the mammal species represented in the assemblages are particularly interesting. Cape buffalo, for example, show attritional profiles, dominated by very young animals and also including relatively large numbers of older adults. Buffalo are known for their aggressiveness, and it is not surprising that Middle Stone Age hunters were unable to kill many of the healthy, prime-age individuals. They could take the more vulnerable juveniles and aging members of the herd, and these animals were brought to the camps to be consumed [Klein, 1982].

Other species such as eland show catastrophic profiles, in which successively older age classes contain smaller numbers of individuals. Here prime-age adults are represented in proportion to their frequency in a stable living population. Eland are relatively docile, less dangerous animals, and they can be driven into traps or over cliff faces. Middle Stone Age people may well have hunted in this fashion. If whole groups were killed, catastrophic death assemblages would result. In the case of other, smaller bovids, which also show catastrophic profiles, driving is less likely to have been used. These smaller nongregarious species may have been taken individually in snares. Like driving, this strategy would net large numbers of prime-age adults, as well as juveniles. But since such practices could reduce the reproductive potential of a prey species, it is probable that Middle Stone Age hunters were relatively inefficient. There is no evidence that these animals were hunted to extinction, and apparently the people could kill only a small proportion of the total number of adult bovids available [Klein, 1981].

If Middle Stone Age people were not clearly more effective hunters than their Acheulian predecessors, they were able to exploit marine resources for the first time. Shells of limpets and tortoises, and the bones of seals and penguins are common in sites along the southern African coast. However, caves such as Klasies River Mouth and Die Kelders in the Cape have yielded relatively few remains of fish or flying birds. It appears that while Middle Stone Age people were quite capable of collecting and scavenging marine foods, they were less skilled as fishermen. Fish and bird bones are more plentiful in Later Stone Age occupations, and these later Upper Pleistocene and Holocene populations were clearly more practiced at both fishing and hunting [Klein, 1977].

Unlike the Neanderthals of Europe, Middle Stone Age groups in sub-Saharan Africa apparently did not bury their dead regularly. As a consequence, human skeletal remains are not often found in caves and shelters occupied early in the Upper Pleistocene. Only in a few cases are there firm associations of fossils with a Middle Stone Age cultural record, and the earliest sites known are in southern Africa. One well-studied Middle Stone Age sequence occurs at Klasies River Mouth, situated on the Cape coast. At Klasies, cave deposits have yielded a wealth of material including stone artifacts, a molluscan fauna, and bird and mammal bones in addition to the hominids. The sediment stratigraphy has been worked out, and a reasonably secure chronological framework has been established. An important monograph on the Klasies excavations has recently been published by Singer and Wymer [1982].

Studies of the sedimentary sequence [Butzer, 1978] together with oxygen isotope analysis of shells collected from the deposits [Shackleton, 1982] place the earliest Klasies Middle Stone Age occupation at the beginning of the Last Interglacial, 120 to 130 ky ago. This conclusion is supported by J.

C. Vogel's [quoted by Partridge, 1982] ionium ages of approximately 100 ky obtained from stalagmite overlying some of the Middle Stone Age levels. People seem then to have inhabited the site intermittently for at least 50 ky. Butzer's stratigraphic work suggests that the last (youngest) Middle Stone Age levels in the caves may date to about 60 ky BP, and Klasies was reoccupied only later in the Holocene.

Human remains have been recovered mainly from the earlier levels, where the associated stone industries are termed MSA I and MSA II. One of the most interesting of the Klasies hominid discoveries was made in MSA I deposits. This relatively complete mandible is somewhat damaged, and all of the anterior teeth are missing. The alveoli show evidence of extensive resorptive change, linked to age and dental disease. Because of these changes, measurements are difficult to take, but the corpus is rather heavily built. Lateral prominence formation is comparable to that in many modern jaws, and there is a strong ("bulging") mental trigon. Protrusion of this chin is accentuated by incurvation of the symphyseal face below the (damaged) alveolar margin. Internally, there is no development of an alveolar planum or superior torus. Neither here nor in the shape of the medial wall of the corpus is there any indication of archaic morphology. Singer and Wymer [1982] comment that this jaw represents fully modern *Homo sapiens*, and there is no reason to question this conclusion.

Other cranial, mandibular and dental remains from MSA II levels are mostly fragmentary. A zygomatic bone is large in comparison to that of recent Africans and is rather flat in facial aspect, but its anatomy is again not noticeably archaic. Another informative fragment consists of part of a frontal bone on which glabella and some of the orbital margin are preserved and to which the upper ends of both nasal bones are still attached. The superciliary eminence is not especially prominent, and the nasal root is broad and flat. This MSA II frontal can be compared to Florisbad, and the difference in robusticity is striking. In the Free State face, glabella is projecting and the supraorbital torus is more heavily constructed [Rightmire, 1978]. Pieces of parietal and loose teeth are also available, along with several broken jaws. One parietal fragment is thick and may have come from a relatively narrow vault. This bone shows few resemblances to Broken Hill or even to Florisbad but could represent a Khoi or San-like cranium [Singer and Wymer, 1982]. With a few exceptions, all of the teeth fall within the size range expected for recent Africans.

The MSA II mandibles vary considerably in size. One is very small and gracile and may have belonged to a female. Others are more massive, and one specimen is especially robust. In this latter individual, much of the anterior corpus is preserved, and the bone is complete on the right side to the level of M_2. The corpus is relatively thick, and Singer and Wymer [1982] note that its upper and lower borders are approximately parallel. However, the anterior height of the body is difficult to measure accurately because of erosion of the tooth sockets. It is likely that the front of the jaw was deeper before this damage occurred. The symphyseal axis is nearly vertical. At the center of this surface there is a blunt swelling that broadens below to form a mental trigon. The trigon itself is not so prominent as in many modern mandibles but still provides evidence of chin formation.

These traits tend to set the Klasies fossil apart from more recent jaws, but at the same time there is little development of an alveolar planum and no expression of a superior transverse torus. A more pronounced internal shelf is present in the Cave of Hearths specimen and in most Neanderthals from localities in Europe and the Middle East. Even the relatively late Neanderthal mandibles from Vindija in central Europe are said to possess distinct transverse tori [Wolpoff et al, 1981]. Therefore, it is fair to conclude that in most aspects of its anatomy, the Klasies jaw is within the range of variation to be expected in a fully modern human population. When the appearance of the other skull and dental remains is also taken into account, there is no real indication that the Klasies people are archaic, despite their association with a Middle Stone Age way of life.

More evidence bearing on this question has been recovered from Border Cave, located in northern Natal Province, South Africa. Here as in the Klasies caves, there is a long record of Middle Stone Age occupation. Studies of the sedimentary sequence and cultural material, radiocarbon dates, and microanalytical data from bone samples are discussed by Beaumont et al [1978] and by Butzer et al [1978]. Results of this work show that the site was first inhabited prior to the beginning of the Last Interglacial, and levels containing Middle Stone Age artifacts are all older than 49 ky BP. Human remains from Border Cave consist primarily of cranial or mandibular parts of three adult individuals and the skeleton of an infant. One of the adult mandibles was recovered in 1974 from deposits unquestionably of the Middle Stone Age. The infant had been excavated earlier, from one of the rare burials recorded at a site of this antiquity. The location of this shallow burial within the sedimentary sequence has been established with some certainty. The remaining adult individuals were discovered by guano diggers, and there is lingering doubt as to their exact provenience in the deposits. However, soil found attached to the partial cranium is best matched by that from levels identified by Butzer with oxygen isotope substage 5d. If the Border Cave cranium is assumed to have come from this part of the sequence, then it may be more than 100 ky old.

All of the Border Cave hominids are demonstrably modern in their anatomy [de Villiers, 1973, 1976]. However, the extent to which one or more of the fossils may resemble living African populations is still not clear. Because of its rugged appearance and well developed superciliary eminences, the adult cranium was compared initially to Florisbad and to other skulls such as Springbok Flats from the Transvaal. The age and cultural associations of the large but very fragmentary Springbok Flats individual have never been established, although "Middle Stone Age" artifacts are said to have accompanied the find. Metric studies have led de Villiers [1973] and Beaumont et al [1978] to emphasize resemblances of Border Cave to this Transvaal specimen and to assign to both the role of "undifferentiated" ancestors from which living populations are derived. Until recently, links between Border Cave and modern Africans have been viewed as fairly remote.

Multivariate statistical treatment of cranial measurements now suggests a rather different conclusion. When measurements relating to frontal form, supraorbital development, and projection of the nasal root are used in discriminant analysis, Border Cave approaches several modern populations and appears to fit best with Khoi or with San males [Rightmire, 1979]. Study of the distances separating the fossil from modern group centroids in the discriminant space confirms that these assignments have biological significance, in that Border Cave does fall within the range of variation expected for the recent populations [Rightmire, 1981b]. De Villiers and Fatti [1982] have conducted a similar analysis and have also concluded that Border Cave is close to living Africans. However, the adult cranium is here said to relate to Nguni males rather than to San.

These Border Cave findings, together with the Klasies fossils, suggest strongly that (some) Middle Stone Age populations of southern Africa were anatomically modern. At Klasies, where the MSA I and MSA II levels are almost certainly of Last Interglacial Age, the cranial and mandibular materials argue for the presence of *Homo sapiens sapiens* at a surprisingly early date. At Border Cave, problems of stratigraphic provenience of several of the hominids have still to be resolved, but the cranium and adult mandibles may also be quite ancient. If so, it can be stated that people very much like living populations (San or Negroes) have been resident in southern Africa for a long time. This evidence is rather sparse and fragmentary, and it may be too soon to base on it any sweeping conclusions concerning the origins of modern humans. But the fossils may indicate that there was little change in local populations during the Upper Pleistocene. It seems unlikely that southern Africa has witnessed any large scale replacement of its peoples during the last 100 ky.

ACKNOWLEDGMENTS

For allowing me to examine fossil material and for offering advice and assistance during the course of research on which this chapter is based, I am grateful to a number of persons, especially the following: H. de Villiers (Johannesburg), R. Hay (Berkeley), Q.B. Hendey (Cape Town), R.G. Klein (Chicago), M.D. Leakey (Olduvai Gorge), R.E. Leakey (Nairobi), C. Magori, F.T. Masao, and A.A. Mturi (all of Dar es Salaam), J.J. Oberholzer (Bloemfontein), R. Singer (Chicago), C.B. Stringer (London), M. Tessema (Addis Ababa), P.V. Tobias (Johannesburg), and S. Waane (Dar es Salaam).

My work in Africa has been supported by National Science Foundation grants BNS 80-04852 and BNS 82-17396.

LITERATURE CITED

Beaumont PB, de Villiers H, Vogel JC (1978): Modern man in sub-Saharan Africa prior to 49,000 years BP: A review and evaluation with particular reference to Border Cave. S Afr J Sci 74:409–419.

Bonde N (1981): Problems of species concepts in paleontology. In Martinelli J (ed): "Concept and Method in Paleontology." University of Barcelona, pp 19–34.

Brain CK (1982): The Swartkrans site: Stratigraphy of the fossil hominids and a reconstruction of the environment of early *Homo*. In de Lumley MA (ed): "L 'Homo erectus et la Place de l'Homme de Tautavel Parmi les Hominidés Fossiles." 1ᵉʳ Cong Int Paleontol Hum Nice 2(preprints):676–706.

Butzer KW (1973): Re-evaluation of the geology of the Elandsfontein (Hopefield) site, southwestern Cape, South Africa. S Afr J Sci 69:234–238.

Butzer KW (1978): Sediment stratigraphy of the Middle Stone Age sequence at Klasies River Mouth. S Afr Arch Bull 33:141–151.

Butzer KW, Beaumont PB, Vogel JC (1978): Lithostratigraphy of Border Cave, Kwa Zulu, South Africa: A Middle Stone Age sequence beginning c. 195,000 BP. J Archaeol Sci 5:317–341.

Butzer KW, Brown FH, Thurber DL (1969): Horizontal sediments of the Lower Omo Valley: The Kibish Formation. Quaternaria 11:15–29.

Butzer KW, Isaac G Ll (eds) (1975): "After the Australopithecines." The Hague: Mouton.

Chavaillon J (1982): Position chronologique des hominidés fossiles d'Ethiopie. In de Lumley MA (ed): "L 'Homo erectus et la Place de l 'Homme de Tautavel Parmi les Hominidés Fossiles." 1ᵉʳ Cong Int Paleontol Hum Nice 2(preprints):766–797.

Chavaillon J, Brahimi C, Coppens Y (1974): Première découverte d 'hominidé dans l 'un des sites acheuléens de Melka-Kunturé (Ethiopie). C R Acad Sci [D] (Paris) 278:3299–3302.

Clark JD (1980): Early human occupation of African savanna environments. In Harris DR (ed): "Human Ecology in Savanna Environments." London: Academic Press, pp 41–71.

Clark JD, Kurashina H (1979): Hominid occupation of the East-Central highlands of Ethiopia in the Plio-Pleistocene. Nature 282:33–39.

Clarke RJ (1976): New cranium of *Homo erectus* from Lake Ndutu, Tanzania. Nature 262:485–487.

Conroy GC, Jolly CJ, Cramer D, Kalb JE (1978): Newly discovered fossil hominid skull from the Afar depression, Ethiopia. Nature 276:67–70.

Cronin JE, Boaz NT, Stringer CB, Rak Y (1981): Tempo and mode in hominid evolution. Nature 292:113–122.

Dagley P, Mussett AE, Palmer HC (1978): Preliminary observations on the paleomagnetic stratigraphy of the area west of Lake Baringo, Kenya. In Bishop WW (ed): "Geological Background to Fossil Man." Edinburgh: Scottish Academic Press, pp 225–235.

Day MH, Leakey MD, Magori C (1980): A new hominid fossil skull (L.H. 18) from the Ngaloba Beds, Laetoli, northern Tanzania. Nature 284:55–56.

Day MH, Stringer CB (1982): A reconsideration of the Omo Kibish remains and the *erectus-sapiens* transition. In de Lumley MA (ed): "L 'Homo erectus et la Place de l 'Homme de Tautavel Parmi les Hominidés Fossiles." 1er Cong Int Paleontol Hum Nice 2(preprints):814–846.

Delson E (1981): Paleoanthropology: Pliocene and Pleistocene human evolution. Paleobiology 7:298–305.

Delson E, Eldredge N, Tattersall I (1977): Reconstruction of hominid phylogeny: A testable framework based on cladistic analysis. J Hum Evol 6:263–278.

de Villiers H (1973): Human skeletal remains from Border Cave, Ingwavuma District, Kwa Zulu, South Africa. Ann Trans Mus 28:229–256.

de Villiers H (1976): A second adult human mandible from Border Cave, Ingwavuma District, Kwa Zulu, South Africa. S Afr J Sci 72:212–215.

de Villiers H, Fatti LP (1982): The antiquity of the Negro. S Afr J Sci 78:321–332.

Eldredge N, Tattersall I (1982): "The Myths of Human Evolution." New York: Columbia University Press.

Gingerich PD (1979): The stratophenetic approach to phylogeny reconstruction in vertebrate paleontology. In Cracraft J, Eldredge N (eds): "Phylogenetic Analysis and Paleontology." New York: Columbia University Press, pp 41–77.

Gould SJ, Eldredge N (1977): Punctuated equilibria: The tempo and mode of evolution reconsidered. Paleobiology 3:115–151.

Hay RL (1976): "Geology of the Olduvai Gorge. A Study of Sedimentation in a Semiarid Basin." Berkeley: University of California Press.

Hendey QB (1974): Faunal dating of the late Cenozoic of southern Africa, with special reference to the Carnivora. Q Res 4:149–161.

Howell FC (1978): Hominidae. In Maglio VJ, Cooke HBS (eds): "Evolution of African Mammals." Cambridge: Harvard University Press, pp 154–248.

Howells WW (1980): *Homo erectus*: Who, when and where: A survey. Yb Phys Anthropol 23:1–23.

Hublin JJ (1978): Quelques charactères apomorphes du crâne nèandertalien et leur interprétation phylogénique. C R Acad Sci [D] (Paris) 287:923–926.

Isaac G Ll (1977): "Olorgesailie: Archeological Studies of a Middle Pleistocene Lake Basin in Kenya." Chicago:University of Chicago Press.

Jelínek J (1980a): European *Homo erectus* and the origin of *Homo sapiens*. In Königsson L-K (ed): "Current Argument on Early Man." Oxford: Pergamon, pp 137–144.

Jelínek J (1980b): Variability and geography. Contribution to our knowledge of European and North African Middle Pleistocene hominids. In Jelínek J (ed): "*Homo erectus* and His Time. Contributions to the Origin of Man and His Cultural Development, Vol 1." Brno: Anthropologie, pp 109–114.

Kalb JE, Jolly CJ, Mebrate A, Tebedge S, Smart C, Oswald CB, Cramer D, Whitehead P, Wood CB, Conroy GC, Adefris T, Sperling L, Kana B (1982): Fossil Mammals and artifacts from the Middle Awash Valley, Ethiopia. Nature 298:25–29.

Kalb JE, Wood CB, Smart C, Oswald EB, Mabrete A, Tebedge S, Whitehead P (1980): Preliminary geology and palaeontology of the Bodo d'Ar hominid site, Afar, Ethiopia. Paleogeogr Paleoclimatol Paleoecol 30:107–120.

Klein RG (1973): Geological antiquity of Rhodesian man. Nature 244:311–312.

Klein RG (1977): The ecology of early man in southern Africa. Science 197:115–126.

Klein RG (1980): Environmental and ecological implications of large mammals from Upper Pleistocene and Holocene sites in southern Africa. Ann S Afr Mus 81:223–283.

Klein RG (1981): Stone Age predation on small African bovids. S Afr Archaeol Bull 36:55–65.

Klein RG (1982): Age (mortality) profiles as a means of distinguishing hunted species from scavenged ones in Stone Age archeological sites. Paleobiology 8:151–158.

Klein RG (1983): The Stone Age prehistory of southern Africa. Annu Rev Anthropol 12:25–48.

Laitman JT, Heimbuch RC, Crelin ES (1979): The basicranium of fossil hominids as an indicator of their upper respiratory systems. Am J Phys Anthropol 51:15–34.

Leakey M, Tobias PV, Martyn JE, Leakey RE (1969): An Acheulian industry with prepared core technique and the discovery of a contemporary hominid at Lake Baringo, Kenya. Proc Prehist Soc 25:48–76.

Leakey MD (1971): Discovery of postcranial remains of Homo erectus and associated artifacts in Bed IV at Olduvai Gorge, Tanzania. Nature 232:380–383.

Leakey MD (1980): Early man, environment and tools. In Königsson L-K (ed): "Current Argument on Early Man." Oxford: Pergamon, pp 114–133.

Leakey MD, Hay RL (1982): The chronological position of the fossil hominids of Tanzania. In de Lumley MA (ed): "L 'Homo erectus et la Place de l 'Homme de Tautavel Parmi les Hominidés Fossiles." 1er Cong Int Paleontol Hum Nice 2 (preprints):753–765.

Mturi AA (1976): New hominid from Lake Ndutu, Tanzania. Nature 262:484–485.

Partridge TC (1982): The chronological positions of the fossil hominids of southern Africa. In de Lumley MA (ed): "L 'Homo erectus et la Place de l 'Homme de Tautavel Parmi les Hominidés Fossiles." 1er Cong Int Paleontol Hum Nice 2(preprints):617–675.

Rightmire GP (1976): Relationships of Middle and Upper Pleistocene hominids from sub-Saharan Africa. Nature 260:238–240.

Rightmire GP (1978): Florisbad and human population succession in southern Africa. Am J Phys Anthropol 48:475–486.

Rightmire GP (1979): Implications of Border Cave skeletal remains for later Pleistocene human evolution. Curr Anthropol 20:23–35.

Rightmire GP (1980): Middle Pleistocene hominids from Olduvai Gorge, northern Tanzania. Am J Phys Anthropol 53:225–241.

Rightmire GP (1981a): Patterns in the evolution of Homo erectus. Paleobiology 7:241–246.

Rightmire GP (1981b): More on the study of the Border Cave remains. Curr Anthropol 22:199–200.

Rightmire GP (1982): Estimating stasis. Reply to Levinton. Paleobiology 8:307–308.

Rightmire GP (1983): The Lake Ndutu cranium and early Homo sapiens in Africa. Am J Phys Anthropol 61:245–254.

Shackleton NJ (1982): Stratigraphy and chronology of the KRM deposits: Oxygen isotope evidence. In Singer R, Wymer J: "The Middle Stone Age at Klasies River Mouth in South Africa." Chicago: University of Chicago Press, pp 194–199.

Shipman P, Bosler W, Davis KL (1981): Butchering of giant geladas at an Acheulian site. Curr Anthropol 22:257–268.

Singer R, Wymer J (1982): "The Middle Stone Age at Klasies River Mouth in South Africa." Chicago: University of Chicago Press.

Stringer CB (1978): Some problems in Middle and Upper Pleistocene hominid relationships. In Chivers DJ, Joysey KA (eds): "Recent Advances in Primatology, Vol 3. Evolution." London: Academic Press, pp 395–418.

Stringer CB (1979): A re-evaluation of the fossil human calvaria from Singa, Sudan. Bull Br Mus Nat Hist 32:77–83.

Stringer CB, Howell FC, Melentis J (1979): The significance of the fossil hominid skull from Petralona, Greece. J Archaeol Sci 6:235–253.

Tallon PWJ (1978): Geological setting of the hominid fossils and Acheulian artifacts from the Kapthurin Formation, Baringo District, Kenya. In Bishop WW (ed): "Geological Background to Fossil Man." Edinburgh: Scottish Academic Press, pp 361–373.

Thorne AG, Wolpoff MH (1981): Regional continuity in Australasian Pleistocene hominid evolution. Am J Phys Anthropol 55:337–349.

Tobias PV (1971): Human skeletal remains from the Cave of Hearths, Makapansgat, northern Transvaal. Am J Phys Anthropol 34:335–367.

Tobias PV (1978): The earliest Transvaal members of the genus *Homo* with another look at some problems of hominid taxonomy and systematics. Z Morphol Anthropol 69:225–265.

Volman TP (1984): Early prehistory of southern Africa. In Klein RG (ed): "Southern African Prehistory and Paleoenvironments." Rotterdam: AA Balkema (in press).

Vrba, ES (1982): Biostratigraphy and chronology, based particularly on Bovidae, of southern hominid-associated assemblages: Makapansgat, Sterkfontein, Taung, Kromdraai, Swartkrans; also Elandsfontein (Saldanha), Broken Hill (now Kabwe) and Cave of Heaths. In de Lumley MA (ed): "L 'Homo erectus et la Place de l 'Homme de Tautavel Parmi les Hominidés Fossiles." 1er Cong Int Paleontol Hum Nice 2(preprints):707–752.

Weidenreich F (1943): The skull of *Sinanthropus pekinensis.* Palaeontol Sin Ser D 10:1–292.

Williams MAJ, Williams FM, Gasse F, Curtis GH, Adamson DA (1979): Plio-Pleistocene environments at Gadeb prehistoric site, Ethiopia. Nature 282:29–33.

Wolpoff MH (1980): Cranial remains of Middle Pleistocene European hominids. J Hum Evol 9:339–358.

Wolpoff MH (1982): The Arago dental sample in the context of hominid dental evolution. In de Lumley MA (ed): "L 'Homo erectus et la Place de l' Homme de Tautavel Parmi les Hominidés Fossiles." 1er Cong Int Paleontol Hum Nice 1(preprints):389–410.

Wolpoff MH, Smith FH, Malez M, Radovcić J, Rukavina D (1981): Upper Pleistocene human remains from Vindija Cave, Croatia, Yugoslavia. Am J Phys Anthropol 54:499–545.

The Origins of Modern Humans: A World Survey of the Fossil Evidence, pages 327–410
© 1984 Alan R. Liss, Inc., 150 Fifth Avenue, New York, NY 10011

A Craniological Approach to the Origin of Anatomically Modern *Homo sapiens* in Africa and Implications for the Appearance of Modern Europeans

Günter Bräuer

Anthropologisches Institut, Universität Hamburg, 2000 Hamburg 13, Federal Republic of Germany

CURRENT VIEWS AND THE PRESENT FOSSIL RECORD

Several important hypotheses on the origin of anatomically modern *Homo sapiens* were advanced on the basis of the fossil record known at the beginning of the 1950s. In spite of the numerous new hominid finds made in the last decades, each of these hypotheses still has its advocates.

Opinions differ widely over the degree to which our considerably increased knowledge supports the individual hypotheses or makes them appear less plausible. These points will be discussed in more detail by other contributors to this volume. Here, we shall only briefly discuss the most popular hypotheses in the light of our current knowledge.

There are three different main hypotheses on the origin of *Homo sapiens* in Europe: The pre-sapiens hypothesis [Boule, 1913; Heberer, 1950; Vallois, 1954, 1958], the pre-Neandertal hypothesis [Howell, 1951, 1957; Sergi, 1953; Breitinger, 1955], and the phase or stage hypothesis [Weidenreich, 1943a; Weinert, 1951; Brace, 1964; Brose and Wolpoff, 1971; Spencer and Smith, 1981], an earlier form of which was advanced by Hrdlička [1927].

Today, in many parts of Europe, many still regard the pre-sapiens hypothesis as the most plausible explanation for the origin of Cro-Magnon man [eg, Gieseler, 1974; Heim, 1977; Henke and Rothe, 1980; Schröter, 1982]. As is well known, this hypothesis assumes that two parallel lines existed in Europe since the Middle Pleistocene. One of these led to the classic Neandertals, the other to anatomically modern *Homo sapiens*. In the light of the many new finds and research results, however, the basis of the central assumption of the pre-sapiens hypothesis has become increasingly weakened. This assumption postulated a pre-sapiens—Cro-Magnon line that is represented by the 200–300 ky old hominids from Swanscombe, and possibly also from Steinheim.

However, no compelling proof for such a long and separate line has been found in the last decades. In addition, doubts regarding what were seen to be the principal supports of this hypothesis—the Fontéchevade hominids—have increased. The frontal fragment "Fontéchevade I," which has no supraorbital torus and possesses a very gracile morphology, may be of a subadult age. Whether the final adult condition would have fallen outside of the range of variation of the female pre-Neandertals thus remains open to question. There are also grave doubts about the exact stratigraphic provenance of the fragment [Howell, 1958; Vandermeersch et al, 1976].

In contrast, the "Fontéchevade II" calotte is considerably more robust and possesses a thick cranial vault. Newer descriptive and multivariate analyses have revealed clear affinities to the Neandertals and pre-Neandertals [Howell, 1958; Trinkaus, 1973; Stringer, 1974; Corruccini, 1975]. Furthermore, Tillier's [1977] work on the variability of the frontal sinus indicates that the extant part of the frontal does not support Vallois' [1949] assumption that the torus was absent.

The modern features of the presumed ancestors of the pre-sapiens line (Steinheim and Swanscombe) have been emphasized often. Yet, they must also be seen in the context of the present fossil record. Specifically, both hominids, not only possess quite archaic traits, but also Neandertaloid features (eg, occipitomastoid crests and supra-iniac fossae (Steinheim/Swanscombe), prognathism, a thick supraorbital torus (Steinheim), and a wide and thick-walled occipital (Swanscombe)) [Weinert, 1936; Hublin, 1978; Stringer, 1978; Wolpoff, 1980]. A houselike form (posterior view) similar to that of anatomically modern man may not only be found with Steinheim (although the considerable deformation of this find leaves this assessment somewhat open to question), but also with such ante- and pre-Neandertal specimens as Arago 21/47 and Ehringsdorf B and H [Grimaud, 1982; Behm-Blancke, 1959–1960]. In contrast, Swanscombe possesses a rather slightly rounded houselike form, as can also be found among classic Neandertals (Spy 1) (Fig. 1), yet Swanscombe's greatest breadth is located somewhat lower. Fontéchevade II also exhibits a rounded form (see Fig. 1). Apart from the obvious individual variability of this trait, Hublin [1982] holds that the houselike form does not represent a derived, modern feature, but rather may have arisen as the parietal area expanded out of the plesiomorphic, tentlike form. This phenomena can be observed among certain representatives of the developed *Homo erectus* or archaic *Homo sapiens* (eg, Petralona, Salé, Ngandong).

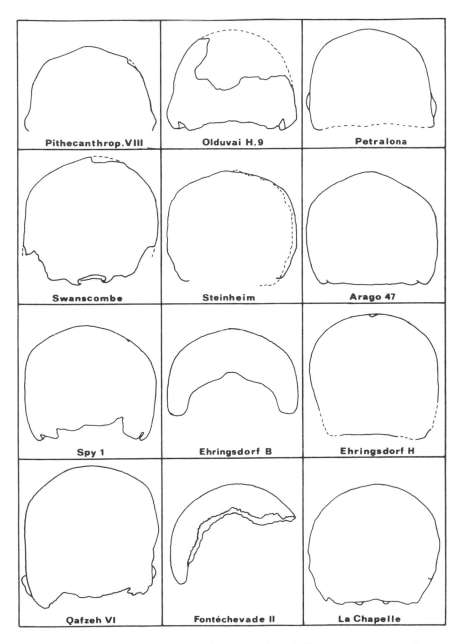

Fig. 1. Examples of parietal variability (occipital view). Drawings are not to scale.

The facial regions of the ante- and pre-Neandertals known today, as well as those from specimens of the non-European developed *Homo erectus* and archaic *Homo sapiens* (cf p 375), show that there was a considerable degree of variation in prognathism and in maxillary morphology during this period. This variation also extends in the gracile direction. The presence of a canine fossa in Steinheim may most likely be seen as an expression of the great polymorphism typical of Middle Pleistocene hominids [cf De Lumley, 1978]. In addition to the evolutionary causes of the heterogeneity present among these hominids, sexual dimorphism apparently also played a role [Wolpoff, 1980]. The newer hominid finds from the Riss and the Last Interglacial, (eg, Biache St. Vaast [Vandermeersch, 1978a], La Chaise "Suard" and "Bourgeois-Delaunay" [Thoma, 1975; Genet-Varcin, 1979; Trinkaus, 1981a; Piveteau et al, 1982]) appear to be steadily closing the gap between such hominids as Steinheim/Swanscombe and the late Neandertals [cf Stringer et al, this volume].

If we consider the substantially greater number of European finds presently available, including in particular, Arago 21/47 and Petralona, then the picture of a morphologically very heterogeneous group of ante-Neandertals is reinforced. But when this group is considered as a whole, it can be assumed that it led to the pre-Neandertals of the Last Interglacial and, ultimately, to the classic European Neandertals as well as the Near Eastern and, in all likelihood, the North African Neandertaloid variants.

Aside from the morphological evidence, considerations of evolutionary biology also make it unlikely that populations of *Homo sapiens* were living in central and western Europe for some 200 ky and were so isolated from one another that two different lineages could form. The pre-sapiens hypothesis can, thus, no longer be considered as a plausible hypothesis for explaining the origins of *Homo sapiens sapiens*. Newer findings suggest that the Swanscombe, Steinheim, and Fontéchevade finds are associated with the line that led to the classic Neandertals and do not represent the beginnings of a pre-sapiens line [Stringer, 1974; Howells, 1978; Trinkaus, 1981a; Hublin, 1982].

The second main model, the pre-Neandertal hypothesis, assumes that the lineages first split during the Eem-Interglacial. Following this event, the southwest Asian progressive pre-Neandertals developed into anatomically modern (a.m.) humans, while the European pre-Neandertals were developing into the more robust classic Neandertals during the beginning of the Würm. Yet, the number of finds from southwest Asia has also grown since the 1950s. Especially important are the newer Qafzeh hominids. Vandermeersch and Tillier, who examined this material, regard all of the Skhūl/

Qafzeh hominids as proto-Cromagnoid and assigned them to a.m. *Homo sapiens* [Vandermeersch, 1978b, 1981a; Vandermeersch and Tillier, 1977; Thoma, 1978]. Although the hominid finds from both sites exhibit a morphological heterogeneity that can hardly be overlooked [McCown and Keith, 1939; Howell, 1958; Howells, 1970], ranging from individuals who are undoubtedly anatomically modern (Qafzeh 9) to individuals with obvious archaic and Neandertaloid features (Skhūl 5, Qafzeh 6), there still remains a substantial morphological gap between these specimens and the Neandertaloids from Tabūn, Shanidar, and Amud, which are probably only slightly older [Vogel and Waterbolk, 1963; Farrand, 1979]. A diachronic trend toward a reduction in size cannot be observed in these Neandertaloids. On the contrary, Trinkaus [1981a] determined that midfacial prognathism increases from Zuttiyeh (?), via Shanidar 2 and 4 (and, perhaps, Amud 1 and Tabūn C1), to Shanidar 1 and 5. An increase towards the more modern group can also be observed when the Neandertaloids (Shanidar, Tabūn, Amud) and the Skhūl/Qafzeh group are compared in terms of tooth sizes, especially those of the front teeth [Stringer, 1982].

The results of Trinkaus' [1981b] analysis of limb proportions also speak against a local evolution towards a.m. humans having occurred in the Near East. He found that the Near Eastern Neandertaloids clearly resemble the European Neandertals in their brachial and crural indices, while the Skhūl/Qafzeh groups, with their relatively longer distal limb segments, are more similar to the European hominids of the Upper Paleolithic.

Thus, the origins of the modern Near Eastern populations are as uncertain as the origins of the modern Europeans, unless one assumes that European a.m. *Homo sapiens* developed autochthonously out of the Neandertals. This is the central assumption of the third model, the phase or stage hypothesis.

An essential prerequisite for assuming that a gradual, continuous evolution occurred is that there was a sufficient span of time between the late Neandertals and early a.m. humans [Brace, 1979]. This assumption has come to be questioned (at least as regards western Europe) because of a series of new finds, especially those from Hahnöfersand (near Hamburg) and Saint-Césaire (Charente). The Hahnöfersand frontal was recovered together with glacial sands during dredging of the Elbe river bed. R. Protsch determined a collagen-based C-14 date of about 36 ky BP. A second C-14 date has recently been obtained through the courtesy of R. Berger and yields a very similar age of 35 ± 2 ky BP (UCLA-2363) [Protsch, personal communication]. The morphology of this fragment indicates that this specimen was not entirely anatomically modern; rather, it exhibits affinities with the

Neandertals, especially with regard to its metric features [Bräuer, 1980a,b, 1981a]. The cranial remains of the Saint-Césaire hominid, whose age probably lies between 31 and 34 ky BP, proves that classic Neandertals existed in France at this late date [Levêque and Vandermeersch, 1980, 1981]. Both of these finds, along with those early Upper Paleolithic hominids that are anatomically modern, do not so much indicate that a gradual evolution occurred, but rather that, in Europe, there was a period of coexistence and, probably, hybridization between the Neandertals and the modern populations [Bräuer, 1982a]. Moreover, this took place at a time when a.m. *Homo sapiens* had already replaced the Neandertaloids in the Near East. While Neandertals still existed in Europe as late as the early Upper Paleolithic, the southwest Asian transition to a.m. *Homo sapiens* had already occurred during the Mousterian. This also shows us that the relationships between cultural stage and morphotype are, by no means, so simple as was long assumed.

Some arguments essential to the phase hypothesis relate to an evolutionary, gradual reduction in tooth and jaw size [Wolpoff, 1971, 1980; Frayer, 1978]. Here, we may not overlook a number of fundamental difficulties having to do with the determination and verification of gradual changes. For example, the uncertain dating of many Würm period hominids causes problems when composing sample populations. In addition, we cannot assume that the differences that can be observed between samples of European Neandertals and a.m. hominids from the Upper Paleolithic could only result from a gradual, unilinear evolution. Gene flow and migration can also lead to these types of changes, or at least figure decisively in bringing them about. Chronologically defined, temporally successive populations do not, in other words, necessarily reflect an evolutionary sequence. Without wishing to go into more detail about general models of evolution at this point (cf p. 391), it can hardly be considered likely that a similar "in situ" evolution towards a.m. *Homo sapiens* occurred more or less independently, in all of the various regional populations of Neandertals [cf Stringer, 1982; Bräuer, 1982a].

South central Europe appears to be a region for which the diachronic changes within the Neandertals (in the direction towards a.m. *Homo sapiens*) is especially well documented [cf Smith, 1982, and this volume]. Here, numerous hominid remains are available, especially from Krapina and Vindija. The more than 80 hominid fragments that have been found at Vindija since 1974 are especially important, not only because they form three stratigraphic groups, but also because the group that is best represented, "G$_3$" (c 35 specimens), may be younger than most of the Neandertals from Krapina [Wolpoff et al, 1981; Smith, 1982]. According to Wolpoff et al

[1981], the numerous though fragmentary remains from the G_3 level, which has been dated to a stadial between c 40 and 59 ky BP, doubtlessly belong to Neandertals. We have only very meager remains from the more recent G_1 level. These fragments cannot be morphologically differentiated from those of the G_3 group. Both Wolpoff et al [1981] and Smith [1982] consider the Vindija Neandertals to be morphologically intermediate between the older Krapina Neandertals of the "hominid zone" (levels 3 and 4) and early a.m. humans from the Upper Paleolithic (eg, Velica Pećina, Mladeč, Brno). In comparison to Krapina, the Vindija Neandertals exhibit a tendency towards smaller anterior teeth, reduced mid-facial prognathism, reduced facial size, thinner and less projecting supraorbital tori, and more pronounced mental eminences. They also differ in several additional characteristics [cf Wolpoff et al, 1981; Smith and Ranyard, 1980].

These indicators of diachronic change merit special attention, particularly since they are based on the largest sample of Neandertals known from any region. Still, these differences do not prove that a.m. *Homo sapiens* did develop here, directly and locally, out of Neandertals. Although the G_3 sample obviously appears to be morphologically more similar to the south central European a.m. *Homo sapiens* than is the Krapina sample, there still remain no doubts about the considerable differences in the total morphological pattern between the Neandertals and currently known representatives of early a.m. *Homo sapiens*. According to Smith [1982], this is also true for such especially robust specimens as Mladeč 5 and 6. The Vindija Neandertals apparently represent a more gracile variant of the Neandertals, one which is more similar to a.m. humans. The Near Eastern Neandertaloids are also more gracile than the west European Neandertals and more similar to the a.m. morphotype [Stringer and Trinkaus, 1981]. Yet, the former also exhibit an increase in mid-facial prognathism from earlier to later specimens (see above).

Should the data on the Krapina and Vindija material be most plausibly interpreted as indicating that a gradual evolution did indeed take place primarily as the result of a reduction in the functional stress of the facial region [Wolpoff et al, 1981], then this would reflect a trend that has not been conclusively documented for the Neandertals from other regions [Howells, 1978; Trinkaus and Howells, 1979]. Consequently, generalizations are not yet warranted. Other possible causes of the differences that can be observed among these samples must also be taken into consideration. Even the authors themselves concede that perhaps a part of the "advanced" characteristics may be the result of a possible preponderance of females and juveniles in the sample [Wolpoff et al, 1981]. Other possible causes are gene flow and/or migration[1], most probably coming into this area from South West Asia, especially since a.m. populations may have existed earlier there than in South Central Europe [Vandermeersch, 1981b]. We could hardly discriminate between the effects of external influences and in situ evolutionary changes, as morphological continuity can be regionally preserved even when "archaic" genes flow into a "newer" gene pool.

In summary, it appears that the current fossil record indicates that during the Upper Pleistocene there were both diachronic, but not only unidirectional, trends as well as periods in which little change can be observed [Howells, 1974] among the Neandertals of the various regions. This statement, however, does not imply anything about causes.

Thus, it appears that none of the three main hypotheses are unequivocally supported by the available evidence, even though the fossil record has continued to grow. While the pre-sapiens hypothesis can probably be regarded as the least plausible of the three [cf Bräuer, 1981–83], the unilinear model of gradual change is also beset by numerous problems. Finally, the most recent findings also speak against a southwest Asian origin of a.m. *Homo sapiens*, as postulated by the pre-Neandertal and phase hypotheses, although the Near East definitely appears to be a region from which a.m. *Homo sapiens* might have spread into the north and played a decisive role in replacing the Neandertals (cf p 395). With the Near East, however, the central question as to the source of the "proto-Cromagnoids" still remains unanswered.

We know from the distribution of the Neandertals that there were contacts and gene flow with North Africa and that the Neandertaloids there were also replaced by a.m. populations during the Würm. Thus, the question arises whether the sources of a.m. *Homo sapiens* might not lie in Africa. Based on the finds and dates available for Africa until roughly the end of the 1960s, most workers proceeded from the assumption that wide parts of the African continent were still settled by archaic Rhodesoid populations—often referred to as African Neandertals—until about 30 to 40 ky ago. Modern humans were often thought to have evolved from the Rhodesoids, and so it was assumed that modern man arose in Africa much later than in Europe [Galloway, 1937a; Keith, 1938; Tobias, 1956, 1961; Brace, 1964]. This also appeared to agree with the archaeochronology generally accepted into the 1960s, which indicated a very late date for the

---

[1] Migrations should not be understood as the short-term movements of large populations, but rather as the territorial changes of groups or tribes, primarily due to ecological factors [see also Howells, 1976, 1982].

beginning of the African Later Stone Age [Clark, 1959, 1970; Bishop and Clark, 1967].

Furthermore, the appearance of a.m. humans in Africa was explained by immigrations from the North, ie, the Mediterranean region, just as Coon [1962] had assumed for the Khoisanoids. In contrast, Thoma [1973] held the view that the Khoisanoids represented an African line separate since the archanthropine stage. He thought that the Negroids first arose during the early Holocene, from immigrant European populations subjected to heavy selective pressures.

Leakey [1953] advanced a hypothesis based on the hominid remains from Kanjera, which were assumed to date from the Middle Pleistocene. Comparable to the European pre-sapiens hypothesis, it held that two lineages had also developed in Africa. One of these led, via Kanjera, to a.m. *Homo sapiens*, the other, parallel line led to the archaic Rhodesoids. Later, Leakey [1972] modified this view and came to consider the Rhodesoids as more the product of hybridization between the *"sapiens"* and *"erectus"* lines. Wells [1957] also assumed the existence of two lines, but he thought that they proceeded from a common basic *Homo sapiens* stock.

But since there have long been uncertainties about the dating of the Kanjera remains, and since such a high age for such a modern morphology completely failed to fit the traditional picture, these suggestions were usually ascribed only a slight significance [Leakey, 1972]. Thus, the "dark continent" of Africa was hardly considered to be a serious candidate for the cradle of *Homo sapiens sapiens*.

Over the last decades, however, this traditional picture has proved to be false in some important points [cf Bräuer, 1981b,c]. Fundamental dating revisions, a number of important new African hominid finds, and extensive new analyses show that recent evolution and the origins of a.m. *Homo sapiens* in Africa followed completely different courses than previously accepted. The new results also appear to offer new perspectives on the obscure source of *Homo sapiens sapiens* in southwest Asia and Europe.

A NEW ANALYSIS OF THE EVOLUTION OF *HOMO SAPIENS* IN AFRICA

The Changed Chronology of the African Stone Age

For decades, the relative chronology of the African Stone Age was primarily determined by the pluvial/interpluvial model [cf Zeuner, 1952; Clark, 1979a]. It was not until the critical research of Cooke [1958] and Flint [1959] that it was shown that this hypothetical scheme was no longer tenable. Radiocarbon datings for sub-Saharan Africa have been carried out since the beginning of the 1950s, especially by W. Libby of the University of Chicago. The few dates that were available around the end of the 1950s constituted the principal bases of the first framework of an absolute chronology for the Middle Stone Age (MSA), the Later Stone Age (LSA), and various typological stages and variants [cf Clark, 1979a]. Based on these dates, J.D. Clark introduced the first radiocarbon chronology—primarily for the MSA and LSA—at the 1959 Pan-African Congress in Leopoldsville (Kinshasa). According to this scheme, the MSA began about 40 ky BP and ended roughly 10 ky ago. In spite of various later modifications, this chronology retained its validity up to 1970, the year in which most of the model was adopted into Clark's *The Prehistory of Africa.*

Following this model, it appeared that sub-Saharan Africa was clearly technologically "retarded" in comparison to Europe. When the transition from the "developed Acheulian" to the MSA had just been completed in Africa, the European Upper Paleolithic had already begun. In other words, the MSA and the Upper Paleolithic appeared to be more or less contemporaneous.

Yet, it was only 2 years later, beginning with Vogel and Beaumont's [1972] publication of a number of new C-14 dates, that a radical revision in the entire chronology of the sub-Saharan Stone Age began. The interface between the MSA and the LSA now appeared to have occurred much earlier—>37 ky BP—while the boundary between the MSA and the Early Stone Age (ESA) was seen as lying at more than 100 ky BP [Beaumont and Vogel, 1972]. In the more than 10 years that have followed, further findings pertaining to the chronology have been supplemented by faunal, sedimentological, and archaeological studies, particular in connection with the new excavations at Border Cave, Klasies River Mouth, Nelson Bay, Die Kelders, Montague Cave, Florisbad, and several other sites [cf the compilation in Singer and Wymer, 1982]. Some of the deposition sequences extend back far beyond the Last Interglacial.

The two oldest sedimentation cycles at Border Cave may correspond to the oxygen isotope stage 6. This suggests an age of c 195 ky BP for the beginning of the MSA sequence at this site. A similarly high age has been estimated through extrapolation of the sedimentation rates on the basis of the available radiocarbon framework [Butzer et al, 1978; Butzer, 1979; Beaumont, 1973, 1980]. Butzer [1978] was able to show that the MSA at Klasies River Mouth (Caves 1, 1A) spans the deep-sea isotope stages 5e to 4. This corresponds to the period betwen c 125 and 60 ky BP. The faunal remains of both vertebrates [Klein, 1976] and molluscs [Voigt, 1982] also confirm the interglacial age [cf Klein, 1979; Singer and Wymer, 1982]. A ^{230}Th/^{234}U age of 174 ± 20 ky BP has been obtained for a Fauresmith facies (final ESA) from Rooidam, near Kimberley [Butzer, 1979].

Evidence is also available that supports a high age for the east African MSA industries. Wendorf et al [1975] presented K/Ar dates of 181 ± 6 and 149 ± 13 ky BP for a MSA sequence near Lake Ziway (Ethiopia). The MSA industry found in the Ngaloba-Beds of the Laetoli region may also possess an age of some 120 ± 30 ky BP [M. Leakey, cited in Magori, 1980; Leakey and Hay, 1982], while a late Acheulian industry with a well-developed Levallois and blade technology found in Kapthurin (northern Kenya) dates to c 200 ky BP [Isaac, 1972; Clark, 1975]. Taken together, the present findings indicate that the transition from the latest Acheulian to the early MSA occurred between 130 and 200 ky BP [Beaumont et al, 1978; Clark, 1979a, 1981].

It is much more difficult to determine when the transition from the MSA to the LSA took place, as a long hiatus follows the MSA settlements at most sites (eg, Nelson Bay, Die Kelders 1). The reasons for this gap, which generally covers a period of more than 10 ky, are mostly vague [Klein, 1973; Deacon, 1978]. The investigations at Border Cave indicate that the final stage of the MSA may date to about 50 ky BP [Butzer, 1978]. The so-called Early LSA of this site follows at c 45 ky BP [Beaumont and Vogel, 1972]. Some of its characteristics are convex-edged scrapers, microbladelets, ground bone points, ostrich eggshell beads, and small bored stones [Beaumont et al, 1978].

Thus, we can see that both the MSA and the LSA have undergone drastic temporal expansions on the order of several times their previously accepted durations. This, in turn, has resulted in extensive revisions in the chronology of the various cultures and culture complexes; a more thorough treatment of these changes would, however, exceed the limits of this work [cf Clark, 1975, 1979a, 1981; Klein, 1974, 1975; Deacon, 1974, 1978; Beaumont et al, 1978; Butzer et al, 1978; Sampson, 1974; Shackleton, 1975; Singer and Wymer, 1982].

One of the most important results of these revisions, and one that should be emphasized, is that sub-Saharan Africa can no longer be considered technologically backwards and stagnated in comparison to North Africa and Europe. Rather, the important technological changes occurred more or less synchronically both north and south of the Sahara [Beaumont and Vogel, 1972; Clark, 1979a]. There are even some indications that Southern Africa lay in the forefront of cultural invention and innovation during the MSA. For example, a number of "final MSA" sites have provided us with small unifacial or bifacial points that could possibly be arrowheads [Beaumont and Vogel, 1972; Beaumont et al, 1978]. Similarly, the Upper Paleolithic "punched blade" traditions which appeared in Europe and the Near East some 35 ky ago, can no longer be considered to be a revolutionary invention of Cro-Magnon man [Clark, 1979a]. They appear as a new but well-developed stage of an old tradition that could be found in various parts of Africa, including the sub-Saharan area, at least as far back as the last interglacial. An example is the "Howieson's Poort" tradition "that makes use of both flake and small blade technologies to produce various tools that, in Europe, would be considered as early Upper Paleolithic" [Clark, 1975].

The Expanding Fossil Record and Hominid Dating

It is obvious that the drastic changes in the archaeochronology of Africa have also resulted in corresponding temporal shifts of the fossil hominids. Thus, those hominds that can be connected in any way with the transition phase leading to a.m. *Homo sapiens* in Africa no longer appear to date into the late Upper Pleistocene—as was assumed as recently as the 1960s—but rather, reach at least as far back as the late Middle Pleistocene. The outline of the Middle and Upper Pleistocene hominid finds dating from c 500 to 30 ky BP given in Table I shows what important discoveries have been made, especially during the last 15 years.

Before turning to a comparative morphological consideration of these hominids, it is necessary to give a short sketch of both the hominids known today and their more or less secure chronological positions [see also Rightmire, this volume].

Until the mid-1960s, the picture for East Africa was based solely on the fragments from Kanjera, Diré-Dawa, and Eyasi (cf Fig. 2). It was indeed difficult to reconstruct the course of hominid evolution in this area using just these fragments. The substantially a.m. character of the Kanjera remains did not fit the high Middle Pleistocene age that was assumed on the basis of the fauna; this, together with the doubts concerning the association of the hominid and faunal material [Boswell, 1935], usually led to these finds being rejected as inconclusive [cf Leakey, 1972]. Even though cranial fragments from two of the four individuals were found in situ with the Middle Pleistocene fauna [Leakey, 1935], Oakley [1974], who performed radiometric measurements (eU308), also considered the hominid material to be significantly younger. It is possible that the hominid remains may thus have an early Upper Pleistocene age, but a greater age, up to perhaps 200 ky can also not be ruled out [Leakey, 1981a; Clark, 1981].

The poorly preserved mandibular fragment from the Cave of "Porc-Épic" near Diré-Dawa not only permits few morphological comments, but its

TABLE I. The Middle and Upper Pleistocene Hominids
From Africa Which Date Between
c 500–30 ky BP[a]

Hominid specimens	Country	Date of discovery
Broken Hill 1–3	Zambia	1921
Singa	Sudan	1924
Florisbad	South Africa	1932
Kanjera 1–5	Kenya	1932/1935
Diré-Dawa	Ethiopia	1933
Rabat	Morocco	1933
Eyasi 1–3	Tanzania	1935/1938
Mugharet el'Aliya	Morocco	1939
Border Cave 1–3	South Africa	1940/1942
Cave of Hearths	South Africa	1947
Haua Fteah	Libya	1952/1955
Hopefield	South Africa	1953
Ternifine 1–4	Algeria	1954/1955
Sidi Abderraham	Morocco	1955
Témara 1	Morocco	1956
Jebel Irhoud 1,2	Morocco	1961/1963
Olduvai H.11	Tanzania	1962
Omo (Kibish) 1–3	Ethiopia	1967
Olduvai H.23	Tanzania	1968
Klasies River Mouth (24 specimens)	South Africa	1968
Jebel Irhoud 3	Morocco	1968
Thomas Quarries 1	Morocco	1969
Salé	Morocco	1971
Thomas Quarries 2	Morocco	1972
Ndutu	Tanzania	1973
Border Cave 5	South Africa	1974
Dar-es-Soltan 2	Morocco	1975
Témara 2	Morocco	1975
Laetoli (Ngaloba) 18	Tanzania	1976
Bodo	Ethiopia	1976/1978
Mumba Rock Shelter	Tanzania	1977

[a]Only two of the Olduvai Hominids have been included in this table. The maxillary fragment OH11 was a surface find; the adhering matrix suggests that it should be associated with the Lower Ndutu Beds (400–60 ky BP). The mandibular fragment OH23 was found in situ in the Lower Masek Beds (600–c 500 ky BP) [Leakey and Hay, 1982]. The remaining jaw and dental material found at Olduvai [discussed in Rightmire, 1980] date prior to the period here under consideration [cf Rightmire, this volume]. In all likelihood, the mandible BK 67, found at Kapthurin (Baringo), also dates prior to this period.

Fig. 2. The Middle and Upper Pleistocene hominid material used in this study (from c 500 ky BP).

chronological position can also be only very inexactly delimited. The association with Stillbay-like artifacts is uncertain [Breuil et al, 1951], as is Vallois' [1951] "Neandertaloid" morphological diagnosis of the specimen, which he himself formulated with great reservations.

Finally, there are also uncertainties regarding the chronological position of the Eyasi hominids discovered by L. Kohl-Larsen. While newer amino acid dates of 34 and 35.6 ky BP indicate that this archaic, possibly Rhodesoid hominid possesses a quite young age [Protsch, 1976, 1981], the associated cultural remains point to a handaxe-poor late Acheulian and/or a Sangoan facies [Grahmann and Müller-Beck, 1967; Schröter, 1978]. Both cultural and stratigraphic comparisons based on new excavations at the Mumba Rock Shelter (only 3–4 km away from the Eyasi site) reveal that a higher early Upper Pleistocene age for the Eyasi deposits is probable [Mehlman, 1979, in press; Bräuer and Mehlman, in press].

The age of the Singa skull, which was found some 60 years ago and which the first workers characterized as "Proto-Bushman" [Smith Woodward, 1938; Wells, 1951], is uncertain. Little trust can be placed in the one available C-14 date of 17.3 ± 2 ky BP (Abu Hugar) because of possible contamination of the crocodile tooth upon which this measurement is based as well as uncertainties pertaining to its stratigraphic origin [Whiteman, 1971; Stringer, 1979]. The artifactual remains, in which Lacaille [1951] recognized strong affinities to the proto-Stillbay (MSA), point more towards an earlier Upper Pleistocene age. This is also supported by the fauna, which contains several extinct species [Bate, 1951]. New examinations of the sedimental deposits of Singa and Abu Hugar have enabled H. Ziegert [1981] to more accurately determine the stratigraphical provenance of the skull. On the basis of the artifacts found in the corresponding layer at Abu Hugar, it is possible that the skull may even be connected with the Final Acheulian. This would give it a still higher age [Ziegert, personal communication]. These investigations have not yet been concluded.

Since the end of the 1960s, a number of important discoveries have been made that have added to this extremely sparse East African fossil record. The remains of three hominids were discovered in the Kibish formation, lower Omo basin, in 1967. Omo 1 was found in situ in Member I, the oldest part of the Kibish formation, together with a few flakes and some faunal material. A ^{230}Th/^{234}U dating of molluscs is available, which gives an age of 130 ky BP [Butzer et al, 1969; Merrick et al, 1973]. Omo 2 is a surface find without any clear faunal or cultural context. Geologically, both sites, which lie some 2.5 km apart [Leakey et al, 1969], probably belong to the same horizon. Nitrogen and uranium measurements also indicate that there are no substantial age differences between the sites. The cranial fragments designated as Omo 3 were assigned to Member III. A radiocarbon date for Member IV suggests that Omo 3 is older than 37 ky BP [Day and Stringer, 1982]. The relatively well-preserved hominid remains from Omo, especially Omo 1 and 2, are of considerable importance, as they already possessed a.m. characteristics in spite of their high age.

The cranial fragments that were recovered near Lake Ndutu in 1973 are most probably older than the Omo remains [Mturi, 1976]. Because of the presumed mineralogical similarities with the uppermost portion of the Masek Beds (Olduvai Gorge), M. Leakey and Hay [1982] consider an age of c 400 ky as likely, but a correlation with the lower unit of the Ndutu Beds (c 200–400 ky BP) cannot be ruled out.

Three years later, another important hominid was found in the Laetoli area of Tanzania [Day et al, 1980]. The skull, whose facial region was also well preserved, was found together with a MSA industry in situ in the Ngaloba Beds. This tuff-containing sediment has been correlated with a similar tuff of the lower section of the Ndutu Beds of Olduvai Gorge. This indicates an estimated age of 120 ± 30 ky BP [Leakey and Hay, 1982].

Another important discovery, this time from the Afar region, was reported in the same year (1976). Cranial remains, including a well preserved facial region, were found together with faunal and archaeological material in layer B of the "Upper Bodo Beds" [Conroy et al, 1978]. The hominid remains are very robust and exhibit a very archaic appearance. The findings presently available, including those resulting from another investigation at the site in 1981, permit this hominid to be dated no more exactly than to within the limits of the middle and upper Middle Pleistocene [White, personal communication].

Finally, the dental remains found at the Mumba Rock Shelter (near the Eyasi site) in 1977 probably date from the early Upper Pleistocene. The Mumba Rock Shelter became known, following the important excavations by the Kohl-Larsens in the 1930s. They recovered human skeletal remains from the LSA and Iron Age up to a depth of c 2 m below the surface [Kohl-Larsen, 1943; Bräuer, 1980c]. The new dental remains come from MSA horizons, 7.00–7.10 m below the surface. Uranium dates (USGS-82/19) point to an age between 130 and 110 ky BP for these horizons [Bräuer and Mehlman, in press].

The number of new finds from the period under consideration in South Africa is smaller, yet the drastic changes in the chronology have cast new light on some long-known hominids. The picture of Upper Pleistocene evolution prevalent until the 1960s was determined by the hominids from Broken Hill/Kabwe, Florisbad, Border Cave, Cave of Hearths, and Hopefield (see Fig. 2). Our current knowledge indicates that the dates previously accepted for all of these hominids are no longer tenable; all of them appear to be considerably older. The well-preserved Broken Hill skull was for more than 50 years the principal support for the assumption that quite archaic Homo erectus-like populations still existed in Southern Africa at a time when the a.m. humans of the Upper Paleolithic were already living in Europe. The Eyasi hominids, whose ages may, however, also be greater (see above), also appeared to lend support to the view that this morphotype was widely diffused during the Upper Pleistocene.

The contemporaneity of the Broken Hill skull and the other hominid remains with the artifacts and fauna from the site, is problematic, as the hominid material was found more or less isolated at the end of a cave passage [Leakey, 1935; Partridge, 1982]. Clark [1970] has described the artifacts as Sangoan (final ESA). Because of the radical temporal expansion of the MSA, the cultural remains today suggest an age no younger than the late Middle Pleistocene [Butzer et al, 1978; Butzer, 1979]. An age of > 125 ky may also be assumed for the faunal remains, which contain a number of extinct species [Klein, 1973; Gentry and Gentry, 1978; Vrba, 1982]. Current findings concerning the stratigraphy of the Broken Hill deposits have led Partridge [1982] to assume that the skull apparently predates the fauna and artifact assemblages. All of the currently available evidence suggests that the Broken Hill remains should most probably be assigned to the upper Middle Pleistocene.

Because of what was usually considered to be a morphology intermediate between Broken Hill and the Khoisanoids, the cranial remains discovered at Florisbad in 1932 were long held to provide evidence that a late, autochthonous evolution towards a.m. humans had occurred in southern Africa [eg, Galloway, 1937b; Keith, 1938; Tobias, 1956].

The associated cultural remains, which were generally seen as "Hagenstad variant" of the early MSA (pre-MSA) or which were assigned to the "Pietersburg Complex" (MSA) [Dreyer, 1953; Oakley, 1954; Sampson, 1972], long suggested an age of c 40 ky BP, an assumption that was supported by several C-14 datings performed in the 1950s [Barendson et al, 1957]. Protsch [1974] arrived at a similar C-14 date of c 39 ky BP.

Not only do we, today, accept a substantially greater age for the early MSA, but we also have new C-14 dates (Pretoria Laboratory) for Peat 2 of > 42.6 ky BP [Rightmire, 1978]. The hominid remains probably originated from the upper levels of the still older Peat 1 [Oakley, 1957; Hoffmann, 1955]. The present evidence suggests a date for Peat 1 lying between the late Middle Pleistocene and the early Upper Pleistocene, an assumption that is also supported by the partially extinct fauna it contains [Beaumont, 1979; Butzer, 1979; Partridge, 1982]. The results, available from R. Clarke's current excavations at Florisbad, also agree with these findings [Partridge, 1982].

The new and detailed investigations into the stratigraphy and chronology of the Border Cave deposits [Beaumont, 1973; Beaumont et al, 1978] have also resulted in a radical revision in the dating of the almost completely a.m. hominids, 1–3 [Butzer et al, 1978]. According to the new chronological framework, the Pietersburg horizons with which the hominids 1 and 2 were most likely associated [Cooke et al, 1945; De Villiers, 1973] have been dated to c 90–115 ky BP [Beaumont, 1980]. Amino acid datings [Protsch, 1975] also suggest a high age—up to 90 ky BP. A similarly great age may be presumed for the juvenile remains (B.C.3) [Beaumont, 1979, 1980]. A mandible (B.C.5) was discovered in 1974 in situ together with artifacts from a middle phase of the Epi-Pietersburg during this new work at Border Cave. An age of c 90 ky BP has been suggested for this find [De Villiers, 1976; Beaumont et al, 1978].

The Cave of Hearths mandible has also had to be redated as being substantially older. There can be no doubts about its direct association with the late Acheulian, in particular the Fauresmith [Tobias, 1962, 1968; Mason, 1971], which thus indicates an upper Middle Pleistocene date [Beaumont, 1979; Tobias, 1982; Partridge, 1982]. New work on the faunal material [Hendey, 1974; Vrba, 1982] underscores this conclusion; the MSA fauna appears to lie quite close to the beginning of the MSA.

The traditional picture that *Homo erectus*-like populations were widely spread throughout Africa as late as 40 ky ago was reinforced by the discovery of the Hopefield calotte at the beginning of the 1950s. This find bears striking affinities to the Broken Hill skull [Singer, 1957; Tobias, 1962]. Singer and Crawford [1958] accepted an association with the predominantly "Cape Coast Fauresmith" (late ESA) cultural remains. Klein [1973] was able to duplicate Singer and Wymer's [1968] findings that the Hopefield fauna (19 of the total of 50 species are extinct) exhibits strong affinities to the fauna from Olduvai Bed IV [cf Hendey, 1969]. Although the cranial remains were found on the surface, there presently appears to be little doubt that this hominid should be placed in the middle or upper Middle Pleistocene [cf Howell, 1978; Beaumont, 1979; Partridge, 1982]. Vrba [1982] has suggested an age between 300 and 600 ky BP.

It was not until the late 1960s that further early Upper Pleistocene hominid remains were discovered in southern Africa. These came from the deposits in Caves 1, 1A, and 1B at Klasies River Mouth, situated on the southern coast of the continent. The thick MSA layers at this site span the Last Interglacial and date between c 120 and 70 ky BP [Butzer, 1978; Klein, 1975; Singer and Wymer, 1982]. Numerous human skeletal remains were recovered from horizons of the early MSA I and MSA II deposits, unfortunately, in a very fragmentary condition. Various techniques have been used to ascertain an age between 95 and 125 ky BP [Butzer, 1982]. For the most part, the hominid remains reveal an astonishingly modern morphology. Singer and Wymer's [1982] recently published monograph has shown that these remains are some of the best stratified and dated specimens of South African early Upper Pleistocene humans.

Finally, a series of important finds come from the region along the northern coast of Africa (see Fig. 2). Some of these date from the Upper Pleistocene. According to the morphological analyses currently available, a part of these finds exhibit Neandertal affinities, while others are more modern.

In 1939, C. Coon discovered several juvenile remains, including a maxillary fragment showing Neandertaloid affinities, in the High Cave near Mugharet el'Aliya. The maxillary fragment, like the permanent molar found later, probably dates to the late Aterian [Coon, 1962].

Two mandibular fragments were discovered in the 1950s in "Levalloiso-Mousterian" deposits of the cave at Haua Fteah, Cyrenaica [McBurney, 1967]. Carbon-14 datings of associated burnt bones indicate that a more or less identical age of c 47 ± 2.3 ky BP may be assumed for the fragments, both of which were found in the same layer [McBurney, 1958; Vogel and Waterbolk, 1963].

The most well-preserved remains of Neandertaloids known from Northern Africa are those from Jebel Irhoud, where a juvenile mandible was recovered several years after the discovery of two adult skulls. The associated "Levalloiso-Mousterian" artifacts [Ennouchi, 1963] and the faunal remains [Ennouchi, 1962] suggest an age lying between 60 and 40 ky BP [cf Briggs, 1968; Jaeger, 1975].

The Témara remains, which were first dated to the beginning of the Last Interglacial [Biberson, 1961], may also be counted among the more recent group of North African hominids. Later reinvestigations into the chronostratigraphy of this site [Roche, 1976], which resulted in the discovery of a quite modern-appearing occipito-parietal fragment [Roche and Texier, 1976; Ferembach, 1976a], led to the hominid remains being assigned to the upper Aterian (c 30 ky BP). A similar age should be attached to the cranial remains discovered at Dar-es-Soltan in 1975 [Ferembach, 1976b].

A considerable temporal gap exists between this younger "Upper Pleistocene group" and the older finds from North Africa (cf Fig. 3). This permits only limited analyses of diachronic changes, a situation contrasting to that for Eastern and Southern Africa. In addition, most sites have only yielded mandibles. The "Middle Pleistocene group" dates between c 200 and 500 ky BP.

The cranial and mandibular fragments from Rabat were discovered in the 1930s. Stratigraphically, they belong to the middle Tensiftian of the Moroccan continental cycle [Biberson, 1963, 1964; Saban, 1977], which roughly corresponds to the beginning of the European Riss. Further confirmation comes in the form of ^{230}Th/^{234}U datings of molluscs from an above-lying horizon. These yielded an age of > 200 ky BP [Stearns and Thurber, 1965].

The two small mandibular fragments from Sidi Abderrahman (Grotte des Littorines) probably possess a similar age of c 200–250 ky BP. Biberson [1956, 1961] classified the cultural remains as "upper Middle Acheulian, Stage VI" and assigned the site to the early Tensiftian.

The oldest known North African hominid remains are from Ternifine. Three mandibles and a parietal fragment were discovered there along with a great number of artifacts (Moroccan Acheulian Stage I, after Biberson [1961, 1964]) and an abundant "upper Villafranchian" fauna [Jaeger, 1975]. The deposits, which have been assigned to the lower Middle Pleistocene [Arambourg and Hofstetter, 1963], probably date from c 500–400 ky BP [for a discussion on the problems of correlation with the Amirian continental cycle, see Jaeger, 1975; Howell, 1978].

Since the end of the 1960s, this group has grown through the addition of the important hominid remains from Thomas Quarries and Salé. Sausse [1975] has discussed the situation concerning the stratification of the mandible known as "Thomas 1." She claims that it is currently not possible to obtain a dating more exact than late Mindel or Riss. The cranio-facial fragment "Thomas 2" probably also dates to between c 200 and 300 ky BP [Genet-Varcin, 1979]. According to Jaeger [1975], the fauna associated with Thomas 2 corresponds to the Tensiftian.

The skull discovered in 1971 near Salé is also of special importance. Although found isolated, faunal material was recovered from the same layer. In contrast to the Riss-period Tensiftian age originally suggested [Jaeger, 1975], the most recent work indicates a greater age of around 400 ky BP [Jaeger, 1982, personal communication].

Figure 3 represents an attempt at placing the hominids we have briefly discussed into a form reflecting their possible dating-spans. Although obvious uncertainties in the dates of most of the hominids plainly exist, there are in most cases more narrowly delimitable timespans which can be considered as most likely. In addition, the material available today is extensive enough to allow an attempt to reanalyse, in the light of the drastic dating revisions, the course recent evolution has taken in Africa.

COMPARATIVE ANALYSES OF AFRICAN CRANIAL REMAINS

In the sections that follow, we will be unable to go into the same degree of detail for all of the African hominid finds previously mentioned. Instead, the emphasis will be placed on those hominids that are more closely connected to the origins of a.m. *Homo sapiens*.

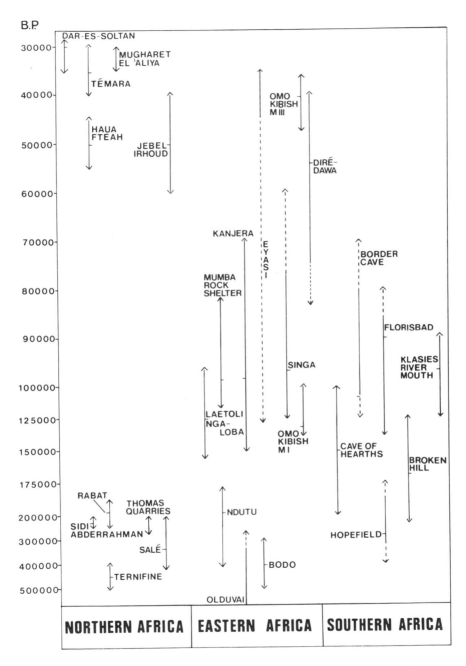

Fig. 3. The chronological positions of the African fossil hominids (up to c 30 ky BP).

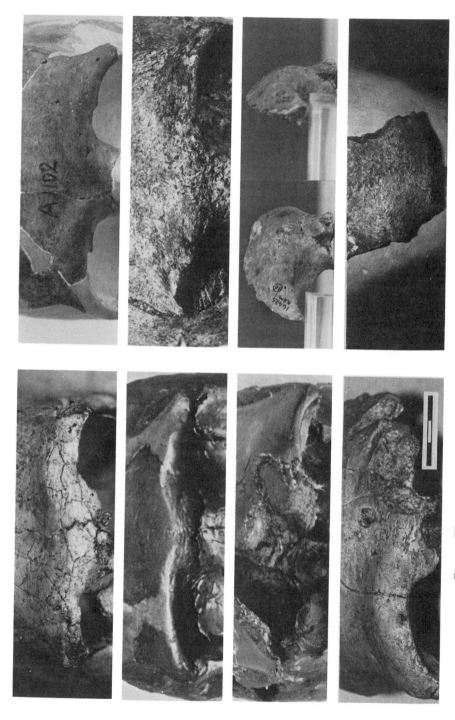

Fig. 4. The supraorbital regions of representatives of late archaic and early anatomically modern *Homo sapiens* (left: Laetoli 18, Omo 1, Omo 2, Florisbad; right: Border Cave 1, Singa, Klasies 16425, Kanjera 1).

This raises the question of the taxonomic definition of *Homo sapiens*. As Howell [1978] points out, there is no inclusive definition of our own polytypic species. Most definitions regard only a single feature, such as cranial capacity, or limit themselves to a.m. *Homo sapiens* (ie, *Homo sapiens* sensu stricto). The problems in classifying archaic forms, such as the Rhodesoids or the Neandertal(o)ids, are the source of the major difficulties in defining *Homo sapiens*. Do these forms belong to the species *Homo sapiens*, do they represent their own species, or should they be assigned to *Homo erectus*?

The difficulties in establishing morphological definitions of taxa have grown as the number of finds has grown. Thus, we know of a large number of finds that cannot be clearly designated as either *Homo erectus* [cf Weidenreich, 1943b] or as a.m. *Homo sapiens*. This group of hominids, which probably dates as far back as the middle of the Middle Pleistocene [cf Wolpoff, 1980], will be referred to in this chapter as "archaic *Homo sapiens*." In doing this, we must bear in mind that it is impossible to draw a distinct line, either temporally or morphologically, between this group and the developed *Homo erectus*.

Because of these taxonomic problems [cf also Kurth, 1968; Vogel, 1970; Thoma, 1973; Jelínek, 1978; Knussmann, 1980; De Lumley, 1981; Trinkaus, 1981a; Howell, 1982], which also apply at the primarily geographically determined subspecies level, Stringer et al [1979] have suggested dividing the species *Homo sapiens* into three "grades" based solely on morphological criteria. This scheme implies no assertions as to the phylogenetic and chronological relationships between the hominids subsumed within a grade. Very archaic specimens, such as Bodo, Broken Hill, and Petralona, are assigned to "grade 1." "Grade 2" encompasses the early Neandertals and comparable finds. "Grade 3" is divided into two groups: the late Neandertals, and a.m. *Homo sapiens*.

Such a graduated breakdown is based on a continuum of evolutionary changes and recognizes the taxonomic difficulties associated with this phase of evolution. It thus appears to more adequately correspond to the actual circumstances than does a rigid taxonomic division. Yet this scheme does not provide a satisfactory solution for the problem of taxonomic assignments. Thus, future interpretations of fossil finds will continue to require definitions or working definitions of the various species and subspecies, especially for *Homo erectus* and a.m. *Homo sapiens* [Day and Stringer, 1982; Howell, 1978, 1982].

If we assume that the Rhodesoids and Bodo date to the middle and/or upper Middle Pleistocene, then we can see that a very robust morphotype was probably spread from South Africa to at least as far as the Horn.[2] Next, a number of morphologically quite heterogeneous finds from eastern and southern Africa (eg, Omo, Laetoli 18, Florisbad, Border Cave, Singa) appear to date to the end of the Middle Pleistocene and/or the early Upper Pleistocene. These possess various constellations of archaic and anatomically modern features. In view of the chronological sequence currently considered as likely, we must ask whether the archaic-Rhodesoid morphotype sensu lato can be considered as an ancestral stock of a.m. *Homo sapiens*. This would require us to determine that those heterogeneous hominids which are closest to a.m. *Homo sapiens* (the "late archaic *Homo sapiens*") possessed the corresponding plesiomorphic features or Rhodesoid affinities.

For interpreting the degree of morphological variability present among the archaic and early a.m. *Homo sapiens*, a number of final Upper Pleistocene finds (c 20–10 ky BP) have been included in the comparative analyses. These finds from various parts of Africa are unquestionably anatomically modern.[3]

Major Cranial Vault Dimensions and Endocranial Capacity

There is considerable variation in the cranial index among both archaic and a.m. hominids. For the 32 final Upper Pleistocene individuals, the spectrum ranges from ultra-dolichocranial (63.3: Mumbwa IV, 3) to brachycranial (80.9: Afalou 40).

While Kanjera 1, with an index value of 64.4, lies in the lower limits of the range of "modern variation,"[4] Singa (81.6), with its unusually short skull, lies slightly above the modern sample.[5] In addition to Kanjera 1, both Omo 2 and Laetoli 18, with values less than 70, are also especially dolicho-

---

[2]The strong affinities between these hominids and such Middle Pleistocene European hominids as Petralona also indicate that there may have been close relationships between the archaic *Homo sapiens* of both continents (cf p 377).

[3] The northern African finds consist of 14 individuals from Afalou-bou-Rhummel, 11 from Taforalt, and one from Dar-es-Soltan; the Eastern African of Olduvai 1, Lukenya Hill, Naivasha, and Gebel Sahaba 18685; the southern African of three individuals from Mumbwa, one from Matjes River, and the Boskop and Tuinplaas finds; and, finally from western Africa, Iwo Eleru.

[4]Both "modern variation" and "modern spectrum," as used here, refer to the a.m. sample from the final Upper Pleistocene.

[5]Some early Holocene skulls are even shorter, eg, Mechta-el-Arbi No. 5.

cranial. The early archaic hominids from Hopefield, Broken Hill and Ndutu, which, like Border Cave 1 and Omo 1[6], possess relatively high index values, fall within the range of variation shown by the Afalou/Taforalt sample (68.9–80.9). This is also true for the *Homo erectus* H.9 from Olduvai Gorge.

The situation is somewhat different when we consider the relationship between cranial length and height.[7] Here, we find a clear increase in height as we proceed from the low Broken Hill vault (whose value of slightly less than 50 places it outside of the modern spectrum, which ranges from 54.1 to 66.1) to Omo 2 and Laetoli 18 (values of c 55) and then to the vaults of Border Cave 1, Singa and Omo 1, which have values around 60 (middle height). This last group, thus, falls in the middle of the range of variation of the Afalou/Taforalt group. The Jebel Irhoud hominids lie within the lower limits of the modern spectrum.

The breadth/height index[7] reveals a spectrum of variation that is comparatively smaller. Salé and Broken Hill, with values around 70, clearly lie below the range of the a.m. sample (74.3–92.9). The somewhat higher skulls of Singa (74.2) and Jebel Irhoud 1 (73.9) follow. In contrast to the intermediate position of Omo 2 and Laetoli 18, with respect to their length-height indices, when the breadth-height index is considered these two hominids are quite similar to the more modern crania of Border Cave 1 and Omo 1. In this trait, they show no differences to the Afalou/Taforalt sample.

Table II shows the cranial capacities of the African hominids here under consideration, as well as several comparative finds from Asia and Europe. Salé, with a cranial volume of less than 1,000 cm[3], falls within the range of variation of the Asiatic *Homo erectus* from Sangiran and Choukoutien, although a number of morphological features of the cranial vault also indicate affinities to *Homo sapiens* (see below). This is also true for Ndutu, which possesses a similarly small cranial capacity.

The early archaic representatives of *Homo sapiens*—Hopefield, Broken Hill 1, and Eyasi 1—possess cranial capacities greater than 1,200–1,250 cm[3] and thus lie above the range of variation of Asiatic *Homo erectus*. The similarities between the three Rhodesoids are remarkable, as it is possible

---

[6]The data for Omo 1 are based on an earlier reconstruction by M.H. Day. These values, however, are little different from those based on a new reconstruction [Day and Stringer, 1982].

[7]Because the cranial basis was fragmented in most of the finds, I used the alternate measurement of auriculo-bregmatic height.

TABLE II. Endocranial Capacities of African *Homo erectus* and *Homo sapiens* Specimens (Including Some Hominids From Asia and Europe)

Hominid	cc	Reference
Africa		
Olduvai H.9	1,067	Holloway [1975]
Salé	930–960	Jaeger [1975]
Hopefield	(1,200–1,250)	Drennan [1953]
Broken Hill/Kabwe	1,280	Day [1977]
Eyasi 1	1,285	Protsch [1981]
Ndutu	(c1,050)	Bräuer[a]
Laetoli 18	1,367	Stringer [personal communication][b]
Omo 2	1,435 ± 20	Day [1977]
Omo 1	c1,430	Day [1969]; Stringer [personal communication]
Kanjera 1	1,350–1,400	Coon [1962]
Singa	1,500	Howell [1978]
Border Cave 1	1,507	De Villiers [1973]
Jebel Irhoud 1	1,480	Ennouchi [1962]
Jebel Irhoud 2	1,430–1,470	Ennouchi [1968]
Asia		
Pithecanthropus I, II IV, VII, VIII range	813–1,059	Holloway [1982]
Choukoutien II, III, X, XI, XII range	915–1,225	Tobias [1973]
Amud 1	1,740	Suzuki and Takai [1970]
Skhūl IV	1,554	McCown and Keith [1939]
Skhūl V	1,450	Snow [1953]
Qafzeh VI	1,568	Vallois and Vandermeersch [1972]
Qafzeh IX	1,508–1,554	Genet-Varcin [1979]
Europe		
Arago 21/47	1,150–1,160	De Lumley [1981]
Petralona	1,190–1,210 (min)	Stringer et al [1979]
Steinheim	1,150–1,175	Howell [1960]

[a]Based on the reconstruction by R. Clarke [1976].
[b]Determined by R. Holloway.

that there may be considerable differences in their dates. Furthermore, it can also be assumed that Eyasi 1 is female.

The hominids from the final Middle and early Upper Pleistocene exhibit greater cranial capacities. The spectrum runs from Kanjera 1 and Laetoli 18, which have cranial capacities of around 1,350 cm^3, to Singa and Border Cave, with capacities of more than 1,500 cm^3. The individuals here under consideration, thus, indicate that there was a trend towards an increase in the braincase during the Middle and early Upper Pleistocene. Yet, there are two basic problems in interpreting the size of the individual endocranium: the wide range of variation within *Homo sapiens sapiens*, which ranges from some 1,000 to 2,000 cm^3, as well as our ignorance of its population of origin. Thus, the differences in cranial capacity exhibited by isolated finds from various populations of archaic *Homo sapiens* can only have a theoretical application when considering the degree of cerebralization. It is only when other analyses, especially of the exocranial morphology, are included that one can more precisely determine the phylogenetic position of a hominid.

Frontal Morphology

The eastern and southern African hominids Hopefield, Broken Hill 1, Bodo, Eyasi 1, and Ndutu (Figs. 5c, 14) possess well-defined supraorbital tori [cf Cunningham, 1980; Gieseler, 1974; Smith and Ranyard, 1980]. The supraorbital regions of the first three of these are massive. Their torus thicknesses (measured above the middle of the orbit) vary between 18 and 21 mm, with Broken Hill possessing the highest value. Thickness decreases towards the lateral processes. The torus divides into two slightly arched segments, similar to the pattern among various European Middle Pleistocene hominids (eg, Petralona, Arago 21, Steinheim). Broken Hill and Petralona also show similarity in exhibiting large frontal sinuses, which extend far into the squama. In contrast, the pneumatisation of Hopefield is assymmetric and restricted primarily to the torus [cf Tillier, 1977; Stringer et al, 1979].

The tori of Eyasi 1 and Ndutu, which possess thicknesses (at the midorbit) of 13 and 9.5–10 mm, respectively, are clearly more gracile. Clarke's [1976] reconstruction of the Ndutu skull, however, features an especially projecting supraorbital region (cf Fig. 14). Eyasi 1's thickness decreases only slightly towards the lateral process, while with Ndutu, it remains almost constant. The frontal sinus of the Eyasi specimen appears quite large [Protsch, 1981]. Both of these hominids—like the three more robust finds—exhibit considerable postorbital constriction. Because of the general affinities, especially those between Eyasi 1 and Broken Hill 1 in further characteristics [cf Wells, 1957], it may be that the differences in torus morphology are also partially due to the sexual dimorphism found among the Rhodesoids and the Middle Pleistocene populations.

The East and South African hominids Laetoli 18, Omo 1, Omo 2, Florisbad, and Border Cave 1 possess no clearly defined tori (Fig. 4). Among these hominids, Laetoli possesses the most toruslike morphology. The me-

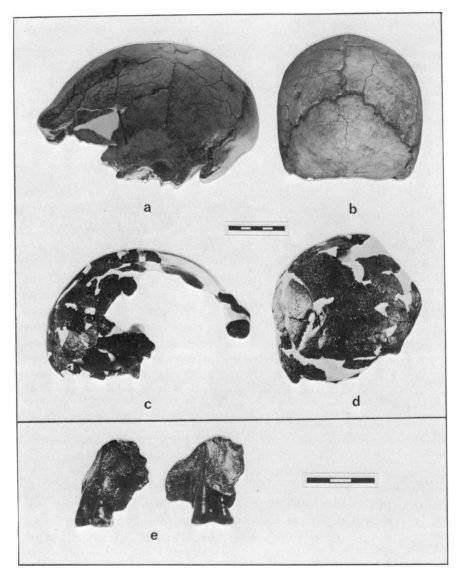

Fig. 5. a and b. Laetoli 18 (Ngaloba Beds) c–e. Eyasi 1. e. Two views of the maxillary fragment (reconstructed by R. Protsch, photographs by G. Unrath).

dial and lateral thicknesses of the relatively strong, slightly curved supraorbital arches are a little more salient than those of Eyasi 1. Both of these hominids possess a widely rounded supraorbital margin, and their tori flatten a little towards the sides. Although the fragmentary character of Eyasi 1 enables only restricted interpretation, there are certain affinities between these two hominids. With Laetoli 18 (Fig. 5) the less projecting supraorbital region and what appears to be slight postorbital constriction point in a modern direction. The superciliary arch and the supraorbital trigone can be distinguished from one another, although not as easily as in modern humans. The frontal sinuses consist of two quite small chambers, located relatively far apart. This is a feature that, although rare, can also be found in a.m. humans [Magori, 1980].

The supraorbital regions of Omo 1 and 2, which are only partially preserved, more closely resemble the modern form with its easily definable supraorbital trigone. Omo 1's glabella region is quite prominent, with a deeply depressed nasal root. In spite of this prominent glabella region, the superciliary arches are separated from one another. In their robustness, they even correspond to the modern situation, such as can be found in the Afalou population. The thickness of the lateral supraorbital region (10 mm) is less than the corresponding values for Eyasi and Laetoli. The lateral view (Fig. 6) also shows us a flattened supraorbital trigone, which can not be differentiated from modern forms. According to the new reconstruction by Day and Stringer [1982], the postorbital constriction is slight. Altogether, the supraorbital region of Omo 1 can, despite a certain robusticity, be described as almost completely anatomically modern.

Although the Omo 2 skull appears much more archaic in many features, certain similarities between its supraorbital region and that of Omo 1 are present. The strongly developed superciliary arch can be clearly differentiated from the equally heavily developed supraorbital trigone. The degree of postorbital constriction is completely modern. In addition, the supraorbital margin is strikingly small (Fig. 6). The preserved portion of the left frontal sinus is well developed both laterally and vertically. Thus, the supraorbital region—in so far as it is present—exhibits clear affinities with the a.m. spectrum, although its archaic robusticity cannot be ignored.

We can also find archaic features in the supraorbital morphology of the South African hominids Florisbad and Border Cave 1 (Fig. 4). Above the Florisbad orbit is a slightly arched and vertically relatively wide toruslike development. It becomes progressively flatter, without noticeable interruption, as one moves from glabella to the zygomatic process. Although its total contours exhibit certain similarities to Laetoli 18, Florisbad is consid-

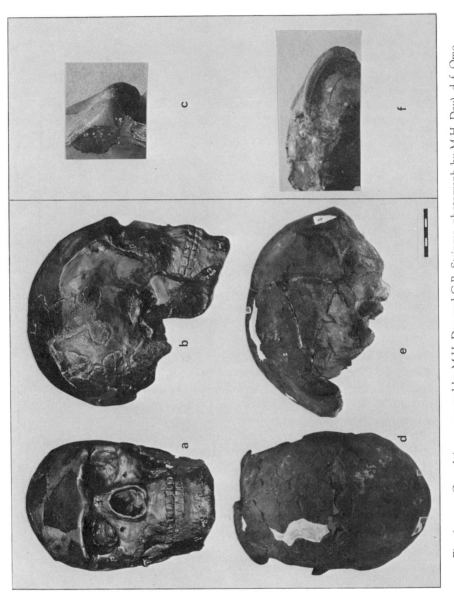

Fig. 6. a–c. Omo 1 (reconstructed by M.H. Day and C.B. Stringer, photograph by M.H. Day). d–f. Omo 2. c shows the left supraorbital trigone, the relatively small supraorbital margin.

erably more gracile in its supraorbital thickness. Florisbad possesses a higher medial value than do the Omo hominids; laterally it roughly equals Omo 1. Florisbad's supraorbital region is not prominent and its postorbital constriction only slight [cf Rightmire, this volume; Fig. 3]. What is remarkable is the heavily rounded supraorbital margin. Rightmire [1978] examined this find and arrived at the conclusion that the supraorbital morphology shows clear similarities to Broken Hill and Hopefield.

If we consider the entire ensemble of features, we find that there are several characteristics of Florisbad's supraorbital region that cannot be found among the modern final Upper Pleistocene hominids from Southern Africa. Indeed, this region—seen purely descriptively—may be assigned a position between the archaic Rhodesoid form and the completely modern form, falling closer to the modern pattern.

The supraorbital region of Border Cave 1 (Fig. 4) has certain affinities with Florisbad, although the former's is much more gracile. With Border Cave 1, it is not possible to clearly separate the arch and the trigone [cf also Cooke et al, 1945]. Medially, the "supraorbital arches" can be clearly distinguished from one another. Further similarities to Florisbad—although again more weakly expressed—can be found in the plainly rounded supraorbital margin. Thus, the supraorbital region cannot be characterized as fully modern, even though unquestionably strong affinities with the recent spectrum have been diagnosed [cf De Villiers, 1973].

As both hominids may possibly date from the same period, they may provide an indication of the amount of variability among the populations living in Southern Africa during the transition from late archaic to a.m. *Homo sapiens*. The differences between the supraorbital structures of these two hominids and the a.m. spectrum of North African forms are remarkable. With the latter, the arch is generally more prominent and is clearly set off from the flat trigone.

The frontal find No. 16425 from Cave 1 at Klasies River Mouth, which belongs to the MSA II, provides evidence that hominids with completely modern supraorbital regions lived in South Africa at almost the same time. Even though the fragment is quite small, it does include a number of diagnostically important features (Fig. 4). The glabella region is not prominent, but rather exhibits only a slight convexity above the very flat nasal root. This convexity certainly falls within the limits of modern variation. Both it and the nasal bones, which show only a little angulation towards one another, suggest connections with recent South African forms. In so far as an evaluation of the supraorbital region is permitted, both the margin and the superciliary arches present a modern appearance [cf also Singer and Wymer, 1982].

In Eastern Africa, the early Upper Pleistocene spectrum of variation of supraorbital morphology also plainly extends into the range of modern variation. Besides the Omo hominids, we find additional evidence from the Sudan (Singa) and from Kenya (Kanjera). A look at the supraorbital region of the Singa specimen (Fig. 4) shows that the superciliary arches are well defined both medially and laterally. The thickness of the supraorbital region, measured above the middle of the orbits, lies intermediate between the Afalou sample and the Omo hominids. Laterally, the supraorbital margin is substantially thinner; the measurements lie below the corresponding values for Boskop and various Afalou specimens. Yet in spite of this lateral thinning, the process is quite prominently developed. The result is a slight amount of postorbital constriction, which, however, when seen in relation to the overall dimensions of the vertical view (Fig. 11), should not be considered as an archaic feature. Singa's supraorbital margin is also well rounded. Altogether, the supraorbital morphology of the Singa hominid can be described as almost completely anatomically modern, with what may be some archaic reminiscences in the region of the processes.

Finally, the Kanjera remains, especially in their supraorbital form, have long appeared to indicate that a.m. proportions were present at a very early date. Both Kanjera 1 and 3 possess frontal fragments that permit evaluation (Fig. 4). The glabella region of Kanjera 1 is quite flat and exhibits no convexity, even in relation to the position of the nasal root. The small extant portion of the left superciliary arch is also very gracile. Thus, in terms of its descriptive features, Kanjera 1's supraorbital region is not only completely modern, but also tends towards the more gracile, apparently female side of the modern spectrum.

The small frontal fragment of Kanjera 3 likewise indicates that there was no torus. The glabella and adjoining arch region are more strongly developed than with Kanjera 1, but the degree of prominence still falls within the a.m. spectrum.

The variability between the unequivocal tori and the a.m. form, with its clearly defined superciliary arches, can lead one to assume that there was a considerable degree of heterogeneity among the populations of the African late Middle and early Upper Pleistocene. However, due to uncertain dating, it remains unknown whether the more robust forms (eg, Laetoli, Omo 2, and Florisbad) are also older than the more gracile forms (such as Omo 1, Kanjera, and Border Cave 1). These differences in robusticity may also be partially due to sexual differences. It is also interesting that similarly broad morphological spectrums have been determined for both Eastern and Southern Africa, in spite of the limited number of finds. The extent to which we may surmise that the changes in both areas are related will be discussed later.

Our knowledge of North African supraorbital morphology is based on just three individuals. All three possess a well-defined supraorbital torus. The torus of Thomas 2 is especially strong and projecting. The obvious postorbital constriction also underlines the quite archaic affinities of this individual, which may be a subadult [cf Ennouchi, 1972]. The supraorbital region of the Jebel Irhoud hominids, which are probably some 200 ky younger, has its greatest affinities with the torus morphology of the Neandertal(o)ids of the Near East and Europe. This is especially true for Jebel Irhoud 1; while in Jebel Irhoud 2 the torus becomes laterally thinner and is somewhat more arched. There are clear differences between this type of torus development and the a.m. proportions in the Afalou/Taforalt populations. Of special interest in this connection is the supraorbital morphology of the skull discovered in 1975 in the grotto "Dar-es-Soltan 2." This find, dated to the upper Aterian [Debénath, 1975], possesses a strongly developed supraorbital region that appears to have certain affinities to the Jebel Irhoud hominids [Ferembach, 1976a; cf also p 395].

In addition to the supraorbital morphology, we can also obtain further information about the variability of archaic and a.m. *Homo sapiens* from an examination of the frontal profile. Figure 7 gives an indication of the variability in the mid-sagittal curves of some final Upper Pleistocene hominids. While the curves of the North African specimens generally have a greater posterior inclination, the East African representatives are characterized by a somewhat steeper inclination in the supraglabellar region. This trend is even more pronounced with the three Southern African finds.

When we compare the curves of the archaic Rhodesoid specimens (Fig. 8a) with those from the modern spectrum, then we can see that the former exhibit a considerably greater inclination. There are clear affinities between Broken Hill, Hopefield, and Eyasi 1. A comparison of Hopefield, Florisbad, and Border Cave 1 (Fig. 8b) reveals certain similarities in the profiles of the first two. In contrast, Border Cave's profile is characterized by a much steeper supraglabellar inclination, which in turn resembles some modern forms, especially those from Southern Africa.

We obtain especially remarkable results when the profiles of Kanjera 1, Singa, and the two modern hominids from Naivasha and Lukenya Hill are superimposed on one another (Fig. 8c). Kanjera possesses a curve almost identical to that of Naivasha; Singa and Lukenya Hill also reveal considerable conformity. Figure 8d shows how the supratoral inclination of the two Jebel Irhoud hominids (especially No. 2) are even steeper than that of the much younger Afalou representatives.

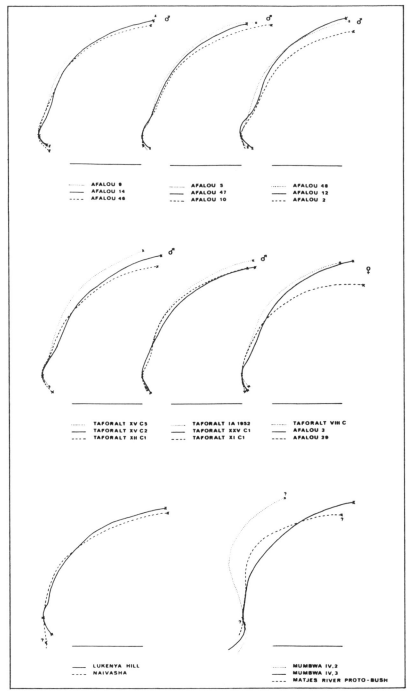

Fig. 7. The mid-sagittal curves of the frontals of some final Upper Pleistocene hominids from various parts of Africa (Frankfurt Plane).

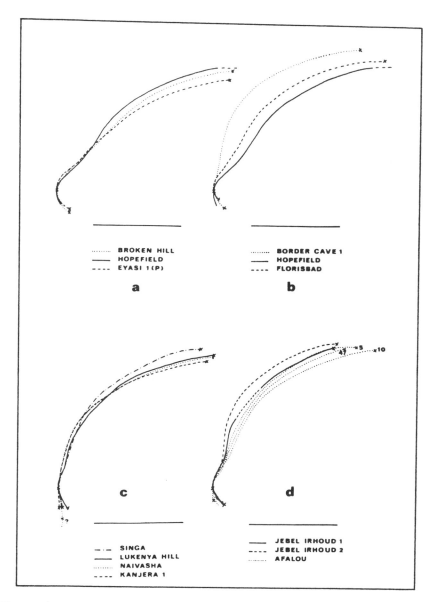

Fig. 8. A comparison of the mid-sagittal curves of the frontals of some fossil hominids (Frankfurt Plane).

When we compare the lateral views of the frontals of several other important finds[8], the very similar profiles of Bodo and Broken Hill are evident. The profile of the Ndutu specimen is extremely hypothetical, but it must, in any case, be somewhat steeper than the profile of the Rhodesoids. Although it is not possible to precisely orient the curve of Laetoli, one cannot fail to notice the similarities to Eyasi 1 in both the profile and the glabella region. The curve of Laetoli, however, follows a steeper course above the supraglabellar fossa. There are also certain similarities to Hopefield.

Because the supraglabellar region of Omo 2 is missing, only limited comparisons are possible. In addition, the proper orientation of this find is uncertain. Characteristic for Omo 2 is the almost straight course of the profile in the first section and the late onset of the curvature. The profiles of Broken Hill and Bodo exhibit their greatest similarities to that of Omo 2 but the angle of inclination to the Frankfurt Plane is much greater with Omo 2 than with the Rhodesoids. There are also obvious similarities between Omo 2 and Laetoli 18. When placed in their most probable orientations, the two lines almost coincide. Omo 1 exhibits clear affinities in frontal profile to the Afalou/Taforalt group. In contrast to the more heavily curved profiles of Singa and Kanjera, the profile of Omo 1 is more similar to the North African spectrum of a.m. humans.

Finally, we also achieve some interesting results through a comparison of the lateral views of the heterogeneous hominids from eastern and southern Africa [cf Bräuer, in press]. Florisbad and Laetoli share some obvious similarities to curvature and inclination, although the supraorbital prominence is somewhat more pronounced with the latter. Omo 1's frontal is steeper than Florisbad's. There are also strong affinities between Border Cave and some of the a.m. hominids from Southern Africa (eg, Tuinplaas = Springbok Flats and Boskop).

The variability of the frontal profiles, considered in its complex entirety, reveals no large gaps between the Rhodesoids and the a.m. form. We find rather, just as with the supraorbital region, that there is a broad spectrum of forms lying between the two poles.

Turning from the descriptive analyses of frontal morphology, we shall now discuss the results of a principal components analysis (Fig. 9). This included six variables measuring both the curvature and the breadth of the

---

[8]As seen simply in profile.

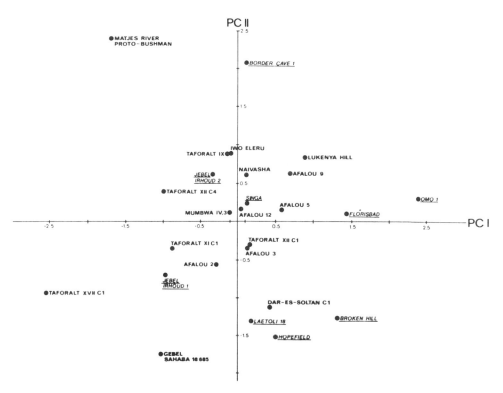

Fig. 9. A principal components analysis based upon six frontal variables. PC I represents 46.5% and PC II 35.6% of the total variance.

frontal.[9] The first principal component appears to primarily express the differences in size, while the second reflects more the differences in frontal curvature. Here, the spectrum ranges from such extremely flat forms as Hopefield and Broken Hill 1 to the heavily curved form of Matjes River (Khoisanoid). The affinities between two of the a.m. individuals (Dar-es-Soltan, Gebel Sahaba) and the very archaic forms are striking. Laetoli 18 also reveals strong affinities to the Rhodesoids. The second principal component places Omo 1 in the middle of the Afalou/Taforalt spectrum. Florisbad's position intermediate to the Rhodesoids and Border Cave 1 is also of interest. The latter hominid possesses obvious affinities to the Proto-Bushman from Matjes River. This result supports the affinities between

---

[9] See Appendix for a list of the variables.

Border Cave and the modern Khoisanoids that were determined by Right-mire [1979]. Singa shows clear affinities with the a.m. spectrum. This is also true of Jebel Irhoud 2's frontal form; in contrast, Jebel Irhoud 1 occupies a more marginal position with regard to the modern spectrum.[10] Further multivariate analyses of frontal morphology [Bräuer, unpublished manu-script], also indicate that Kanjera 1 has modern affinities.

Parietal Morphology

The parietal bones of *Homo erectus* are, when compared to a.m. *Homo sapiens*, longitudinally flatter and shorter. In the occipital view, they are more narrow at the top and diverge as they go down. The maximum breadth is generally found at a low point, near the auricularia [Howell, 1978]. Figure 10 gives a view of the variability in sagittal curvature. In the upper part of the very broad range lie the representatives of archaic *Homo sapiens* (eg, the eastern and southern African hominids Broken Hill, Omo 2, and Singa). At the periphery, yet outside of the a.m. sample, lies Laetoli, followed by Eyasi 1 and Kanjera 1. The strong affinities that these three hominids show to one another, and their close connections to Naivasha (Kenya), are worthy of mention.

Within the Rhodesoids, there are clearly great differences between Bro-ken Hill and Hopefield [cf also Drennan, 1953]; both of Hopefield's values, like those of Omo 1, fall in the middle of the modern spectrum. The two North African specimens Salé[11] and Jebel Irhoud 1, with their extremely slight parietal curvatures, take up very extreme positions. This is principally due to the very short parietal arches, whose lengths lie below the range of variation of the other hominids. The arch-lengths for Omo 2, Kanjera 1, Hopefield, and Broken Hill lie around the lower limits of variation found for the Afalou/Taforalt sample (c 120 mm).[12]

The occipital views of the a.m. North African sample reveal a substantial amount of individual variability. Those forms with lateral walls that are either relatively vertical or show only a slight tendency to diverge upwards are dominant. Besides these main forms, there are also single forms that are either laterally somewhat rounded (eg, Afalou 14) or display relatively

---

[10] It should be emphasized that no specific torus measurements were included.

[11] Pathological changes of Salé's skull shape cannot be ruled out [Hublin, personal communication].

[12] Some East African Holocene finds have values well below 120 mm [Bräuer, in preparation].

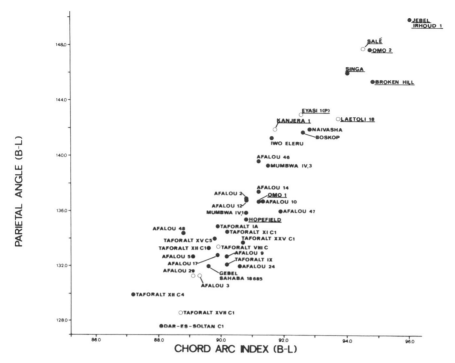

Fig. 10. A bivariate comparison based upon measurements of the mid-sagittal parietal curvature (●, male; ○, female).

straight contours that show only a slight convergence towards the top (Afalou 2). From the a.m. hominids from western and eastern Africa, we find that Lukenya Hill, Naivasha, and Olduvai 1 possess the types of forms dominant in the north. There are uncertainties with the Iwo Eleru specimen. These result from the fragmentary condition of the find, whose reconstruction necessarily includes some supplementary guesswork. Yet, it is possible that the lateral sides are somewhat rounded. The few a.m. South African specimens that permit an assessment possess a broad spectrum of forms, including those with diverging, rounded, and parallel sides.

   If we now compare the older hominids to these "modern" forms, the following finds appear to possess either diverging or parallel sides: Omo 1, Laetoli 18, Kanjera 1, Singa, Border Cave 1, Eyasi 1, and possibly, Hopefield. However, the very wide and low form exhibited by Singa, with its diverging sides, stands alone among those finds which date back to the final Upper Pleistocene [Bräuer, unpublished manuscript].

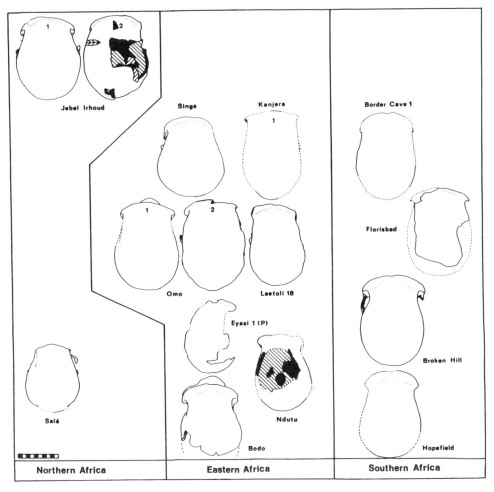

Fig. 11. The vertical views of some African fossil hominids (> 30 ky BP) arranged in a rough chronological order.

Both Omo 2 and Broken Hill possess a straight or very slightly converging contour, and thus, have certain similarities to Afalou 2. The lateral sides of Ndutu are slightly convergent, but also somewhat rounded. The Jebel Irhoud hominids possess lateral sides that are vertical or slightly rounded. They differ in this trait from most of the European classic Neandertals, which exhibit more transverse-oval forms (cf Fig. 1). Finally, Salé also possesses, above the prominent supramastoid region, slightly converging sides. These are not dissimilar from those of Jebel Irhoud 1, Ndutu, and Omo 2, although the vault of the latter is considerably higher.

In summary, the occipital views of the archaic hominids exhibit close relationships to the a.m. spectrum. There are none of the tentlike forms so typical with *Homo erectus* (eg, Olduvai H.9). Both the Rhodesoids and Salé, with their expanded parietal regions, exhibit closer affinities to the modern spectrum than to the *Homo erectus* forms, a fact that supports assigning them to "archaic *Homo sapiens*."

When viewed from above, it appears that pronounced parietal bosses are quite common among the a.m. individuals from northwest Africa. The two lateral sides are equally rounded on a few of the crania (Taforalt 11C1, 18C2, 17C1, Afalou 29, 47). Pronounced parietal bosses may also be found among the few individuals from West and East Africa (Iwo Eleru, Lukenya). There are also specimens with rounded, less prominent forms (Gebel Sahaba 18685, Olduvai 1); and one specimen with more or less parallel sides (Naivasha). Forms in which the lateral sides exhibit only a slight outward curvature are also present in the South African group (Mumbwa IV,1; IV, 3; Tuinplaas). Both Matjes River and Boskop possess strongly pronounced parietal bosses, which in the latter are larger.

In degree of postorbital constriction (Fig. 11), the Rhodesoids, Bodo, and Ndutu are plainly different from those of the modern spectrum. These four hominids are more or less equally curved in the tubera region which is without any special pronouncement. We may assume that Eyasi 1 probably possessed a similar form. The Omo hominids, Border Cave, and possibly Florisbad (following the reconstruction) also possess equally curved parietal regions. The postorbital regions of these crania, are, however, modern. The Laetoli 18 cranium, which is quite small, reveals a certain pronouncement in its parietal bosses. Its overall form exhibits similarities to Kanjera 1. Some of Kanjera 1's more extreme features, unmatched by any other specimen studied, may be the result of the reconstruction work.

The vertical contour of the Singa cranium is of special interest. The very pronounced parietal bosses were one of the reasons that Coon [1962] and others considered this hominid to be an early Khoisanoid. Yet such pronounced parietal bosses are by no means found only with this specimen. Iwo Eleru exhibits some observable similarities to this find, as do a number of modern, northwest African hominids (Afalou 5, 40, Taforalt 1A). Thus, Singa's vertical form, like its frontal morphology (see above), can no longer be considered as evidence of Khoisanoid affinities. If we compare Singa with Boskop, or with the Matjes River "Proto-Bushman," we can see that the three are clearly dissimilar with regard to the position and shape of their parietal bosses.[13]

In sum, the late archaic and early modern hominids from eastern and southern Africa exhibit an obvious heterogeneity, but they show no substantial deviations from the equally heterogeneous a.m. forms. The form of Kanjera 1 is the only one that does not quite conform to this picture.

Figure 12 presents the results of a multivariate comparison based upon six parietal variables.[14] These deal with both the curvature shown in the mid-sagittal profile and the maximum bi-parietal breadth. The distribution of the finds, as based on the first two principal components, reveals an interesting separation: the older finds are all on the left side, and the final Upper Pleistocene finds on the right. Laetoli has a marginal position with respect to the modern spectrum. This holds for Hopefield, Omo 1, and Kanjera 1 as well. The two Omo hominids exhibit affinities with one another. Broken Hill is plainly dissimilar to the modern spectrum; Salé and Singa are even more so.

Other multivariate analyses of parietal morphology, including some that considered the minimum frontal breadth and the overall cranial size [Bräuer, unpublished manuscript], have also shown that Singa holds a very isolated position. This most probably reflects this hominid's extremely short parietals, yet pathological changes may also account for some of Singa's uniqueness.

Occipital and Temporal Morphology

Our considerations here will primarily concentrate on three aspects: the curvature and angulation of the occipital profile, the morphological variation at the boundary of the nuchal plane, and the features of the mastoid and supramastoid regions.

As with the supraorbital region, the occipitals of both archaic and a.m. *Homo sapiens* reveal a broad spectrum of variation. This ranges from clear torus formations, via more or less toruslike developments, to gracile nuchal lines. Hopefield and Broken Hill possess prominent occipital tori. Inion and the opisthocranion coincide on both crania. The occipital torus of Hopefield follows an almost horizontal line and is limited to a medial position some 6 cm long. A flat but well-developed supratoral sulcus is located above it. In

---

[13] A new preparation of the Singa skull has shown that the parietal bosses are abnormally thickened, probably due to pathological alterations [Stringer, personal communication].

[14] See Appendix.

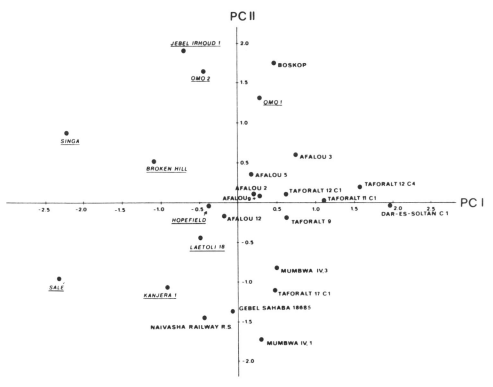

Fig. 12. A principal components analysis based upon six parietal variables. PC I represents 61.7%, and PC II 19.0% of the total variance.

profile, the squamous and the nuchal parts (the latter are only very fragmentarily preserved) follow slightly curved lines to their conjunction, where they form a marked angle.

A similar angle is present on Broken Hill, but the marked torus proceeds, bilaterally and continuously, from the missing inion region to the occipito-mastoid suture. There, it merges into the mastoid crest. Altogether, the torus reveals a greater downward curvature. The mastoid process is large in comparison to the final Upper Pleistocene hominids from southern Africa, but small when compared to the Afalou/Taforalt group. The supramastoid crest is prominent and is separated from the mastoid crest by a marked fossa.

Eyasi 1 resembles the Rhodesoids in both contour and angularity (Fig. 5c). The torus, however, is only slightly developed and is restricted to a medial position. Protsch [1981] determined that the *lamina externa* had been abraded along the length of the torus. The mastoid process is relatively small, but might fall within the range of "modern" variation. The isolated occipital fragment known as "Eyasi 2" possesses a somewhat larger angle. Medially, the occipital torus is slightly more prominent than with Eyasi 1, but clearly less than with Broken Hill and Hopefield.

Ndutu has an occipital profile somewhat more rounded than those of the hominids already discussed. Although Ndutu's profile shows comparatively modern affinities, this find also possesses a torus thickening that reaches a height of c 1.6 cm on the midsagittal line. Above this lies a distinctive supratoral depression. The torus becomes much thinner towards the sides. It has been conjectured that Eyasi 1 (and perhaps 2) and Ndutu may be females. This could partially explain why these hominids have less marked tori.

The length of Ndutu's mastoid process is also moderate, being somewhat less than with Broken Hill. Ndutu's value falls in the lower regions of the female Afalou spectrum. Clarke [1976] has pointed out that the form of the process, with its very marked mastoid crest, resembles those of the hominids 9 and 12 from Olduvai. This crest runs dorsally at roughly the level of the occipital torus. The supramastoid crests are also markedly developed.

Laetoli 18 does not only show modern affinities in its supraorbital region, but also in its occipital morphology. Its profile follows a uniform curve below the lambda region. The opisthocranion lies well above the inion (cf Fig. 5). The nuchal plane is bordered by an undercut occipital torus that is limited to the central portion of the occipital and of relatively slight prominence. A prominent, downwards projecting occipito-mastoid crest runs along the occipitomastoid suture. Although the flattened curvature of the frontal and parietal regions point towards affinities with Eyasi 1 (cf Fig. 5), Laetoli 18's occipital profile is less angular. Laetoli's mastoid processes are very small. With lengths (below the Frankfurt Plane) of 20 mm, they lie outside of the observed range of final Upper Pleistocene variation. The supramastoid crests are well developed.

Substantial portions of Omo 1's occipital have been preserved (cf Fig. 6). This specimen exhibits a slight, perhaps "toruslike," thickening in the central region that becomes laterally thinner and then disappears. Anatomically modern humans (eg, Olduvai 1 and various individuals of the Afalou/Taforalt populations) also exhibit such prominently developed superior nuchal lines.

A comparison of Omo 1's contours with those of the Afalou/Taforalt hominids shows its similarities to certain of these hominids (eg, to Afalou 5). The opisthocranion lies well above the inion. Although we are unsure of the position of porion, we can still say that the mastoid process was quite long, probably equal to the average for the male Afalou hominids. The mastoid crest can be described as moderately to markedly developed.

The occipital morphology of Omo 2 clearly differs from this modern form. The opisthocranion coincides with the inion on the highly set torus (Fig. 6). In profile, the squamous and nuchal parts follow an almost straight course and form a clear angle with one another. The toral thickening is primarily restricted to a central region, above which a slight supratoral depression is located. Prominent occipitomastoid crests are present. The occipital traits of Omo 2 find their greatest affinities with the Rhodesoids, although there are clear differences in the form of the torus. Omo 2's mastoid region is, in its entirety, very robust. Both the mastoid process and the mastoid and supramastoid crests are strongly developed. The mastoid crest runs dorsally towards the occipital torus.

There is no torus on the occipital of the Singa hominid; both the nuchal lines and the nuchal plane have only a relatively weak relief. In profile, the occipital is rounded; here again it appears completely modern. The same is true for the strongly curved occipital of Kanjera 1. Singa and Laetoli 18 bear certain resemblances to one another in the overall contours of their occipitals. Singa's mastoid processes are short (close to the lower limit of the female Afalou spectrum) and relatively small, but the supramastoid and mastoid crests are well developed.

Only a small fragment from the right side of Border Cave 1's occipital has been preserved. This provides no indication that its form deviated from that of the modern spectrum. The quite small mastoid processes (both of which are present), which feature relatively prominent supramastoid crests, also indicate modern affinities.

The Middle Pleistocene North Africans also exhibit modern tendencies in their occipital morphology. Salé, for example, has well-rounded supratoral and nuchal regions. The toruslike thickening is, vertically, relatively wide, but not especially prominent. A weakly developed supratoral sulcus lies above. The opisthocranion is located above the inion. The mastoid processes are quite strongly developed in relation to the small cranium. They are longer than those of Broken Hill.

The well-rounded occipital from Rabat (Kébibat), which is probably a juvenile specimen, also exhibits obvious modern affinities. There is no occipital torus, but only a slightly emphasized superior nuchal line [cf Saban, 1975].

The Jebel Irhoud hominids possess a relatively low (with reference to the Frankfurt Plane) occipital torus that is well-developed centrally and tapers off laterally. Both hominids possess an occipital bun (Jebel Irhoud 2's is more prominent) like that found with most of the European Neandertals. None of the other African hominids here under discussion share this characteristic. A comparison of these profiles with those of several of the Afalou/Taforalt individuals reveals that the occipital of Jebel Irhoud 1 has a greater overall dorsal extension. Opisthocranion and inion do not coincide. The mastoid processes of the Jebel Irhoud finds are relatively small; their length lie in the lower regions of the female Afalou variation. Both hominids exhibit strongly developed mastoid crests; the supramastoid crest is more pronounced with Jebel Irhoud 2 [cf Ennouchi, 1962, 1968].

The occipital fragment of Témara 2 also lacks an occipital bun. Its profile, dimensions, and gracile nuchal plane reveal clear affinities to the Afalou/Taforalt spectrum and are plainly different from those of the Irhoud specimens [Ferembach, 1976b].

To summarize, the occipital morphology reveals an extraordinary amount of variability in angulation and delimitation of the nuchal plane. This is true of both those forms which can be classified as tori and the toruslike formations. The latter are not only found among archaic but also among a.m. hominids. A clear, definitive demarcation is not possible in many cases. Sexual dimorphism also contributes considerably to the spectrum of variation of this complex of features as well as to that of the mastoid region.

Because many of the older hominids possess fragmentary occipitals, only a few of them could be included in the metric description of the occipital curvature. The bivariate spectrum (Fig. 13) ranges from forms with low index and angle values, ie, with a greater dorsal projection, to flatter forms with higher values. Although the modern individuals exhibit a large amount of variation, especially as regards the occipital angle, most of the archaic hominids clearly distance themselves from this spectrum.

Omo 2 has an especially extreme position, as the descriptive comparison also suggested. Salé, Ndutu, and Eyasi 1 follow. Singa, with its very long occipital arc, lies between this group and the adjacent modern spectrum. Omo 1, in profile, possesses entirely modern proportions.

Upper Facial Skeleton

Changes in vault and face morphology are the result of a complex interaction between evolutionary and biomechanical forces [Biegert, 1957; Moss and Young, 1960]. The transition from early archaic to a.m. *Homo sapiens* was accompanied by a reduction in the area of cranial muscle attachments as well as a diminution of the massivity of the masticatory apparatus [Wolpoff, 1980].

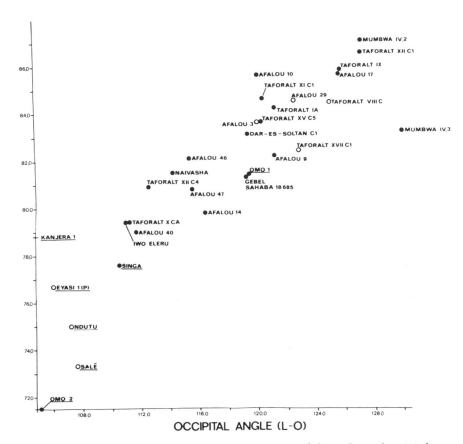

Fig. 13. A bivariate comparison based upon measurements of the midsagittal occipital curvature (●, male; ○, female).

Bodo and Broken Hill 1 provide an impression of the robust upper facial morphology of early archaic *Homo sapiens* (Fig. 14). The nasion-alveolare heights of these two hominids (Bodo, 88 mm; Broken Hill 1, 96 mm) lie well above the variation shown by the a.m. samples of this study, which extends to 78 mm. The length from the nasion to the nasospinale is quite similar in Bodo (60) and Broken Hill 1 (59), but the subspinal part is longer in the latter. The mid-facial region presents a contrasting picture. Here Bodo, with its massively developed zygomatic processes and malar bones, is considerably broader than Broken Hill 1 (bimaxillary breadth: Bodo, c 132; Broken Hill 1, 107). Both hominids lack a canine fossa, but Broken Hill 1 does

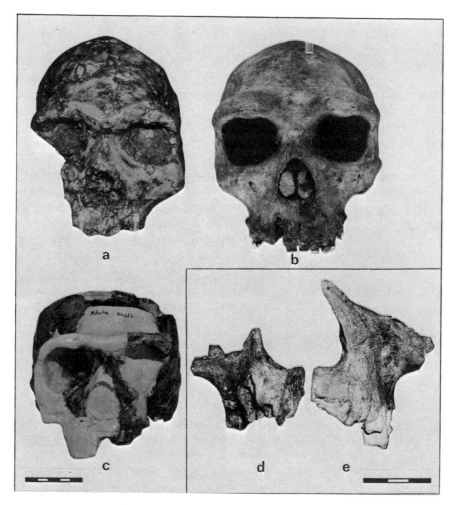

Fig. 14. (a) Bodo, (b) Broken Hill 1, (c) Ndutu, (d) Broken Hill 2, (e) the left half of Laetoli 18's maxilla (photograph provided by C.B. Stringer).

exhibit a slight concavity in the region below the infraorbital foramina. In addition, the curvature of the zygomatic process (*Incurvatio inframalaris*) of this hominid is more highly located and more strongly curved than is the case with either Bodo or the European Neandertals.

An attempt at orientating the fragmentary Bodo skull to the Frankfurt Plane reveals that the entire upper facial skeleton is more prognathous than Broken Hill's, whose prognathism is primarily restricted to the alveolar process. Bodo's zygomatic bones (Fig. 14) are situated more anteriorly, while

Broken Hill 1's are transversely arranged and located more posteriorly. In their facial morphology, these two hominids bear striking similarities to the Middle Pleistocene hominids Petralona and Arago 21 [Conroy, 1980; Stringer, 1981], although all four specimens differ widely in extent of prognathism. The two African hominids lie between Petralona and Arago 21 [following the reconstruction of M.-A. De Lumley, 1981]. With the exceptions of Laetoli 18 and Jebel Irhoud 1, the facial skeletons of the other representatives of archaic and early a.m. *Homo sapiens* consist of only small fragments.

Also remarkable is the maxillary morphology of Broken Hill 2. In contrast to Broken Hill 1, hominid 2 (an isolated maxillary fragment) possesses an obvious canine fossa (Fig. 14). The conditions of the find preclude a clear determination of whether the two hominids were contemporaneous [cf Wells, 1947]. If both hominids do possess a similarly high Middle Pleistocene age, this would indicate that there was a considerable variation in maxillary morphology (evidence that this was the case in Europe is provided by Steinheim, Arago 21, and Petralona); but it should be noted that Broken Hill 2, despite its modern affinities, is a rather massive jaw whose nasal breadth surely lay above the range of modern variation.

Ndutu's maxillary region around the nasal aperture has been partially preserved (Fig. 14). The small, strongly curved and laterally directed zygomatic process indicates a rather gracile middle face. The area around the infraorbital foramen appears to indicate the presence of a trace of an infraorbital fossa or slight concavity. The preserved subnasal section suggests some alveolar prognathism.

It is very difficult to make any assessment of the infraorbital region in the Eyasi 1 hominid, from which only a small fragment of the alveolar process (with the C and P1 and a portion of the alveolus of P2) exists (Fig. 5e). This piece does show general affinities to Broken Hill 1 [Wells, 1947].

Florisbad possesses an obvious canine fossa and a malar bone with a modern morphology. These traits were probably also present in both Border Cave 1 (only a gracile malar fragment has been preserved) and the maxillary-malar fragment of Kanjera 1.

Much of both sides of Laetoli 18's maxilla were preserved (Fig. 14). The zygomatic processes are oriented laterally some 90°. The *incurvatio inframalaris* has a high position and is curved, as is also the case with a.m. humans. The left side, which is the better preserved of the two, exhibits a slightly developed canine fossa. A clear concavity can be seen in the infraorbital region. In its entirety, the facial morphology shows unquestionable affinities to a.m. *Homo sapiens*. The subnasal region is characterized by marked prognathism.

There appear to be general similarities in the upper facial morphology of Laetoli and Jebel Irhoud 1. Although the latter does not possess a clear canine fossa, its *incurvatio inframalaris* and the gracility of its zygomatic processes and malar bones exhibit obvious affinities to the a.m. maxillary form. The differences between these features and the heavily developed maxillary sinuses and mid-facial prognathism of the European Neandertals are obvious. The maxillae of the Shanidar Neandertaloids are also relatively gracile, but lack a clear canine fossa [Stringer and Trinkaus, 1981]. Coon [1962] has claimed that the maxillary fragment from Mugharet el 'Aliya provides evidence that this infant had a relatively large and long face for its age. Importantly, this find possesses no canine fossa.

This primarily descriptive discussion shall now be complemented with a few metric comparisons. These were, however, severely limited by the fragmentary character of most of the hominids.

The peculiar robustness and the archaic character of Bodo's upper face is also shown by the unusually large interorbital breadth (mf-mf) of 34 mm. In contrast, Broken Hill 1's breadth is only 27 mm, a value lying just above the Afalou spectrum (19.0–26.0). The measurement for Petralona, which because of its similarities to Bodo and Broken Hill, has been included in this comparison of upper facial morphologies, lies between these two. The other African hominids that could be measured (Ndutu, Florisbad, Border Cave 1, Jebel Irhoud 1) fall within the spectrum of modern final Upper Pleistocene variation. Of these four, Florisbad most resembles Broken Hill. The biorbital breadths of Bodo, Petralona, and Florisbad all possess very similar values between 122 and 124 mm; the highest value from an a.m. individual (110 mm) is found in the Afalou sample. Of interest is the fact that Border Cave possesses a value close to this relatively high figure, while the two a.m. skulls from Southern Africa (Tuinplaas and the Matjes River "Proto-Bushman") have values only slightly above 100. Both Jebel Irhoud 1, whose value (113.0) is slightly less than Broken Hill's (116.5) and the latter lie between the upper limit of the modern spectrum and Bodo/Florisbad.

Bodo, Broken Hill 1, Florisbad, Border Cave 1 (reconstruction), and Jebel Irhoud 1 all permitted measurements of their orbital widths (mf-ect). All of the values lie well away from those of the a.m. sample (39.5–46.5). Border Cave 1 also possesses a high value (49.0) in this measurement, one which corresponds to that of Broken Hill 1. Yet it cannot be ruled out that the very large orbital width is at least partially a result of uncertainties resulting from the extensive reconstruction work. Although Bodo and Broken Hill clearly differ in their orbital indices (67.9/79.6), both of these values fall into the Martin/Saller [1957] category of "low orbits." Florisbad possesses very low orbits similar to those of Bodo (and Petralona). Jebel Irhoud 1 and Border Cave 1 lie between Bodo and Broken Hill.

When the nasal aperture is considered, Bodo is extremely wide (38), with a value well over that of Jebel Irhoud 1 (33.5). The nasal widths of the other hominids fall within the final Upper Pleistocene spectrum, which ranges from c 25 to 32 mm. Broken Hill 1 and Florisbad have positions in the upper regions of this spectrum, Laetoli (28.5) roughly corresponds to the "modern" mean, and Ndutu possesses a value of only 27 mm. Bodo and Jebel Irhoud 1 greatly resemble one another in their nasal indices. Their values lie somewhat above the Afalou/Taforalt spectrum (46.7–59.6). The nasal forms of Broken Hill 1 (52.5) and Florisbad (50.8) can be described as mesorrhine. Petralona (55.9), in contrast to Bodo (63.3), falls within the Afalou spectrum.

Another interesting feature of the nasal region is the position of the nasal bones with regard to one another. Bodo and Broken Hill 1 possess relatively large simotic angles that fall into the upper regions of the modern spectrum (75.4–140.0) and most resemble that of Lukenya Hill (127.0) and Mumbwa IV, 3 (140.0). The angle between the nasal bones of Florisbad is also relatively wide. In contrast, Ndutu possesses an especially pronounced nasal bridge. Its simotic angle (83.6) lies in the lower regions of the Afalou spectrum (75.4–123.8).

Determination of the cheek height provides a measurement of the size of the zygomatic bone. Bodo, with a value of 36 mm, lies well above the modern spectrum (as does Petralona, with 34.5 mm). Broken Hill 1, on the other hand, possesses a value that falls in the upper reaches of the modern spectrum (21.0–30.0), near that of Dar-es-Soltan C1. The cheek height of Jebel Irhoud 1 falls in the middle of the Afalou/Taforalt sample, thereby stressing the modern mid-facial affinities of this Neandertaloid. The cheek heights of Florisbad and Border Cave 1 are even less. Kanjera 1's extremely low value of 20 mm is of some interest; this figure lies below the spectrum of the a.m. finds. The early Holocene individual "Gamble's Cave 4" (Kenya) possesses similarly gracile malar bones. More detailed metrical comparisons of the upper facial morphology are not possible here [see Bräuer, unpublished manuscript].

Additional information on the morphological affinities of the well-preserved specimens of archaic *Homo sapiens* is provided by a principal components analysis (Fig. 15). It is based on eight facial measurements listed in the Appendix and, for purpose of comparison, includes the Petralona find. The first principal component already covers 82.4% of the total variance. The distribution of the individuals reveals the clear differences between the a.m.

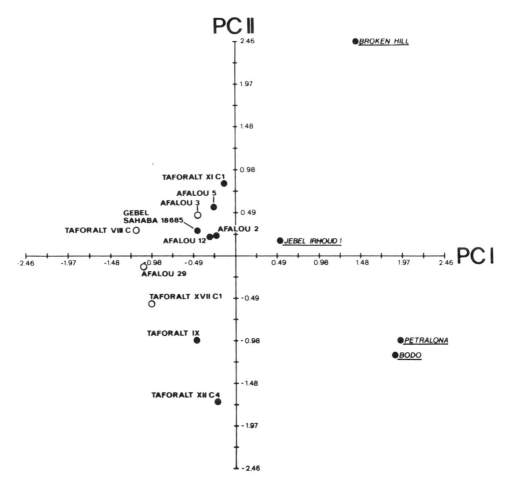

Fig. 15. A principal components analysis based upon eight upper facial variables. PC I represents 82.4% and PC II 7.9% of the total variance.

forms and the early archaic specimens from Bodo, Broken Hill, and Petra-lona. An additional result is the great similarities between these three hominids. When the second principal component (which, however, repre-sents only a very small amount of variance) is also included, the affinities between Bodo and Petralona are greater than those between Bodo and Broken Hill. As it appears unlikely that these resemblances are primarily the results of parallel developments, one could conclude that there were close phylogenetic relationships between the early archaic *Homo sapiens* of Africa and Europe. The multivariate analysis indicates that the facial morphology of Jebel Irhoud 1 is plainly more modern than those of the Middle Pleistocene hominids, even though we can discern that the former is also clearly separated from the modern spectrum. For at least those few individuals that could be included in this statistical analysis of facial morphology, the results serve as a good summary of the previous discussion of descriptive and metric characteristics.

Mandible

African hominid mandibles dated between c 500 and 30 ky BP can be divided geographically and chronologically into four different groups: 1) The Middle Pleistocene finds from southern (and eastern[15]) Africa. The South African exponents of this group are the ramus fragment from Hopefield and the body fragment from the Cave of Hearths. 2) Those southern and eastern African finds dating from the early Upper Pleistocene (Border Cave 2,3,5, Klasies River 41815, 13400, 21776, 14695, 16424, Omo 1, Diré-Dawa (?)). 3) The Middle Pleistocene finds from northwest Africa (Ternifine 1-3, Thomas 1, Sidi Abderrahman, Rabat). 4) Those northern African finds which date to between c 50 and 30 ky BP (Haua Fteah 1,2, Jebel Irhoud 3, Témara 1).

As the previous morphological analyses of the vault and face have shown, the second group may be of special importance for an understanding of the transitional phase leading to a.m. *Homo sapiens*. Our information on the ancestors (group 1) of these southern African hominids is relatively limited. This is due to the very fragmentary nature and/or subadult age of the specimens, but the two fragments composing this group possess a morphology different from that of the modern spectrum.

The Hopefield ramus fragment was discovered some 450 m from the calotte fragments. Yet both could possibly belong to the same individual, as the ramus was found together with a fragment of the posterior end of the right parietal bone that fits precisely into the reconstruction of the lambdoid suture of the Hopefield vault [Singer, 1954; Drennan and Singer, 1955]. The intact portion of the ramus includes the mandibular foramen and the posterior wall of the alveolus of the third molar. This permits a relatively reliable reconstruction of the ramus breadth and a determination of the coronoid height above the alveolar plane. This reconstruction reveals a very

---

[15] The mandibular remains from Olduvai and Baringo (see Table I) are probably older than 600 ky BP. The sole exception is OH 23, which is probably c 500 ky old. These finds will, thus, not be herein discussed [cf Rightmire, 1980].

broad ramus (c 49 mm), according to Drennan and Singer [1955]. The Hopefield ramus greatly resembles the Mauer mandible in this feature as well as in the height and curvature of its anterior border and its flat *incisura mandibulae* (sigmoid notch). The ramus form of Hopefield clearly differs from those of Ternifine and Thomas. The attempts that have been made to reconstruct the mandible from Broken Hill 1 [Dart, 1955; Drennan, 1955] indicate that this ramus was probably less broad and higher than can be assumed for Hopefield.

The second Middle Pleistocene mandible fragment comes from the Cave of Hearths and belonged to a child of some 12 years. Both the body fragment and the teeth possess a number of archaic features. Tobias [1971], when discussing the low and robust body, pointed out that robustness decreases as the height increases during growth. Thus, it is almost impossible to reliably assess this feature of this find. Yet the angle of inclination of the *symphysis menti*[16] indicates a marked degree of prognathism, such as can also be found in other Middle Pleistocene hominids. The slight subalveolar depression[17] and the moderately developed mental trigone[18] indicate that the chin was slightly developed, similar to the chins of various Neandertals and Rabat. A well-developed *planum alveolare*[19] and a superior transverse torus may be found on the posterior surface of the symphysial region [Tobias, 1971]. All in all, the Cave of Hearths mandible may be assigned to early archaic *Homo sapiens*, whose southern African representatives appear to belong to the Rhodesoid variant [see also Dart, 1948; Tobias, 1971, 1982].

Much more obviously, modern morphological features are found in those mandibles found in southern or eastern Africa and appearing to date to the early Upper Pleistocene (group 2).

In their robustness, the two adult mandibles Border Cave 2 and 5 (Fig. 16) fall within the modern spectrum. The low absolute sizes of the two bodies are demonstrated by their small heights and thicknesses. De Villiers' [1973, 1976] investigations indicated that Border Cave 2's dimensions lie close to the means of female South African Negroids and close to the female

---

[16] The angle between the alveolar plane and the incision-gnathion line [Weidenreich, 1936].

[17] This corresponds to the *Incurvatio mandibularis anterior* [Weidenreich, 1936] and is a component of the *mentum osseum*.

[18] Mental protuberance and mental tubercles.

[19] The inclination of the bone surface towards the dorsal and inferior directions.

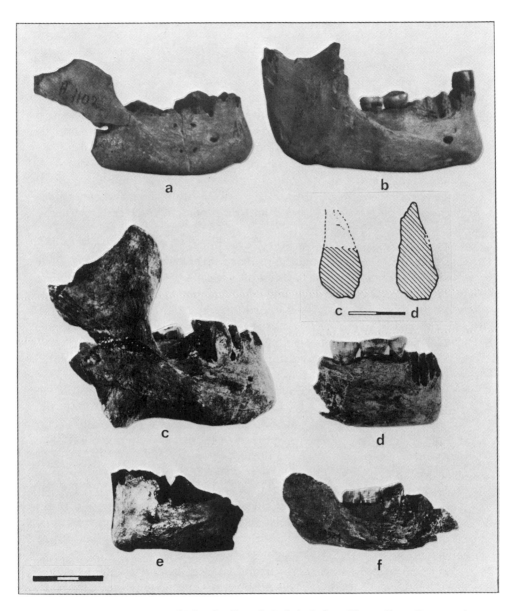

Fig. 16. (a) Border Cave 2, (b) Border Cave 5, (c–f) finds from Klasies [from Singer and Wymer, 1982], (c) No. 41815 (including a cross-section view of the symphysis), (d) No. 13400 (including a cross-section view of the symphysis), (e) No. 21776, (f) No. 16424.

San (the latter affinity is based on a multivariate analysis). The same multivariate analysis showed that Border Cave 5 possesses affinities with the male South African Negroids.

Both Border Cave 2 and 5 possess well-developed chins formed by a subalveolar depression and a pronounced mental protuberance. The posterior surface of the symphysial region follows an almost vertical course with both hominids. Neither has an alveolar plane. Border Cave 3, a mandible from a 3–6-month-old infant, cannot be metrically distinguished from Negroid infant mandibles of the same age [De Villiers, 1973]. The degree of development of the mental protuberance also falls within the observed range of recent southern African variation. Overall, the mandible exhibits affinities to both recent Negroid and San infant mandibles [De Villiers, 1973].

The mandibular fragments from Caves 1 and 1B at Klasies River Mouth (Fig. 16) are especially significant. A relatively well-preserved mandible (41815) was found in direct association with a MSA I (c 120 ky BP). The mandibular remains recovered with a MSA II industry (> 95 ky BP) include two body fragments, each with a portion of the chin region (13400, 21776); a symphysial fragment (14695); a body fragment with a portion of the ramus (16424); and several isolated teeth (some of which belong to 13400).[20]

The mandible 41815 is robustly built and features well-developed gonia and a large ramus. The dimensions of the body, however, fall within the modern spectrum. The robusticity index at M1 (49.0) is similar to that of Tuinplaas (48.6) and smaller than that of Mumbwa IV, 2 (57.7). Only the coronoid height, as seen in the published photographs, may possibly lie in the upper limits of, or even slightly exceed, the modern spectrum. The better-preserved portion of the right ramus suggests that the *incisura mandibulae* was not shallow. The chin region is well developed and possesses a prominent mental protuberance. There is no rearwards projecting alveolar plane on the posterior surface of the symphysis, but a modern formation instead. The tooth crowns are relatively small. Overall, there appears to be no reason to question assigning this mandible to a.m. *Homo sapiens* [see Singer and Wymer, 1982].

The four slightly more recent fragments exhibit a considerable amount of morphological variability. The cross-section of the symphysis of fragment

[20] The author was only able to examine a portion of these remains; most of the data discussed here were taken from Singer and Wymer [1982].

13400 greatly resembles that of 41815 (Fig. 16), although the *mentum osseum* and the mental protuberance are more weakly developed. Despite the fragmentary condition of the outer alveolar border, a slight subalveolar depression can be observed. A small but prominent mandibular torus is present on the basal portion of the body (P2/M1). The symphysial height (32.5) and the height at M1 (c 31 mm) fall within the lower regions of the final Upper Pleistocene variation and reveal similarities to the mandible 5 from Border Cave. The thickness of the body also falls in the modern spectrum. Some archaic affinities are revealed by the relatively slight prominence of the chin region and the approximately parallel upper and lower margins of the body.

Specimen 21776 possesses a somewhat more strongly developed chin. The robusticity of the body at the M1 lies within the a.m. spectrum and resembles those of final Upper Pleistocene specimens from Southern Africa. The author's examination of this fragment indicated that, in its overall size and form, 21776 also bears a great resemblance to the Fish Hoek mandible, which possibly dates from the early Holocene [Rightmire, 1978].

Only a few features of the small symphysial fragment known as 14695 can be ascertained with any certainty, as the upper surface of the bone shows signs of erosion. It is probable that the chin was at least slightly developed. The posterior surface of the symphysial region appears to have possessed no archaic features.

The last of these four mandibular fragments, 16424, is extremely small. The body height at M_1 is only about 20 mm, a value far below the observed range of final Upper Pleistocene variation (26.0–43.0). The thickness of 13.2 mm falls within the lower regions of this variation (12.0–18.0). The extant portion of the ramus also indicates that the ramus was low and gracile. Singer and Wymer [1982] see strong similarities to the Bushman mandibles in these various traits. The teeth are also unusually small (M2: M-D 9.3/B-L 8.8; M3: 8.3/8.0), lying at the lower end of the variation exhibited by recent African populations [see also Brabant, 1965; Bräuer and Mehlman, in press.].

Of the Klasies River Mouth mandibles, No. 41815 provides evidence that unequivocally a.m. forms were present at a very early date. From those 4 fragments which date to between c 95 and 120 ky BP, Nos. 21776 and 16424 should in all likelihood be considered as anatomically modern, while 13400 appears to possess some archaic traits and thus probably comes from a population living during the transitional period from late archaic to a.m. *Homo sapiens*. Number 14695 may also possibly belong to this transitional spectrum. The distinct gracility of 16424 could be taken as an indication that there was also a considerable sexual dimorphism during the early Upper Pleistocene. If we include the Border Cave mandibles in this spectrum, then it seems that a primarily a.m. morphotype was already present in southern Africa at that time.

In contrast to the south, eastern Africa has yielded few mandibular remains. The Omo 1 mandible (Fig. 6) includes most of the right ramus, an adjacent portion of the body, and the chin region. The ramus, in both its height and its breadth-height index, falls well within the Afalou/Taforalt spectrum. The same is true of the chin region, which possesses a well-developed mental trigone.

The poorly preserved body fragment from Diré-Dawa, whose measurements and robusticity fall just within the modern range of variation, nevertheless possesses certain archaic affinities. The external symphysial surface runs vertically. There is no indication of any chin development, although the fragmentary condition of this find limits an exact analysis [Vallois, 1951; Briggs, 1968]. The posterior surface features a *planum alveolare*. A marked inferior transverse torus may be observed. A postmolar space—a trait common to most Neandertals—is not present with Diré-Dawa. Despite these archaic, perhaps Neandertaloid affinities, Vallois [1951] did not definitively assign this find to *Homo sapiens neanderthalensis*. It is just as possible that this hominid is related to the spectrum of non-Neandertaloid forms of the archaic *Homo sapiens* of East Africa [see also Howell, 1978].

Most of the African mandibles come from the North African coastal region. The oldest and most archaic of these are the three Ternifine specimens. The mandibular fragments of Thomas Quarry I and Sidi Abderrahman possess clear affinities to the three specimens of "evolved *Homo erectus.*"

Ternifine III, a completely preserved mandible, is unusually large and massive. The broad and high rami show a great upward divergence and possess a narrow and deep *incisura mandibulae*. Number I also possesses a very massive body. Number II, with its smaller dimensions, is probably female. All three possess a symphysial region that follows a rearwards course; none of them possess a *mentum osseum*. However, both individuals II and III have a *trigonum basale*.[21] All three exhibit a strongly inclined *planum alveolare* and superior and inferior transverse tori [Arambourg, 1963].

The much younger Thomas 1 demi-mandible resembles these finds, especially Ternifine II. Most of the important features that Arambourg

---

[21] A triangular prominence in the symphysial axis, located on the lower margin of the bone.

[1963] considered as characteristic of the Ternifine specimens may also be found on Thomas 1. Its teeth are also quite large and the M3 is reduced. While the body heights and thicknesses fall within the a.m. ranges of variation, the robusticity-indices have values even greater than those of the Ternifine mandibles.

Two mandibular fragments have been recovered at Sidi Abderrahman: a right corpus fragment with an intact lower margin and three molars in position, and a smaller left body fragment with the P1 in position. Both, the absolute body height and thickness (M2) of the main fragment (34.5/17.0) and the index value fall within the Afalou/Taforalt spectrum. The upper and lower margins of the body are parallel. The symphysis was probably inclined slightly backwards. The molars are large and resemble those from Ternifine; the M3 is reduced.

Finally, the juvenile body fragment from Rabat also belongs to the group of northwest African Middle Pleistocene mandibular finds. We have already seen that the Rabat occipital possesses advanced, *sapiens*-like features. The same is true of the mandible. The robusticity of the body (M1/M2), however, is high (comparable to that of the Ternifine specimens and Thomas 1) yet we must remember that the Rabat mandible belongs to a c-17-year-old subadult [Arambourg and Biberson, 1956]. The anterior surface of the symphysis is almost vertical and exhibits an even rearward rounding in its lower portion. Although there is no clear *mentum osseum*, there are the beginnings of a mental trigone. A *planum alveolare* is present; it runs toward the superior transverse torus [Saban, 1975]. The dental features exhibit strong affinities to the Ternifine finds, especially No. II [Arambourg, 1963].

In sum, the Middle Pleistocene mandibles present a picture in which archaic characteristics are disappearing. To be more precise, they represent a period in which forms with the constellation of features typical for *Homo erectus* are being superseded by forms with mosaics of *erectus*-like and *sapiens*-like features [see Howell, 1978].

Aguirre and De Lumley [1977] have compared the Middle Pleistocene mandible finds from Europe, Africa, and China. They found that the northwest African finds have greater affinities to the southern European ante-Neandertals (Atapuerca, Arago, Montmaurin) than to the east Asian finds. The Choukoutien mandibles exhibit less robustness in several important features. Based on these findings, as well as other evidence, the authors considered it possible that there were close general ties between southern Europe and northwest Africa during this period.

Between the northwest African Middle Pleistocene finds and the last group of mandibles (group 4) lies a span of no less than 150 ky. The two mandibular fragments from Haua Fteah, which were examined by Mc-Burney et al [1953] and Tobias [1967], were found to possess strong affinities to the Tabūn/Shanidar Neandertaloids in both their metric and nonmetric features. Haua Fteah I (18–25 years old) and II (c 12–14 years old) both possess low and broad rami; the indices of the two finds (I, 70.2; II, 68.4) lie above the final Upper Pleistocene spectrum (44.6–65.6).[22] Both also possess a relatively shallow *incisura mandibulae*, although the values of their incisura indices (I, 27.0; II, 28.1) fall into the lower regions of the great variation shown by the Afalou/Taforalt sample (25.6–54.1).

Tobias [1967] found special similarities of ramus morphology between Haua Fteah II and the juvenile mandible from Ksar'Akil as well as several of the Skhūl finds. In contrast to the west Asian Neandertaloids, the Haua Fteah remains lack the retromolar space. All in all, these two mandibular fragments appear to indicate that there was also a Neandertaloid population spread throughout North Africa, a population, however, that had "at least some of its genetic roots in the earlier population of Africa" [Tobias, 1968].

The Témara mandible also possesses a number of archaic traits and exhibits certain affinities to Rabat. The body is very low and robust; some of the values of its indices even exceed those of Rabat. The fragmentary ramus has a broad and low appearance. The depth of the *incisura mandibulae*—expressed as the incisura index (30.5)—falls in the lower regions of the Afalou/Taforalt spectrum (see above). The symphysis is vertical. There is a slight subalveolar depression that leads to a slightly developed *mentum osseum*. The beginnings of a *trigonum mentale* are also present. The incision-gnathion line is less oblique than with Rabat. The *planum alveolare* is obviously reduced; the superior and inferior transverse tori are only very weakly developed. In its overall appearance, the symphysial region, thus, tends in the a.m. direction. The M3 is partially concealed by the anterior margin of the ramus. The crown area of the molars is very large, similar to that of the Middle Pleistocene hominids [see also Howell, 1978; Tobias, 1968]. Témara 1 may expand the spectrum of hominids that possess varying constellations of "archaic-African," Neandertaloid, and a.m. traits.

These tendencies also appear to have found expression in the robust child's mandible (8–9 years old) known as Jebel Irhoud 3. The anterior surface of the symphysis follows a vertical course. Nevertheless, the slight subalveolar depression, together with the mental tubercles and the slight

---

[22] However, many Holocene specimens from all parts of Africa possess values similar to those of the Haua Fteah finds.

protuberance, suggest the beginnings of a *mentum osseum*. Both the *planum alveolare* and the *torus transversus superior* are slightly developed. The teeth are unusually large, even greater than those of the European and west Asian Neandertals.

Hublin and Tillier [1981] have maintained that this hominid finds its greatest affinities in the hominids from the Near East, especially in Skhūl/Qafzeh. They found no support for Ennouchi's [1969] claim that this mandible is associated with the African Neandertaloids. Yet the present author's opinion is that this mandible cannot be assigned to *Homo sapiens sapiens* with any greater assurance than can the adult Jebel Irhoud hominids. Rather, the different archaic-plesiomorphic and partially Neandertaloid features indicate a spectrum like that previously described for Témara and Haua Fteah. This spectrum certainly suggests that there were connections with the Near East, but the role of the older northwest African group remains uncertain. The great temporal differences between these two groups and the small number of finds to compare leave this question largely open.

CONCLUSIONS AND PHYLOGENETIC CONSIDERATIONS

As a modification of the "grade" model advanced by Stringer et al [1979], I would like to propose the following classification and assignments for the African finds that have been analysed in this chapter:

Homo sapiens "grade 1" (early archaic *Homo sapiens*): Bodo, Broken Hill 1, Hopefield, Eyasi, Ndutu, Cave of Hearths, Rabat, Salé (?), Sidi Abderrahman (?), Thomas (?).

Homo sapiens "grade 2" (late archaic *Homo sapiens*): Omo 2, Laetoli 18, Florisbad, Diré-Dawa (?), Broken Hill 2 (?).

Homo sapiens "grade 2a" (Neandertaloid *Homo sapiens*[23]): Jebel Irhoud, Haua Fteah, Mugharet el'Aliya, Témara 1 (?).

Homo sapiens "grade 3" ([early] anatomically modern *Homo sapiens*): Klasies River, Border Cave, Omo 1, Kanjera, Mumba, Témara 2, Singa (?).

The representatives of 'early archaic *Homo sapiens*' may be geographically divided into a northwestern group, an eastern group, and a southern group. As the morphological analyses have shown, there are obvious similarities between the eastern and the southern groups. We must remember, however,

---

[23] Here, the term 'Neandertaloid' is used to classify specimens which resemble Neandertals. In contrast, the European Neandertals are designated as 'Neandertalids', ie, Neandertals *sensu stricto*.

that the dating uncertainties, the large geographic distances, and what appears to be a considerable degree of sexual dimorphism all counsel that any attempt to phylogenetically rank these archaic hominids be done with the greatest caution.

Nevertheless, the Hopefield, Broken Hill 1, and Bodo hominids provide the basis for the assumption that such a robust morphotype was spread from the southern tip of Africa to at least as far as the Horn. As is assumed for the European ante-Neandertals (cf p 330), Eyasi and perhaps Ndutu indicate that a great degree of sexual dimorphism may also have existed in Africa. However Broken Hill 1 and Eyasi 1 suggest that both sexes shared a general resemblance in terms of their overall morphology.

In comparison to *Homo erectus* (eg, Olduvai H.9), all of the representatives of early archaic *Homo sapiens* are considerably more evolved. Their cranial vaults are more expanded and, in many features, already possess affinities to a.m. *Homo sapiens*. In this respect, Ndutu exhibits an especially pronounced mosaic of archaic traits and traits that approach the a.m. spectrum. In spite of the small number of finds, it currently appears likely that the further development towards a.m. *Homo sapiens* proceeded via a morphotype similar to the Rhodesoid form. This, thus, appears to further confirm the view early held by Tobias [1956] that an autochthonous evolution, via Rhodesoid forms, had occurred in sub-Saharan Africa—although, we see this today within a completely different chronological framework.

The pronounced morphological similarities between the male representatives of early archaic *Homo sapiens* and various southern European ante-Neandertals (especially Petralona), allow us to conclude that there was a close phylogenetic relationship between the populations of Europe and Africa. It is much more likely that the European Middle Pleistocene hominids had an African ancestry than an Asiatic origin [Stringer et al, 1979; Stringer, 1981]. The mandibular morphology of the Northwest group also suggests that there were closer contacts with Europe [Aguirre and De Lumley, 1977].

While Thomas 1 and Sidi Abderrahman show strong affinities to the Ternifine mandibles (developed *Homo erectus*), Salé and, to a greater extent, Rabat, reveal certain affinities to the a.m. spectrum. The few cranial remains and the unusual morphology of the Salé find[24] are not sufficient to determine whether, and if yes, to what degree, there were connections between the developments which took place in the Northwest region of the African continent and those which occurred south of the Sahara.

---

[24] See the footnote on p 364.

The various analyses indicate that the hominids Laetoli 18, Omo 2, Florisbad, and (probably) Diré-Dawa, all of which appear to date to the final Middle or early Upper Pleistocene, possess affinities to both early archaic and a.m. *Homo sapiens*. It is possible that the robust maxillary fragment Broken Hill 2, with its canine fossa, may also belong to this group. These hominids exhibit various individual mosaics of archaic-plesiomorphic, anatomically modern, and intermediate traits, whereby the overall affinities to the a.m. morphotype are much more pronounced than is the case with the early archaic forms. The hominids that have been classified above as late archaic *Homo sapiens* ("grade 2") include both the less archaic *Homo sapiens*, which evolved from parts of the early archaic spectrum, and those from the spectrum resulting from the hybridization between the early a.m. *Homo sapiens* ("grade 3") and the late archaic populations, which is assumed to have occurred during the replacement phase (see below). It does not seem possible to distinguish between these two sections of the late archaic spectrum. *"Homo sapiens* grade 3" includes those hominids that are either completely anatomically modern or bear only a few archaic reminiscences.

The comparative analyses discussed in this chapter, when seen against the backdrop of the datings currently considered likely, suggest that early a.m. *Homo sapiens* was widely spread throughout Africa as far back as the early Upper Pleistocene. During this period, modern humans were almost certainly present in southern Africa (Klasies River, Border Cave), and most probably in eastern Africa as well (Singa, Kanjera, Mumba[25]). Omo 1 may even provide evidence that a.m. populations were present as early as the final Middle Pleistocene.

Thus, the late archaic finds that we have assigned to "grade 2" were probably more or less contemporaneous with a number of a.m. specimens from eastern and southern Africa. It is precisely for this reason that we must assume that a broad morphological spectrum existed in these regions during the late Middle and early Upper Pleistocene. The presently available representatives of grades 2 and 3 appear to confirm this assumption. The morphological spectrum suggested by the two Omo hominids is especially remarkable. While Omo 1 exhibits only a very few archaic reminiscences, Omo 2 (which was possibly Omo 1's contemporary) clearly possesses archaic (probably "Rhodesoid") traits [see also Rightmire, 1976]. Laetoli 18, in its

---

[25] Both the metric dimensions and the crown pattern of the molar remains from the Mumba Rock Shelter (Tanzania) fall within the a.m. spectrum [Bräuer and Mehlman, in press].

overall morphology, strongly resembles Eyasi 1. The supraorbital morphology of Florisbad also reveals obvious reminiscences to its ancestral Rhodesoid spectrum. Even the cranium of Kanjera 1, which is usually viewed as being very modern, exhibits affinities to Laetoli and Eyasi 1 when we consider the curvature of its parietals and occipital. Those of the mandibular remains discovered at Klasies River Mouth that have been dated to the early Upper Pleistocene also possess a number of archaic affinities, while the contemporaneous frontal fragment from the same site falls within the range of recent variation.

In summary, there is a relatively large body of evidence from eastern and southern Africa that supports the conclusion that a.m. *Homo sapiens* originated in those areas during the late middle and/or early Upper Pleistocene. It is difficult to precisely evaluate the extent to which these developments proceeded independently in eastern and southern Africa, but the fact that the observed changes most probably occurred at roughly the same time, taken together with certain morphological affinities between finds from both regions, tends to support the assumption that there were connections during this phase. Yet, as has been shown, the early a.m. humans of eastern and southern Africa also exhibit specific similarities to the final Upper Pleistocene finds from their respective regions [see also De Villiers, 1973; Rightmire, 1979; De Villiers and Fatti, 1982]. Thus, autochthonous regional differentiations probably also played an important role in the changes that took place during the Upper Pleistocene.

The developments that led to the appearance of a.m. *Homo sapiens* in southern and eastern Africa appear to have run their course by the time of the early Upper Pleistocene—some 100 to 70 ky ago. At that same time, evolution in Europe and western Asia was just completing the steps that resulted, respectively, in the classical and late Neandertals.

It appears likely that the North African Neandertaloids (*Homo sapiens* "grade 2a") developed out of the spectrum of the pre-Neandertals and Near Eastern Neandertals, an assumption that has received support from a number of multivariate analyses [eg, Stringer, 1974, 1978]. Moreover, it seems apparent that gene flow between the populations of southwest Asia and those of North Africa also existed during the Würm. Other uncertainties remain regarding the extent to which local roots reach back into the Middle Pleistocene (Salé, Thomas, Rabat) and the roles these might have played in the development of the specific local morphologies. These are due to the broad chronological gap (based on the lack of finds) between the two fossil groups. The morphological analyses, in any case, indicate that local factors may very well have played a role.

The great amount of morphological heterogeneity among archaic and a.m. *Homo sapiens* and the postulated hybridization between differentially evolved groups necessitates a few comments on the question about the most plausible evolutionary model.

Gould and Eldredge [1977], as well as Stanley [1979, 1981], have maintained that a model of punctuated evolution and species formation (based primarily on observations of invertebrates) can also be applied to hominid evolution. They dispute whether gradual evolution occurred within the hominid taxa. Cronin et al [1981] have taken a clearly contradictory position as a result of their analysis of the fossil hominids dating from roughly the last 4 million years. They found that the most reasonable model for explaining hominid evolution is one of phyletic gradualism with a varying rate of evolution. There are many examples of this that may be cited, including the progressive cerebralization that occurred in the species *Homo erectus*, or the line that led from the ante-Neandertals, via the pre-Neandertals, to the late Neandertals. Furthermore, Gould and Eldredge [1977] oversimplify matters in assuming that early archaic *Homo sapiens* corresponds to a.m. *Homo sapiens*. The arguments against this have been laid out in this chapter.

If we consider the process of evolutionary change on a more fundamental level, then it seems that the determination of what appears to have been a punctuated or sporadic origin of species does not rule out the assumption that gradual change took place, as population genetics postulates. Instead of considering this punctuated model as a tool for precisely comprehending the basic course of evolution, we should rather see the hypothesis of "punctuated equilibrium" as a less-detailed way of describing portions of the fossil record.

Evolutionary biology provides the framework for assuming that new developments almost always occur on a microevolutionary basis within subpopulations that have become isolated or are somewhat marginal. Thus, such developments only occur in a portion of the spectrum of variation exhibited by an entire population. Depending on the duration and speed of evolution, the isolated subpopulations will become more or less differentiated from the original parent population. If the isolation ends, then the subpopulation and the parent population will, in all likelihood, interbreed. In hominid evolution, these events normally might have taken place on the subspecies or population level. One of the implications of such a course of evolutionary change is that, at any particular time, a species was probably composed of a varying number of allopatric populations that were isolated from one another by differentially permeable barriers to gene flow [Pilbeam, 1975].

When there are not great morphological differences between a subpopulation and its parent population, then the fossil record will tend to give the impression of a gradual continuum. In contrast, the concurrence of more greatly differing populations will result in what appears to be breaks, or jumps. When we consider phylogenetic change in this way, then we see that both models actually represent an artificial polarization. It remains uncontested that the rate of evolution can vary in accordance with different conditions of selection and is also dependent on the size of the population. This, moreover, also provides us with the basis for accelerated development.

However the opposite rule, namely that substantial changes and especially speciation can only occur during periods of accelerated evolution, cannot be derived from these axioms of evolution. There is simply no biological basis for doing so. Evolutionary changes are the result of an interplay between natural selection, gene flow, migration, and drift; any realistic reconstruction of human phylogeny must therefore include a complicated network reflecting these factors [Bilsborough, 1976]. Moreover, hominid evolution is fundamentally connected with the evolution of culture.

According to the fossil record of the African Middle and Upper Pleistocene, the transition from the developed *Homo erectus* to the archaic *Homo sapiens* does not appear to have occurred quickly, but rather relatively slowly and continuously. For this reason, it is hardly possible to draw a clear taxonomic boundary between these two species (cf p 349).

The hominid finds from the upper Middle and early Upper Pleistocene reveal a considerably greater amount of morphological heterogeneity, which is apparently related to the appearance of the modern morphotype. This morphological change—"the modernization" of the cranium from late archaic ancestors—appears to have taken place at a relatively faster tempo than did, for example, the transition from *Homo erectus* to archaic *Homo sapiens* [see also Rightmire, 1981; Delson, 1981].

Although the dates for the finds from the turn of the Upper Pleistocene are quite inexact, the evidence that we do possess seems to indicate that a.m. humans originated with a comparatively accelerated rate of evolution.

Thus, the role Africa has played in the discussion about the origins of the early moderns has undergone decisive change. Today, Africa appears to be the place from which we possess the earliest and best documentation of the origins of a.m. *Homo sapiens*. At the present time, his debut in Europe and Asia remains unclear. Could it be that the key to this problem will also be found in Africa?

In the first section of this chapter, the main hypotheses dealing with the origins of Cro-Magnon man in Europe were subjected to a critical discussion. It is my opinion that the substantially increased number of Middle and Upper Pleistocene hominid finds currently available indicates that there is little likelihood that Homo sapiens sapiens originated in Europe. Questions about the provenance of a.m. humans also ultimately arise in connection with the heterogeneous complex of Near Eastern finds (Skhūl/Qafzeh).

The initial dating revisions for the African Stone Age led Beaumont and Boshier in 1972 to consider Africa as a potential cradle of a.m. humans. De Villiers' [1973, 1976] subsequent investigations of the Border Cave hominids provided additional support for the possibility of an early, South African origin of a.m. humans [cf Beaumont et al, 1978]. Following this, the great age of the Border Cave hominids led Protsch [1975] to conclude that early a.m. humans had, in the span of a few thousand years (between 40 and 35 ky BP), expanded from southern Africa into all of the continents. In his model, the Rhodesoids represent a "recent" evolutionary sideline which developed parallel to the line of a.m. humans and that was still spread throughout southern and eastern Africa as late as 35–40 ky ago. This line eventually became extinct.

While it currently appears very probable that a.m. Homo sapiens was present in southern Africa during the early Upper Pleistocene, the present study provides evidence of a much more complex picture of the origins of a.m. humans in the various regions of Africa as well as of the possibility of a northward expansion. Moreover, no firm evidence for a swift and global expansion of a "Border Cave-type" has emerged. Many research findings from other areas of the world also contradict the idea that the archaic populations of these regions were replaced in such a radical fashion [eg, Thorne and Wolpoff, 1981; Bräuer, in press]. In addition, the Rhodesoids scarcely seem to represent a recent developmental sideline of basic a.m. humans.

At the "1er Congrès International de Paléontologie Humaine" in Nice, I introduced a phylogenetic model designated as the "Afro-European sapiens hypothesis" [Bräuer, 1982b]. This model (Fig. 17) is primarily based on a series of morphological analyses, described both here and elsewhere [Bräuer, unpublished manuscript], and the new African chronological framework. The main elements of this scheme concern the evolution from archaic to a.m. Homo sapiens in eastern and southern Africa. A detailed discussion of these aspects of the model has already been given. Now we shall look more closely at that portion of the hypothesis which is related to the postulated northward expansion.

As the morphological analyses have shown, there are especially striking similarities between the final Pleistocene Afalou/Taforalt sample and early

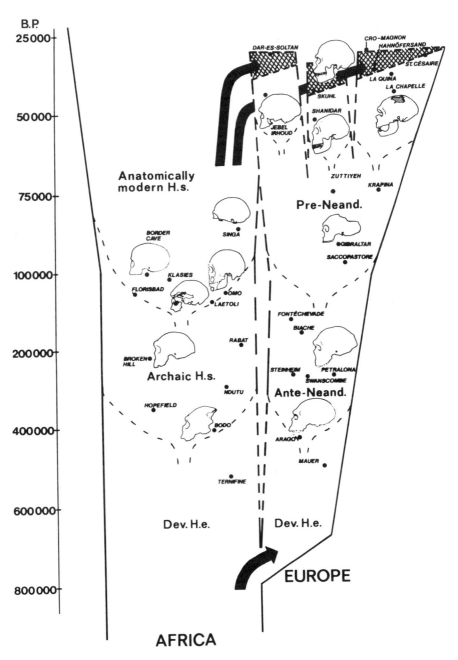

Fig. 17. A new phylogenetic model of hominid evolution in Africa and Europe during the Middle and Upper Pleistocene (the "Afro-European *sapiens* hypothesis").

a.m. hominids from East Africa, especially Omo 1. Such close connections are not found between the a.m. humans from Northern or Southern Africa. Furthermore, Stringer [1978] was able to show that there are strong affinities of cranial morphology between Omo 1, the Qafzeh/Skhūl hominids, and the late Paleolithic Europeans [see also Day and Stringer, 1982]. The current fossil record indicates that the expansion of a.m. *Homo sapiens* into the north had apparently reached at least as far as Sudan (Singa) by the early Upper Pleistocene (cf Fig. 3).

Besides the evidence against a Near Eastern local evolution that was discussed at the beginning of this chapter, it also seems very unlikely that such an evolutionary process from a Neandertaloid to a substantially Cro-Magnoid morphology occurred in this area between c 50 to 40 ky BP, while, somewhat further to the south, a very similar a.m. cranial form had already developed, probably more than 50 ky earlier, out of the non-Neandertaloid ancestors. Thus, the assumption that a.m. humans spread from the East African area into the North also appears to conform to premises of evolutionary theory.

The replacement of the North African Neandertaloids also appears to have occurred in the course of this postulated expansion. The more recent human remains from the Upper Aterian (c 30 ky BP) of Morocco may possibly represent evidence of this replacement. While the cranium of Dar-es-Soltan 2 possesses certain affinities to the Jebel Irhoud hominids (although it is practically anatomically modern), the occipital from Témara has as modern an appearance as those of the Afalou/Taforalt hominids [Ferembach, 1976a,b; Thoma, 1978].

Finally, a.m. humans seem to have "spread" further, via the Near East, into Europe, increasingly absorbing and replacing the Neandertals that were living there.[26] This period of replacement probably lasted for several thousand years, and it can also be assumed that various degrees of hybridization between Neandertalid and modern populations occurred during this time (cf p 332). This would at least partially explain the numerous Upper Paleolithic finds that appear to feature Neandertalid reminiscences [Brace, 1964; Smith and Ranyard, 1980; Smith, 1982]. Furthermore, this would mean that Cro-Magnon people were not—contrary to the traditional, widespread assumption—of European origin and also did not have their roots among marginal Neandertaloids, but rather go back to the same line from which the modern African populations also descended.

---

[26] Cf footnote p 334.

It is unclear why Thoma [1978], who considers such a possibility to be likely, attached less importance to Omo 1. Rather he saw the more archaic Omo 2 hominid as possibly representing a non-Neandertaloid ancestral form, out of which the Cro-Magnoids developed in two independent lines. One of these was held to have led to the European "Cro-Magnoids," the other branch to the African "Mechtoids." According to Thoma [1973], the early a.m. humans of the Near East and western Asia are primarily of a "Europoid" type, yet, they also exhibit various features (alveolar prognathism, prominent cheek bones, flat nasal root/bones) with "Negroid" tendencies [McCown and Keith, 1939]. This "Europoid-Negroid spectrum" seems, however, to be rather generally characteristic of early a.m. *Homo sapiens* [cf Schwidetzky, 1970, 1979].

One of the tasks of future research is to further probe the hypothesis of such an Afro-European connection. Comparative analyses of the Upper Paleolithic finds from Eurasia and Africa would be especially valuable in further testing this question.

Here, only the Staroselye infant shall be mentioned in this context. This hominid, from the Crimean Peninsula (Black Sea), dates from the late Mousterian and may provide evidence of a possible connection between the a.m. *Homo sapiens* of the Near East and the a.m. humans of Western Russia. Roginski [1954] considered this infant skull (6–7 months), with its high, dolichocranial vault, low face, angular orbits, and alveolar prognathism, to have its strongest resemblances with the Skhūl hominids, especially the infant skull Skhūl I. A reexamination of this find led Alexeyev [1976] to conclude that Staroselye provides additional evidence to support the assumption that contacts between the hominids of eastern Europe and populations from more southern regions existed as early as the late Mousterian. Alexeyev also sees evidence of connections with southern regions in the early Upper Paleolithic skull from Markina Gora (USSR).

The postulation of gene flow and population migrations from eastern Africa into the North leads to the question of paleoclimatology. Specifically, to what degree does the present paleoclimatic knowledge indicate that such an expansion was possible and probable? In addition, we must ask about archaeological evidence of potential routes for such a spreading, especially along the Nile valley. The time span in question lies between c 70 and 40 ky BP.

Boaz et al [1982] have recently discussed the association between the current knowledge about paleoclimatic conditions during the Upper Pleistocene and hominid evolution. On the basis of an oxygen isotope curve the authors assumed that the Neandertals apparently relatively quickly made room for a.m. *Homo sapiens* as the climate became colder some 34 ky ago. They attributed the Neandertal replacement to migrations and "swamping" by the numerically superior populations of a.m. humans, which were coming out of western Asia and North Africa. A wide land bridge along the Dardanelles arose during isotope stage 2, when the sea level lay more than 80 m below that of today.

Boaz et al [1982] saw a repeated, increasing desertification and the shortage of resources it caused as the main reasons for these migrations into the north and west. This desertification probably resulted in population pressures among the hunter-gatherer populations in the stricken regions of western Asia and what is today the Sahara. This lead to an expansion into the colder, but more productive steppe and tundra regions of western Asia and Europe.

The expansion of the deserts did not reach its peak until c 18 to 20 ky BP [Sarnthein, 1978]. Large portions of the Sahara were probably uninhabitable during this period [Clark, 1979b]. Yet in the time preceding this period, ie, since the beginning of the Last Glacial, there were alternate phases of relative aridity and humidity in the Sahara region. In particular, a cooler and more humid climate prevailed in the Sahara during the early phases of the Last Glacial (stages 2 to 4) until about 40 ky BP, and again between c 32 and 24 ky BP [Clark, 1980]. Although the period most relevant for this postulated expansion lies above the range of the conventional C-14 method, there can be little doubt that the area between eastern Africa and the Mediterranean was passable, at least for longer spans of time [Sarnthein, personal communication]. Migrations and expansions of populations of greater magnitude can be generally considered to be quite likely, contingent upon the changing climatic conditions during the glacial.

The strong general affinities between the Mousterian of the Maghreb, the eastern Sahara, and the Nile Valley is seen by Clark [1980] as evidence of movements from east to west or west to east, eg, from Maghreb to the Egyptian Nile, via the West Sahara, or vice versa [see also Crew, 1975]. The Nile Valley may be considered as a likely area through which this expansion from East Africa into the north proceeded. In spite of the increasing desertification that occurred in northern Africa and the Sahara during the Last Glacial, the Nile Valley appears to have been the only region of North Africa that was more or less continuously inhabited since the Middle Pleistocene. "When large parts of the Sahara were too arid to sustain human settlements, the Nile Valley was never sufficiently inhospitable to prevent man from being able to occupy it successfully" [Clark, 1980]. Although there is still no evidence of Middle Paleolithic settlements from the more northern regions of the Nile, the Mousterian is well represented between Luxor and Dongola (Sudan) [Clark, 1980; Wendorf and Schild, 1976].

If we summarize this short digression on both the paleoclimatic conditions prevalent during the first half of the Last Glacial and the archaeological picture [cf also p 336; Beaumont and Vogel, 1972; Beaumont et al, 1978], then it appears that the hypothesis advanced in this chapter—although primarily based upon morphology—is also possible in the light of these other considerations. An African origin of a.m. humans would, as R. Leakey [1981b] so carefully put it, fit certain facts very well. Moreover, such a model also possesses the greatest likelihood when all of the currently available facts are considered and the alternatives weighed. One question remains: how can the developments which occurred in Eastern Asia be brought into relation to the "Afro-European *sapiens* hypothesis," which only deals with the western part of the Old World [cf Bräuer, in press]. Did modern humans originate monocentrically or polycentrically?

ACKNOWLEDGMENTS

I am exceedingly grateful to the following individuals for their kind permission to study the fossil hominid remains under their care and for the various forms of assistance they have provided during the course of my research: C.K. Brain, Y. Coppens, M.H. Day, H. De Lumley, E. du Pisani, H. de Villiers, K.A. Eisa, M. Fredman, J.-L. Heim, Q.B. Hendy, A.R. Hughes, E.N. Keen, R. Kruszynski, M.D. Leakey, R.E. Leakey, C. Magori, F. Masao, E. Mbua, J.K. Melentis, A. Morris, A.T. Nkini, H. Oberholzer, P. Schröter, B. Senut, C.B. Stringer, P.V. Tobias, E. Vrba, J. Wallace, and T.D. White.

I would also like to thank the persons listed below for information on the dating of particular hominids or for providing photographs or other information on the hominids: P. Beaumont, J.D. Clark, G. Conroy, J.-J. Hublin, H. Müller-Beck, R. Protsch, M. Sarnthein, R. Singer, G. Unrath, H. Ziegert.

For their assistance with graphics and photography, I am grateful to T. Görnemann, S. Müller, and C. Stegemann. I am also grateful to U. Lüttmann, who assisted with the computations, and to J. Baker, who translated the manuscript into English.

To R. Knussmann, F.H. Smith, and C.B. Stringer, I express my sincere appreciation for critically commenting on an earlier draft of this paper.

Finally, I am most grateful to the Deutsche Forschungsgemeinschaft for their past and continuing support of my research.

LITERATURE CITED

Aguirre E, de Lumley M-A (1977): Fossil men from Atapuerca, Spain: Their bearing on human evolution in the Middle Pleistocene. J Hum Evol 6:681–688.

Alexeyev VP (1976): Position of the Staroselye find in the hominid system. J Hum Evol 5:413–421.

Arambourg C (1963): Le gisement de Ternifine. Arch Inst Paleont Hum 32:1–190.

Arambourg C, Biberson P (1956): The fossil human remains from the Paleolithic site of Sidi Abderrahman (Morocco). Am J Phys Anthropol 14:467–490.

Arambourg C, Hofstetter R (1963): Le gisement de Ternifine, pt. 1: Historique et géologie. Arch Inst Paleont Hum 32:9–36.

Barendsen GW, Deevey ES, Gralenski LJ (1957): Yale natural radiocarbon measurements III. Science 126:908.

Bate DMA (1951): The mammals from Singa and Abu Hugar. In Arkell AJ, Bate DMA, Wells LH, Lacaille AD (eds): "The Pleistocene Fauna of Two Blue Nile Sites." Fossil Mammals of Africa. London: Brit Mus Nat Hist, Vol 2, pp 1–28.

Beaumont PB (1973): Border Cave—a Progress Report. S Afr J Sci 69:41–46.

Beaumont PB (1979): Comment to: Rightmire GP (1979): Implications of Border cave skeletal remains for later Pleistocene Human Evolution. Curr Anthropol 20:26–27.

Beaumont PB (1980): On the age of Border Cave Hominids 1–5. Palaeontol Afr 23:21–33.

Beaumont PB, Boshier A (1972): Some comments on recent findings at Border Cave, Northern Natal. S Afr J Sci 68:22–24.

Beaumont PB, De Villiers H, Vogel JC (1978): Modern man in sub-Saharan Africa prior to 49000 years BP: A review and evaluation with particular reference to Border Cave. S Afr J Sci 74:409–419.

Beaumont PB, Vogel JC (1972): On a new radiocarbon chronology for Africa south of the Equator. Afr Stud 31:65–89, 155–182.

Behm-Blancke G (1959-60): Altsteinzeitliche Rastplätze im Travertingebiet von Taubach, Weimar-Ehringsdorf. Alt Thüringen 4:1–245.

Biberson P (1956): Le Gisement de l'Atlanthrope de Sidi Abderrahman (Casablanca). Bull Archeol Maroc 1:37–92.

Biberson P (1961): Le Cadre Paléogéographique de la Préhistoire du Maroc atlantique. Publ Serv Antiq Maroc 16:116.

Biberson P (1963): Human evolution in Morocco in the framework of the paleoclimatic variations of the Atlantic Pleistocene. In Howell FC, Bourlière F (eds): "African Ecology and Human Evolution." Chicago: Viking Fund Publ. Vol 36, Anthropology, pp 417–447.

Biberson P (1964): La place des Hommes du paléolithique marocain dans la chronologie du Pleistocène atlantique. l'Anthropologie 68:475–526.

Biegert J (1957): Der Formwandel des Primatenschädels. Morphol Jb 98:77–199.

Bilsborough A (1976): Patterns of Evolution in Middle Pleistocene Hominids. J. Hum Evol 5:423–439.

Bishop WW, Clark JD (eds) (1967): "Background to Evolution in Africa." Chicago: University of Chicago Press.

Boaz NT, Ninkovich D, Rossignol-Strick M (1982): Paleoclimatic setting for Homo sapiens neanderthalensis. Naturwissenschaften 69:29–33.

Boswell PGH (1935): Human remains from Kaman and Kanjera, Kenya Colony. Nature 135:371.

Boule M (1913): "L'Homme Fossile de la Chapelle-aux-Saints." Ann Paléont 8:1–70.

Brabant H (1965): Observations sur la denture des Pygmées de l'Afrique Centrale. Bull Groupe Eur Rech Sci Stomatol Odontol 8:27–49.

Brace CL (1964): The fate of the "Classic" Neanderthals: A consideration of hominid catastrophism. Curr Anthropol 5:3–43.

Brace CL (1979): "The Stages of Human Evolution, 2nd ed." Englewood Cliffs, NJ: Prentice-Hall.

Bräuer G (1980a): Die morphologischen Affinitäten des jungpleistozänen Stirnbeines aus dem Elbmündungsgebiet bei Hahnöfersand. Z Morphol Anthropol 71:1–42.

Bräuer G (1980b): Nouvelles analyses comparatives du frontal Pleistocène Supérieur de Hahnöfersand, Allemagne du Nord. l'Anthropologie 84:71–80.

Bräuer G (1980c): Human skeletal remains from Mumba Rock Shelter, Northern Tanzania. Am J Phys Anthropol 52:71–84.

Bräuer G (1981a): New evidence on the transitional period between Neanderthals and modern man. J Hum Evol 10:467–474.

Bräuer G (1981b): Neuere Hypothesen zur Hominiden evolution während des späten mittleren und oberen Pleistozäns in Afrika. Schriften Uŕ Frühgeschichte Akad Wissenschaft DDR 41 (in press).

Bräuer G (1981c): Current problems and research on the origin of Homo sapiens in Africa. Human biologia Budapestinensis. 9:69–78.

Bräuer G (1982a): A comment on the controversy "Allez Neanderthal" between M.H. Wolpoff, A. ApSimon, C.B. Stringer, R.G. Kurszynski, and R.M. Jacobi in Nature 289 (1981). J Hum Evol 11:439–440.

Bräuer G (1982b): Early anatomically modern man in Africa and the replacement of the Mediterranean and European Neanderthals. I. Cong Int Paleont Hum Nice.

Bräuer G (1981–83): Der Stirnbeinfund von Hahnöfersand—und einige Aspekte zur Neandertaler-Problematik. Hammaburg 6 (in press).

Bräuer G (in press): The "Afro-European sapiens hypothesis," and hominid evolution in East Asia during the late Middle and Upper Pleistocene. In Andrews P, Franzen J (eds): The Early Evolution of Man, with special emphasis on Southeast Asia and Africa. A Memorial Symposium in honor of GHR von Koenigswald. Frankfurt.

Bräuer G, Mehlman MJ (in press): Hominid molars from a Middle Stone Age level at Mumba Rock Shelter, Tanzania. Am J Phys Anthropol.

Breitinger E (1955): Das Schädelfragment von Swanscombe und das 'Präsapiens-Problem'. Mitt Anthropol Ges Wien 84/85:1–45.

Breuil H, Teilhard de Chardin P, Wernert P (1951): Le Paléolithique du Harrar. l'Anthropologie 55:219–230.

Briggs LC (1968): Hominid Evolution in Northwest Africa and the Question of the North African "Neanderthaloids." Am J Phys Anthropol 29:377–386.

Brose DS, Wolpoff MH (1971): Early upper palaeolithic man and late middle palaeolithic tools. Am Anthropol 73:1156–1194.

Butzer KW (1978): Sediment stratigraphy of the Middle Stone Age sequences at Klasies River Mouth, Tzitzikama Coast, South Africa. S Afr Archaeol Bull 33:141–151.

Butzer KW (1979): Comment to Rightmire GP (1979). Curr Anthropol 20:28.

Butzer KW (1982): Geomorphology and sediment stratigraphy. In Singer R, Wymer J (eds): "The Middle Stone Age at Klasies River Mouth in South Africa." Chicago: University of Chicago Press, pp 33–42.

Butzer KW, Beaumont PB, Vogel JC (1978): Lithostratigraphy of Border Cave, Kwa Zulu, South Africa: A Middle Stone Age sequence beginning ca. 195.000 B.P. J Archaeol Sci 5:317–341.

Butzer KW, Brown FH, Thurber DL (1969): Horizontal sediments of the lower Omo Valley: Kibish formation. Quaternaria 11:15–29.

Clark JD (1959): Carbon-14 chronology in Africa South of the Sahara. Actes IVeme Cong Panafr Préhist Étude Quat, pp 303–311.

Clark JD (1970): "The Prehistory of Africa." New York: Praeger.

Clark JD (1975): Africa in Prehistory: Peripheral or Paramount? Man 10:175–198.

Clark JD (1979a): Radiocarbon dating and African archaeology. In Berger R, Suess HE (eds): "Radiocarbon Dating." Berkeley: University of California Press, pp 7–31.

Clark JD (1979b): Early human occupation of Savanna environments in Africa with particular reference to the "Sudanic zone." Wenner Gren Conference, 1978.

Clark JD (1980): Human populations and cultural adaptations in the Sahara and Nile during prehistoric times. In Williams MAJ, Faure H (eds): "The Sahara and the Nile." Rotterdam: Balkema, pp 527–582.

Clark JD (1981): Prehistory in southern Africa. In Ki-Zebro J (ed): "General History of Africa. I. Methodology and African Prehistory." Berkeley: University of California Press, pp 487–529.

Clarke R (1976): New cranium of Homo erectus from Lake Ndutu, Tanzania. Nature 262:485–487.

Conroy GC (1980): New evidence of Middle Pleistocene hominids from the Afar desert, Ethiopia. Anthropos (Athens) 7:96–107.

Conroy GC, Jolly CJ, Cramer D, Kalb JE (1978): Newly discovered fossil hominid skull from the Afar depression, Ethiopia. Nature 276:67–70.

Cooke HBS (1958): Observations relating to Quaternary Environments in East and South Africa. Geol Soc S Afr 60:73 pp.

Cooke HBS, Malan BD, Wells LH (1945): Fossil man in the Lebombo-Mountains, South Africa: The 'Border Cave', Ingwavuma District, Zululand. Man 45:6–13.

Coon CS (1962): "The Origin of Races." New York: Knopf.

Corruccini RS (1975): Metrical analysis of Fontéchevade II. Am J Phys Anthropol 42:95–97.

Crew HL (1975): An evaluation of the relationship between the Mousterian complexes of the eastern Mediterranean: A technological perspective. In Wendorf F, Marks A (eds): "Problems in Prehistory, North Africa and the Levant." Dallas: Southern Methodist University Press, pp 427–438.

Cronin JE, Boaz NT, Stringer CB, Rak Y (1981): Tempo and mode in hominid evolution. Nature 292:113–122.

Cunningham DJ (1908): The evolution of the eyebrow region of the forehead, with special reference to the excessive supra-orbital development in the Neanderthal race. Trans R Soc, Edinburgh 46:283–311.

Dart RA (1948): The first human mandible from the Cave of Hearths, Makapansgat. S Afr Archaeol Bull 3:96–98.

Dart RA (1955): Extinct and extant human mandibles. S Afr J Sci 51:258–267.

Day MH (1969): Early Homo sapiens remains from the Omo River Region of South-west Ethiopia. Nature 222:1135–1138.

Day MH (1977): "Guide to Fossil Man, 3rd ed." London: Cassell.

Day MH, Leakey MD, Magori C (1980): A new hominid fossil skull (L.H. 18), from the Ngaloba Beds, Laetoli, northern Tanzania. Nature 284:55–56.

Day MH, Stringer CB (1982): A reconsideration of the Omo Kibish remains and the erectus-sapiens transition. I. Cong Int Paleont Hum Nice, pp 814–846.

Deacon J (1974): Patterning in the radiocarbon dates for the Wilton/Smithfield complex in Southern Africa. S Afr Archaeol Bull 29:3–18.

Deacon J (1978): Changing patterns in the Late Pleistocene/Early Holocene Prehistory of Southern Africa as seen from the Nelson Bay Cave Stone Artifact Sequence. Q Res 10:84–111.

Debenath A (1975): Découverte de restes humains probablement atériens à Dar-es-Soltane (Maroc). CR Acad Sci [D] (Paris) 281:875–876.

Delson E (1981): Paleoanthropology: Pliocene and Pleistocene human evolution. Paleobiology 7:298–305.

De Lumley M-A (1978): Les Anténeandertaliens. In Bordes F (ed): "Les Origines Humaines et les Époques de l'Intelligence." Paris: Masson, pp 159–182.

De Lumley M-A (1981): L'homme de Tautavel. Critères morphologiques et stade évolutif. In de Lumley H, Labeyrie J (eds): "Datations Absolues et Analyses Isotopiques en Préhistoire. Methodes et Limites." Coll Intern, Tautavel: CNRS pp 259–264.

De Villiers H (1973): Human skeletal remains from Border Cave, Ingwawuma District, KwaZulu, South Africa. Ann Transvaal Mus 28:229–256.

De Villiers H (1976): A second adult human mandible from Border Cave, Ingwawuma District, KwaZulu, South Africa. S Afr J Sci 72:212–215.

De Villiers H, Fatti LP (1982): The Antiquity of the Negro. S Afr J Sci 78:321–332.

Drennan MR (1953): A preliminary note on the Saldanha skull. S Afr J Sci 50:7–11.

Drennan MR (1955): The special features and status of the Saldanha skull. Am J Phys Anthropol 13:625–634.

Drennan MR, Singer R (1955): A mandibular fragment, probably of the Saldanha skull. Nature 175:364.

Dreyer TF (1953): The Origin and Chronology of the Fauresmith Culture. Res Nat Mus Bloemfontein I:56–76.

Ennouchi E (1962): Un Néandertalien: L'homme du Jebel Irhoud (Maroc). l'Anthropologie 66:279–299.

Ennouchi E (1963): Les Néanderthaliens du Jebel Irhoud (Maroc). CR Acad Sci [D] (Paris) 256:2459–2460.

Ennouchi E (1968): Le deuxième crâne de l'homme d'Irhoud. Ann Paleont (Vert) 54:117–128.

Ennouchi E (1969): Présence d'un enfant néanderthalien au Jebel Irhoud (Maroc). Ann d Paleont (Vert) 55:251–265.

Ennouchi E (1972): Nouvelle découverte d'un archanthropien au Maroc. CR Acad Sci [D] (Paris) 274:3088–3090.

Farrand WR (1979): Chronology and palaeoenvironment of Levantine prehistoric sites as seen from sediment studies. J Archaeol Sci 6:369–392.

Ferembach D (1976a): Les Restes humains atériens de Témara (Campagne 1975). Bull Mem Soc Anthropol Paris 13:175–180.

Ferembach D (1976b): Les Restes humains de la grotte de Dar-es-Soltane 2 (Maroc), Campagne 1975. Bull Mem Soc Anthrop Paris 3(Ser 13):183–193.

Flint RF (1959): Pleistocene Climates in Eastern and Southern Africa. Bull Geol Soc Am 70:343–374.

Frayer DW (1978): Evolution of the dentition in Upper Paleolithic and Mesolithic Europe. Univ Kansas Publ Anthropol 10:1–201.

Galloway A (1937a): Man in Africa in the light of recent discoveries. S Afr J Sci 23:89–120.

Galloway A (1937b): The nature and status of the Florisbad skull as revealed by its nonmetrical features. Am J Phys Anthropol 23:1–16.

Genet-Varcin E (1979): "Les Hommes Fossiles. Découvertes et Travaux Depuis Dix Années." Paris: Boubée.

Gentry AW, Gentry A (1978): The Bovidae (Mammalia) of Olduvai Gorge, Tanzania. Parts 1 and 2. Bull Br Mus (Nat Hist) 29:289–446, 30:1–83.

Gieseler W (1974): Die Fossilgeschichte des Menschen. In Heberer G (ed): "Die Evolution der Organismen, III." Stuttgart: Fischer, pp 171–517.

Gould SJ, Eldredge N (1977): Punctuated equilibria: The tempo and mode of evolution reconsidered. Paleobiology 3:115–151.

Grahmann R, Müller-Beck H (1967): "Urgeschichte der Menschheit. 3rd ed." Stuttgart: Kohlhammer.

Grimaud D (1982): Le pariétal de l'Homme de Tautavel. I. Cong Int Paleont Hum Nice, pp 62–88.

Heberer G (1950): Das Präsapiens-Problem. In Grüneberg H, Ulrich W (eds): "Moderne Biologie. Festschrift zum 60. Geburtstag von Hans Nachtsheim." Berlin, pp 131–162.

Heim J-L (1977): Un nouveau regard sur l'évolution humaine. ANDES 23:2–10.

Hendey QB (1969): Quaternary vertebrate fossil sites in the south-western Cape Province. S Afr Archaeol Bull 24:96–105.

Hendey QB (1974): The late Cenozoic Carnivora of the Southwestern Cape Province. Ann Transvaal Mus 63:1–369.

Henke W, Rothe H (1980): "Der Ursprung des Menschen." Stuttgart: Fischer.

Hoffmann AC (1955): Important contributions of the Orange Free State to our knowledge of primitive man. S Afr J Sci 51:163–168.

Holloway R (1975): Early hominid endocasts: Volumes, morphology and significance for hominid evolution. In Tuttle RH (ed): "Primate Functional Morphology and Evolution." The Hague: Mouton, pp 393–416.

Holloway R (1982): Homo erectus brain endocasts: Volumetric and morphological observations, with some comments on cerebral asymmetries. I. Cong Int Paleont Hum Nice, pp 355–369.

Howell FC (1951): The place of Neanderthal man in human evolution. Am J Phys Anthropol 9:379–416.

Howell FC (1957): The evolutionary significance of variation and varieties of 'Neanderthal' man. Q Rev Biol 32:330–347.

Howell FC (1958): Upper Pleistocene men of the Southwestern Asian Mousterian. In Koenigswald GHR von (ed): "Hundert Jahre Neandertaler." Köln: Böhlau.

Howell FC (1960): European and northwest African Middle Pleistocene hominids. Curr Anthropol 1:195–232.

Howell FC (1978): Hominidae. In Maglio VJ, Cooke HBS (eds): "Evolution of African Mammals." Cambridge: Harvard University Press, pp 154–248.

Howell FC (1982): Diversity and Variability in Homo erectus and the question of the presence of Homo erectus in Europe. I. Cong Int Paleont Hum Nice.

Howells WW (1970): Mount Carmel man: Morphological relationships. Int Cong Anthropol Ethnol Sci Tokyo 1968 I:269–272.

Howells WW (1973): "Cranial Variation in Man." Cambridge: Harvard University Press.

Howells WW (1974): Neanderthals: Names, hypotheses and scientific method. Am Anthropol 76:24–38.

Howells WW (1976): Explaining modern man: Evolutionists versus migrationists. J Hum Evol 5:477–495.

Howells WW (1978): Position phylétique de l'homme de Néanderthal, In Bordes F (ed): "Les Origines Humaines et les Époques de l'Intelligence." Paris: Masson, pp 217–237.

Howells WW (1982): Comment to Smith FH: Upper pleistocene hominid evolution in south-central Europe: a review of the evidence and analysis of trends. Curr Anthropol 23:688–689.

Hrdlička A (1927): The Neanderthal phase of man. J R Anthropol Inst 67:249–269.

Hublin J-J (1978): "Le Torus Occipital Transverse et les Structures Associées: Évolution Dans le Genre Homo." Thèse de 3ᵉ Cycle, Université de Paris VI.

Hublin J-J (1982): Les Anténéandertaliens: Présapiens ou Prénéandertaliens. Geobios Mem Spec 6:345–357.

Hublin J-J, Tillier A-M (1981): The Mousterian juvenile mandible from Irhoud (Morocco): A phylogenetic interpretation. In Stringer CB (ed): "Aspects of Human Evolution." London: Taylor and Francis, pp 167–185.

Isaac GL (1972): Chronology and the tempo of cultural change during the Pleistocene. In Bishop WW, Miller JA (eds): "Calibration of Hominid Evolution." Edinburgh: Scott Academic Press, pp 381–430.

Jaeger J-J (1975): The Mammalian Faunas and Hominid Fossils of the Middle Pleistocene of the Maghreb. In Butzer KW, Isaac G (eds): "After the Australopithecines." The Hague: Mouton, pp 399–418.

Jaeger J-J (1982): Position chronologique des Hominidés fossiles d'Afrique du Nord. I. Cong Int Paléont Hum Nice.

Jelinek J (1978): Comparison of mid-Pleistocene Evolutionary Process in Europe and in South-East Asia. Proc Symp Nat Select Liblice 1978, Praha, pp 251–267.

Keith, A (1938): The Florisbad skull. Nature 141:1010.

Klein RG (1973): Geological antiquity of Rhodesian man. Nature 244:311–312.

Klein RG (1974): Environment and subsistence of prehistoric man in the Southern Cape Province, South Africa. World Archaeol 28:238–247.

Klein RG (1975): Middle Stone Age Man—Animal Relationships in Southern Africa: Evidence from Die Kelders and Klasies River Mouth. Science 190:265–267.

Klein RG (1976): The mammalian fauna of the Klasies River Mouth Sites, southern Cape Province, South Africa. S Afr Archaeol Bull 31:75–98.

Klein RG (1979): Stone age exploitation of animals in southern Africa. Am Sci 67:151–160.

Knussmann R (1980): "Vergleichende Biologie des Menschen. Lehrbuch der Anthropologie und Humangenetik." Stuttgart: Fischer.

Kohl-Larsen L (1943): "Auf den Spuren des Vormenschen, 2 Vols." Stuttgart: Strecker and Schröder.

Kurth G (1968): Zur Abgrenzung von Speziationen unter den gegenwärtig bekannten Hominiden, vor allem des Mittelpleistozäns. In Kurth G (ed): "Evolution und Hominisation." Stuttgart: Fischer, pp 204–229.

Lacaille AD (1951): The stone industry of Singa-Abu Hugar. In Arkell AJ, Bate DMA, Wells LH, Lacaille AD (eds): "The Pleistocene Fauna of two Blue Nile Sites." Fossil Mammals Afr 2:43–50.

Leakey LSB (1935): "Stone Age Races of Kenya." London: Brit Mus Nat Hist, Oxford Univ Press.

Leakey LSB (1953): "Adam's Ancestors." London: Methuen.

Leakey LSB (1972): Homo sapiens in the Middle Pleistocene and the evidence of Homo sapiens' evolution. In Bordes F (ed): "The Origin of Homo sapiens." Paris: UNESCO, pp 25–29.

Leakey MD, Hay RL (1982): The chronological position of the fossil hominids of Tanzania. I. Cong Int Paleont Hum Nice, pp 753–765.

Leakey REF (1981a): African fossil man. In Ki-Zebro J (ed): "General History of Africa. I. Methodology and African Prehistory." Berkeley: University of California Press, pp 437–451.

Leakey REF (1981b): "The Making of Mankind." London: Rainbird.

Leakey REF, Butzer K, Day MH (1969): Early Homo sapiens remains from the Omo River region of South-West Ethiopia. Nature 222:1132–1138.

Levêque F, Vandermeersch B (1980): Découverte de restes humains dans un niveau castel-perronien à Saint-Césaire (Charente-Maritime). CR Acad Sci [D] (Paris) 291:187–189.

Levêque F, Vandermeersch B (1981): Le néandertalien de Saint-Césaire. Recherche 12:242–244.

Magori CC (1980): "Laetoli Hominid 18: Studies on a Pleistocene Fossil Human Skull From Northern Tanzania." Ph.D. thesis, London.

Martin R, Saller K (1957): "Lehrbuch der Anthropologie. 3rd ed." Stuttgart: Fischer.

Mason RJ (1971): "Cave of Hearths in Prehistory." Johannesburg: Witwatersrand University Press.

McBurney CBM (1958): Evidence for the distribution in space and time of Neanderthaloids and allied strains in northern Africa. In von Koenigswald GHR (ed): "Hundert Jahre Neanderthaler." Köln: Böhlau.

McBurney CBM (1967): "Haua Fteah and the Stone Age of the Southeast Mediterranean." Cambridge: Cambridge University Press.

McBurney CBM, Trevor JC, Wells LH (1953): The Haua Fteah fossil jaw. J R Anthropol Inst 83:71–85.

McCown TD, Keith A (1939): "The Stone Age of Mount Carmel, Vol II." Oxford: Claredon.

Mehlman MJ (1979): Mumba-Höhle revisited: The relevance of a forgotten excavation to some current issues in East African prehistory. World Archaeol 11:80–94.

Mehlman MJ (in press): A reassessment of the associations and the age of the Lake Eyasi (Tanzania) hominid fossil crania. The Afr Archaeol Rev.

Merrick HV, de Heinzelin J, Haesaerts P, Howell FC (1973): "Archaeological Occurrances of Early Pleistocene Age From the Shungura Formation, Lower Omo Valley, Ethiopia." Nature 242:572–575.

Moss ML, Young RW (1960): A functional approach to craniology. Am J Phys Anthropol 18:281–292.

Mturi AA (1976): New hominid from Lake Ndutu,Tanzania. Nature 262:484–485.

Oakley KP (1954): Study tour of early hominid sites in southern Africa. S Afr Archaeol Bull 9:75–87.

Oakley KP (1957): The dating of the Broken Hill, Florisbad, and Saldanha skulls. "Proc 3rd Pan Afr Cong Prehist, Livingstone 1955." London: Chatto and Windus, pp 76–79.

Oakley KP (1974): Revised dating of the Kanjera hominids. J Hum Evol 3:257–258.

Partridge TC (1982): The chronological positions of the fossil hominids of Southern Africa. I Cong Int Paleont Hum Nice, pp 617–675.

Pilbeam DR (1975): Middle Pleistocene hominids. In Butzer KW, Isaac GI (eds): "After the Australopithecines." The Hague: Mouton, pp 809–856.

Piveteau J, de Lumley M-A, Debenath A (1982): Les hominides de la chaise. I Cong Int Paleont Hum Nice, pp 901–917.

Protsch R (1974): Florisbad: Its palaeoanthropology, chronology, and archaeology. Homo 25:68–78.

Protsch R (1975): The absolute dating of Upper Pleistocene sub-Saharan fossil hominids and their place in human evolution. J Hum Evol 4:297–322.

Protsch R (1976): The position of the Eyasi and Garusi Hominids in East Africa. In Tobias PV, Coppens Y (eds): "Les Plus Anciens Hominidés." Paris: CNRS, Coll VI, pp 207–238.

Protsch R (1981): The palaeoanthropological finds of the Pliocene and Pleistocene. In Müller-Beck H (ed): "Die Archäologischen und Anthropologischen Ergebnisse der Kohl-Larsen-Expeditionen in Nord-Tanzania 1933–1939, Vol 3." Tübingen: Archaeol Venatoria.

Rightmire GP (1976): Relationships of Middle and Upper Pleistocene hominids from sub-Saharan Africa. Nature 260:238–240.

Rightmire GP (1978): Florisbad and human population succession in southern Africa. Am J Phys Anthropol 48:475–486.

Rightmire GP (1979): Implications of Border Cave skeletal remains for Later Pleistocene human evolution. Curr Anthropol 20:23–35.

Rightmire GP (1980): Middle Pleistocene hominids from Olduvai Gorge, northern Tanzania. Am J Phys Anthropol 53:225–241.

Rightmire GP (1981): Patterns in the evolution of Homo erectus. Paleobiology 7:241–246.

Roche J (1976): Chronostratigraphie des restes atériens de la grotte des contrebandiers à Témara (Province de Rabat). Bull Mem Soc Anthropol Paris Ser 13(3):165–173.

Roche J, Texier P (1976): Découverte de restes humains dans un niveau atérien supérieur de la grotte des Contrebandiers, à Témara (Maroc). CR Acad Sci [D] (Paris) 282:45–47.

Roginsky Ya (1954): Morphological features of the skull of a child from the late Mousterian layer of the Staroselye cave. Sov Ehtnogr 1:27–39.

Saban R (1975): Les restes humains de Rabat (Kébibat). Ann Paléont Vert 61:153–207.

Saban R (1977): The place of Rabat Man (Kébibat, Morocco) in Human Evolution. Curr Anthrop 18:518–524.

Sampson CG (1972): The Stone Age industries of the Orange River Scheme and South Africa. Nat Mus Bloemfontein Mem 6:1–288.

Sampson CG (1974): "The Stone Age Archaeology of Southern Africa." New York: Academic Press.

Sarnthein M (1978): Sand deserts during glacial maximum and climatic optimum. Nature 272:43–46.

Sausse F (1975): La Mandibule Atlanthropienne de la Carrière Thomas I (Casablanca). L'Anthropologie 79:81–112.

Schröter P (1978): Die archäologische Einordnung der Fundstelle. In Müller-Beck HJ (ed): "Die Archäologischen und Anthropologischen Ergebnisse der Kohl-Larsen-Expeditionen in Nord-Tanzania 1933–1939, Vol 2." Tübingen: Archaeol Venatoria, pp 66–68.

Schröter P (1982): Präsapiens- und Sapiens-Menschen der Spätzeit. Kindlers Enzyklopäd Der Mensch 2:95–146.

Schwidetzky I (1970): Rassengeschichte. In Heberer G, Schwidetzky I, Walter H (eds): "Das Fischer-Lexikon: Anthropologie." Frankfurt: Fischer.

Schwidetzky I (1971): "Hauptprobleme der Anthropologie." Freiburg: Rombach.

Schwidetzky I (1979): "Rassen und Rassenbildung beim Menschen." Stuttgart: Fischer.

Sergi S (1953): Morphological position of the 'Prophaneranthropi' (Swanscombe and Fontéchevade). In Howells WW (ed): "Ideas on Human Evolution." Cambridge: Harvard University Press.

Shackleton NJ (1975): The stratigraphic record of deep-sea cores and its implications for the assessment of glacials, interglacials, stadials, and interstadials in the mid-Pleistocene. In Butzer KW, Isaac GI (eds): "After the Australopithecines." The Hague: Mouton, pp 1–24.

Singer R (1954): The Saldanha Skull from Hopefield, South Africa. Am J Phys Anthropol 12:345–362.

Singer R (1957): Investigation at the Hopefield Site. In Clark JD (ed): "Proc 3rd Pan Afr Cong Prehist, Livingstone 1955." London: Chatto and Windus, pp 175–182.

Singer R, Crawford JR (1958): The significance of the archaeological discoveries at Hopefield, South Africa. J R Anthropol Inst 88:11–19.

Singer R, Wymer JJ (1968): Archaeological investigations at the Saldanha Skull site in South Africa. S Afr Archaeol Bull 23:63–74.

Singer R, Wymer JJ (1982): "The Middle Stone Age at Klasies River Mouth in South Africa." Chicago: University of Chicago Press.

Smith FH (1982): Upper Pleistocene hominid evolution in southcentral Europe: A review of the evidence and analysis of trends. Curr Anthropol 23:667–703.

Smith FH, Ranyard GC (1980): Evolution of the supraorbital region in Upper Pleistocene fossil hominids from South-Central Europe. Am J Phys Anthropol 53:589–609.

Smith Woodward A (1938): A fossil skull of an ancestral Bushman from the Anglo-Egyptian Sudan. Antiquity 12:193–195.

Snow CE (1953): The ancient Palestinian Skhūl V reconstruction. Am Sch Prehist Res Bull 17:5–12.

Spencer F, Smith FH (1981): The significance of Aleš Hrdlička's "Neanderthal Phase of Man:" A historical and current assessment. Am J Phys Anthropol 56:435–459.

Stanley SM (1979): "Macroevolution: Pattern and Process." San Francisco: Freeman.

Stanley SM (1981): "The New Evolutionary Timetable." New York: Basic Books.

Stearns CE, Thurber DL (1965): Th230-U234 dates of late Pleistocene marine fossils from the Mediterranean and Moroccan littorals. Quaternaria 7:29–42.

Stringer CB (1974): Population relationships of Later Pleistocene hominids: A multivariate study of available crania. J Arch Sci 1:317–342.

Stringer CB (1978): Some problems in Middle and Upper Pleistocene hominid relationships. In Chivers D, Joysey K (eds): "Recent Advances in Primatology, 3." London: Academic Press, pp 395–418.

Stringer CB (1979): A re-evaluation of the fossil human calvaria from Singa, Sudan. Bull Br Mus Nat Hist (Geol) 32:77–83.

Stringer CB (1981): The dating of European Middle Pleistocene hominids and the existence of Homo erectus in Europe. Anthropologie (Brno) 19:3–14.

Stringer CB (1982): Towards a solution to the Neanderthal problem. J Hum Evol 11:431–438.

Stringer CB, Howell FC, Melentis JK (1979): The significance of the fossil hominid skull from Petralona, Greece. J Archaeol Sci 6:235–253.

Stringer CB, Trinkaus E (1981): The Shanidar Neanderthal crania. In Stringer CB (ed): "Aspects of Human Evolution." London: Taylor and Francis, pp 129–165.

Suzuki H, Takai F (1970): "The Amud Man and his Cave Site." Tokyo: The University of Tokyo.

Thoma A (1973): New evidence for the polycentric evolution of Homo sapiens. J Hum Evol 2:529–536.

Thoma A (1975): Were the spy fossils evolutionary intermediates between classic Neanderthal and modern man? J Hum Evol 4:387–410.

Thoma A (1978): L'origine des Cro-magnoides. In Bordes F (ed): "Les Origines Humaines et les Époques de l'Intelligence." Paris: Masson, pp 261–282.

Thorne AG, Wolpoff MH (1981): Regional continuity in Australasian Pleistocene hominid evolution. Am J Phys Anthropol 55:337–349.

Tillier A-M (1977): La pneumatisation du massif craniofacial chez les hommes actuels et fossiles. Bull Mem Soc Anthropol Paris Ser 13(4):177–189, 287–316.

Tobias PV (1956): Evolution of the Bushmen. Am J Phys Anthropol 14:384.

Tobias PV (1961): New evidence on the evolution of man in Africa. S Afr J Sci 57:25-38.

Tobias PV (1962): Early members of the Genus Homo in Africa. In Kurth G (ed): "Evolution und Hominisation." Stuttgart: Fischer, pp 194-196.

Tobias PV (1967): The hominid skeletal remains of Haua Fteah. In McBurney CBM (ed): "The Haua Fteah (Cyrenaica) and the Stone Age of the South-East Mediterranean." Cambridge: Cambridge University Press, pp 338-352.

Tobias PV (1968): Middle and Early Upper Pleistocene members of the Genus Homo in Africa. In Kurth G (ed): "Evolution und Hominisation, 2nd ed." Stuttgart: Fischer, pp 176-194.

Tobias PV (1971): Human skeletal remains from the Cave of Hearths. Makapansgat, Northern Transvaal. Am J Phys Anthropol 34:335-368.

Tobias PV (1973): "The Brain in Hominid Evolution." New York: Columbia University Press.

Tobias PV (1982): The fossil hominids of the African continent and their comparison with Tautavel Man and the other Anteneanderthals. I Cong Int Paleont Hum Nice.

Trinkaus E (1973): A Reconsideration of the Fontéchevade fossils. Am J Phys Anthropol 39:25-36.

Trinkaus E (1981a): Evolutionary continuity among Archaic Homo sapiens. In Ronen A (ed): "The Transition From Lower to Middle Paleolithic and the Origin of Modern Man." University of Haifa.

Trinkaus E (1981b): Neanderthal limb proportions and cold adaptation. In Stringer CB (ed): "Aspects of Human Evolution." London: Taylor and Francis, pp 187-224.

Trinkaus E, Howells WW (1979): The Neanderthals. Sci Am 241:118-133.

Valoch K (1968): Evolution of the Paleolithic in central and eastern Europe. Curr Anthropol 9:351-390.

Vallois HV (1949): The Fontéchevade fossil men. Am J Phys Anthropol 7:339-362.

Vallois HV (1951): La mandibule humaine fossile de la grotte du Porc-Épic près Diré-Daoua (Abyssinie). l'Anthropologie 55:231-238.

Vallois HV (1954): Neanderthals and Praesapiens. J R Anthropol Inst 84:111-130.

Vallois HV (1958): La Grotte de Fontéchevade II: Anthropologie. Arch Inst Paleontol Hum 29:1-164.

Vollois HV, Vandermeersch B (1972): Le crâne mousterian de Qafzeh (Homo VI). l'Anthropologie 76:71-96.

Vandermeersch B (1978a): "Le Crâne Pré-Wurmien de Biache-Saint Vaast (Pas-de-Calais)." In Bordes F (ed): "Les Origines Humaines et les Époques de l'Intelligence." Paris: Masson, pp 153-157.

Vandermeersch B (1978b): Quelques aspects du problème de l'origine de l'homme moderne. In Bordes F (ed): "Les Origines Humaines et les Époques de l'Intelligence." Paris: Masson, pp 251-260.

Vandermeersch B (1981a): Les hommes fossiles de Qafzeh (Israël). Paris: CNRS.

Vandermeersch B (1981b): Les premiers Homo sapiens au Proche-Orient. In Ferembach D (ed): "Les Processus de l'Hominisation." Paris: CNRS, pp 97-100.

Vandermeersch B, Tillier A-M, Krukoff S (1976): Position chronologiques des restes de Fontéchevade. Congr UISPP, Coll 9, Le peuplement anténeandertalien de l'Europe, Nice, pp 19-26.

Vandermeersch B, Tillier A-M (1977): Étude préliminaire d'une mandibule d'adolescent provenant des niveaux Moustériens de Qafzeh, Israël. Eretz Israel 13:177-183.

Vogel C (1970): Gegenwärtige Probleme der Morphologie in der Stammesgeschichte von Primaten und Mensch. Z Morphol Anthropol 62:185-206.

Vogel C (1970): Gegenwärtige Probleme der Morphologie in der Stammesgeschichte von Primaten und Mensch. Z Morphol Anthropol 62:185–206.

Vogel JC, Beaumont PB (1972): Revised radiocarbon chronology for the Stone Age in South Africa. Nature 237:50–51.

Vogel JC, Waterbolk HT (1963): Groningen radiocarbon dates IV. Radiocarbon 5:163–202.

Voigt EA (1982): The Molluscan Fauna. In Singer R, Wymer JJ (eds): "The Middle Stone Age at Klasies River Mouth in South Africa." Chicago: Chicago University Press, pp 155–186.

Vrba E (1982): Biostratigraphy and chronology, based particularly on Bovidae, of southern hominid-associated assemblages: Makapansgat, Sterkfontein, Taung, Kromdraai, Swart-krans. I. Cong Int Paleontol Hum Nice, pp 707–752.

Weidenreich F (1936): The mandibles of Sinanthropus pekinensis: A comparative study. Paleontol Sin Ser D 7:1–134.

Weidenreich F (1943a): The 'neanderthal man' and the ancestors of 'Homo sapiens'. Am Anthropol 45:39–48.

Weidenreich F (1943b): The skull of Sinanthropus pekinensis: A comparative study on a primitive hominid skull. Palaeontol Sin Ser D 10:1–291.

Weinert H (1936): Der Urmenschenschädel von Steinheim. Z Morphol Anthropol 35:463–518.

Weinert H (1951): "Stammesentwicklung der Menschheit." Braunschweig: Vieweg.

Wells LH (1947): A note on the broken maxillary fragment from the Broken Hill cave. J R Anthropol Inst 77:11–12.

Wells LH (1951): The fossil human skull from Singa. In Arkell AJ, et al (eds): "The Pleistocene Fauna of two Blue Nile Sites. Fossil Mammals of Africa Vol 2." London: Brit Mus Nat Hist, pp 29–42.

Wells LH (1957): The place of the Broken Hill skull among human types. In Clark JD (ed): "Proc 3rd Pan Afr Cong Prehist." Livingstone (1955). London: Chatto and Windus, pp 172–174.

Wendorf F, Laury RL, Albritton CC, Schild R, Haynes CV, Damon PE, Shafiqullah M, Scarborough R (1975): Dates for the Middle Stone Age of East Africa. Science 187:740–742.

Wendorf F, Schild R (1976): "Prehistory of the Nile Valley." New York: Academic Press.

Whiteman AJ (1971): "The Geology of the Sudan Republic." Oxford: Oxford University Press.

Wolpoff MH (1971): Metric trends in hominid dental evolution. Case Western Reserve Univ. Stud. Anthropol 2:244 pp.

Wolpoff MH (1980): Cranial Remains of Middle Pleistocene European Hominids. J Hum Evol 9:339–358.

Wolpoff MH, Smith FH, Malez M, Radovcić J, Rukavina D (1981): Upper Pleistocene Human Remains from Vindija Cave, Croatia, Yugoslavia. Am J Phys Anthropol 54:499–545.

Woodward AS (1938): A fossil skull of an ancestral Bushman from the Anglo-Egyptian Sudan. Antiquity 12:193–195.

Zeuner F (1952): "Dating the Past, 3rd ed." London: Methuen.

Ziegert H (1981): Abu Hugar Palaeolithic Site (Blue Nile Province, Sudan). A preliminary report. Paper UISPP Cong, Mexico.

APPENDIX: VARIABLES USED IN THE MULTIVARIATE ANALYSES

I. Analysis of frontal morphology (cf Fig. 9)

Six variables:	Glabella-bregma arc
	Glabella-bregma chord
	Glabella-bregma subtense
	Glabella subtense fraction
	Frontal angle (determined after H - FRA, yet using the glabella-bregma chord)
	Minimum frontal breadth (M - 9)

Analysis of parietal morphology (cf Fig. 12)

Six variables:	Maximum cranial breadth (H - XCB, M - 8)
	Parietal arch (M- 27)
	Parietal chord (H - PAC, M - 30)
	Parietal subtense (H - PAS)
	Bregma subtense fraction (H - PAF)
	Parietal angle (H - PAA)

III. Analysis of upper facial morphology (cf Fig. 15)

Eight variables:	Outer biorbital breadth (M - 43)
	Biorbital breadth (M - 44)
	Nasion-prosthion length (H - NPH, M - 48)
	Cheek height (H - WMH)
	Orbital breadth (M - 51)
	Orbital height (H - OBH, M - 52)
	Nasal breadth (H - NLB, M - 54)
	Nasal height (H - NLH, M - 55)

H, abbreviations used by Howells [1973].
M, numbers used by Martin and Saller [1957].

The Origins of Modern Humans: A World Survey of the Fossil Evidence, pages 411–483
© 1984 Alan R. Liss, Inc., 150 Fifth Avenue, New York, NY 10011

Modern *Homo sapiens* Origins: A General Theory of Hominid Evolution Involving the Fossil Evidence From East Asia

Milford H. Wolpoff, Wu Xin Zhi, and Alan G. Thorne

Department of Anthropology, University of Michigan, Ann Arbor, Michigan
48104 (M.H.W.), Institute of Vertebrate Palaeontology and Paleoanthropology,
Beijing, People's Republic of China (W.X.Z.), and Department of Prehistory,
Research School of Pacific Studies, Australian National University, Canberra,
Australia (A.G.T.)

> Our knowledge of the fossil history of non-human subspecies is almost
> negligible and *Homo* will probably furnish the first well-studied case of
> the evolution of several subspecies in geological time [Van Valen,
> 1966, p 382].

The east Asian hominid fossil sequence presents an unequaled opportunity
for the development and testing of hypotheses about human evolution. For
at least a million years [Matsu'ura, 1982; Sémah et al, 1981; Liu and Ding,
1983; Zhou et al, 1982], or approximately one quarter of human evolutionary time, and most of the time span of the genus *Homo*, east Asia was the
easternmost edge of the inhabited world. Sites yielding hominid fossils
representing that entire time span have been recovered from the whole of
the region, from Zhoukoudian in the north to southern most Australia in
the south. This extraordinary fossil record includes the majority of known
Homo erectus remains. Indeed, this species was first discovered, defined, and
described in east Asia. For a considerable part of the time period of human
paleontological studies, the east Asian fossil record provided data for the
interpretation of human fossils from other major geographic regions (for
example, when the first of the earliest known hominids was recovered from
Laetoli, the specimen was initially placed in the genus "*Meganthropus*"
[Weinert, 1950], originally named and described in east Asia). Finally, east
Asia is in the source area for two of the major groups of modern humanity.
Attempts to deal with the wealth of fossil evidence from east Asia have
resulted in several major schemes for interpreting human evolution, partic-
ularly the origin of modern *Homo sapiens*.

Thus, there is ample reason to believe that east Asia is of great importance in understanding human evolution. It is our contention that the

region is critical for this understanding, since it provides an independent means of critically examining the two competing hypotheses about modern *Homo sapiens* origins that were developed in order to account for the evolutionary situation in Europe, at the opposite end of the human range.

THEORIES OF MODERN *HOMO SAPIENS* ORIGINS

One of the most important problems in human paleontology concerns the origin of modern populations. It is widely perceived, although not necessarily accurately, that living populations of *Homo sapiens* differ from the more archaic Upper Pleistocene representatives of the species in a number of substantial ways. At the same time, these modern populations exhibit a number of important similarities. The problem revolves around the question of whether all living populations have a single recent (Late Pleistocene) origin, or whether they evolved in many different regions from local archaic populations of *Homo sapiens*.

Rapid Replacement

The contention of a recent single origin for modern populations is essentially a punctuation model, which provides for evolutionary change as a consequence of isolation, speciation, and subsequent rapid replacement. This outward migration, or "Noah's Ark" hypothesis [Howells, 1976a], explains the variation of recent people everywhere as the result of dispersion from a common source population that was already morphologically modern.

The paleontological basis. The source area for the original population of modern *Homo sapiens* has been treated a number of ways by the various workers who support this hypothesis. The source region may be unstated or unknown [Howells, 1976a, 1980, 1983], or placed alternatively in Europe [Hrdlička, 1927], China [Macintosh and Larnach, 1976; Weckler, 1957], the Near East [Vandermeersch, 1970, 1972, 1981], or more generally western Aisa [Bodmer and Cavalli-Sforza, 1976; Guglielmino-Matessi et al, 1979; Howell, 1951], sub-Saharan Africa [Protsch, 1978; Rightmire, 1979], or, in the most recent suggestion, Australia [Gribbin and Cherfas, 1982].

The punctuation model denies that there is any significant evidence for regional continuity spanning the Middle and Late Pleistocene, apart from adaptive similarities that might be shared because they respond to the same sources of selection (climate, altitude, etc). Even when the possibility of admixture with existing local populations is admitted (coincidentally denying by implication that modern *Homo sapiens* origins was a *speciation* event), the process of admixing could only be important

... to the extent that such "introgressive hybridizations" were not sufficient to destroy the similarity pattern created by the previous history. If they were important enough to have created the observed similarity pattern, then the pattern must have originated earlier, and the problem is only shifted backwards in time [Guglielmino-Matessi et al, 1979, p 563].

If modern populations diverged recently from a single basal line, the similarities that they share would be explained. There would be no need to account for parallel evolutionary trends in different parts of the world, because only one of the earlier lines gave rise to the modern populations. The rapid replacement model suggests that most earlier morphologies disappeared as a consequence of this replacement. Modern population differences are alternatively accounted for by drift (often the geneticists' explanation) or local selection (often the biological anthropologists' explanation).

The contention that modern populations have a single recent origin, and that they rapidly replaced the preceding archaic people of regions throughout the inhabited world, was essentially a European explanation for the appearance of modern people in Europe without recourse to a Neandertal ancestry for modern Europeans [Brace, 1964a]. Most developments of the Neandertal replacement theme seem to regard the rest of the world as an afterthought, although recognizing that if there was rapid replacement in Europe, there might have also been rapid replacement everywhere else. The main interest in other regions stems from the search to find the source area where the modern populations evolved, or didn't evolve, as the case may be.

In more peripheral areas of human occupation, such as parts of Africa, and the islands of southeast Asia . . . Middle and early Upper Pleistocene local populations may have become extinct without contributing to the modern sapient gene pool [Kennedy, 1980, p 366].

Non-European specimens are rarely presented in any detail. For instance, in the recent text by Genet-Varcin [1979]—a book in which the only two hypotheses about modern *Homo sapiens* origins discussed at length are those of Thoma and Protsch—the 159-page chapter on Late Pleistocene *Homo sapiens* includes only 21 pages dealing with the world beyond the Greater Mousterian Culture Area. It is as though Boule and Vallois [1957, p 370] were correct in asserting that beyond Europe "human paleontology can present only a relatively poor catalogue."

Apart from general statements about the replacement of archaic populations throughout the world [such as Kennedy's above, Howell and Washburn, 1960; Protsch, 1978; and others of similar opinion], the rapid replacement hypothesis has only occasionally been applied specifically to the east Asian region. For instance, in his last paper Macintosh [Macintosh and Larnach, 1976] suggested a south China origin for the modern populations of the area, and Howells [1967] has regarded the Niah burial as a more modern contemporary of the Ngandong sample and therefore a more likely ancestor of the modern populations of Australasia. Similarly, Thoma [1964] considers the living Mongoloids as descendants of Siberian populations that in turn are descendants of the Near Eastern Neandertals, and therefore does not derive modern Chinese from the Middle Pleistocene remains of China. Finally, Protsch [1978] argues that modern humans appeared first in South Africa and very rapidly replaced *all* archaic populations within the last 40 ky years. This is probably the most extreme of the replacement positions.

However, not all of the workers advocating the interpretation of rapid replacement in Europe support this extreme position for other regions. Thus, Boule and Vallois [1957, p 521] conclude:

> We know that Neandertal Man existed in Asia as he did in Europe . . . there he no longer appears in isolation; some of his forms are linked with the Prehominians, others with the Men of the Upper Paleolithic . . . whereas the Neandertal Man of Europe occupies the position of a type apart . . . the little we know about this type in Asia shows it as included within a regular evolutionary sequence.

The genetic basis. The other basis for the punctuation model is found in the formulations of Cavalli-Sforza and Edwards [Edwards, 1971; Cavalli-Sforza and Edwards, 1965]. The argument for rapid replacement is based largely on the genetic traits of living populations and the computation of evolutionary trees leading to a common ancestral population at a period variously estimated at 30 to 100 ky ago. The proponents contend that while drift is not an exclusive agent of differentiation, it is much more important than selection. Calculations and trees are based on calibration of blood systems [Cavalli-Sforza et al, 1964], anthropometric traits [Cavalli-Sforza and Edwards, 1965], craniometrics [Howells, 1973a; Guglielmino-Matessi et al, 1979], and most recently on mitochondrial DNA sequencing studies by Wilson [Gribbin and Cherfas, 1982].

Problems in the Paleontological Interpretation

The rapid replacement model thus seems to rest on both morphological (ie, paleontological) and genetic foundations. However, we suspect that its

popularity among paleoanthropologists has more to do with the fact that it
provides a solution to the "Neandertal problem" in Europe, than the general
insight this model may provide for the origin of modern humans elsewhere.
Indeed, outside Europe this model seems to be contradicted in many regions
by evidence for local continuity between Middle Pleistocene, Late Pleistocene, and modern populations [Rightmire, 1976, 1978, 1981; Ferembach,
1973, 1979; Jaeger, 1975; Saban, 1977; Thorne and Wolpoff, 1981; Tobias,
1968; Thoma, 1975; Weidenreich, 1947a; Coon, 1962; Jelínek, 1980b, 1982]—
a point recognized by some supporters of the rapid replacement model
applied within Europe [such as Boule and Vallois, 1957].

The rapid replacement hypothesis leaves a widely recognized problem in
the relation of fossil and modern regional variation. As Coon [1959, p
1,399] said of Howells, in a review of *Mankind in the Making*:

> When Howells comes to explain the modern distribution of these races
> in the world, he gets into serious traffic trouble, and no gendarme - *ex
> machina* drops out of the sky to straighten it out.

The extent of this traffic problem might better be appreciated through a
consideration of how some of the modern groups came to inhabit the areas
in which they are found, according to proponents of a single recent origin
for modern populations. For instance, consider the populating of Indonesia,
according to Hooton [1949, p 641]. The original inhabitants, "Proto-Austra-
loids" mixed with a "minority of Negritos," were followed by a wave of
"Mediterraneans" from Europe, who in turn were followed by the "upper
crust mixed Mongoloid (Malay)." Seemingly, every major "race" has contributed to the peopling of these islands, implying that races were able to
form, differentiate, migrate great distances, and subsequently mix within a
period that according to Protsch [1978] may have been as short as 40 ky!

Such an explanation is required by the rapid replacement hypothesis,
but when it is necessary to look at its implications in regional detail, it lacks
a certain degree of evolutionary credibility. Thus we contend that the
paleontological underpinnings for rapid replacement as the explanation for
modern *Homo sapiens* origins may not be valid if this model is applied on a
worldwide basis.

Problems Surrounding the Genetic Argument

We also contend that there are problems involved in the genetic underpinnings of this model [see also Weiss and Maruyama, 1976]. Briefly put, we
have reason to question whether genetic data for living populations provide

independent information for examining any historical model [Livingstone, 1973]. Tree (dendrogram) analyses of genetic distance relations have been widely published for human populations [for instance, see the recent summary by Nei and Roychoudhury, 1982], and tend to be self-verifying since most branching analyses of distance relations give approximately the same information [Harpending, 1974]. Branching analyses, and most other forms of genetic distance analyses [for instance, Piazza et al, 1981] necessarily assume that population differentiation occurred through population splitting from an original parental form, and that the main mechanisms resulting in the subsequently accumulating differences are constant, random mutations and genetic drift. Thus, when used to support the rapid replacement model, the method must assume the very point it is used to establish—the common origin of modern populations and their subsequent divergence through branching. Moreover, although similar adaptations raise the possibility that genetic similarity could reflect common selection after divergence, natural selection and the effects of gene flow must necessarily be discounted.

We have other questions about the relevance of genetic distance studies in reconstructing evolutionary relationships and mechanisms. As Morton and Lalouel [1973] put it, the problem with branching analysis is that

> . . . its relevance to evolution is unclear, although it is also combined with the pretension that the dendrogram is in every detail a cladogram of populations evolving at a constant rate. Without this assumption, the method has not been shown to have any desirable properties.

Moreover, we question whether genetic distance, even if it is a valid reflection of the evolutionary process, actually corresponds to the morphological differences observable in the fossil record. Genetic distance analysis is based on the simple multiallelic systems normally measured in branching studies. These systems greatly contrast with the genetic complexities underlying polygenic traits (ie, morphology). Thus, observable patterns of genetic variability and morphological variability may have little direct relationship to each other because the sources of variability are fundamentally different. For instance, consider the example of *Pan* and *Homo*, a case in which dramatic morphological variation reflecting at least 4 million years of evolutionary and adaptive divergence is contrasted with an extraordinary genetic similarity [King and Wilson, 1975]. In their review of worldwide genetic relations, Nei and Roychoudhury [1982] conclude that genetic distance between populations is not always correlated with morphological difference, suggesting:

. . . evidently, evolution at the structural gene level and at the morphological level do not obey the same rule.

To some extent (at least with respect to the first major "split"), the differences between genetic and morphological trees have been reconciled through manipulations of the morphological data to remove the effects of climate [Guglielmino-Matessi et al, 1979]. Yet, these manipulations may remove some of the most important sources of differences in selection between human groups, and it is unreasonable to expect that accurate evolutionary information will result from studies that systematically remove the consequences of change due to natural selection from the data they propose to interpret.

In sum, without discounting the important role of modern populational genetic studies in the understanding of past and present evolutionary phenomena, we contend that because of the very nature of what is studied, and the assumptions necessary for the sorts of analyses that have been published, these studies have not been an appropriate basis for supporting the rapid replacement (or any other historical) hypothesis. We agree with Livingstone [1973, p 48] that

> if the problem is to reconstruct population history, there are more relevant data and better ways to do it . . . it would be highly unlikely that the genetic differences at any locus are solely due to, and therefore reflect accurately, the population history of the groups, or that the gene frequency differences would reflect the history.

Local Regional Continuity

The alternative model of modern *Homo sapiens* origins is based on the contention that modern populations evolved in different geographic areas from already differentiated ancestral groups of archaic *Homo sapiens* (or *Homo erectus*). The classic interpretation of this model is accurately described by Weiss and Maruyama [1976, p 32].

> Human races have been produced over a long time in which local population areas (such as Asia, Africa, Europe) were more or less isolated.

Historically, the hypothesis stemmed from the observations of Weidenreich [1938, 1939a, 1943, 1946], especially with regard to the evolutionary sequence in North China and Australasia (see discussion below). However, he also interpreted the fossil sequence in Africa and Europe in the same manner [1947a].

The static model. Weidenreich's model was largely observational. Weidenreich was aware of the potential contradiction between local evolutionary continuity and worldwide evolutionary changes. In one of his final publications he attempted to resolve this problem through the contention that orthogenesis was the primary orienting factor in human evolution, thus accounting for the evolution of different isolated groups in a single common direction [1947b].

Because this explanation was ultimately unacceptable, Coon [1962] was motivated to interpret Weidenreich's ideas within the synthetic theory of evolution, with special emphasis on the problem of how one polytypic species can evolve into another. Although focusing on the role of evolutionary processes in the orientation of morphological change, he became mired in two problems that have contributed greatly to the widespread dissatisfaction with his monumental work. The first of these, previously recognized by Weidenreich, involves the contradiction between worldwide evolutionary trends and local continuous sequences. Coon [1962, p 37] sidesteps this by stating: "We cannot hope to settle the question of parallel evolution versus peripheral gene flow in the evolution of each race." The second results from his application of Linnean taxonomy to the evolution of the polytypic genus *Homo*. Here he concludes that even before *Homo sapiens*, there were distinct races. Each of these was subject to different evolutionary histories, to different patterns of selection, and to important grade-changing mutations at different times. This leads to the contentions of separate evolutionary rates and the differently timed crossing of "species thresholds" that most workers have found unacceptable. Eventually, Weidenreich's model was dismissed along with Coon's explanation—an action that might have been a case of throwing out the baby with the bath water. Thus, in the past two decades only a few workers (see references above) have emphasized the evidence for local regional continuity between modern and ancient populations of *Homo sapiens*.

However resurrected, the regional continuity model contains a basic contradiction. In our view, this stems from a static model of evolutionary change. As long as morphological continuity, over long periods of time and involving different characteristics in a number of regions, is assumed to imply that these regions were genetically isolated, two sets of observations are rendered contradictory to each other: the regional continuity observation that implies *differences* in selection and regional isolation, and the observation of worldwide evolutionary trends (such as the appearance of modern *Homo sapiens* in each region) that implies *similarities* in selection, and gene flow between regions. Unless the initial gene pools were identical *and* the local adaptations were the same, isolation should result in increasing differences between regions and the appearance of separate evolutionary pathways. Gene flow, on the other hand, should result in just the opposite— the loss of differences between regions and the appearance of homogenous modern populations. Long-lasting regional differences, as perceived by Weidenreich, therefore seem to require an explanation that rests on both gene flow, and the absence of gene flow, as well as on both differences in selection from region to region and similarities in selection acting throughout the inhabited world.

According to this static model of evolutionary change, regional continuity throughout the Pleistocene would therefore seem to require a long-lasting and very fortuitous ratio between local differences in selection and widespread similarities in selection: just enough local difference to allow long-lasting regional distinctions to be maintained, but just enough widespread similarity to allow the species to evolve in the same direction, all across its range. The maintenance of these evolutionary forces at serendipitous levels must go on long enough to account for widespread populations of *Homo* evolving in the same general direction over a long period of time (in taxonomic terms changing from *Homo erectus* to *Homo sapiens*), with the populations becoming neither more significantly divergent nor more significantly similar, and without one population replacing all the others. This is the observation that Weidenreich perceived but could not explain in modern evolutionary terms, and that Coon tried to explain but in a manner that incorporated this contradiction.

Coon's solution was to propose an explanation that rested heavily on isolation. Thus, he posited that the degree of genetic continuity within each region over time greatly exceeded the degree of genetic interconnections between regions, concluding [1962, p 37] that

> ... the races of man differ more from each other in a quantitative genetic sense than *Homo erectus* and *Homo sapiens* did, and our races are older than our species.

Resolving the contradiction. In the decades following Weidenreich's work, two approaches to the attempted resolution of this contradiction developed: the model of regional continuity with isolation, and the model of rapid replacement by populations of anatomically modern *Homo sapiens* from a single source during the Late Pleistocene. As mentioned above, the modern version of the regional continuity with isolation model is essentially Coon's [1962]. Although Coon closely followed Weidenreich's interpretations, he was never able to incorporate into his reasoning Weidenreich's lateral infrastructure (which we would now call the lines of gene flow) connecting the main evolutionary lineages that Weidenreich perceived. Yet, Weidenreich was clear about this. For instance, summarizing his position on the relation between Sangiran *Homo sapiens*, Ngandong, and modern Australians [1943, pp 249–250], he wrote:

> . . . at least one line leads from *Pithecanthropus* and *Homo soloensis* to the Australian aborigines of today. This does not mean, of course, that I believe all the Australians of today can be traced back to *Pithecanthropus* or that they are the sole descendants of the *Pithecanthropus-Homo soloensis* line.

In 1962, Coon argued that only genetic isolation could account for the development and maintenance of regional differentiation, and thus was left with the contradiction. If worldwide evolutionary trends (changes in grade) were not a consequence of gene flow, the required explanation must lie elsewhere. Thus, authors such as Sarich [1971] and Krantz [1980] attempted to resolve the contradiction through the argument that many cultural innovations and other aspects of cultural behavior could be expected to spread much more rapidly than genetic information. Since culture is the main adaptive mechanism in hominids, one might expect worldwide communalities in selection and a rapidly spreading impetus of adaptive change that would orient a common direction to human evolution, even in the absence of significant gene flow (a requirement of Coon's model). In this role, culture was envisioned as the great leveler of adaptive differences in the human populations.

> For most of man's million-year existence, the way of life of one group strongly resembled the lifeway of another. Life in Paleolithic England and stone-age Tanganyika were much alike . . . One may question whether it took more brains to succeed in Neolithic Ireland or in the Indus Valley of the New Stone Age [Garn, 1962, p 224].

Krantz [1980] focused specifically on the contention of a recent very rapid spread of modern linguistic ability to account for the worldwide similarities in selection resulting in the appearance of modern populations without invoking gene flow.

Unfortunately, as Coon developed his model of genetic isolation and differing evolutionary rates, and formalized its consequences within a taxonomic system, the different rates of change that resulted in different times for crossing species (ie, grade) boundaries had implications that resulted in the widespread rejection of his model. The cultural communality aspect of this model has remained in favor, although in the absence of an evolutionary mechanism to account for how it could actually operate, especially in the absence of gene flow.

Only a few workers have realized that in the absence of gene flow, even the widespread dissemination of cultural ideas and innovations that had selective advantage for all populations would not necessarily result in either a common direction to morphological evolution across the hominid range, or to the maintenance of genetic unity as a single biological species. The reaction of morphologically different populations with genetically divergent gene pools to the same source of selection may differ dramatically [Dobzhansky, 1960], since selection can only modify the variation that is present [Lewontin, 1974; Wolpoff, 1980a].

> The rate and direction of phenotypic change depend crucially on the pattern of genetic variation available for natural selection or random genetic drift to act upon. But to a considerable extent the patterns of genetic and phenotypic variation are themselves shaped by selection [Charlesworth et al, 1982, p 478].

Thus, if populations differ phenotypically and are genetically divergent, morphological changes that follow the spread of selectively important technological changes, such as the ability to produce fire or knowledge of the prepared core technique for the manufacture of stone tools, *could* under some circumstances increase populational diversity in a way that would contradict Weidenreich's and Coon's observations of regional continuity, and Weiss and Maruyama's [1976] contentions about interpopulational genetic similarity. For instance, the same technological change could improve the sub-Arctic adaptation of northern peoples and the forest adaptation of central Africans, each with very different morphological consequences as the result of divergent shifts in balancing selection. Increasing divergence and the lack of substantial gene flow would lead to speciation according to virtually any model [Sokal, 1973].

We find it difficult to believe that the spread of ideas was not accompanied by a spread of genes (witness the genetic "consequences" following the spread of the Bible and the steel axe), and therefore the contradiction does not seem to be resolved by the supposed effects of the worldwide similarities in selection that presumably result from the spread of cultural innovations and information. Clearly, cultural information can and does spread *more rapidly* than genes. However, technological innovations, at least insofar as they can be discerned from the archaeological record, comprise only a small percentage of the totality of cultural innovations. In our view it is likely that the spread of archaeological innovations is a conservative estimator for the magnitude of gene flow in the past. With regard to the question of whether communalities in the sources of selection *alone* (ie, in the absence of gene flow) can account for the single direction of evolutionary change all across the human range, what we find most important is that ultimately cultural information does not spread *without* the spread of genes. We hardly wish to deny the absolutely critical role of culture in orienting the direction of human evolution. What we contend is that by itself, culture does not resolve the contradiction in Weidenreich's thinking. Humans, after all, are not the only evolving species to show long-term polytypic variation.

Coon's last attempt to grapple with this problem [1982] involves a rather different approach to avoiding the contradiction. He posits that even in the face of local differences in selection acting on geographically distinct *Homo erectus* populations, a low level of gene flow could account for the worldwide transformation of the species into *Homo sapiens*, if this transformation was the result of a macromutation doubling the surface area of the brain. Thus the predominating effect of local selection was not disturbed by what was in effect a one-time event, the rapid spread of a single macromutation. Coon's use of a saltation explanation would seem to be contradicted by his earlier contention that *Homo sapiens* did not evolve at the same rate everywhere. He addresses this problem through the claim that the saltation process left some populations with more neurons than they needed. In these cases, brain size fell back to *Homo erectus* levels.

We find this explanation unlikely, unsupported, and as we will discuss later, unnecessary. Coon posits a major energy crisis during the Middle Pleistocene, resulting in faunal changes that leave a niche for hominids to fill. This provided the basis for a dramatic level of selection promoting the fortunate brain doubling macromutation that resulted in the *Homo sapiens* transformation. However, we contend that under these conditions of intense selection the genetic basis for much more than brain size would rapidly spread, just as during intensive dog breeding for a limited range of breed standards, many unrelated (and often undesirable) characteristics are also fixed in canine populations. If the circumstances that Coon describes were correct, we believe that along with the doubled brain size a large number of common features would have rapidly spread over the human range, obliterating the evidence for regional continuity. Effectively, then, Coon's latest hypothesis is a rapid replacement hypothesis and should not predict the presence of evidence for regional continuity, evidence that Coon still accepted.

Solution unsatisfactory. In sum, the attempts to resolve the contradiction in the regional continuity model have not been generally acceptable. Thus, however convincing the morphological observations of Weidenreich and Coon may be, the lack of an acceptable theoretical basis has marred the continuity model, and has contributed to the success of the alternative (rapid replacement) hypothesis.

THE HISTORY OF INTERPRETATIONS OF THE EAST ASIAN FOSSIL RECORD

Beginning with Dubois, several major schemes of human evolutionary interpretations have had their origin in the east Asian fossil record. Dubois [1894] argued that hominids originated in the region—a theme reflected in later works by Weidenreich [1946] and Woo [1962] among others.

Another important interpretive scheme developed for the region involves an evolutionary continuity between its ancient and modern inhabitants. The notion that there is an evolutionary and morphological sequence embracing east Asian hominids throughout the Pleistocene is of long standing. It is through the publications of Weidenreich [especially 1939a, 1943, and 1947a] that this proposal first received widespread scientific recognition. In his concluding portion of the Zhoukoudian cranial monograph, Weidenreich argued that

> . . . the various racial groups of modern mankind took their origin from ancestors already differentiated. . . . There are clear evidences that *Sinanthropus* is a direct ancestor of *Homo sapiens* with closer relations to certain Mongolian groups than to any other races [1943, p 276].

This was a constant theme through Weidenreich's works—one that he applied to the inhabited world of the Pleistocene. Thus,

> The Australian natives have some of their characteristics in common with the fossil Wadjak-Keilor man and with Homo soloensis. Homo soloensis himself appears as an advanced pithecanthropus phase. Some of the characteristic features of Sinanthropus reappear in certain Mongolian groups of today. The same relations exist between Rhodesian

man and certain fossil South African forms of modern man. The Skhūl group of Palestine presents forms intermediate between the typical Neanderthal man from Tabūn and fossil modern man from Europe [1947a, p 201].

The hypothesis of early geographic differentiation became known as the "Polycentric School"—a phrase coined by Weidenreich himself [1939a]. Few hypotheses have been more consistently misrepresented [for instance, see Howells's [1967, p 241] "candelabra" rendition of Weidenreich's thoughts].

Apart from the interpretive scheme advanced by Weidenreich, and following him Coon, prior to the 1980s only Howells [especially 1967] has dealt with the east Asian region as a whole. Howells's ideas, discussed in more detail below, are quite different in that he envisions no connection between the archaic and modern populations in any part of the region, although admitting considerable antiquity for the regional differentiation of east Asia's modern populations [1976c, p 648]. To the south, Ngandong is interpreted as a perhaps slightly more advanced form of *Homo erectus*, lacking any clear connection to the early modern specimens from Wajak (Wadjak) and Niah (although in later publications Kow Swamp and other similarly robust Pleistocene Australians are found to be somewhat problematic). To the north, no special connection is recognized between the Middle Pleistocene Zhoukoudian people and later Asians [1983, p 298]: "because of their basic similarities, all modern men have a common ancestor who was later than Peking man." Maba is regarded as a "faraway Neanderthal" [1967, p 204], lacking clear connection to (what he regarded as) the undifferentiated early Chinese specimens from Ziyang, Liujiang, and the Upper Cave at Zhoukoudian. It has remained Howells's conviction that modern populations have a single recent origin.

There has been no lack of publications about the region subsequent to Coon and Howells. These have been authored mainly by local scholars, and with few exceptions [for instance Casiño, 1976] have dealt with evolutionary issues within subportions of the region. Because the approaches to the fossil material have come increasingly to diverge with a rapidly expanding fossil record and a greater focus on local problems, north and south portions of the east Asian region are best discussed separately.

North Asia

Weidenreich dealt mainly with the Zhoukoudian remains, detailing the relationship he perceived between this fossil sample and certain living Asian groups. In the Zhoukoudian cranial monograph he suggested 12 features that he believed indicated morphological continuity in the region. These are: mid-sagittal crest and parasagittal depression; metopic suture; Inca bones; "Mongoloid" features of the cheek region; maxillary, ear and mandibular exostoses (the mandibular exostoses forming a mandibular torus); a high degree of platymerism in the femur; a strong deltoid tuberosity in the humerus; shovel-shaped upper lateral incisors; and the horizontal course of the nasofrontal and frontomaxillary sutures. While Weidenreich considered the special similarities connecting the Zhoukoudian sample and living north Chinese, he admitted some problem in placing the Upper Cave remains within his scheme [1939b].

After the initial propositions about north Asian regional continuity made by Weidenreich, and later Coon, only a few western scholars have dealt with hypotheses regarding the full Pleistocene sequence of human evolution in north Asia. Aigner has consistently adhered to a scheme emphasizing local continuity [1976, 1978; Aigner and Laughlin, 1973]. Aigner, like Coon, argues from the morphological evidence (although including specimens not earlier available such as the Lantian cranium and mandible) that there is a special communality of features extending through the Pleistocene of China that indicate continuous local evolution. In addition to features discussed by Weidenreich, Aigner [1976] mentions the profile contour of the nasal saddle and of the nasal roof, pronounced frontal orientation of the malar facies and the frontosphenoidal processes of the maxilla, and a rounded infraorbital margin, even with the floor of the orbit. This local continuum is seen as the consequence of isolation, and much of Aigner's work has been involved in attempts to establish isolation from the archaeological record and from fauna. This developed in more detail an earlier proposal of cultural and morphological continuity in China made by Chang [1963], and has helped lead to the realization that the Asian archaeological sequence is regionally unique and probably should not be interpreted in a European framework [Yi and Clark, 1983, and comments therein].

Thoma [1964, 1973, 1975] also claims a polycentric approach to north Asian Pleistocene evolution, and to the origin of the Mongoloids. However, his hypothesis is quite unlike Weidenreich's. Thoma derives modern Mongoloids through a European lineage, involving (what he regards as) generalized Neandertal forms such as Amud and Teshik Tash which he believes evolved into Paleosiberians who in turn gave rise to modern north Asians.

After liberation, scholars in the People's Republic of China clearly accepted Weidenreich's interpretive framework, and they expanded on the morphological evidence supporting it [Woo, 1956, 1982; Wu, 1961, 1981; Wu and Zhang, 1978].

Weidenreich mentioned a series of characteristics (discussed above) of Sinanthropus to demonstrate the continuity from the Zhoukoudian folk to modern Mongoloids. But his contention was not widely accepted. One of the important reasons was that the role of discrete (nonmetric) traits in establishing evolutionary relationships and differences was unclear, and in fact is still disputed [Corruccini, 1974; Livingstone, 1980; Stringer, 1980]. We believe the criticisms can be dismissed when appropriate cautions and limitations in nonmetric analysis are adhered to. It has long been recognized that discrete traits are heritable in the same sense that metric traits are heritable; mainly, their expression represents the interaction of genotype and environment [Corruccini, 1976; Trinkaus, 1978; Cheverud et al, 1979]. Populational distinctions based on discrete traits involve frequency differences [Hertzog, 1968; Cadien, 1972]. In comparisons of living populations many of these differences have been found to be significant, and most workers have concluded that variation is best analyzed through a combination of metric and nonmetric features. A number of earlier analyses attempted interregionally broad comparisons based on discrete variables, with results that engendered much of the recent debate over their validity. We propose that this problem stems from comparing widely diverse groups, and agree with Ossenberg [1976, p 707] that the validity of nonmetric analysis is limited by the diversity of the groups analyzed. Comparisons are likely to be misleading or invalid unless they are between populations that are fairly closely related. In sum, we believe that by limiting the geographic region over which these comparisons are made (as Weidenreich did), combining metric and discrete variables, focusing on the distributions of frequencies rather than of "types," and considering the potential role of behavior and environment in the expressions of the traits, observations such as those made by Weidenreich are phenetically valid.

Another obstacle to accepting Weidenreich's interpretation was the fact that there were not enough well-dated human fossils from the region to fill the long gap between these two terminals. It took the better part of four decades to improve this situation. From the time Weidenreich first proposed regional continuity to the early 1950s, no more human fossils were found in China except a child's incisor from the Ordos region [Woo, 1956], five teeth (including a large shovel-shaped central maxillary incisor) and fragments of the tibia and humerus from the rubble left on, but not necessarily derived from, level 29 from the earlier excavations at Zhoukoudian [Woo and Chia, 1954], and the Upper Cave remains from Zhoukoudian, dated by radiocarbon to 10.5–18.3 ky BP (the most recent review of all Chinese dates is given by Zhou et al [1982]—unless otherwise indicated all of the Chinese dates are from this source). The Ordos tooth was too scanty to be useful and the

Upper Cave remains were considered by Weidenreich, as well as by all of the other anthropologists who dealt with the specimens during that period, as far from a definite ancestor for the modern Chinese.

In 1951, a female skull was found in Ziyang County of Sichuan Province [Woo, 1958a]. It was considered Late Paleolithic in age but because it was in secondary sediments with vertebrate fossils from different periods, the date is still in dispute. There is a sagittal keel and a prenasal fossa on it but no maxillary torus. The supraciliary arches are extraordinarily well developed compared with a modern Chinese female. They are separated from the frontal squama by a supratoral sulcus.

Three teeth of a child of late Middle or early Upper Pleistocene age were found near a village named Dingcun in Shanxi Province [Woo, 1958b] in 1954. Both of the incisors are shovel shaped. The wrinkle pattern of the molar shows an intermediate position between the Zhoukoudian Lower Cave condition and modern Chinese.

In 1956, a fragmentary maxilla was discovered from near Changyang, Hubei [Chia, 1957]. It was associated with *Ailuropoda-Stegodon* fauna including *Hyaena sinensis* and therefore was considered Middle Pleistocene in age. According to its morphology and stratigraphy it might belong to the late part of the Middle Pleistocene or the early part of the Late Pleistocene (the discovery of Liujiang in 1959 showed that this fauna persisted later than once thought). It is so fragmentary that it cannot add morphological evidence relating to the hypothesis of continuous development from Sinanthropus through Mongoloids. No maxillary torus exists on this specimen.

Also in 1956, a crushed palate and cranial base were found at Chilinshan, Kwangxi [Chia and Woo, 1959]. The fauna is recent, and the specimen is either latest Pleistocene or earlier Holocene (on the basis of its fossilization, it is unlikely to be extremely recent). The facial portion is of interest, because while the palate is only moderate in size, the zygomatic process of the maxilla is quite large and somewhat puffed, resulting in the lack of a canine fossa. The masseter attachment on the zygomatic process is rugose, and broadens posteriorly onto the short remaining stub of the zygomatic bone. The nose is broad, and there is a very sharp inferior nasal border. Below the nose there is distinct alveolar prognathism. While the degree of prognathism is marked, its exact expression is difficult to judge because of the curved alveolar plane (the maxillary correspondence to the curve of Spee); if the anterior third of the toothrow is used as the base the angle of the maxilla's anterior seems less acute, while if the posterior two-thirds is used the angle is more acute. The occipital plane is much longer than the nuchal plane, but there is a distinct nuchal torus extending almost 3 cm to either side of the low triangular inion prominence.

Maba is represented by a skull cap unearthed from Guandong Province [Woo and Peng, 1959]. It is from the latest part of the Middle Pleistocene or the early part of the Late Pleistocene according to its faunal associations. The specimen possesses some features showing its intermediate position between Sinanthropus and Mongoloids such as a weak sagittal keel on the frontal bone, a frontal sinus which is restricted to the interorbital area, the frontonasal and frontomaxillary sutures form a nearly horizontal straight line, the malar surface of the frontosphenoidal process of the zygomatic bone faces more forward than those in Neandertals, seen from above the supraorbital torus slants posterolaterally, and there is a rather large profile angle of the nasal bone. Given the wide interorbital area the nasal bones are surprisingly narrow (as in Dali), and the nasals support a strongly developed central ridge along their join (resembling the late Lower Cave cranium H3 as well as the Upper Cave crania). This configuration is still particularly characteristic of some northern Mongoloid populations [Oschinsky, 1964].

A fairly complete fossil skull of anatomically modern type was found in 1959 within a cave in Liujiang, Kwangxi with *Ailuropoda-Stegodon* fauna [Woo, 1959]. It is considered to be the approximate age of the Upper Cave remains, and based on the faunal comparison it is clearly older than Chilinshan (also from Kwangxi). The upper lateral incisors are shovel shaped. We can reasonably infer that the central incisors must have had the same feature although one of them was lost and the other was severely worn. The third molars had not erupted, although the skull belonged to a middle-aged person. The flatness of the upper and middle face and of the nasal saddle, the lower margin of the pyriform orifice, and the contour of the sutures between the frontal bone and the nasal plus maxillary bones indicates its affinity with other Chinese fossil specimens. The protruding zygomatics and the orientation of the frontosphenoidal process are also Mongoloid features. The vault can be regarded as male if the association with the postcranial remains (including a pelvis) is correct. It is less robust in most features than the Upper Cave male no. 101, and in particular has a much flatter nasal saddle and a lower facial height.

In 1959, both Woo Rukang and Cheboksarov hypothesized the continuous development of physical type in Paleolithic China. They were of the opinion that Upper Cave specimens were proto-Mongoloids, but no detailed evidence was presented to support this. From 1959 to 1961, one of us [Wu Xinzhi, 1961] published articles to demonstrate this notion in detail and pointed out that, as a sample, the Upper Cave remains are particularly close to Chinese, Eskimos and Amerindians [see also Neumann, 1956].

The three skulls from that cave have a number of common characters, as mentioned earlier by Weidenreich [1939b, 1943]. They are lowness of upper face and angular shape of the orbit, great interorbital breadth, moderate prognathism, a wide pyriform orifice, and prenasal fossae. There are also other common features such as the low cranial index, more inclined frontal bones, moderate length of the palate, a variable but generally high degree of facial flatness, and the convex sagittal contour of the anterior surface of alveolar process of the maxilla.

All of these common features are in accordance with Mongoloid affinities except the low orbit and broad nose, and these can be placed in the context of the regional evolutionary sequence. The lower orbits are different from those of modern Mongoloids. Weidenreich considered this a European Paleolithic and/or Melanesoid feature. In fact, low orbits are common in specimens from that period all over the world. The orbital index of the Liujiang Upper Paleolithic skull found in China is also rather low (68.7) and the orbits of this specimen are also rectangular in shape. Chinese Mesolithic (Djalainor found in Inner Mongolia) and Neolithic (as from Baoji, Shaanxi) skulls have orbital indices (77.5 and 77.3 for Djalainor, 78.0 for the Baoji specimens) intermediate between Chinese Paleolithic and modern populations. These data indicate a transitional development of the orbital index from Upper Cave through modern Chinese. Besides, although Melanesians have lower orbits compared with other modern populations, their average orbital index is still higher than Upper Cave no. 102 (72.3).

Weidenreich had envisioned the higher nasal index (55.9) of no. 102 as one of its Melanesoid features. With regard to nasal aperture shape, the nasal indices of the Liujiang skull (58.5) and the Djalainor Mesolithic skull as well as Chinese Neolithic (Baoji male mean 52.5, female mean 52.43; Banpo 50.0) show that there is a pattern of decreasing nasal index from upper Paleolithic through Recent times in China. Moreover, some populations of modern Mongoloids, such as southern Chinese and Paltacalo-Indians, still retain broad noses. Finally, cranium no. 102 is young, perhaps as little as 13 to 14 years in age. If growth were complete, the adult nasal index would be lower.

In addition to the common features of the sample, it is relevant to also review some of the unique features of each cranium. The total facial height index of the no. 101 skull is lower than that in modern Chinese but it is close to Santa Rosa Indians. The naso-malar angle (135°) is low for a Mongoloid vault from Asia, although it falls within the low end of the Amerind range as reported by Oschinsky [1964]. The zygo-maxillary angle of 128° is not as low compared with Asian and American Mongoloids. The marked supraorbital development and the supratoral sulcus of no. 101 is very unusual for modern Asians, but easily exceeded by other Late Pleistocene specimens from north Asia such as the 6-ky-old Siberian cranium from Serovo [Laughlin et al, 1976].

Weidenreich mentioned many characters of no. 102 as evidence for attributing it to the Melanesoid type. Some of these features were discussed above as common features in all the Upper Cave specimens; and others will be discussed below. The sloping forehead and extremely high vault is one of the features he cited. No. 102 has two broad depressions on each side of the frontal squama, lateral to the midline at about its sagittal center. These depressions cross the squama transversely, and are connected by a shallow groove. Seemingly similar depressions in the Zhoukoudian Lower Cave crania are not equivalent; in the Lower Cave specimens only one of these appears on each side of the squama, and they are associated with the frontal keel (a structure absent in Upper Cave cranium no. 102). We believe that the morphology of the no. 102 cranium corresponds to the lateral depressions described by Magitot [1885] as marking the mechanism of deformation in artificially deformed European crania. Other features indicating artificial deformation [see Brown, 1981] in no. 102 include the long, flat frontal, the marked parietal curvature in the sagittal plane, the great cranial height, and the flattened occiput with its lack of angulation at the superior nuchal line. We are certain that the no. 102 vault shows the effects of artificial deformation. Correcting for this influence, the vault was probably lower and thus its vault height and forehead slope are not necessarily Melanesian features. In fact many populations of Mongoloids also have high vaults.

Weidenreich also cited the total facial index as similar to Melanesoid form. But mandible no. 104, which he believed belonged to skull no. 102, actually belongs to another individual. The maximum width of the dental arch of this mandible is larger than that on the skull by 10 mm (the difference of the greatest widths between upper and lower dental arches of the same skull does not exceed 4 mm as measured on 100 Chinese skulls). Moreover the second and third molars on mandible no. 104 were heavily worn while the second molars on skull no. 102 had just erupted and the third molars were unerupted. The upper and lower dentitions cannot fit. Therefore the total facial index of this skull cannot be calculated.

Some features of no. 102 are different from those in Melanesians. Palatal and maxillo-alveolar indices of no. 102 (85.1 and 126.2, respectively) are much higher than those in Melanesians (means of 63.6 and 108.2, respectively). The latter index of no. 102 even exceeds the range of variation in Melanesians (100–118.2). Weidenreich explained this incongruity by the fact that the last molar of this skull had not erupted. He argued that after the eruption of this tooth the length of the maxillary alveolar process would increase so that the index would be comparable with the Melanesians. But although unerupted, the third molars had already occupied their spaces on the alveolar process of this Upper Cave skull. The maxillo-alveolar index exceeds the Melanesian range so greatly that only if maxillo-alveolar length was enlongated by 10 mm could the indexes match each other. This much additional growth in no. 102 would be very unlikely.

Weidenreich focused on sagittal keeling, low cranial index, extremely high vault, and large transverse cranio-facial index as his basis for attributing no. 103 to Eskimos. In fact none of these are unique to Eskimos; they also exist on the skulls of certain Mongoloid groups. There are 11 out of 71 northern Chinese skulls examined by one of us (Wu Xinzhi) with a transverse cranio-maxillary index higher than the average value of Eskimos (100.8 according to Oetteking [1930]). In 15 skulls of Tibetan B Group measured by Morant [1923, 1924] there are six with an index higher than 101. The largest value Morant reports is 105.7 (no. 30). Similarly, while the very high naso-malar and zygo-maxillary angles of the specimen (148° and 131°) resemble Eskimo mean values, they are below the mean values for Neolithic Siberian crania, and fall within the Arikara Amerind range. The nasal index of no. 103 (50.0) is not concordant with those in Eskimos, who have average populational values ranging between 43.5 and 46.1 for females [Hrdlička, 1942; Oschinsky, 1964]. The absolute width of the anterior nasal aperture of this skull accounts for the difference in indexes. This width is much larger than those reported for Eskimo females (21.5–25.0 [data from Hrdlička, 1942; Oschinsky, 1964]). Thus it is more reasonable to envision this skull as belonging to a broad category—the Mongoloid line—instead of a narrower one such as the Eskimoid type.

Finally, the tori on maxilla no. 110 and mandible no. 101 strengthen the evidence linking them with Mongoloids.

Weidenreich overstated and misinterpreted some features of the Upper Cave skulls, instead of diagnosing them according to their total morphological pattern. He had no opportunity to examine the Mesolithic and earlier Neolithic human specimens in China, instead comparing his fossils only with modern populations in his analysis. But morphological characters should be envisioned in view of their historical development. The various populations on the same evolutionary line did not necessarily have common features in all respects because the earlier populations may have experienced only intermediate changes in the evolutionary process giving rise to later populations.

Coon [1962] utilized all of the Chinese human fossils then known to support the notion of continuity. Coon's analysis of the Zhoukoudian remains drew on a wider sample for its comparisons than was available for Weidenreich's analysis. Moreover, he had new Middle and Late Pleistocene specimens that could be fitted into the interpretive framework (Dingcun, Changyang, Maba, Ziyang, Liujiang, Ordos, and Chilinshan). He was unable to examine these firsthand, however, and casts were not available so the discussions relied on brief Chinese publications. In the conclusions of his analysis he wrote: "the Upper Cave skulls from Choukoutien approach the end of the Sinanthropus-Mongoloid line . . . the sooner we forget about the Ainu-Melanesian-Eskimo label the better." Coon also discussed some features of Maba Man as further evidence for continuity, for instance stating: "the lower border of the orbit projects forward, as in Sinanthropus and Mongoloids." However, his notion about the lower border of the orbit is unconvincing. There is no inferior orbital margin preserved in Maba and one can hardly judge whether or not this margin is protruding on the Sinanthropus specimens available. He agreed with Woo that "Liujiang man was a Mongoloid form of *Homo sapiens* still in the process of evolution," contending that "the skull deviates somewhat from the Mongoloid line in an Australian direction."

In 1963 and 1964, a mandible and a skullcap of *Homo erectus* were found respectively from two sites in Lantian County, Shaanxi Province [Woo, 1964, 1966]. They are estimated to be about 0.65 and 0.75 million years in age, respectively, on the basis of paleomagnetic studies. The mandible from Chenjiawo shows very ancient evidence of third molar agenesis (a not uncommon condition in modern China), since the third molar was not erupted in this woman of advanced age, and radiographs show it was not present in the alveolus. There are multiple mental foramina, also a common characteristic at Zhoukoudian. The Gongwangling skull shows a series of similarities with its peers at Zhoukoudian. The skull has traces of sagittal keeling and *eminentia cruciata*. The frontonasal and frontomaxillary sutures form a more or less straight line, and the degree of flattening of the nasal saddle is intermediate between that of Sinanthropus and Mongoloids. The widths of upper and middle parts of the nasal bones probably approximate each other. The anterior sagittal contour of the alveolar process of the maxilla is slightly convex. The maxilla is rather small, especially low, with very anterior cheeks. It is comparable with the three small maxillae from Zhoukoudian, as well as with the low Dali face (see below) and the Changyang fragment. Together, these maxillae show that facial reduction was quite early in north Asia, compared with other regions. The Lantian maxilla has no torus.

Since 1965 a number of fossil incisors of Homo erectus and Homo sapiens were found at different places, namely Yuanmou [Zhou and Hu, 1979], Yunxian [Woo and Dong, 1980], Yunxi, Xichuan [Wu and Wu, 1982a], Tonzi [Wu et al, 1975], and Xijiayao [Chia et al, 1979; Wu, 1980]. Without exception, they are shovel shaped.

In 1966 a new discovery of Sinanthropus was made at Locus H, Zhoukoudian, from layer 3, one of the very highest levels in the Lower Cave. This layer has been variously estimated at 230 ky and 370 ky in age. The remainder of a vault, known to Weidenreich as the H3 (cranium no. 5) specimen on the basis of a large occipitotemporal fragment and a small portion of the mastoid, was recovered [Chiu et al, 1973]. This allowed a complete reconstruction of the H3 cranial vault, a specimen that is much later in the Zhoukoudian sequence than the other cranial remains. The specimen, almost certainly a male, was found to clearly resemble the other Zhoukoudian crania in many details as well as in its gross size and shape. On the other hand, there are a number of differences that are of some interest, although of limited value because they only describe a single individual. Its cranial capacity (1,140 cm^3) is larger than the mean for the four adult crania (1,075 cm^3) from the eighth and ninth layers, an expansion which Wu and Lin [1983] consider significant. While the supratoral sulcus is very well excavated, the anterior of the frontal is less distinctly bossed and the frontal keel is less prominent than in the earlier crania. The posterior edge of the temporal squama is steeper and higher. The supraorbital torus is vertically thinner than any of the earlier specimens, the frontal is sagittally longer but its squama is thinner (as is the parietal, except at its posterior), and the nasal root is more depressed. Moreover, there are a number of differences regarding the temporals that Weidenreich [1943] noted, but did not remark on or attribute to the much higher stratigraphic position of the specimen. These include a gentle curvature to the parietal margin of the zygomatic process (the form of the process is triangular in the earlier specimens and both the parietal and the much shorter sphenoidal margins are straight), the temporal squama is higher (more closely approaching the modern condition), and the floor of the tympanic plate is thinner.

A new excavation of Dingcun site in 1976 yielded a fragment of a child's parietal that seemed to possess an Inca bone, judging from the contour of its superoposterior corner. It is interesting that one of the Xujiayao parietals (approximately 100 ky in age) has the same feature.

In 1978, a fairly complete skull of the archaic type of Homo sapiens was found in Dali, Shaanxi [Wu, 1981]. According to the stratum and faunal correlations it is from the terminal part of the Middle Pleistocene, 100 ky or more younger than the top of the Zhoukoudian Lower Cave. It has a series of common features shared with other human fossils in China, including a frontal boss, sagittal keeling, *eminentia cruciata* near bregma, a posterolateral slant to the supraorbital torus (as seen from above), an angular torus, and an Inca bone. The profile angle of the nasal bone is large, the face and nasal saddle flat. Contrasting with the robustness of the vault, the face is rather low. The frontosphenoidal process of the zygomatic bone faces more forward. The junction between the lower margin of the maxilla and the zygomatic bones is more angular. Thus, the face can be characterized as quite flat; in fact, the naso-malar angle (145°) approximates the Eskimo mean. The frontonasal and frontomaxillary sutures combine to form an almost straight horizontal line. All of these features are different from those in Neandertals [Wu and Wu, 1982b]. In addition, the shape of the orbit and the anterior position of the greatest cranial breadth also differentiates Dali and Neandertals into different geographical groups. In the preliminary announcement of the skull [Wu, 1981], it was concluded that the new evidence firmly supported the contention that

> . . . the human evolutionary line is continuous in China from Yuanmou Man to modern Chinese. This is in opposition to the replacement model of human evolution on the development of Chinese fossil man.

In 1980, another skull of *Homo erectus* was found from near Hexian, Anhui [Wu and Dong, 1982]. The age is thought to be contemporary with the later part of the Zhoukoudian Lower Cave. This skull has a metopic suture that is not obliterated, as in Sinanthropus skull L2. Sagittal keeling extends from the level of the frontal tubercle through the vicinity of bregma. This region of the forehead exhibits a frontal boss similar to that in the Zhoukoudian crania. The frontal sinus is small, and does not extend to the roof of the orbits. As seen from above, the superior borders of the orbits are rounded. Interestingly, while many of the Hexian features resemble the Sinanthropus crania, certain details such as the poorly expressed supratoral sulcus of the frontal, the well-excavated *sulcus supratoralis* of the occiput, and the general configuration of vault shape as seen from the rear, more closely resemble the Indonesian forms.

Based on the observations mentioned above we can make a new evaluation of the various features indicating morphological continuity of human fossils in this region.

Incisor shovel shaping is the most consistent feature characterizing Pleistocene fossil humans in China, beginning with the earliest specimen known (from Yuanmou—now considered to be early Middle Pleistocene in age [Liu and Ding, 1983]). It still maintains very high frequency in living Mongoloids [Hrdlička, 1920; Cadien, 1972]. Although it does appear in other regions, it is either at low frequencies or limited to certain periods of time or to certain smaller areas. This feature has been considered as an adaptive hereditary structure [Portin and Alvesalo, 1974; Cadien, 1972]. Although the face is not as well represented in the region as we wish, the known material shows a number of distinctly regional features. One of these is size (mainly height) reduction. Compared with other areas, facial and posterior dental reduction is earlier in north China than anywhere else. Indeed, posterior tooth size is smaller in the Zhoukoudian Lower Cave sample than it is in Holocene fossil Australians. Third molar agenesis is a related feature that appears early, and persists through the fossil record of the region. Its frequency is still highest in Mongoloid populations [Cadien, 1972]. The flatness of the upper middle face and particularly of the nasal saddle is also a frequently found feature in the Pleistocene human populations of China. The frontosphenoidal process of the zygomatic faces more forward than in other regions. The junction of the lower margins of the maxilla and zygomatic bones is more angular in shape when viewed from beneath, and Wu and Wu [1982b] contend it is in the lower face that regional features are most distinct for the Middle Pleistocene Chinese specimens.

Some regional features clearly change through time. Certain of these characteristics that have become less frequent in the region still occur more regularly in Mongoloids than in groups from any other area. Thus, sagittal keeling appeared more regularly in the earlier skulls and became gradually fainter in later specimens. The frequency of Inca bones is higher in earlier specimens and became rarer in later ones. The presence of a relatively small frontal sinus (restricted to the interorbital area) is consistent, but the sinus reduces in absolute size throughout the Pleistocene. In earlier crania frontonasal and frontomaxillary sutures form an almost straight line. In *Homo erectus* from the region the upper and middle part of the nasal bones are more or less similar. Exostoses on the jaw bones are more frequent in Sinanthropus, they have lower frequencies in later specimens from China, but still characterize Mongoloid populations [Oschinsky, 1964].

Although all of these characteristics are not absolutely absent in other populations of the Pleistocene world, they have much lower frequencies and are distributed discontinuously in regions other than China.

In sum, when we consider the possibility of replacement, how can we account for the many similarities between the presumably extinct ancient inhabitants and their modern counterparts of different ancestry? If all of these common features were obviously adaptive, an explanation of common adaptation would be possible. However, this is not the case. Instead, we feel that the evidence firmly supports only one interpretation of the fossil data. That is, the evolutionary sequence in north Asia clearly indicates regional continuity in its pattern of morphological variation. If this pattern of continuity can be shown at all, it can be shown for the full span of habitation for the area.

Southeast Asia

The southern regional sequence has a similar theoretical history. The suggestion of an Indonesian/Australian relationship goes back to Dubois's [1922] analysis of the Wajak remains. The features of the *Homo erectus* remains from Trinil and Sangiran, plus the Ngandong series, led Weidenreich to conclude that the Indonesian hominids

> . . . agree in typical but minor details with certain fossil and recent Australian types of today so perfectly that they give evidence of a continuous line of evolution leading from the mysterious Java forms to the modern Australian bushman [1946, p 83].

Weidenreich was hampered in his analysis of the three Australian fossils then available (Talgai, Cohuna, and Keilor) by the fact that none of these appeared to be well dated [1945]. Like Keith [1931], who described the first two crania at length, he saw dramatic affinities between these specimens (and Wajak) and the modern Australians. Weidenreich, however, also had the Ngandong sample for comparison, and with these time-successive samples he hypothesized a continuous (although not unique) evolutionary line from *Homo erectus* in Indonesia:

> . . . there is now an almost continuous phylogenetic line leading from the Pithecanthropus group through Homo soloensis to the Wadjak man and from there to the Australian aboriginal of today [1945, p 31].

While Weidenreich [1951] did not construct a complete list of regional features that he saw in Indonesian and Australian hominids, he cited the forehead shape, the prebregmatic form of the frontal bone, the obelionic region, and frontal sinus form and development as the basis for his assertion.

Coon [1962] added little to this interpretation, reaching the same conclusion from the same three Australian fossils and the Wajak remains, noting that cranial features "definitely link the Australian aborigine to the succession of . . . populations in Java" [1962, p 410]. More recently, Macintosh expressed this argument by noting that in the Australian fossil hominids "the mark of ancient Java is on all of them" [1965, p 59].

Since Coon reviewed the region, there has been an important expansion of the Pleistocene Indonesian fossil record, including the recovery of at least 23 new *Homo erectus* specimens from Sangiran, and the critical calvarium from Sambungmachan which provides a link between Indonesian *Homo erectus* and the Ngandong sample that follows it in time [Jacob, 1976; Wolpoff, 1980a]. Yet, initial publication of these new materials, including Sangiran 17, the most complete of all known *Homo erectus* crania [Sartono, 1975; Jacob, 1973], has been largely descriptive, with the main focus of attention on the question of taxonomy within the Indonesian sequence [Jacob, 1975, 1976; Sartono, 1975, 1982; Holloway, 1981]. Thus, Jacob recognizes three species of *Pithecanthropus* (*modjokertensis, erectus, soloensis*) as well as a second genus, *Meganthropus*. In his earlier publications Sartono recognized two subspecies of *Homo erectus* in the Indonesian material [1975, p 356], designating these as *ngandongensis* (a large-brained variety) and *modjokertensis* (a small-brained variety), and also assumed that *Meganthropus* was a separate genus. By 1982, he revised his view, coming to believe that there were three separate groups in Java, representing successive stages, in hominization: 1) *Australopithecus*, represented by a *Gigantopithecus*-like form; 2) *Homo paleojavanicus*, represented by an earlier group (ie, *Meganthropus*) and a later one (Weidenreich's *Pithecanthropus robustus*); and 3) *Homo erectus*, represented by an earlier group (Weidenreich's *Pithecanthropus erectus*) and a later one (the Ngandong sample).

Taxonomy is also the focus of the attempted cladistic reevaluation of the Ngandong fossils [Santa Luca, 1980]. Issues that might be relevant to problems of evolutionary progression and regional variation are submerged in a phylogenetic analysis that results in the contention that there are two Indonesian lineages coexisting for a half million years or more. Santa Luca does raise several points of interest in an analysis of the east Asian and Australasian region. He concludes, for instance, that regional similarity is more important than grade in comparing (what we regard as) the extremes of an east Asian cline during the Middle Pleistocene (in the analysis Ngandong was considered to be Middle Pleistocene for morphological reasons and because this earlier date reduces the span of coexistence for the two lineages Santa Luca perceives). However, the study has a number of limitations, not the least of which is its exclusive focus on the Ngandong remains in detailing character states and the exclusion of many of the earlier

Indonesian *Homo erectus* specimens, as well as the (possibly contemporary—[see Xia, 1982]) H3 cranium from the highest levels of the Lower Cave at Zhoukoudian, in the comparative analysis.

With regard to the origin of modern *Homo sapiens*, hypotheses span the entire range of possibilities. Sartono [1982], supporting an earlier suggestion made by von Koenigswald [1973], argues that *Ramapithecus* was the last common ancestor of the completely separated regional sequences of hominid evolution in Africa and Australasia. In his view, there is an unbroken (although overlapping) line of Australasian evolution from an early australopithecine form through regionally distinct early modern *Homo sapiens* populations such as represented at Wajak and Kow Swamp. Only after the appearance of modern *Homo sapiens* do the subsequent population expansions lead to gene flow and interbreeding between the previously isolated groups, resulting in greater similarities between living populations than characterized the isolated lineages of the past. Jacob's [1976] position is somewhat different. He argues that the so-called Australoid features are primitive for all *Homo sapiens*, and therefore the Australoid features of the early modern population represented at Wajak reflect time (ie, grade) and not region. Jacob finds Mongoloid features as well as Australoid ones in the Wajak remains, including the broad face, and the flat nasal root lacking a depression at nasion. Thus, Jacob contends that Wajak was not a direct ancestor of the living Australians. However, Jacob does not support a rapid replacement hypothesis for the region since he regards part of the ancestry of modern Australians as stemming from Indonesia at some point during the time span between Ngandong and Wajak, involving at least some populations that are old enough to not be fully "modern."

The other extreme in interpretations is provided by Santa Luca [1980] and Howells [1967]. Santa Luca contends that "no clear morphological connection can be established between the Ngandong skulls and the Wadjak/Keilor group" (p 120) [see also Stęślicka 1947]. Howells develops this theme further, giving considerable weight to the date and affinities of the Niah remains (but ignoring Brothwell's [1960] admission that the specimen was a burial and that the date was obtained from an "equivalent level" elsewhere in the cave). Howells argues that since Niah and Ngandong may well have been contemporary [1967, p 236], one could hardly have given rise to the other. By 1973, with the evidence from Mungo, he expanded this argument to encompass the simultaneous appearance of people similar to Niah in Australia. Although at the same time he was puzzled by the more archaic appearance of the later Kow Swamp remains (9–12 ky BP), he contended [1973b, p 112] "the Kow Swamp people strongly suggest Solo man, but they do not actually show his special anatomical traits." As recently as 1983, Howells still rejected the Weidenreich interpretation, and therefore remains the only major scholar to argue for sudden replacement in Australasia based on evidence from within the region [1983; Howells and Schwidetzky, 1981].

Unlike the northern and southern ends of east Asia, no new fossil material that could be helpful in understanding the origin of modern *Homo sapiens* has been recovered in Indonesia since even before Weidenreich's time [Glinka, 1981]. The Wajak specimens remain alone in the span between the Ngandong sample and the historic present, and these fossils have defied all attempts at date determination. Although the degree of fossilization has suggested a Late Pleistocene age [Dubois, 1922], a more recent, perhaps Holocene, date is indicated by the lack of nitrogen in the bones [Oakley et al, 1975] and by the similarity of the (presumably associated) fauna with fauna from the Sampung Cave Holocene deposits [Jacob, 1967].

These two specimens have been commonly regarded as Australoid [Dubois, 1922; Weidenreich, 1945; Coon, 1962; Howells, 1967]. Alveolar prognathism and an indistinct nasal margin such as found in both of the Wajak faces are indeed unusual in modern Chinese, although the condition is approached by the Late Pleistocene palate from Hong Kong [Jacob, 1968], and by the Chilinshan palate. These features reflect the Australoid affinities of the specimens, as do the convergence of the temporal lines well behind the orbits, the marked cranial vault thicknesses, and the great breadths of the midface (bizygomatic, bimaxillary, and nasal). Moreover, in the palates and mandibles of the Wajak specimens, there is a degree of robustness reminiscent of the Australian populations. In fact, while far from identical, the palate and dentition of Wajak 2 is astonishingly similar to the much earlier Sangiran 4 specimen, as Dubois [1922] pointed out. The Wajak dentitions are quite large, anteriorly and posteriorly, and would be unusual in a Late Pleistocene sample from continental Asia.

On the other hand, Jacob [1967] is convincing in his contention that a number of the facial and palatal features of the specimens seem to have been influenced by significant gene flow from the north. Both of the faces are quite flat, especially in the middle and upper portions. The nasal root is not depressed, and the nasal bones themselves are nonprojecting and, taken together, form a surface that is very flat transversely along its entire length. In the midface, the zygomatic processes of the maxilla form a flat surface, in an almost straight line between the junctures with the zygomatics, broken only around the nasal apertures, which projects slightly (resulting in a very poorly expressed canine fossa as is found in many modern south Chinese faces, as well as in the Chilinshan fossil face). The orbits are broad and low, basically rectangular in shape, and their inferolateral edge is rounded.

Indeed, many of the Wajak features commonly regarded as Australoid are either not particularly Australian-like or are not unique to Australia. For instance, neither specimen has a supraorbital torus (as is commonly reported). The supraciliary arches are only moderately developed, and only cranium no. 2 has any expression of a lateral toral structure (between the trigone and the supraciliary eminence). The expression of these frontal superstructures in Zhoukoudian Upper Cave cranium no. 101 and Siberian crania such as Serovo exceeds that of either of these alleged Australoids. Other Wajak features interpreted as Australoid are also shared with Late Pleistocene and recent specimens from China (especially, but not uniquely, from south China). The forehead slope can be matched in many modern Chinese crania. The broad interorbital area has its counterpart in the Upper Cave crania as well as in Liujiang. The flatness across the nasal bones is also similar to Liujiang; in this case, both of these more southern specimens contrast with the Upper Cave crania. The lambdoidal flattening and hemi-bun of Wajak 1 are easily matched in the occipital regions of Upper Cave no. 101 and Liujiang.

Thus, while they resemble the Australian populations in a number of features, we concur that the Wajak remains seem to reflect the effects of gene flow from the North. Another Late Pleistocene specimen we regard as reflecting these effects is from the Tabon cave in the Philippines, reported to have an indirect radiocarbon date of about 23 ky BP. Macintosh [Macintosh and Larnach, 1976] considers the specimen Australian-like, and Howells [1976c] calls it "non-Mongoloid," elsewhere [1973a] suggesting Ainu affinities. We agree that some features show similarity to the Australian populations; particularly, the great sagittal length of the frontal, the contrasting narrow biorbital and orbital breadths, and the marked projection of the supraorbital region in front of the anterior face of the frontal squama. At the same time, however, we regard the Tabon frontal as exhibiting a striking resemblance to the Upper Cave specimens, especially to no. 101, but for the features mentioned above and the more delicate and reduced vertical thickness of the lateral portion of the supraciliary arches. Both Tabon and Upper Cave no. 101 are characterized by a well developed glabellar portion of the supraorbital region, with the full extent of the supraorbital area separated from the frontal squama by a shallow, broad supratoral sulcus. The inferior surface of the supraorbitals is uncurved (horizontal). Above this area, the squama is evenly curved between the prominent but low-positioned temporal lines. The postorbital constriction is minimal, and dimentionally the minimum frontal breadth approaches the maximum breadth of the frontal. The sagittal curvatures of the two specimens are about the same. Below this area, the interorbital region is broad, but the nasal bones are quite narrow with a strong central ridge. There is very little depression of the nasal saddle, but both specimens show a moderately high nasal angle (the angulation mainly involves the lower portion of the bone).

The great antiquity of intermediacy for specimens geographically between the north and south extremes, for instance the Middle Pleistocene *Homo erectus* specimen from Hexian, indicates that such gene flow has been long standing in east Asia, and likely was dynamically stable. One might describe this as a morphological cline.

Perhaps not surprisingly, as Jacob [1967] and Santa Luca [1980] have claimed, there are no specific features of the Wajak specimens showing unique continuity with the Ngandong remains (we believe the situation would probably be different if the faces and jaws of these earlier Indonesians were known). Instead, to find clear evidence of continuity with the Ngandong and earlier Indonesians, it is necessary to examine the Late Pleistocene human populations from even further to the south.

The Australian Record

The expansion of the Australian fossil record since the writings of Coon and Weidenreich has been dramatic. The sites at Kow Swamp, Lake Nitchie, Coobool Creek and the Willandra Lakes in southern Australia, and at Cossack in Western Australia, include hominids that in our view substantially reinforce the anatomical basis for links between Indonesia and Australia [Thorne, 1971a,b, 1976, 1980a,b; Thorne and Macumber, 1972]. Moreover, occupation of Australia is now known to span *at least* 40 ky, and well-dated human fossil remains include perhaps the earliest known specimen attributed to anatomically modern *Homo sapiens*, Lake Mungo III, dated to 30 ± 2 ky [Bowler and Thorne, 1976].

In the search for the geographic source of the Australians a number of conflicting hypotheses have been proposed. All assume or argue some Indonesian involvement. The simplest, that of Abbie [1966, 1968, 1975], suggests that all present variation is a result of local adaptation and drift. Abbie's argument is based on examination of living and recent cranial material, and on an assumption of minimal time depth to Australian prehistory. It is important to note, however, that Abbie regarded the Tasmanians as a separate Melanesian issue, as the Tasmanians "never were on the Australian continent" [1968, p 23]. Macintosh [1967] noted that there appeared to be two prehistoric forms, although by the time of his final work [Macintosh and Larnach, 1976] he concluded that the observed prehistoric variation could be subsumed within the modern range of features (a contention also maintained by Howells and Schwidetzky [1981]). As noted above, Macintosh assumed that all Australians had some Indonesian genetic base. Birdsell's trihybrid theory [1949, 1967, 1977] is the most detailed migrational construct, based entirely on his extensive knowledge of contemporary variation. In it, Birdsell perceived three waves of morphologically distinct groups crossing from Indonesia to Australia at different times. Birdsell does not identify the geographical source of the first wave (the Oceanic Negritos), but sees an Ainu link for the second wave (the Murrayians), and India as the evolutionary center for the final migrant group (the Carpentarians).

The most recent proposals to explain the Pleistocene Australian cranial variation [Freedman and Lofgren, 1979a,b; Thorne 1977, 1980a,b] involve two morphological sources, with the migration of people characterized by robust morphologies from Indonesia and a separate movement of groups displaying more gracile features from continental east Asia, followed by admixture within Australia to create the Holocene Aboriginal range of forms.

Australian links to the much earlier Indonesian populations are based on some elements of morphology (see below). No population or fossil exactly resembling the Ngandong hominids has ever been found in Australia. Many of the differences between the fossil Australians and these earlier Indonesians are undoubtedly due to natural selection, as Abbie originally suggested. Yet, it has been difficult to detail the sources of the selection acting on these populations. For instance, while north Asia is characterized by a continuously changing environment, the environment and ecology of the south have been relatively stable through most of the Pleistocene [Chappell, 1976], environmentally disturbed mainly by the dramatically changing sea levels [Keesing, 1950; Chappell and Thom, 1977], and ecologically changing mainly in the decline and extinction of local megafauna. This makes it difficult to propose an argument of response to changing environment as the basis for Australindonesian hominid evolution. Another difficulty in establishing the details of Late Pleistocene hominid evolution in the region, and the details of the peopling of Australia, stems from the virtual absence of a clearly associated archaeological record [Hutterer, 1977; Bartstra, 1982]. Neither Indonesian nor earlier Australian adaptive strategies can be reconstructed in any obvious way from the absence of evidence. Moreover, there is no demonstrable geographic source area for the earliest Australian lithic assemblages. These earliest Australian archaeological sequences do not display clear links to forms in Indonesia [Hutterer, 1977] or to anywhere else for that matter. The fact that few if any stone tools are yet to be clearly associated with Javan *Homo erectus* [Bartstra, 1982; Soejono, 1982] may be significant, if stone working proves to be a relatively minor aspect of Middle and later Pleistocene Indonesian material culture. Later in time, the presence of edge-ground tools in north Australia and in Japan, between 20 and 25 ky ago, and their apparent absence elsewhere at this age, is intriguing. Yet, at the moment, overall attempts to explore the source of the first Australians necessarily rely on skeletal data alone.

General resemblance to Indonesian hominids. A number of the Pleistocene and early Holocene Australian fossils show specific resemblances to the Indonesian fossil samples from both Sangiran and Ngandong. Such similarities of the vault include: the posterior position of the minimum frontal breadth, well behind the orbits; flatness of the frontal in the sagittal plane (even when disregarding those specimens from Coobool Crossing and Kow Swamp that Brown [1981] considers to be definitely or possibly deformed); the horizontal orientation of the supraorbital's lower border; and, the distinct prebregmatic eminence. In fact, however, most of the similarities connecting the earlier Indonesians and the Australian fossil hominids are found in the face. The unfortunate absence of faces from Ngandong allows only comparisons with Sangiran to be made. The facial similarities include the marked prognathism characteristic of the region from at least the time of Sangiran 17 [Thorne and Wolpoff, 1981], the maintenance of large posterior dentitions throughout the Middle and Late Pleistocene, the persistence of the zygomaxillary ridge, eversion of the lower border of the malar, and rounding of the inferolateral border of the orbit.

Specific resemblance to Ngandong. Weidenreich [especially 1943, pp 248–250] was quite emphatic in his contention that living Australian Aborigines show a recognizable heritage from the Ngandong population. He focused particularly on the occasional appearance of a well-developed, supraorbital torus discontinuous over glabella but undivided over each orbit, associated with a flat, long, receding forehead lacking a supratoral sulcus. He also noted the importance of prelambdoidal depressions, sharp angulations between the occipital and nuchal planes, and short (or nonexistent) sphenoparietal articulations in the pterion regions, as additional evidence for this relationship.

Larnach and Macintosh [1974] quantified the comparison, scoring 207 Australian Aborigine crania, 80 New Guinea Aborigines, and comparative samples of Europeans and Asians for 18 characteristics that Weidenreich attributed uniquely to the Solo crania. Of these, six were absent in all of the modern samples. *Nine of the twelve other features attained their highest frequencies in either the Australian or the New Guinea samples.* These were: the large rounded zygomatic trigone; absence of a supraorbital sulcus; suprameatal tegmen; transverse squamo-tympanic fissure; angling of the petrous to tympanic in the petro-tympanic axis; lambdoidal protuberance; marked ridge-shaped occipital torus; external occipital crest emerging from the occipital torus; marked supratoral sulcus on the occiput. The authors did not consider frontal flattening. If they had, this would emerge as a tenth Solo character found at its highest frequency in modern Australians.

Moreover, contrary to claims that have been made by Howells [1976b; Howells and Schwidetzky, 1981], there are unique points of resemblance between a number of the *fossil* Australian specimens and the Ngandong Indonesians. Kow Swamp 9, for example, preserves a nuchal torus configuration that is very common at Solo, with a marked posterior projection of the torus itself combined with a distinct separation of the torus from the occipital with superior and inferior sulci, and an inferior dip at inion. The lateral supraorbital corners of this specimen form a posteriorly facing triangle (lateral frontal trigone) whose apex is the temporal ridge—a condition otherwise unique to the Ngandong specimens.

Cohuna, a member of the Kow Swamp group, shows several of the same features (for instance, a posterior position for the minimum frontal breadth and a (small) lateral frontal trigone). In addition, there are the following resemblances to the Ngandong sample: the coronal suture closely approaches the anterior angle of the temporal squama; the superior border of the temporal is straight, while the posterior border is very low; the maximum cranial breadth is found at the mastoid angle of the parietal; the *torus angularis* is found at the lambdoidal suture, superior to the asterionic region; the temporal line forms a ridge spanning the entire length of the frontal (a characteristic shared by virtually all of the Kow Swamp males).

On the frontal of Kow Swamp 7, the broad supratoral sulcus is continuous with lateral sulci paralleling the temporal ridges on their medial sides. The resulting continuous sulcus thus outlines the entire superior portion of the frontal squama on three sides, resulting in the appearance (but not the actuality) of a very broad, low, frontal boss. This condition characterizes all of the Ngandong males.

The Kow Swamp 5 glenoid fossa closely resembles the very unusual Ngandong condition in which virtually the entire articular surface anterior to the fossa's roof forms a vertical (ie, anterior) wall. The horizontal position of the articular surface, anterior to this wall, is so horizontally short that there is no inferior surface for the articular eminence. Like the Ngandong fossae, the Kow Swamp 5 glenoid fossa in its entirety is deep and particularly similar to the Ngandong fossae in its sagittal narrowness.

These resemblances are not restricted to the Kow Swamp group. In particular, remains from Talgai, Mossgiel, Lake Nitchie, Coobool Creek, and Cossack share these features to a greater or lesser extent. However, these crania are variable. The two sites with rather large samples, Kow Swamp and Coobool Creek, include individuals at the opposite ends of the morphological spectrum, and indeed most of the specimens at these sites represent the intermediates between these extremes.

Thus, we conclude that a very specific case can be made linking features in some of the Pleistocene Australians with the Ngandong fossils. At the same time, the range of variation and the morphological details of that variation indicate that the ancestry of the Australians is more complex than a simple unique line of descent from the Ngandong folk.

Other sources. Weidenreich [1943, p 250], as discussed above, also contended that not all of the Australians are the descendants of the Indonesian (Sangiran → Solo) line. The question of additional sources for the modern Australian gene pool is the basis for a major debate within Australian paleoanthropology.

Weidenreich [1945] regarded the Keilor vault (radiocarbon dated on bone collagen to 12.9 ky BP according to Oakley et al [1975]) as virtually identical to Wajak 1, using the phrase "members of the same family" to describe the degree of their likeness. There is indeed considerable similarity between these two, particulary in size, proportions, and facial flatness. At the same time, some of the differences include the weak angular torus of Keilor (Wajak lacks a torus), Keilor's lack of an occipital bun but its distinct centrally located nuchal torus, the much greater glabellar prominence in Wajak, and the larger but less everted Wajak malars (lacking the zygomaxillary ridge of Keilor and other Australians). At the least, the Keilor/Wajak comparison shows the continued influx of peoples from Indonesia at a time when the gene flow from continental Asia had left its mark. However, the south China specimen from Liujiang can be added to this comparison, because here too the similarities outweigh the differences. Again the similarities include size, proportion, and especially facial flatness. Moreover, Liujiang also shares the wide interorbital area, low square orbits, indistinct nasal margin, and thickened lateral supraorbital corners with the other two. Liujiang, however, is less prognathic, has the most prominent nasals and highest nasal angle, the weakest nuchal area, the least-prominent canine jugae, and the most-rounded anterior alveolar margin of the three. That is to say, it is the most Chinese-like. Nevertheless, there is little doubt that the comparison of these three crania confuses the question of the source area for at least some of the Australian immigrants. While Indonesia has always been a certainty, continental Asia is an additional possibility [Thorne, 1980a].

Another comparison that supports a broader regional source for the Australian populations is that between the Ziyang and Mungo 1 females. Mungo is the less complete of these two, consisting mainly of a calotte. While regional identity is not easily established from the calotte alone, the similarity of these two specimens is interesting. It involves the very small size of both specimens, similar proportions, the similar degree of central supraorbital development (although vertically thicker and much more anteriorly projecting in Mungo 1), the separation of the supraorbital from the frontal squama by a distinct supratoral sulcus, and the vertical separation of internal and external occipital protuberances.

Australian remains ranging from the more gracile variants at Kow Swamp and Coobool Creek to Mungo 1 and 3, Keilor, and Green Gully, show elements that could be explained by ancestry from the north Asian populations described above. Whether this indicates several different geographical sources for the Pleistocene Australians or reflects the changing nature of the gene pool in a single source area (presumably Indonesia) cannot be resolved at present. The fact is, however, that both present and prehistoric Australian populations reflect a dual ancestry that involves the two distinct morphological clades that have existed in east Asia since the earliest hominid habitation. Events in the later Pleistocene shift the preexisting balance between populations at the northern and southern ends of the range in east Asia (see below). Yet, the end result in Australia is a series of populations in which the effects of the Indonesian ancestry (the Sangiran → Solo line) predominate.

If our reasoning and interpretations are correct, east Asia in the Late Pleistocene begins to act as a population source for more peripheral areas to the east (the Americas), southeast, and south [Stewart, 1974; Brace, in this volume; Brace and Hinton, 1981, and references therein], and certainly by the Holocene if not before, for the north and west.

A THEORY OF MULTIREGIONAL EVOLUTION

The multiregional evolution hypothesis [Thorne and Wolpoff, 1981] proposes an explanation for the earlier observations of local regional continuity in east Asia and elsewhere, and the more recently discovered evidence for local differentiation and continuity within this region discussed above. We believe that the observations that were contradictory in Weidenreich's scheme and unconvincingly explained in Coon's can be satisfactorily accounted for.

The multiregional hypothesis was initially developed to account for observations that were made by Weidenreich, and that continue to characterize a vastly expanded fossil sample from east Asia. The hypothesis stemmed from attempts to relate the morphological differences seen in *Homo erectus* remains from northern and southern portions of east Asia to the morphological differences between populations from these extreme ends of the region today. The resulting evolutionary model was applied in detail to a comparison of Sangiran *Homo erectus* remains and the Pleistocene Australian sample from Kow Swamp. However, from a western Pacific perspective, the Australasian sequence that was used in this initial development is merely the southern end of what we view as a long lasting cline stretching northward through what now is Malaysia and Thailand, to east and northeast Asia. Pleistocene fossil remains representing the time span this cline existed have been mostly limited to its extreme ends, resulting in their interpretation as sampling markedly different groups (ie, the Weidenreich/Coon notion of the Sinanthropus → Mongoloids and Pithecanthropus → Australoids dichotomy).

While these ends do represent the extremes of morphological variation over the same delineated area today, and perhaps in the past, we believe the evolutionary situation to be much more complex (and interesting) than Coon's idea of isolated and separate groups of *Homo erectus* independently evolving at different rates into distinct modern populations. Indeed, modern population biology provides no model that would allow this [Dobzhansky, 1963].

Gradual Evolution in Polytypic Species

Observations of regional continuity in the fossil record are not limited to the hominids. For instance, two different geographically dispersed rodent groups, *Cricetodon* species [Freudenthal, 1965] and *Sigmodon* species [Martin, 1970], have maintained geographically distinct subspecies through evolutionary changes regarded as crossing species boundaries. In the case of *Sigmodon*, two temporal species span a time range from the Late Pliocene through the Middle Pleistocene—a range quite equivalent to the span of the genus *Homo*. The history of the European genus *Cricetodon* is somewhat more complex, since cladogenesis also seems to have characterized populations in all of the regions where the geographically distinct subspecies are represented. In neither case is an evolutionary explanation provided.

Perhaps the lack of explanatory attempts outside of paleoanthropology should not be surprising. The fact is that apart from hominids and the two genera discussed above, fossil histories of paleosubspecies are virtually unknown [Van Valen, 1966]. We suspect that this is not because geographic variation within fossil species was rare, but rather because the tendency has been to interpret such variation at the species level because it fits the morphospecies or evolutionary species criteria [Levinton and Simon, 1980]. It is quite possible that long-lasting geographic variation within past species can only be recognized when the interbreeding concept of biological species [Mayr, 1969] is used to generate criteria for recognizing fossil species, and even then polytypism may be difficult to recognize.

> Different groups of organisms are subjected consistently to certain speciation modes because of their genetic, population-structural, and ecological attributes. For instance . . . there are very consistent differences in population structure between frogs and mammals such that mammals are much more likely to display higher degrees of population subdivision . . . this implies that mammals are more subject to rapid population divergence . . . [and] that mammals are more prone to form polytypic species. The result is that . . . the number of fossil mammalian species could actually be overestimated [Templeton, 1982, p 117].

Apart from the multiregional evolution hypothesis, to our knowledge, no other evolutionary model has ever been proposed to relate the long-term evolution of central and peripheral populations in a widespread polytypic species except for the peripatric speciation model of allopatric speciation (not to be confused with the *para*patric model of sympatric speciation) and its various punctuational formulations [for instance Mayr, 1954, 1982; Gould and Eldredge, 1977; Stanley, 1978, 1979]. Effective criticisms of the punctuational model [Levinton and Simon, 1980; Charlesworth et al, 1982; Templeton, 1980, 1982] focus on virtually every aspect of evolutionary theory *except* the problem of gradual change within a polytypic species. Gradual long-term evolution in polytypic species is not unknown, but to date it has not been explained in a convincing way. Indeed, it is our observation that commonplace as they are [cf Mayr, 1969, pp 38–50], polytypic species are mainly discussed in the evolutionary literature with regard to peripatric speciation. In most other contexts, the analysis of geographic variation is simplified with the panmictic assumption.

The reason for this is the widespread perception that there is no fundamental difference between monotypic and polytypic species; ". . . the difference between monotypic and polytypic species . . . is not a profound one" [Carson, 1982, p 419]. We cannot reconcile this perception with the evidence for a completely different parceling of genetic variation in widespread polytypic species, in contrast to monotypic species [see Templeton, 1980, and references therein]. For instance, Ayala [1982] finds between four and 11 times as much genetic distance between subspecies as between local populations in the major vertebrate groups. We believe that the genesis of this situation is not so much in a conviction that polytypic species can only change through peripatric speciation, as in a lack of realization that polytypic species pose a problem.

While the punctuational model provides an alternative explanation for evolutionary change in a polytypic species, it cannot account for regional continuity in evolving populations across the boundary created by speciation events. Since we contend that gradualism [as defined by Bock, 1979; Gingerich, 1979; and others] is the most likely explanation of Pleistocene evolutionary change in *Homo*, we believe that the punctuational model does not validly apply.

Instead, we propose the theory of multiregional evolution as an attempt to deal with gradual evolutionary change in a widespread polytypic species generally regarded as of sufficient magnitude to represent the evolution of one species into another without resorting to a punctuational explanation with subsequent rapid population replacement. While the theory was developed in order to account for regional continuity during the evolution of the genus *Homo*, it does not rely on any features or behaviors unique to the hominids. We suggest that this theory might account for regional continuity across species "boundaries" in any widespread polytypic form, and consequently, is a valid alternative to the punctuational hypothesis whenever regional continuity can be demonstrated.

Three Attributes of the Genus *Homo's* Evolution

The theory of multiregional evolution attempts to explain three phenomena that characterize the Pleistocene evolution of the genus *Homo*: 1) the *initial* contrast of a variable source population of African hominids, with a number of differing monomorphic peripheral populations at the edges of the hominid range; 2) the *early appearance* of features that show evolutionary continuity (ie, regional features) with modern populations at the periphery, contrasted with the much later appearance of features showing regional continuity at the center; 3) the *maintenance* of these center/periphery contrasts through most of the Pleistocene. These three phenomena are related to the development and maintenance of long lasting clines throughout the hominid range, balancing gene flow against opposing selection and (especially at the periphery) drift.

Ours is not the first suggestion that clines maintained by gene flow and selection characterize the geographic distribution of features in *Homo* [Livingstone, 1962, 1964; Brace, 1964b; Birdsell, 1972; Brues, 1972, 1980]. Indeed, this is the generally accepted explanation for polytypism in any species that is widely dispersed geographically [Mayr, 1963; Stanley, 1978; Morris and Nute, 1978; Charlesworth et al, 1982]. We propose the theory of multiregional evolution as an attempt to show how the specific *pattern of evolution* in a polytypic species can be explained as a consequence of modern clinal theory. Multiregional evolution theory is more than an empirically based clinal analysis, and moreover differs from most classic clinal studies or simulations that must account for *both* the initial distribution *and* the subsequent stable or changing clinal pattern with the same evolutionary mechanisms. The theory we propose takes advantage of the known fossil record, allowing much of the initial distribution of variation to be specified, and suggests that different mechanisms are responsible for the establishment of the initial pattern of variation and the maintenance of the pattern for much of subsequent hominid evolution.

Center and Edge

The process of multiregional evolution involves two distinguishable stages. First, the initial polytypic populations must be established. It has been proposed that these differences resulted from the mechanics of the initial spread of hominids from Africa. The establishment of peripheral populations with intraregional homogeneity and interregional heterogeneity during this spread has been described in the "centre and edge" hypothesis by Thorne [1981], whereby the morphological characteristics of polytypic species reveal

> . . . almost invariably that the degree of polymorphism decreases toward the border of the species, and that the peripheral populations are not infrequently monomorphic [Mayr, 1963, p 389].

During the initial habitation, decreased genetic and morphological variability can result from continued drift in small populations, founder, and peninsula (bottleneck) effects. Indeed, a species expansion involving small populations, the most likely model for the initial hominid habitation of Eurasia [Birdsell, 1968, 1972], creates the ideal circumstances for establishing reduced genetic variation within these small populations [Templeton, 1980, 1982; Wright, 1978].

Specific environments encountered during these initial expansions to varying regions and the random aspects of drift, provide the mechanisms for adaptive divergence [Templeton, 1981, 1982], and help establish genetic differences between the peripheral populations. While some of these differences may respond to specific environments or local habitats, the random nature of genetic drift insures that other variations will not initially reflect differences in adaptation. Thus, the "centre and edge" hypothesis provides an explanation for how the contrasting pattern of central variability and peripheral intraregional homogeneity combined with interregional heterogeneity was initially established. In the "centre and edge" model, as applied to the hominids of east Asia, the "edge" is defined as the eastern geographic limit to the spread of early hominids in the Old World. To what extent this edge is ecologically marginal is a point that is yet to be fully determined; the northern edge of this geographic boundary is likely to also have been an ecological margin, while the southern and eastern edges are bounded by ocean.

Early *Homo erectus* specimens from East Asia (a "central" [cf Briggs, 1981] area and possibly at or near the source area for the species) demonstrate high phenotypic variation [Thorne and Wolpoff, 1981; Coon, 1982]. In contrast, we have shown [Thorne and Wolpoff, 1981; Table 4] that the extensive series of Indonesian hominids at one "edge" of the *Homo erectus* range exhibit significantly reduced variability, holding sex constant, over a similar time span. Regional homogeneity also characterizes the Zhoukoudian crania, in spite of a time span that may be as long as 200 ky between the earliest and latest specimens in the Lower Cave. Thus, the evidence suggests that during the initial spread of hominids to the eastern periphery, a pattern developed in which central variable populations were contrasted with peripheral populations showing marked regional distinctions but relative internal homogeneity.

Regional Continuity

The second stage involves the *maintenance* of this contrasting pattern of central and peripheral variation for long periods of time, during which general evolutionary trends characterize the entire polytypic species. It is our experience that this pattern of contrasting variation for the central and peripheral populations of polytypic species is quite common, although often obscured by the predilections of some taxonomists who transform polytypic species into several monotypic ones based on the presence of geographically distinguishable morphotypes. Actually, polytypic species are a universal phenomenon, present in all families, terrestrial and marine [Mayr, 1969, pp 38–50], and are especially characteristic of mammals because of their population structure [Templeton, 1980].

The polytypic center. There are a variety of factors that could explain the persistence of elevated polytypism at the center of a species range [Mayr, 1963, 1970; Carson, 1955, 1959; Cook, 1961]. These are by no means mutually exclusive. 1) Since by center we mean initially the earliest inhabited area of optimal adaptation, perhaps the area where the species evolved, it is likely that a greater range of morphological variation will be tolerated within groups from that region. 2) Central populations tend to be contiguous and to have relatively higher population densities. Potential competition between these populations could increase the range of interpopulational differences, and thus the heterogeneity of the region as a whole. 3) Gene flow through populations will tend to be greatest at a species' geographic center, because gene flow through central populations is multidirectional. 4) Selection is likely to be relaxed for populations undergoing rapid expansion, which would act to increase the range of phenotypes present at the center of the range of a colonizing species [Carson, 1968].

The monomorphic periphery. In the peripheral (and/or ecologically marginal) populations, events associated with both the initial colonization and the subsequent period of local adaptation help develop the combination of *pronounced difference between* allopatric populations and *relative homogeneity within* them. It is our contention that the observed morphological homogeneity (which at its extreme is monotypism) reflects both a reduced number of genetic polymorphisms and a reduction in the average heterozygosity because it was initially established by the partial isolation of numerous small populations with histories of founder effect and continuous drift for long timespans [Wright, 1978; Birdsell, 1968].

> Ancestral populations inhabiting ecologically and/or geographically marginal areas will often tend to be more subdivided and have smaller deme size than populations in the ecological and/or geographic centers of the species range [Templeton, 1982, p 114].

Given the likely pattern of earlier Pleistocene human population movements throughout Eurasia, these are the circumstances under which reduced genetic variation can be expected in the peripheral or marginal regions [Wright, 1977; Templeton, 1980].

> Founder populations will . . . display considerable losses of genetic variation if, after the founder event, the population remains very small for many generations [Templeton, 1982, p 114].

A consideration of the population structure that probably characterized humans during the process of initial habitation, if not throughout most of *Homo's* evolution, is critical for understanding the genetic conditions within the colonizing species. A population structure such as that reported for the Yanomama Amerinds [Neel, 1978; Smouse et al, 1981], but also observed in nonhuman primates [Cheverud et al, 1978], lies between the panmictic and the isolated island population models [Wright, 1940], and is a reasonable prediction for the population structure of Pleistocene hominids. In the Yanomama population structure, villages split along lines of kinship so that they are distinguished by marked genetic divergence in spite of the (often considerable) gene flow between them. What is genetically important about this structure is the fact that the variance effective population size will be smaller than the inbreeding effective size. According to Neel [1978], the genetic consequences of this structure are: 1) A considerable portion of the genetic variability is distributed between villages. 2) The loci that are polymorphic within a village have a high level of heterozygosity. 3) However, the *overall* levels of *homozygosity* are high.

> Therefore, selection for homozygous effects of an allele are undoubtedly more important with this type of population structure as compared to a large panmictic population [Templeton, 1980, p 1,020].

A single village drawn out of this net to found a new population would carry with it a reduced number of polymorphisms.

> . . . an isolated founder population of Yanomama-like individuals would not simply recreate their ancestral situation. The founder effect would greatly increase the stability of the genetic background in the descendants of the founders; a situation very different from the ancestral condition [Templeton, 1980, p 1,021].

Templeton [1980] further argues that such a founder population would accumulate inbreeding rapidly because so many of its members would be closely related. Another variant of these conditions that applies to the Yanomama (and presumably also to our Pleistocene ancestors) would occur when part of the village forms a founder population, since the founding individuals are almost always more closely related to each other than would be expected in a random sample from the village. Thus, there would also be a reduction in the levels of individual heterozygosities, and one would expect the overall level of genetic variability to be reduced. In his discussion of Neel's data, Templeton concludes that such populations would be less likely to speciate at the periphery than small populations drawn at random from a panmictic species. *These are precisely the observations we have made regarding peripheral hominid populations.* Mainly, they vary interregionally, are each monomorphic, and in no recognizable case did they speciate.

As local adaptive responses develop, these initial genetic differences, diverse patterns of selection, and the continued action of drift in the regions with lower population densities would help to retain if not expand the genetic differences initially established [Lande, 1976]. Moreover, drift and reduced gene flow to the peripheral or ecologically marginal areas might help maintain homogeneity within these populations [Templeton, 1982]. This is because at the species' boundaries "the total amount of gene flow is reduced in peripheral populations and near the periphery gene flow becomes increasingly one-way outward" [Mayr, 1970].

We are aware that not all peripheral populations show reduced genetic variation [Lewontin, 1974]. Our point is that not all peripheral populations have the evolutionary history and continued small population size that would create the circumstances that result in reduced genetic variation. However, it is quite credible to contend that these conditions were met during the time that Eurasia was first inhabited, as they are still met by a number of aboriginal populations today, especially if the population structure was of the Yanomama type as we believe was probably the case. Thus, Wright [1978], in discussing Birdsell's study of population structure in western Australian Aborigines, concludes that

> He estimates the average total number in a tribe to be about 500, with a breeding population of about 185 and an effective number of about 100. This is small enough for the building up of considerable differences among large areas at each nearly neutral polymorphic locus merely by sampling drift.

It is the *combination* of population structure and the small size of peripheral populations that provides the circumstances that account for the long-lasting consequences we perceive.

Finally, in understanding the implications of this process, we recognize that a clear distinction between ecological periphery and geographic periphery should be made [Lewontin, 1974]. These are not necessarily the same; not all ecological margins are located at the extremes of the geographic range, and portions of the geographic edge may not be ecologically marginal. However, while the geographic border of a species does not necessarily have markedly suboptimal conditions, the fact that marginal populations received most of their gene flow from more central ones [Sokal, 1973] tends to prevent optimal adaptations from being achieved at the geographic edge of a species range [Mayr, 1954], and the geographic edge in many cases comes to also act as an ecological margin at which fewer diverse polymorphisms may be maintained [Karlin and McGregor, 1972].

Persistence of regional differences. Given an explanation for the maintenance of the center and edge contrast discussed above, an additional characteristic of the evolutionary history of *Homo* must also be considered. This is the persistence of the specific pattern of regional differences (the interregional heterogeneity) established during the geographic spread of the polytypic species *Homo erectus*. Recognizable elements of these geographic differences were maintained at the periphery over a span of at least three-quarters of a million years. Morphological continuity at the center spans less than half this time. We contend that a dynamic model of long-lasting morphological clines that results from a balance of gene flow and opposing selection and drift best accounts for the observed pattern.

Clinal Theory

Clines are, ultimately, the product of two conflicting forces: selection, which would make every population uniquely adapted to its local environment, and gene flow, which would tend to make all populations of a species identical [Mayr 1970, p 215].

Actually, Mayr's definition conforms to the earlier concept of clines, as described by Huxley [1939] and Haldane [1948] in terms of migration/opposing selection equilibria. Much of the subsequent development of the cline concept has been through mathematical modeling, with the main focus on attempts to ascertain the characteristics, limiting factors, and equilibrium ranges for clinal distributions [Wright, 1940; Nagylaki, 1975;

Slatkin, 1973, 1978]. Computer modeling has been a particularly useful tool, and it has been demonstrated that there is much less sensitivity of equilibria to varying relations of opposing selection and gene flow than was once thought [Hiorns and Harrison, 1977; Slatkin, 1978; Nagylaki, 1976]. Moreover, the effects of gene flow are magnified when clines are two-dimensional [Nagylaki, 1975; Sokal, 1973]. This suggests that high levels of gene flow are not necessary for the maintenance of clinal equilibria encompassing a significant amount of total variation—a precept in keeping with most Pleistocene hominid gene flow models [Wright, 1931; Birdsell, 1972]. Finally, various computer models have been satisfactorily compared to actual clinal distributions [Endler, 1977; Mani, 1980]. In sum, the varying relations of clinal equilibria to differing magnitudes of gene flow, differing patterns of opposing selection, and differing levels of heterozygosity have been established in some detail [Endler, 1977].

One point we find crucial in the development of clinal theory is the contention that differentiation along a cline is virtually inevitable, almost regardless of the level of gene flow [Endler, 1973, 1977; Hanson, 1966; Brues, 1972; Weiss and Maruyama, 1976; Ford, 1975]. Indeed, clines tend to reduce the dispersal role of neutral alleles when these are linked to other alleles affected by the selection maintaining the cline (although even moderately advantageous alleles are unaffected). Thus, "local heterozygosity will be reduced, and divergence between regions increased" [Barton, 1979, p 339].

Why population differentiation is almost inevitable when distributions are clinal was first addressed by Wright [1940] through his concept of isolation by distance. One outgrowth of this idea was the demonstration that long-standing clines do not necessarily require opposing selection but rather can develop from a balance between drift and gene flow when contiguous populations are small [May et al, 1975], as might be expected at the periphery of a species. The presence of clinal distributions is not necessarily a marker of corresponding selection differences over the range of the cline. Thus, clinal distributions involving small populations and a balance between gene flow and drift may be expected to produce (or maintain) more homogeneous populations, especially at the edge of the range of a polytypic species where population densities tend to be lower and the potential for drift is greater.

Selection gradient. It has been contended that there is one other mechanism that could produce clines: a selection gradient in the absence of gene flow [Ehrlich and Raven, 1969]. Discounting the importance of gene flow between demes was part of the general resurgence of sympatric speciation hypotheses [Gould and Johnston, 1972; White, 1978] of the last decade.

This view disregards the interbreeding concept of species by denying that gene flow acts, in any important sense, to unite local populations into a larger integrated genetic entity. Mayr [1970, 1982], Stanley [1978], and others argue strongly against this case. In the absence of gene flow, there is inevitably speciation [Ayala, 1982]. Moreover, modern understanding of the low levels of gene flow necessary to promote stable clinal equilibria suggests that the case itself may have been overstated since Ehrlich and Raven [1969, p 1,228] admit

> we would agree that the introduction of genetic novelties into natural populations, even at a low level, may be important in supplying raw material for selection.

Of course, the converse problem, that is the effect of various amounts of gene flow in the presence of selection gradients, is a rather different issue. In a series of computer simulations, Livingstone [1969] has shown that there is a surprising insensitivity of the form of the cline to differing gene flow magnitudes when a clear gradient in selection exists [see also Jain and Bradshaw, 1966; Antonovics, 1971].

Gene Flow

The Ehrlich and Raven [1969] argument about gene flow and the responses to it brings focus on the related questions of the form, magnitude, and role of gene flow in a polytypic species.

The form of gene flow. With regard to form, the Ehrlich and Raven argument, as the earlier argument by Coon [1962], seems to regard the main form of gene flow as spreading mutations that may be promoted if they prove to be advantageous. We propose that this represents only a small part of the form that gene flow normally takes. Such a view seems to assume that most evolutionary change results from the introduction (or appearance) of new mutations. In hominids, at least, this is a very unlikely assumption, given the extraordinary genetic similarity of humans and chimpanzees [King and Wilson, 1975]. In the words of Lewontin [1974, p 179]

> ... the overwhelming preponderance of genetic differences between closely related species is latent in the polymorphisms existing within the species.

If this is the nature of genetic difference *between closely related species*, one can begin to appreciate how much less is the genetic variation between regional populations of the same polytypic species. Certainly, this variation is an order of magnitude less than the variability within these populations [Dobzhansky, 1970; Lewontin, 1974; Nei and Roychoudhury, 1982]. This brings us to our contention that the normal, most frequent, and probably most important form of gene exchange involves shifting frequencies of alleles already present in the populations exchanging genes. Gene flow is better understood in terms of its effect on differential allele frequencies across a cline than in terms of the transport of one new allele spreading throughout the cline (presumably becoming established in each population before it stands a reasonable likelihood of spreading to the next).

The magnitude of gene flow. The magnitude of gene flow is a second issue. As mentioned above, consideration of clines as two-dimensional (rather than one-dimensional) networks magnifies the potential effects of gene flow on equilibria [Nagylaki, 1975; Sokal, 1973]. Moreover, when a gradient in selection exists, the gradient has a predominating effect on the resulting cline's form; the main effect of gene flow magnitude differences is to change the steepness of gradient across the cline, and in some cases to displace the geographic positions of equilibria frequencies. These effects are magnified when the discontinuous distribution of individuals who live in groups is also taken into account. For these reasons, and because clines can exist when there are selection gradients in the absence of gene flow, we propose that in most cases clines will be maintained within a species regardless of the magnitude of gene flow. It follows that the degree of differentiation along a cline, the steepness of the gradient, and the form that the gradient takes are all somewhat independent of the magnitude of genetic interchange. Put another way, *given the presence of gene flow* (because its absence will eventually result in speciation), *clines will invariably form in a polytypic species*. The particular attributes of these clines will be dictated by all of the relevant variables, including the magnitude and direction of gene flow (and the degree to which it is reciprocal) and the intensity and distribution of local selection, but also the distribution of populations, population sizes and breeding structures, the effect of drift, and the pattern of initial genetic variation. *We reject the ideas that selection or drift can "override" the influence of gene flow* (ie, the magnitude of gene flow is "too small"), *or that gene flow can "swamp" the effects of selection* (ie, the magnitude of gene flow is "too large"). These still widely held precepts make no sense in terms of the balance model of clines. Variation in the evolutionary forces underlying clines can alter their form, but cannot erase the clines themselves [Slatkin, 1973, 1978].

This point is important in discussing the magnitude of actual gene flow between populations. The recent series of claims suggesting that this magnitude is rather low [Ehrlich and Raven, 1969; Endler, 1977] seem to rely as much on the argument that the expected effects of higher magnitudes cannot be found, as on actual measurements of gene flow. Although Mayr [1963] makes some reasonable estimates, actual magnitudes of gene flow between living animal populations remain largely unknown. We take no position on the issue of magnitude; both large and small magnitudes for various species have been reported. Moreover, while the gene flow reported between adjacent living human populations is usually anything but miniscule, we admit the possibility that this magnitude may be lower in other species, or in human prehistory (as might be suggested by the emerging understanding of how different Pleistocene archaeological sequences in various regions actually were). It is our contention that over geologic time, this issue is irrelevant because genetic comparisons between human populations [Piazza et al, 1981; Lewontin, 1974; Nei and Roychoudhury, 1982] or between humans and the African apes [King and Wilson, 1975; Brown et al, 1982; Templeton, 1983] show that the actual magnitude of *genetic* change to be accounted for during the evolution of *Homo* is surprisingly small.

The role of gene flow. The third issue, the role of gene flow in a polytypic species, can be discussed on several different levels. In an important sense, without gene flow there will be no polytypic species [Mayr, 1963]—a point even admitted by Ehrlich and Raven [1969] through their claim that because gene flow is rare (or of low magnitude) between many populations, the biological species definition is incorrect. One way or another, polytypic species require gene flow because in its absence there is eventually speciation. However, this does not fully answer the question about its role. The opposite of the Ehrlich and Raven perspective is the species-unifying concept of Mayr [1963] and others, in which gene flow between populations is seen as supporting cohesion and stability within the species as a whole, maintaining the distinction and identifiability of the species. A rather different problem arises from this concept; mainly, the question of how the process responsible for stability can also allow change [Stanley, 1978; Altukhov, 1982]. The solution offered is that major change happens during peripatric speciation. This is an answer that we hardly wish to deny, but our experience with the evolution of the genus *Homo* suggests that it is not the *only* answer.

To begin with, species cohesion is not maintained by gene flow alone [Lande, 1980]. A rather different explanation for species cohesion is provided in the more recent discussions by Mayr [1970], Carson [1975, 1982], Lewontin [1974], and others. In these publications, the idea of coadapted genetic systems representing the maximum fitness under balancing selection achieved by whole organisms has been envisioned as an internal restraint, limiting how much a species can change without a complete reorganization of these genetic systems (Mayr's "genetic revolution"). Carson [1982] considers all populations of a polytypic species as sharing a common coadapted genetic system, with the populational or subspecies differences seen as a consequence of "a set of genes not tied in so tightly with the balances" (p 418). The coadapted genetic system for a species is not dependent on gene flow for its maintenance.

The problem we find in this concept is the nature of normal subspecific or populational differences in a polytypic species. These often tend to reflect local adaptations. Consequently, this model leads to a contradiction in which a single coadapted genetic system is established within an entire polytypic species because of the heterozygotic combinations promoted by balancing selection to maximize the fitness of individuals, and yet the gene combinations that respond to local selection are not part of this system.

For this reason, critical to the understanding of the role of gene flow is the distinction we believe should be made between factors that maintain species cohesion and stability, and factors that maintain the phenetic similarity of populations within species. These are not the same phenomena; we contend that a special explanation for phenetic similarity is hardly required in the face of the extremely small number of genetic differences distinguishing populations enclosed within the protected gene pool of a biological species.

If we allow that the phenetic similarity within biological species is a separate phenomenon, we can envision the stability and cohesion maintaining mechanisms in a somewhat different manner. When not called upon to explain phenetic similarity, we regard it likely that in polytypic species, coadapted genetic systems are not species-wide, but rather can be expected to differ from population to population. Indeed, such differences would magnify the isolation by distance phenomenon. Our model comes much closer to Wright's [1931, 1977, 1980] shifting balance theory, with morphologically distinct intraspecific groups reflecting different adaptive peaks maintained by stabilizing selection. The idea of adaptive peaks weakens the requirement for a genetic revolution concept of species change by attenuating the effects of coadapted genetic systems on the cohesion and stability of a species (a point not ignored by either Wright or Mayr). Indeed, the question has been raised more than once as to whether genetic revolutions in species-wide coadapted genetic systems actually characterize the course of evolution [Charlesworth et al, 1982; Wright, 1980; Lande, 1980; Templeton, 1980, 1982]. The shifting balance model with its multiple adaptive peaks better fits what is known about polytypic species evolution in the absence of particular speciation events [Charlesworth et al, 1982]. On the other hand, it is certainly not an argument against peripatric speciation. As Lande [1980] has emphasized, the peripheral populations in such a scheme (more subject to drift and therefore with fewer polymorphisms), still retain sufficient variation for potentially rapid evolution.

Gene flow is the latticework that connects the populations of a polytypic species, representing multiple adaptive peaks, through its participation in the support of clines. We therefore suggest that the mechanisms that promote stability and cohesion within a species can in fact also provide for a mechanism to allow species-wide evolutionary change; as changes in the factors that produce clinal equilibria vary, the clines change in response. Because the clines represent a balance (albeit differing from area to area), evolutionary responses can be of much greater magnitude than the magnitude of changes in the equilibria-producing forces, just as a small shift in frequency can cause a great change in the interference pattern between two overlapping waves. Thus, evolutionary change in a species constituted as we describe is neither directly related in magnitude to the amount of gene flow between populations, nor directly limited by the inability to alter coadapted genetic systems (if the distribution of these systems is the object of the clinal variation as we believe it must be since both the genetic systems of populations and the clines between them exist in polytypic species). We envision the polytypic species as a *dynamic* system, externally bounded and protected by the limits to gene flow, and internally diversified by the evolutionary forces sustaining clinal relations between differentiated populations. The stability within such a species does not result in its stagnation.

Clinal Theory and Regional Distinctions

Mechanisms causing geographic variation have been studied by theoretical population geneticists . . . . The extent of genetic differentiation between two or more local populations is determined by the balance between the strength of gene flow and natural selection or random genetic drift [Charlesworth, 1982, p 476].

With regard to our model of hominid evolution, we are especially concerned with the pattern of internal diversification within polytypic species. We suggest that clinal theory, as it is understood at present, provides an explanatory basis for the second part of the multiregional evolution hypothesis; the persistence of the "centre and edge" pattern of contrasting homogeneity at the periphery and heterogeneity at the center of the range, and the maintenance of regional distinctions for populations at the periphery of the hominid range. Eventually, regional continuity as indicated by the appearance of populations recognizably ancestral to Late Pleistocene or recent populations also characterizes the original central region. Finally, subsequent hominid adaptations establish multiple regional centers (in the Late Pleistocene) and the earlier contrasting pattern no longer applies. A long-term mechanism of dynamically changing balances between gene flow, and opposing selection for some features and drift for others, resulted in the creation of a latticelike network of clines; some local, some regional, and some perhaps spanning a large portion of the hominid range from the center to the peripheries. This pattern would account for the maintenance of regional distinctions that were established during the colonization of the inhabited Old World. Moreover, if small population effects were more important toward the periphery, as Mayr [1970] suggests often is the case, relative homogeneity might be expected to persist for some features at the periphery, to the extent that gene flow and drift are the mechanisms that establish the clines for them. If so, it would follow that at least some of the homogeneous characteristics of peripheral populations showing long-lasting regional continuity were not especially adaptive.

The clinal distributions of other characteristics would be expected to result from a gene flow and opposing selection balance. These would be more susceptible to change as a consequence of changing selection across the hominid range, and in hominids it is likely that some of these features respond to rapidly spreading cultural innovations and ideas without significant genetic interchange.

Similarly, other factors than a drift-gene flow balance can contribute to the continued pattern of relative homogeneity at the periphery, including reduced and unidirectional gene flow, and interruptions to the effect of selection due to the influence of drift on adaptive characteristics. Moreover, the most realistic models of gene flow/selection balance regard the magnitude and even direction of selection as differing over the range of the cline. Indeed, the initial pattern of genetic variability may be a factor resulting in differing magnitudes of selection since there is an interdependence between selection and genetic polymorphism. Selection can only act on the genetic combinations present; different combinations can affect both the magnitude and conceivably even the direction of selection. For instance, because of anterior tooth loading, strain may be concentrated in the central supraorbital region in both Eskimos and Australian Aborigines. A marked supraorbital torus, however, may only appear in the Australians because of their much lower foreheads [see Russell, 1983].

Summary: The Pattern of Change

The multiregional evolution hypothesis suggests that drift, due to small population effects, and differences in selection resulting from environmental variation as well as the existence of regionally distinct morphotypes, both characterized populations at the marginal or peripheral portions of the hominid range. The initially different gene pools of these populations were established during the process of first habitation. The pattern of regional variation was maintained throughout most of the Pleistocene by a balance between the local forces promoting homogeneity and regional distinction (selection, drift), and multidirectional gene flow. As a consequence, a long-lasting dynamic system of morphological clines came to characterize the multiregional distribution of our polytypic lineage.

If this was the normal pattern guiding the evolution of the genus *Homo*, we contend that the conflict between gene flow and local selection that underlies so many interpretive disputes cannot actually exist. All evolutionary changes require a balance between these potentially opposing forces and must be accounted for by shifts in this balance (except in the extreme cases of latest Pleistocene and Holocene large scale population movements for which the multiregional evolution model does not apply). It makes no sense, in our view, to argue about whether gene flow or selection or drift "predominated" to account for a specific local evolutionary sequence because the multiregional evolution model requires all of these, and focuses on *the balance between them* in explanation of evolutionary change. In the clinal model, shifts in the magnitudes of the opposing forces change the gradient of the cline, and it is this changing gradient, viewed over time in a specific region, that provides the data to characterize regional evolutionary change. This is not an alternative to the gradualist model; as applied to a polytypic species, this *is* the gradualist model!

DISCUSSION

The data we have presented indicate that all across east Asia hominid evolution led to the appearance of modern populations without either a speciation *event* or a dramatic influx of populations from elsewhere as described in Howells' [1976a] "Noah's Ark" model. That is, the evolution of the genus *Homo* in east Asia was gradual (continuous, although not necessarily at a constant rate), and not punctuational. We have detailed more than sufficient evidence to demonstrate regional continuity over this area, beginning with its earliest inhabitants. We have proposed that the early Chinese samples from Zhoukoudian, Lantian, and Hexian are related to modern populations of China through a number of characteristics, and by morphologically intermediate fossils from Dali, Maba, Xujiayao, and other sites with less complete materials. In a parallel fashion, the Sangiran hominids can be linked to the modern Australians through both regionally distinct characteristics and by morphologically intermediate samples from Sambungmachan and Ngandong.

Thus, the interpretation of regional continuity is based on the evidence we have provided that there were two distinct although not geographically discontinuous morphological clades at the north and south ends of the eastern periphery. These morphological clades have been connected by clines of intergradation for as long as we can find evidence of geographically intermediate specimens, and while commonly described as "Mongoloid" and "Australoid," they could as easily and accurately be called northern and southern.

For its entire known history, the northern clade has been characterized by posterior dental reduction, facial height reduction, transverse facial flattening (involving the nasal saddle, the orientation of the frontosphenoidal process of the zygomatic, and the angulation of the lower margin of the maxilla at the maxillozygomatic junction), and relatively high frequencies for a number of discrete features such as incisor shoveling, sagittal keeling, frontal sinus reduction, and exostoses of the jaws. The southern clade is characterized by facial massiveness, marked prognathism, dental megadonty, and a number of different discrete features found at relatively high frequencies such as frontal flatness, the posterior positioning of minimum frontal breadth, the horizontal orientation of the lower border of the supraorbitals, the form of the pterionic region, the zygomaxillary ridge, and the suite of nine additional nonmetric characters common in the Ngandong sample, which Larnach and Macintosh [1974] demonstrated occur at their highest frequencies in living Australians. Thus, we regard the fossil evidence from east Asia as disproving the hypothesis of a single migratory origin for the modern populations of the region.

Modern populations descending from each of these clades and from the clines stretching between them evolved throughout the region. The evolutionary pattern we perceive is complex, but describable. Yet, there have been extraordinary difficulties in attempts to describe it. Historically, taxonomy has been a burden rather than a help in recognizing and accounting for the evolutionary pattern in east Asia.

The Species Problem

There are several taxonomic implications to the approach and interpretation we have suggested here. These are taxonomic and not phylogenetic, since they involve the way names are used, and not the way that hominid evolution is perceived. However, they are problems precisely because the names have traditionally stood in the way of the perception.

We regard the species distinction between *Homo erectus* and *Homo sapiens* as being problematic. The issue we address stems from the difficulty in clearly distinguishing an actual boundary between *Homo erectus* and *Homo sapiens*. We are not the first to raise this point [see Hemmer, 1967; Jelínek, 1978, 1980a,b, 1981]. From a purely cladistic outlook *Homo erectus* should be sunk, since species originating through anagenesis (ie, without branching) are not recognized as separate species [Wiley, 1981, pp 38–41] according to the criteria of phylogenetic systematics.

However, whether or not a cladistic approach to this problem is taken, the question of a valid species distinction arises from the fact that in regions where the appropriate fossils can be found, the difficulties of distinguishing a boundary beween these species of the genus *Homo* has been widely recognized. Thus, problems of classification have arisen for specimens in Europe (Mauer, Vértesszöllös, Bilzingsleben, Petralona, and the Arago sample), in Africa (Broken Hill, Saldanha, Bodo, Rabat, and Ndutu), and Indonesia (Sambungmachan and the Ngandong sample). Because the taxonomic determinations for these particular specimens are problems of classification, and not of the perception of evolutionary process, we regard their solution as arbitrary, and as a matter of taste, rather than as a matter of systematics. What is more interesting, in our view, is the fact that there are any problems in classification at this boundary at all!

We contend that no morphological definition of the *Homo erectus*/*Homo sapiens* boundary can be validly applied everywhere in the world (for instance, see Krantz [1980] and references therein). We believe this is why there has been so much discussion about whether or not *Homo erectus* can be identified in Europe. Since comparisons of early European specimens are necessarily made with *Homo erectus* samples from Africa or Asia, without a clear understanding of what features are regionally specific to Europe, grade is inevitably confused with regional variation in these comparisons. The fact is that if (as we believe under current taxonomic usage) *Homo erectus* did exist in Europe, the theory of multiregional evolution predicts that this European sample should be regionally distinct from *Homo erectus* samples in other areas. Little wonder, then, that a clear boundary cannot be found (ie, distinct morphological criteria for a species boundary cannot be established).

In our view, there are two alternatives. We should either admit that the *Homo erectus*/*Homo sapiens* boundary is arbitrary and use nonmorphological (ie, temporal) criteria for determining it [cf Wolpoff, 1980b], or *Homo erectus* should be sunk [cf Weidenreich, 1943, 1951; Hemmer, 1967; Jelínek, 1978, 1980, 1981]. There are ample precedents for either alternative.

Simpson [1961], in discussing the implications of phyletic evolution, argues that (p 165): "the lineage must be chopped into segments for purposes of classification, and this must be done arbitrarily." In practice, however, phyletic species are rarely defined in a completely arbitrary way. This is because "in most fossil sequences there are convenient breaks between horizons to permit a nonarbitrary delimitation of species" [Mayr, 1969, p 35] because of accidents of discovery, and if "one or more species are found and defined before the more extensive lineage is at hand, those species should be preserved as far as possible" [Simpson, 1961, p 166]. The taxonomy of the genus *Homo* developed in a way that clearly is compatible with these criteria for phyletic species definition. When in the middle decades of the century *Homo erectus* was defined to include Sinanthropus and Pithecanthropus, a temporal and morphological gap was perceived to separate this species from the two representatives of early *Homo sapiens* then recognized, Steinheim and Swanscombe (at the time Broken Hill was considered too late to impinge on this question, and was regarded as an "African Neandertal"). This gap allowed clear definition of the boundary between the two phyletic species, and provided the basis for a definitive morphological distinction.

The problems we now face stem from a vastly larger sample that appears to be "transitional" between *Homo erectus* and *Homo sapiens*; a sample representing several regions, and indicating that the comparison with Steinheim and Swanscombe might have been misleading to begin with because these specimens appear to be females in a region where the corresponding males (Petralona, Bilzingsleben) are much more robust and resemble *Homo erectus* to a much greater degree. In this case, our guidance from Simpson and Mayr is to retain *Homo erectus* and the original boundary between it and *Homo sapiens*, because *Homo erectus* has been widely recognized for some time and on the average shows clear morphological distinctions from *Homo sapiens*, but to admit that this boundary is a historic one and is morphologically arbitrary. This would prevent further fruitless efforts at attempting to find a morphological definition for the boundary.

The alternative, sinking *Homo erectus*, was first discussed by Weidenreich [1943, p 246]:

> it would not be correct to call our fossil *"Homo pekinensis"* or *"Homo erectus pekinensis"*; it would be best to call it *"Homo sapiens erectus pekinensis."* Otherwise it would appear as a proper "species" different from *"Homo sapiens"* which remains doubtful, to say the least.

Including all fossil forms of Homo in Homo sapiens would be more in keeping with the concept of the evolutionary species, as defined by Simpson [1961] and widely utilized in phylogenetic analysis. As long as it is not replaced by a temporal subspecies, sinking Homo erectus would carry the advantages of explicitly recognizing the arbitrariness of the boundary, and eliminating the perceived need to "explain" how a "new" species (Homo sapiens) could have appeared in so many different regions. More importantly, it would eliminate the necessity of relying on dates to determine which species a number of specimens belong to. We suspect that this is the most important advantage, since dates have been notably unreliable for the time span of interest in this problem, and therefore the tendency has been to ascertain taxonomy on the basis of morphology, leading to the problems and contradictions discussed above. There is historic precedence; numerous temporal species have been sunk in Homo (Homo neanderthalensis, Homo primigenius, Homo soloensis, etc). The resulting species, Homo sapiens, would have a longevity of approximately 1.5 million years, not far from the modal duration for Pleistocene mammalian species [Stanley, 1978].

Sinking Homo erectus, however, is not without its drawbacks. The taxon is well known and fairly well described. Whether or not it is "real" may be an arbitrary distinction, but as an element of terminology it carries considerable information about both morphology and age. Sinking the species leaves the problem of how this information can be expressed, either within or outside of a taxonomic framework.

In sum, either alternative in dealing with the species problem has undesirable consequences. Applying a categorical terminology to describe differences when there is continuous variation creates a reality of groups and boundaries that in fact never existed. Rejecting a categorical terminology leaves no acceptable method for referring to the differences.

Grade

The problems involved in use of temporal species to describe variation in the genus Homo reflect the difficulty in dealing with evolutionary grade in an evolving polytypic lineage. Grade is the concept that created the greatest difficulties for Coon's [1962] scheme, and indeed continues to haunt his more recent formulations [1982]. Coon explicitly used species names as referents for grade. Thus, when claiming that some subspecies "crossed the sapiens boundary" later than others, there was more than a terminological inference. The statement could be, and in fact was, interpreted to mean that some human groups reached the modern level of biological organization later than others. It was difficult not to conclude that these populations were less evolved than their contemporaries from other areas (although Coon never said that), and this implication provided much of the basis for rejecting his entire scheme.

We believe that the problem was not so much in his scheme, as in his use of the grade concept to describe it. All living *Homo sapiens* populations are of the same grade, regardless of their morphology, society, or culture. If true today, this precept was equally true yesterday, and generally at any particular time in the past (as long as a single lineage is considered). It follows that the grade concept, like the concept of the temporal species discussed above, cannot be morphologically defined in any useful way within a polytypic species (for instance, Day and Stringer [1982] recently proposed a set of morphological criteria to define an "anatomically modern" *Homo sapiens* grade which would exclude at least one third of the Late Pleistocene and Holocene Australian fossils). The alternatives are the same; drop it, or admit to an arbitrary temporal definition to delineate the grade characteristics. In fact, insofar as grade and species are used synonymously (as in Coon's case), exactly the same arguments and definitions can apply.

Grade is also often applied below the species level, in many cases synonymously with subspecies, creating the specter of a mixed subspecies terminology including regional groups, temporal groups, and their various combinations (witness the unfortunate history of *"Homo sapiens neanderthalensis"*). This usage comes from the desire to distinguish modern populations from their recent ancestors, without creating species names for every discernable difference. While most workers have recognized that time and place represent very different and fully independent sources of variation, in practice the tangled subspecies terminology has proven very difficult to untwine [for instance, see Campbell, 1965]. "Archaic" *Homo sapiens*, for instance, was coined to provide a phrase to describe a premodern *Homo sapiens* grade without using the term "Neandertal" or a variant of it (Neandertaloid, pre-Neandertal, tropical Neandertal, African Neandertal, overseas Neandertal, etc). Thus, at least in the definition of the terminology, grade need not be potentially confused with region—reflecting a precept in keeping with our own thinking.

Stringer et al [1979] have attempted to even further subdivide *Homo sapiens* using a numbered grade system, arguing (correctly) that a simple system of worldwide grades does not characterize the fossil record sufficiently to deal with what they regard as the "Neandertal problem" in Europe. They seem to implicitly grasp that changing "Neandertal grade" into "archaic sapiens grade" changes terminology without really changing concepts (although there is no explicit recognition of this in their presentation).

However, we perceive no clear advantage in the numbered grade system they present. The categories (with criteria that were never stated) confound variation due to grade with regional variation, and still result in a large residual of unclassifiable specimens. The use of subspecies, or numbered stages, to represent grade may even create more problems than the use of grade to delineate species [Bonde, 1977]. Moreover, from the vantage point of a gradualistic model, a deeper objection lies once again in the delineation of boundaries, for we wonder whether a boundary between grade 2 and grade 3b (or grade 3a and grade 3b) of Stringer et al [1979], that is between "archaic" and "modern" *Homo sapiens*, can be drawn any better than the *Homo erectus/Homo sapiens* boundary. We particularly question whether there are any morphological criteria that can be applied to the definition of this archaic/modern boundary on a worldwide basis, in view of the fact that even regionally this boundary has proven to be excruciatingly difficult to draw. If criteria for this boundary are difficult to establish in regions as far apart as south-central Europe [Jelínek 1976; Wolpoff et al, 1981; Smith, 1982, and this volume] and Australia (here, what might be perceived as grade differences are actually due to different source populations), where fossil remains are well represented, the question that comes to mind is how one can discuss modern *Homo sapiens* origins as a distinguishable event?

Our answer to this rhetorical question is that such an event cannot be distinguished *because there was no event*. We feel that we are trapped by our terminology. The identification of differences resulted in using names to describe them, the names became entities (ie, the map became the territory), the entities required boundaries to remain separate, and finally the crossing of a boundary became an event. The categorical terminology is our best means of expressing differences in grade, and yet it is this terminology that *creates* the boundaries and therefore the need to explain how they are crossed. Evolutionary change in *Homo* is real, grade is real, and therefore changes in grade are real. Whether we express the continuity of the evolutionary process by removing temporal subdivisions or by admitting to arbitrary ones, we should not forget that there never were any boundaries except of our own terminological making. Coon should never have had to explain how five allopatric subspecies crossed a species boundary but still maintained their identity. It was his own terminology that created the need for an explanation.

Multiregional Evolution and Modern *Homo sapiens* Origins

At one time, the "modern" vs "archaic" (=primitive, =Neandertal, etc) *Homo sapiens* distinction seemed very clear in Europe, where Neandertals were perceived to be replaced by Cro Magnons. Earlier in the century other regions lacked sufficient fossil remains to effectively question the European framework. Instead, from a European perspective it became common to interpret the worldwide appearance of Homo sapiens within this framework, characterized by Howells as the Noah's Ark model. With a vastly increased fossil record, proper consideration of earlier discoveries, and a developing geochronological framework, the modern/archaic distinction has become difficult to define in many regions. If anything, the problem is most acute at the extreme margins of the human range, Europe and Australia, as one might expect from the theory of multiregional evolution, since it is at the peripheries that the evidence for regional continuity is strongest, and the continuity has the greatest antiquity.

What makes the problem so acute is the *comparison* of the European and the Australian peripheries. In Europe, where replacement is often considered to have been a dramatic event, morphologically and archaeologically, the earlier population is presumed to have vanished without archaeological trace or morphological remnant. In contrast, in Australia, where the populating of the continent is *known* to have involved population movements in the Late Pleistocene, and the migrants derive from *recognizably different* source populations—indeed from historically different morphological clades—the resulting populations of the continent clearly reflect a meld of genetic heritages, with the morphology of the earlier inhabitants predominating. This is not the place to discuss the European situation in detail, but this comparison of peripheral regions surely indicates that something is very wrong in the classic interpretation of European prehistory. Clear archaeological evidence for a migratory origin of more modern populations has always been lacking in this archaeologically rich region. Indeed, the major adaptive changes associated with the European Upper Paleolithic do not occur at its onset. Thus, in his evaluation of subsistence changes in the southern Franco-Cantabrian region, Straus [1983, p 103] concludes:

> . . . the limited subsistence base of the Middle Paleolithic had broadened considerably by the end of the Upper Paleolithic . . . . These changes did not occur instantaneously at c 35,000 BP. In fact, it is hard to find radical differences between the Mousterian and the early Upper Paleolithic (Chatelperronian and archaic Aurignacian) in terms of site numbers and locations or subsistence activities.

Moreover, the morphological similarities connecting the latest European Neandertals with the *early* so-called modern populations (especially in central and eastern Europe) make a simplistic migratory replacement scheme unsupported by the totality of the evidence [Hrdlička, 1927; Weidenreich, 1947a; Brace, 1964a; Brose and Wolpoff, 1971; Jelínek, 1976; Wolpoff et al, 1981; Smith, 1982, and this volume].

We believe that, properly interpreted, the European situation parallels the Australian one. These are both peripheral regions in the range of a polytypic species. In both cases evolutionary change reflects a shifting gene flow/selection (and drift) balance. In both cases, the elements of this balance change through the Late Pleistocene, with technological advances promoting skeletal gracilization, and increasing amounts of gene flow from the more central areas reflecting the Late Pleistocene improvements in human adaptation and the consequent population expansions. In Australia the gene flow was primarily from the southern part of east Asia and Indonesia (themselves changing genetically over this period), while in the Greater Mousterian Culture Area gene flow from both Asia and (perhaps in western Asia) also from Africa helped influence subsequent evolutionary changes. In both regions continuity with earlier populations can be shown, and in particular the morphological continuity with the original clade in the region predominates in the living populations.

Continental east Asia does not fully parallel these two regions, although the Pleistocene evidence for regional continuity is just as persuasive. However, in the Late Pleistocene continental east Asia began to contrast with Europe and Australindonesia, becoming more the *source* than the *recipient* of gene flow. Simply put, continental east Asia does not remain a periphery. Instead it increasingly comes to act as a center in the Late Pleistocene, with gene flow extending from it to the east (the New World), west, and south (although this undoubtedly remained multidirectional, as the Late Pleistocene specimens from Indonesia and south China show). The regional features marking the northern morphological clade were long established, and as the region became a center, many of these were not lost in spite of the increased polytypism that subsequently evolved. Thus, the spread of these features has helped trace the path of population expansions and other forms of gene flow.

We contend that this is the normal pattern of evolution in *Homo*. It is in situ in the only sense that gradual evolution in a polytypic species can ever be in situ, and it involves gene flow in a manner that evolution in a polytypic species always must. The origins of modern *Homo sapiens* are to be found in the evolutionary process we have described as multiregional evolution, and not in Noah's Ark.

ACKNOWLEDGMENTS

We are sincerely grateful for the opportunity to accomplish this work together. Funding for our participation in this project came from a variety of sources, and we acknowledge our indebtedness for the support given by the National Science Foundation (Grant BNS 76-82729 and the Committee for Scholarly Exchange with the People's Republic of China program grant INT-8117276), the Australian National University (especially in support of the conference on "Bones, Molecules, and Man"), and the support of the Academy of Sciences of the People's Republic of China.

For their permission to examine specimens in their care, and their courtesy extended during our visits, we thank the late G.H.R. von Koenigswald of the Senckenberg Museum, T. Jacob of the Gadjah Mada University Medical School, D. Kadar of the Bandung Geological Museum, C.B. Stringer of the British Museum (Natural History), I. Tattersall of the American Museum of Natural History, and P. Brown of the Department of Prehistory, University of New England.

For their helpful comments during the development of this manuscript, we are indebted to C. Childress, A. Mann, K. Rosenberg, M. Russell, and V. Vitzthum. We especially thank Frank Livingstone, who proved to be an enduring but encouraging critic of our work.

LITERATURE CITED

Abbie AA (1966): Physical characteristics. In Cotton BC (ed): "Aboriginal Man in South and Central Australia." Adelaide: Government Printer, Part I, pp 9–45.

Abbie AA (1968): The homogeneity of Australian Aborigines. Archaeol Phys Anthropol Oceania 3:221–231.

Abbie AA (1975): "Studies in Physical Anthropology II." Canberra: Australian Institute of Aboriginal Studies.

Aigner JS (1976): Chinese Pleistocene cultural and hominid remains: A consideration of their significance in reconstructing the pattern of human bio-cultural development. In Ghosh AK (ed): "Le Paléolithique Inférieur et Moyen en Inde, en Asie Centrale, en Chine et Dans le Sud-est Asiatique." Paris: UISPP Colloque VII, CNRS, pp 65–90.

Aigner JS (1978): Important archaeological remains from North China. In Ikawa-Smith F (ed): "Early Paleolithic in South and East Asia." The Hague: Mouton, pp 163–232.

Aigner JS, Laughlin WS (1973): The dating of the Lantian man and his significance for analyzing trends in human evolution. Am J Phys Anthropol 39:97–110.

Altukhov YP (1982): Biochemical population genetics and speciation. Evolution 36:1168–1181.

Antonovics J (1971): The effects of a heterogeneous environment on the genetics of natural populations. Am Sci 59:593–599.

Ayala FJ (1982): Gradualism versus punctualism in speciation: reproductive isolation, morphology, genetics. In C Barigozzi (ed): "Mechanisms of Speciation." New York: Alan R. Liss, pp 51–66.

Barton NH (1979): Gene flow past a cline. Heredity 43:333–339.

Bartstra GJ (1982): Homo erectus erectus: The search for his artifacts. Curr Anthropol 23(3):318–320.

Birdsell JB (1949): The racial origin of the extinct Tasmanians. Records Queen Victoria Museum, Launceston 2(3):105–122.

Birdsell JB (1967): Preliminary data on the trihybrid origin of the Australian Aborigines. Archaeol Phys Anthropol Oceania 2:100–155.

Birdsell JB (1968): Some predictions for the Pleistocene based on equilibrium systems among recent hunter-gatherers. In Lee RB, DeVore I (eds): "Man the Hunter." Chicago: Aldine, pp 229–240.

Birdsell JB (1972): The problem of the evolution of human races: Classification or clines? Social Biology 19:136–162.

Birdsell JB (1977): The recalibration of a paradigm for the peopling of Greater Australia. In Allen J, Golson J, Jones R (eds): "Sunda and Sahul: Prehistoric Studies in Southeast Asia, Melanesia and Australia." New York: Academic Press.

Bock WJ (1979): The synthetic explanation of macroevolutionary change—A reductionist approach. Bull Carnegie Museum Nat Hist 13:20–69.

Bodmer WF, Cavalli-Sforza LL (1976): "Genetics, Evolution, and Man." San Francisco: Freeman.

Bonde N (1977): Cladistic classification as applied to vertebrates. In Hecht M, Goody P, Hecht B (eds): "Major Patterns of Vertebrate Evolution." New York: Plenum, pp 741–804.

Boule M, Vallois HV (1957): "Fossil Men." New York: Dryden.

Bowler JM, Thorne AG (1976): Human remains from Lake Mungo: Discovery and excavation of Lake Mungo III. In Kirk RL, Thorne AG (eds): "The Origin of the Australians." Canberra: Australian Institute of Aboriginal Studies, pp 127–138.

Brace CL (1964a): The fate of the "Classic" Neanderthals: A consideration of hominid catastrophism. Curr Anthropol 5:3–43 and 7:204–214.

Brace CL (1964b): A nonracial approach towards the understanding of human diversity. In Ashley Montagu MF (ed): "The Concept of Race." New York: Free Press, pp 103–152.

Brace CL (1980): Australian tooth size clines and the death of a stereotype. Curr Anthropol 21(2):141–164.

Brace CL, Hinton RJ (1981): Oceanic tooth-size variation as a reflection of biological and cultural mixing. Curr Anthropol 22(5):549–569.

Briggs JC (1981): Do centers of origin have a center? Paleobiology 7:305–307.

Brose DS, Wolpoff MH (1971): Early Upper Paleolithic man and late Middle Paleolithic tools. Am Anthropol 73:1156–1194.

Brothwell DR (1960): Upper Pleistocene human skull, from Niah caves, Sarawak. Sarawak Museum J 9:323–349.

Brown P (1981): Artificial cranial deformation: A component in the variation in Pleistocene Australian Aboriginal crania. Archaeol Phys Anthropol Oceania 16:156–167.

Brown WM, Prager EM, Wang A, Wilson AC (1982): Mitochondrial DNA sequences of primates: Tempo and mode of evolution. J Mol Evol 18:225–239.

Brues A (1972): Models of clines and races. Am J Phys Anthropol 37:389–399.

Brues A (1980): Comment on "Sapienization and speech," by G. Krantz. Curr Anthropol 21:779–780.

Cadien JD (1972): Dental variation in man. In Washburn SL, Dolhinow P (eds): "Perspectives on Human Evolution, Vol 2." New York: Holt, Rinehart, and Winston, pp 199–222.

Campbell BG (1965): The nomenclature of the Hominidae. Occasional Paper 22, Royal Anthropological Institute of Great Britain and Ireland.

Carson HL (1955): The genetic characteristics of marginal populations of *Drosophila*. Cold Spring Harbor Symp Quant Biol 20:276–287.

Carson HL (1959): Genetic conditions which promote or retard the formation of species. Cold Spring Harbor Symp Quant Biol 24:87–105.

Carson HL (1968): The population flush and its genetic consequences. In Lewontin RC (ed): "Population Biology and Evolution." Syracuse: Syracuse University, pp 123–137.

Carson HL (1975): The genetics of speciation at the diploid level. Am Nat 109:83–92.

Carson HL (1982): Speciation as a major reorganization of polygenic balances. In Barigozzi C (ed): "Mechanisms of Speciation." New York: Alan R Liss, pp 411–433.

Casiño ES (1976): Looking for missing links in missing lands. In Tobias PV, Coppens Y (eds): "Le Plus Anciens Hominidés." Paris: UISPP Colloque VI, CNRS, pp 409–424.

Cavalli-Sforza LL, Barrai I, Edwards AWF (1964): Analysis of human evolution under random genetic drift. Cold Spring Harbor Symp Quant Biol 29:9–20.

Cavalli-Sforza LL, Edwards AWF (1965): Analysis of human evolution. In Geerts SJ (ed): "Genetics Today." Proc XIth Int Cong Genet, 1963, Vol 3, pp 923–933.

Chang Kwangchi (1962): New evidence on fossil man in China. Science 136:749–760.

Chang Kwangchi (1963): "The Archeology of Ancient China." New Haven: Yale University.

Chappell J (1976): Aspects of late Quaternary paleography of the Australian-East Indonesian Region. In Kirk RL, Thorne AG (eds): "The Origin of the Australians." Canberra: Australian Institute of Aboriginal Studies, pp 11–22.

Chappell J, Thom BG (1977): Sea levels and coasts. In Allen J, Golson J, Jones R (eds): "Sunda and Sahul: Prehistoric Studies in Southeast Asia, Melanesia, and Australia." London: Academic Press, pp 175–291.

Charlesworth B, Lande R, Slatkin M (1982): A Neo-Darwinian commentary on macroevolution. Evolution 36:474–498.

Cheboksarov NN (1959): On the continuous development of physical type, economic activities, and Paleolithic culture of the people in the territory of China. Soviet Ethnog 4:1–25.

Cheverud JM, Buettner-Janusch J, Sade D (1978): Social group fission and the origin of intergroup genetic differential among the Rhesus monkeys of Cayo Santiago. Am J Phys Anthrop 49:449–456.

Cheverud JM, Buikstra JE, Twichell E (1979): Relationships between non-metric skeletal traits and cranial size and shape. Am J Phys Anthropol 50:191–198.

Chia Lanpo (1957): Notes on the human and some other mammalian remains from Changyang, Hupei. Vertebrata PalAsiatica 3:247–257.

Chia Lanpo, Wei Qi, Li Chaorong (1979): Report on the excavation of Xujiayao (Hsuchiayao) man site in 1976. Vertebrata PalAsiatica 17:277–293.

Chia Lanpo, Woo Jukang (1959): Fossil human skull base of Late Paleolithic stage from Chilinshan, Leipin District, Kwangsi, China. Vertebrata PalAsiatica 3:37–39.

Chiu Chunglang, Gu Yümin, Zhang Yinyun, Chang Shenshui (1973): Newly discovered Sinanthropus remains and stone artifacts at Choukoutien. Vertebrata PalAsiatica 11:109–131.

Cook LM (1961): The edge effect in population genetics. Am Nat 95:295–307.

Coon CS (1959): Review of "Mankind in the Making," by WW Howells. Science 130:1399–1400.

Coon CS (1962): "The Origin of Races." New York: Knopf.

Coon CS (1982): "Racial Adaptations. A Study of the Origins, Nature, and Significance of Racial Variations in Humans." Chicago: Nelson-Hall.

Corruccini RS (1974): An examination of the meaning of cranial discrete traits for human skeletal biological studies. Am J Phys Anthropol 40:425–446.

Corruccini RS (1976): The interaction between nonmetric and metric cranial variation. Am J Phys Anthropol 44:285–294.

Day MH, Stringer CB (1982): A reconsideration of the Omo Kibish remains and the erectus-sapiens transition. In "L'*Homo erectus* et la Place de l'Homme de Tautavel Parmi les Hominidés Fossiles, Vol 2." Nice: Louis-Jean Scientific and Literary Publications, pp 814–846.

Dobzhansky T (1951): "Genetics and the Origin of Species, 3rd edition." New York: Columbia University.

Dobzhansky T (1960): Individuality, gene recombination, and nonrepeatability of evolution. Aust J Sci 23:71–74.

Dobzhansky T (1963): The possibility that *Homo sapiens* evolved independently 5 times is vanishingly small. Sci Am 208(2):169–172.

Dobzhansky T (1970): "Genetics of the Evolutionary Process." New York: Columbia University.

Dubois E (1894): "*Pithecanthropus erectus*: Eine Menschenähnliche Übergangsform von Java." Batavia: Landes Drucherei.

Dubois E (1922): The proto-Australian fossil man of Wadjak, Java. Koninklijke Akad Wetenschappen Amsterdam [B] 23:1013–1051.

Edwards AWF (1971): Mathematical approaches to the study of human evolution. In Hodson FR, Kendall DG, Tautu P (eds): "Mathematics in the Archaeological and Historical Sciences." Edinburgh: University Press, pp 347–355.

Ehrlich PR, Raven PH (1969): Differentiation of populations. Science 165:1228–1232.

Endler JA (1973): Gene flow and population differentiation. Science 179:243–250.

Endler JA (1977): "Geographic Variation, Speciation, and Clines." Princeton, NJ: Princeton University Press.

Ferembach D (1973): L'évolution humaine au Proche-Orient. Paleorient 1:213–221.

Ferembach D (1979): L'émergence du genre *Homo* et de l'espèce *Homo sapiens*. Les faits. Les incertitudes. Biometrie Hum 14:11–18.

Ford EB (1975): "Ecological Genetics." London: Chapman and Hall.

Freedman L, Lofgren M (1979a): The cossack skull and a dihybrid origin of the australian aborigines. Nature 282:298–300.

Freeman L, Lofgren M (1979b): Human skeletal remains from cossack, western Australia. J Hum Evol 8:283–299.

Freudenthal M (1965): Betrachtungen über die Gattung *Cricetodon*. Koninklijke Akad Wetenschappen [B] 68:293–305.

Garn SM (1962): "Human Races, revised edition." Springfield: Thomas CC.

Genet-Varcin E (1979): "Les Hommes Fossiles." Paris: Boubee.

Gingerich PD (1979): The stratophenetic approach to phylogeny reconstruction in vertebrate paleontology. In Cracraft J, Eldredge N (eds): "Phylogenetic Analysis and Paleontology." New York: Columbia University Press, pp 41–77.

Glinka J (1981): Racial history of Indonesia. In Schwidetzky I (ed): "Rassengeschichte der Menschenheit. Asien I: Japan, Indonesian, Oceanien." Munich: R. Oldenbourg, pp 79–113.

Gould SJ, Eldredge N (1977): Punctuated equilibria: The tempo and mode of evolution reconsidered. Paleobiology 3:115–151.

Gould SJ, Johnston RF (1972): Geographic variation. Annu Rev Ecol Systematics 3:457–498.

Gribbin J, Cherfas J (1982): "The Monkey Puzzle: Reshaping the Evolutionary Tree." Bodley Head, London.

Guglielmino-Matessi CR, Gluckman P, Cavalli-Sforza LL (1979): Climate and the evolution of skull metrics in man. Am J Phys Anthropol 50:549–564.

Haldane JBS (1948): The theory of a cline. J Genet 48(3):277–284.

Hanson WD (1966): Effects of partial isolation (distance), migration and fitness requirements along with environmental pockets upon steady state gene frequencies. Biometrics 22:453–468.

Harpending H (1974): Genetic structure of small populations. Ann Rev Anthropol 3:229–243.

Hemmer H (1967): "Allometrie-Untersuchungen zur Evolution des Menschlichen Schädels und Seiner Rassentypen." Stuttgart: Fischer.

Hertzog KP (1968): Associations between discontinuous cranial traits. Am J Phys Anthropol 29:397–404.

Hiorns RW, Harrison GA (1977): The combined effects of selection and migration in human evolution. Man 12:438–445.

Holloway RL (1981): The Indonesian *Homo erectus* brain endocasts revisited. Am J Phys Anthropol 5:503–522.

Hooton EA (1949): "Up From the Ape, revised edition." New York: MacMillan.

Howell FC (1951): The place of Neanderthal man in human evolution. Am J Phys Anthropol 9:379–416.

Howell FC, Washburn SL (1960): Human evolution and culture. In Tax S (ed): "Evolution After Darwin, Vol III." Chicago: University of Chicago, pp 33–56.

Howells WW (1967): "Mankind in the Making, revised edition." Garden City: Doubleday.

Howells WW (1973a): Cranial variation in man. A study by multivariate analysis of patterns of difference among recent human populations. Papers Peabody Museum Archaeol Ethnol 67:1–259.

Howells WW (1973b): "Evolution of the Genus *Homo*." Reading: Addison-Wesley.

Howells WW (1976a): Explaining modern man: Evolutionists, *versus* migrationists. J Hum Evol 5:477–496.

Howells WW (1976b): Metrical analysis in the problem of Australian origins. In Kirk RL, Thorne AG (eds): "The Origin of the Australians." Canberra: Australian Institute of Aboriginal Studies, pp 141–160.

Howells WW (1976c): Physical variation and prehistory in Melanesia and Australia. Am J Phys Anthropol 45:641–650.

Howells WW (1980): *Homo erectus*—who, when and where: A survey. Yrbk Phys Anthropol 23:1–23.

Howells WW (1983): Origins of the Chinese people: interpretations of the recent evidence. In Keightley DN (ed): "The Origins of Chinese Civilization." Berkeley: University of California, pp 297–319.

Howells WW, Schwidetzky I (1981): Oceania. In Schwidetzky I (ed): "Rassengeschichte der Menschenheit. Asien I: Japan, Indonesian, Oceanien." Munich: R. Oldenbourg, pp 115–166.

Hrdlička A (1920): Shovel-shaped teeth. Am J Phys Anthropol 3:429–465.

Hrdlička A (1927): The Neanderthal phase of man. J Roy Anthropol Inst 57:249–274.

Hrdlička A (1942): Catalogue of human crania in the United States National Museum collections. Eskimo in General. Proc US Natl Museum 91:169–429.

Hutterer KL (1977): Reinterpreting the Southeast Asia Paleolithic. In Allen J, Golson J, Jones R (eds): "Sunda and Sahul: Prehistoric Studies in Southeast Asia, Melanesia and Australia." London: Academic Press, pp 31–71.

Huxley JS (1939): Clines: An auxiliary method in taxonomy. Bijdragen Dierkunde (Leiden) 27:491–520.

Jacob T (1967): "Some Problems Pertaining to the Racial History of the Indonesian Region." Utrecht: Drukkerij Neerlandia.

Jacob T (1968): A human Wadjakoid maxillary fragment from China. Koninklijke Akad Wetenschappen Amsterdam [B] 71:232–235.

Jacob T (1973): Paleoanthropological discoveries in Indonesia with special reference to the finds of the last two decades. J Hum Evol 2:473–485.

Jacob T (1975): Morphology and paleoecology of early man in Java. In Tuttle RH (ed): "Paleoanthropology, Morphology, and Paleoecology." The Hague: Mouton, pp 311–325.

Jacob T (1976): Early populations in the Indonesian region. In Kirk RL, Thorne AG (eds): "The Origins of the Australians." Canberra: Australian Institute of Aboriginal Studies, pp 81–93.

Jaeger JJ (1975): The mammalian faunas and hominid fossils of the Middle Pleistocene of the Maghreb. In Butzer K, Isaac G (eds): "After the Australopithecines." The Hague: Mouton, pp 399–418.

Jain SK, Bradshaw AD (1966): Evolutionary divergence among adjacent plant populations. Heredity 21:407–441.

Jelínek A (1982): The Tabūn cave and Paleolithic man in the Levant. Science 216:1369–1375.

Jelínek J (1976): The Homo sapiens neanderthalensis and Homo sapiens sapiens relationship in Central Europe. Anthropologie 14:79–81.

Jelínek J (1978): Homo erectus or Homo sapiens? Rec Adv Primatol 3:419–429.

Jelínek J (1980a): European Homo erectus and the origin of Homo sapiens. In Kønigsson LK (ed): "Current Argument on Early Man." Oxford: Pergammon, pp 137–144.

Jelínek J (1980b): Variability and geography. Contribution to our knowledge of European and North African Middle Pleistocene hominids. Anthropologie 18:109–114.

Jelínek J (1981): Was Homo erectus already Homo sapiens? Les Processus de L'Hominisation. CNRS Int Colloq 599:91–95.

Karlin S, McGregor J (1972): Polymorphisms for genetic and ecological systems with weak coupling. J Theor Populat Biol 3:210–238.

Keesing FM (1950): Some notes on early migrations in the southwest Pacific area. Southwest J Anthropol 6:101–119.

Keith A (1931): "New Discoveries Relating to the Antiquity of Man." London: Williams and Norgate.

Kennedy GE (1980): "Paleoanthropology." New York: McGraw-Hill.

King MC, Wilson AC (1975): Evolution at two levels in humans and chimpanzees. Science 188:107–116.

Krantz G (1980): Sapienization and speech. Curr Anthropol 21:773–792.

Lande R (1976): Natural selection and random genetic drift in phenotypic evolution. Evolution 30:314–334.

Lande R (1980): Genetic variation and phenotypic evolution during allopatric speciation. Am Naturalist 116:463–479.

Larnach SL, Macintosh NWG (1974): A comparative study of Solo and Australian Aboriginal crania. In Elkin AP, Macintosh NWG (eds): "Grafton Elliot Smith: The Man and his Work." Sydney: Sydney University, pp 95–102.

Laughlin WS, Okladnikov AP, Derevyanko AP, Harper AB, Atseev IV (1976): Early Siberians from Lake Baikal and Alaskan population affinities. Am J Phys Anthropol 45:651–660.

Levinton JS, Simon CM (1980): A critique of the punctuated equilibria model and implications for the detection of speciation in the fossil record. Systematic Zool 29:130–142.

Lewontin RC (1974): "The Genetic Basis of Evolutionary Change." New York: Columbia University.

Livingstone FB (1962): On the non-existence of human races. Curr Anthropol 3:279–281.

Livingstone FB (1964): On the nonexistence of human races. In Ashley Montagu MF (ed): "The Concept of Race." New York: Free Press, pp 46–60.

Livingstone FB (1969): An analysis of the ABO blood group clines in Europe. Am J Phys Anthropol 31:1–10.

Livingstone FB (1973): Gene frequency differences in human populations: Some problems of analysis and interpretation. In Crawford MH, Workman PL (eds): "Methods and Theories of Anthropological Genetics." Albuquerque: University of New Mexico, pp 39–67.

Livingstone FB (1980): Comment on "Sapienization and speech," by G. Krantz. Curr Anthropol 21:784.

Lui Tungsheng, Ding Menglin (1983): Discussion on the age of "Yuanmou Man." Acta Anthropol Sinica 2:40–48.

Macintosh NWG (1965): The physical aspect of man in Australia. In Berndt RM, Berndt CH (eds): "Aboriginal Man in Australia." Sydney: Angus and Robertson, pp 29–70.

Macintosh NWG (1967): Fossil man in Australia. Aust J Sci 30:86–98.

Macintosh NWG, Larnach SL (1976): Aboriginal affinities looked at in world context. In Kirk RL, Thorne AG (eds): "The Origin of the Australians." Canberra: Australian Institute of Aboriginal Studies, pp 113–126.

Magitot E (1885): Essai sur les Mutilations ethniques. Bull Soc Anthropol Paris 8:21–25.

Mani GS (1980): A theoretical study of morph ratio clines with special reference to melanism in moths. Proc Roy Soc London [Biol] 210:299–316.

Martin RA (1970): Line and grade in the extinct medium species of Sigmodon. Science 166:1504–1506.

Matsu'ura S (1982): A chronological framing of the Sangiran hominids. Bull Natl Sci Museum Tokyo [D] 8:1–53.

May RM, Endler JA, McMurtne E (1975): Gene frequency clines in the presence of selection opposed by gene flow. Am Naturalist 109:659–676.

Mayr E (1954): Changes of genetic environment and evolution. In Huxley JS, Hardy AC, Ford EB (eds): "Evolution as a Process." London: Allen and Unwin, pp 188–213.

Mayr E (1963): "Animal Species and Evolution." Cambridge: Belknap.

Mayr E (1969): "Principles of Systematic Zoology." New York: McGraw-Hill.

Mayr E (1970): "Populations, Species, and Evolution." Cambridge: Belknap.

Mayr E (1982): Speciation and macroevolution. Evolution 36:1119–1132.

Morant GM (1923): A first study of the Tibetan skull. Biometrika 14:193–260.

Morant GM (1924): A study of certain Oriental series of crania including the Nepalese and Tibetan series in the British Museum. Biometrika 16:1–104.

Morris LN, Nute PE (1978): Effects of isolation on genetic variability: Macaque populations as model systems. In Meier RJ, Otten CM, Adel-Hameed F (eds): "Evolutionary Models and Studies in Human Diversity." The Hague: Mouton, pp 105–125.

Morton NE, Lalouel J (1973): Topology of kinship in Micronesia. Am J Hum Genet 25:422–432.

Nagylaki T (1975): Conditions for the existence of clines. Genetics 80:595–615.

Nagylaki T (1976): Clines with variable migration. Genetics 83:867–886.

Neel JV (1978): The population structure of an Amerind tribe, the Yanomama. Annu Rev Genet 12:365–413.

Nei M, Roychoudhury AK (1982): Genetic relationship and evolution of human races. In Hecht MK, Wallace B, Prace GT (eds): "Evolutionary Biology, Vol 14." New York: Plenum, pp 1–59.

Neumann GK (1956): The Upper Cave skulls from Choukoutien in the light of Paleo-Amerind material. Am J Phys Anthropol 14:380.

Oakley KP, Campbell BG, Molleson TI (1975): "Catalogue of Fossil Hominids, Part III: Americas, Asia, Australia." London: British Museum (Natural History).

Oetteking B (1930): Craniology of the north Pacific coast. Mem Museum Nat Hist 11:1–391.

Oschinsky L (1964): "The Most Ancient Eskimos." Ottawa: Canadian Research Centre for Anthropology, University of Ottawa.

Ossenberg NS (1976): Within and between race distances in population studies based on discrete traits of the human skull. Am J Phys Anthropol 45:701–716.

Piazza A, Monozzi P, Cavalli-Sforza LL (1981): Synthetic gene frequency maps of man and selective effects of climate. Proc Natl Acad Sci 78:2638–2642.

Portin P, Alvesalo L (1974): The inheritance of shovel shape in maxillary central incisors. Am J Phys Anthropol 41:59–62.

Protsch R (1978): "Catalogue of Fossil Hominids of North America." New York: Fischer.

Rightmire GP (1976): The relationships of Middle and Upper Pleistocene hominids from sub-Saharan Africa. Nature 260:238–240.

Rightmire GP (1978): Florisbad and human population succession in southern Africa. Am J Phys Anthropol 48:475–486.

Rightmire GP (1979): Implications of the Border Cave skeletal remains for later Pleistocene human evolution. Curr Anthropol 20:23–35.

Rightmire GP (1981): Later Pleistocene hominids of eastern and southern Africa. Anthropologie 19:15–26.

Russell MD (1983): "The Functional and Adaptive Significance of the Supraorbital Torus." Ann Arbor: University Microfilms.

Saban R (1977): The place of Rabat man (Kébibat, Morocco) in human evolution. Curr Anthropol 18:518–524.

Santa Luca AP (1980): "The Ngandong Fossil Hominids." Yale University Publications in Anthropology, No. 78, New Haven.

Sarich V (1971): Human variation in an evolutionary perspective. In Dolhinow P, Sarich V (eds): "Background for Man." Boston: Little Brown, pp 182–191.

Sartono S (1975): Implications arising from *Pithecanthropus* VIII. In Tuttle RH (ed): "Paleoanthropology. Morphology and Paleoecology." The Hague: Mouton, pp 327–360.

Sartono S (1976): The Javanese Pleistocene hominids: a re-appraisal. In Tobias PV, Coppens Y (eds): "Les Plus Anciens Hominidés." Paris: UISPP Colloque VI, CNRS, pp 456–464.

Sartono S (1982): Characteristics and chronology of early men in Java. In: "L'*Homo erectus* et la Place de L'Homme de Tautavel Parmi les Hominidés Fossiles." Vol 2." Nice: Jean-Louis Scientific and Literary Publications, pp 491–541.

Sémah F, Sartono S, Zaim Y, Djubiantono T (1981): A palaeomagnetic study of Plio-Pleistocene sediments from Sangiran and Simo (Central Java): Interpreting first results. Mod Quaternary Res Southeast Asia 6:103–110.

Simpson GG (1961): "Principles of Animal Taxonomy." New York: Columbia.

Slatkin M (1973): Gene flow and selection in a cline. Genetics 75:733–756.

Slatkin M (1978): Spatial patterns in the distribution of polygenic characters. J Theor Biol 70:213–228.

Smith FH (1982): Upper Pleistocene hominid evolution in South-Central Europe: A review of the evidence and analysis of trends. Curr Anthropol 23:667–703.

Smouse PE, Vitzthum VJ, Neel JV (1981): The impact of random and lineal fission on the genetic divergence of small human groups: A case study among the Yanomama. Genetics 98:179–197.

Soejono RP (1982): New data on the Paleolithic industry of Indonesia. In "L'*Homo erectus* et la Place de L'Homme de Tautavel Parmi les Hominidés Fossiles, Vol 2." Paris: CNRS, pp 578–592.

Sokal RR (1973): The species problem reconsidered. Systematic Zool 22:360–374.

Stanley SM (1978): Chronospecies' longevities, the origin of genera, and the punctuational model of evolution. Paleobiology 4:26–40.

Stanley SM (1979): "Macroevolution. Pattern and Process." San Francisco: Freeman.

Steślicka W (1947): The systematic position of Ngandong-Man. Ann Univ Mariae Curie (Lublin) [C] 2:37–109.

Stewart TD (1974): Perspectives on some problems of early man common to America and Australia. In Elkin AP, Macintosh NWG (eds): "Grafton Elliot Smith: The Man and his Work." Sydney: Sydney University, pp 114–135.

Straus LG (1983): From Mousterian to Magdalenian: Cultural evolution viewed from Vasco-Cantabrian Spain and Pyrenean France. In Trinkaus E (ed): "The Mousterian Legacy. Human Biocultural Change in the Upper Pleistocene." Oxford: BAR International Series 164, pp 73–111.

Stringer CB (1980): Comment on "Sapienization and speech," by G. Krantz. Curr Anthropol 21:788.

Stringer CB, Howell FC, Melentis J (1979): The significance of the fossil hominid skull from Petralona, Greece. J Archaeol Sc 6:235–253.

Templeton AR (1980): The theory of speciation via the founder principle. Genetics 94:1101–1138.

Templeton AR (1981): Mechanisms of speciation—A population genetic approach. Ann Rev Ecol 12:23–48.

Templeton AR (1982): Genetic architectures of speciation. In Barigozzi C (ed): "Mechanisms of Speciation." New York: Alan R. Liss, pp 105–121.

Templeton AR (1983): Phylogenetic inference from restriction endonuclease cleavage site maps with particular reference to the evolution of humans and the apes. Evolution 37:221–244.

Thoma A (1964): Die Entstehung de Mongoliden. Homo 15:1–22.

Thoma A (1973): New evidence for the polycentric evolution of *Homo sapiens*. J Hum Evol 2:529–536.

Thoma A (1975): L'origine de l'homme moderne et de ses races. Recherche 6:328–335.

Thorne AG (1971a): Mungo and Kow Swamp: Morphological variation in Pleistocene Australians. Mankind 8(2):85–89.

Thorne AG (1971b): The racial affinities and origins of the Australian Aborigines. In Mulvaney DJ, Golson J (eds): "Aboriginal Man and Environment in Australia." Canberra: Australian National University, pp 316–325.

Thorne (1976): Morphological contrasts in Pleistocene Australians. In Kirk RL, Thorne AG (eds): "The Origin of the Australians." Canberra: Australian Institute of Aboriginal Studies, pp 95–112.

Thorne AG (1977): Separation or reconciliation? Biological clues to the development of Australian society. In Allen J, Golson J, Jones R (eds): "Sunda and Sahul: Prehistoric Studies in Southeast Asia, Melanesia, and Australia." London: Academic Press, pp 187–204.

Thorne AG (1980a): The longest link: Human evolution in Southeast Asia and the settlement of Australia. In Fox JJ, Garnaut RG, McCawley PT, Maukie JAC (eds): "Indonesia: Australian Perspectives." Canberra: Research School of Pacific Studies, pp 35–43.

Thorne AG (1980b): The arrival of man in Australia. In Sherratt A (ed): "The Cambridge Encyclopedia of Archaeology." Cambridge: Cambridge University Press, pp 96–100.

Thorne AG (1981): The centre and the edge: The significance of Australian hominids to African paleoanthropology. In Leakey RE, Ogot BA (eds): "Proceedings of the 8th Panafrican Congress of Prehistory and Quaternary Studies, Nairobi, September 1977." Nairobi: TILLMIAP, pp 180–181.

Thorne AG, Macumber PG (1972): Discoveries of late Pleistocene man at Kow Swamp, Australia. Nature 328:316–319.

Thorne AG, Wolpoff MH (1981): Regional continuity in Australasian Pleistocene hominid evolution. Am J Phys Anthropol 55:337–349.

Tobias PV (1968): Middle and early Upper Pleistocene members of the genus *Homo* in Africa. In Kurth G (ed): "Evolution und Hominization, 2nd ed." Stuttgart: Fischer, pp 176–194.

Trinkaus E (1978): Bilateral asymmetry of human skeletal nonmetric traits. Am J Phys Anthropol 49:315–318.

Vandermeersch B (1970): Les origines de l'homme moderne. Atomes 25(272):5–12.

Vandermeersch B (1972): Récentes découvertes de squelettes humains à Qafzeh (Israël): essai d'interprétation. In Bordes F (ed): "The Origin of *Homo sapiens*." Paris: UNESCO, pp 49–54.

Vandermeersch B (1981): Les Premiers *Homo sapiens* au Proche-Orient. In Ferembach D (ed): "Le Processus de L'Hominisation, No. 599." Paris: CNRS, pp 97–100.

Van Valen L (1966): On discussing human races. Perspect Biol Med 9:377–383.

von Koenigswald GHR (1973): *Australopithecus, Meganthropus,* and *Ramapithecus.* J Hum Evol 2:487–492.

Weckler JE (1957): Neanderthal man. Sci Am 197(6)89–97.

Weidenreich F (1938): *Pithecanthropus* and *Sinanthropus.* Nature 141:376–379.

Weidenreich F (1939a): Six lectures on *Sinanthropus pekinensis* and related problems. Bull Geol Soc China 19:1–110.

Weidenreich F (1939b): On the earliest representatives of modern mankind recovered on the soil of East Asia. Peking Nat Hist Bull 13:161–174.

Weidenreich F (1943): The skull of *Sinanthropus pekinensis:* A comparative study of a primitive hominid skull. Palaeontol Sinica, n.s. D, No. 10 (whole series No. 127).

Weidenreich F (1945): The Keilor skull. A Wadjak skull from southeast Australia. Am J Phys Anthropol 3:21–33.

Weidenreich F (1946): "Apes, Giants, and Man." Chicago: University of Chicago.

Weidenreich F (1947a): Facts and speculations concerning the origin of *Homo sapiens*. Am Anthropol 49:187–203.

Weidenreich F (1947b): The trend of human evolution. Evolution 1:221–236.

Weidenreich F (1951): Morphology of Solo man. Anthropological Papers of the Am Museum Nat Hist 43:(part 3):205–290.

Weinert H (1950): Über die neuen Vor- und Frühmenschenfunde aus Afrika, Java, und Frankreich. Z Morphol Anthropol 42:113–148.

Weiss KM, Maruyama T (1976): Archaeology, population genetics, and studies of human racial ancestry. Am J Phys Anthropol 44:31–50.

White MJD (1978): "Models of Speciation." San Francisco: Freeman.

Wiley EO (1981): "Phylogenetics. The Theory and Practice of Phylogenetic Systematics." New York: John Wiley and Sons.

Wolpoff MH (1980a): "Paleoanthropology." New York: Knopf.

Wolpoff MH (1980b): Cranial remains of Middle Pleistocene European hominids. J Hum Evol 9:339–358.

Wolpoff MH, Smith FH, Malez M, Radovčić J, Rukavina D (1981): Upper Pleistocene human remains from Vindija Cave, Croatia, Yugoslavia. Am J Phys Anthropol 54:499–545.

Woo Jukang (Wu Rukang) (1958a): Tzeyang Paleolithic man—earliest representative of modern man in China. Am J Phys Anthropol 16:459–471.

Woo Jukang (Wu Rukang) (1958b): Investigation of human teeth. In Pei Wenchung (ed): Report on the Excavation of Paleolithic sites at Tingtsun, Hsiangfenhsien, Shanxi Province, China. Institute of Vertebrate Paleontology Academia Sinica Mem.

Woo Jukang (Wu Rukang) (1959): Human fossils found in Liujiang, Kwangsi, China. Vertebrata PalAsiatica 3:109–118.

Woo Jukang (Wu Rukang) (1962): The mandibles and dentition of Gigantopithecus. Palaeontol Sinica, New Series D, Number 11.

Woo Jukang (Wu Rukang) (1964): A newly discovered mandible of Sinanthropus type—Sinanthropus lantianensis. Sci Sinica 13:801–812.

Woo Jukang (Wu Rukang) (1966): The hominid skull of Lantian, Shensi. Vertebrata PalAsiatica 10:1–22.

Woo Jukang (Wu Rukang) (1982): Paleoanthropology in China, 1949–79. Curr Anthropol 23:473–477.

Woo Jukang (Wu Rukang), Chia Lanpo (1954): New Discoveries about Sinanthropus pekinensis in Choukoutien. Sci Sinica 33:335–352.

Woo Jukang (Wu Rukang), Dong Xingren (1980): The fossil human teeth from Yunxian, Hubei. Vertebrata PalAsiatica 18:142–149.

Woo Jukang (Wu Rukang), Dong Xingren (1982): Preliminary study of Homo erectus remains from Hexian, Anhui. Acta Anthropol Sinica 1:2–13.

Woo Jukang (Wu Rukang), Peng Ruce (1959): Fossil human skull of early Paleolithic stage found at Mapa, Shaoquan, Kwangtung Provence. Vertebrata PalAsiatica 3:176–182.

Wright S (1931): Evolution in Mendelian populations. Genetics 16:97–159.

Wright S (1940): Breeding structure of populations in relation to speciation. Am Nat 74:232–248.

Wright S (1968): "Evolution and the Genetics of Populations. Volume 1. Genetic and Biometric Foundations." Chicago: University of Chicago.

Wright S (1977): "Evolution and the Genetic of Populations, Vol 3. Experimental Results and Evolutionary Deductions." Chicago: University of Chicago.

Wright S (1978): "Evolution and the Genetics of Populations, Vol 4. Variability Within and Among Natural Populations." Chicago: University of Chicago.

Wright S (1980): Genic and organismic selection. Evolution 34:825–843.

Wu Maoling (1980): Human fossils discovered at Xujiayao site in 1977. Vertebrata PalAsiatica 18:229–238.

Wu Maoling, Wang Linghong, Zhang Yinyun, Chang Senshui (1975): Fossil human teeth and associated cultural remains from Tonzi County, Guizhou Provence. Vertebrata PalAsiatica 13:14–23.

Wu Rukang (Woo Jukang), Lin Shenglong (1983): Peking man. Sci Am 248 (6):86–94.

Wu Rukang (Woo Jukang), Wu Xinzhi (1982a): Human fossil teeth from Xichuan, Henan. Vertebrata PalAsiatica 20:1–9.

Wu Rukang (Woo Jukang) (1982b): Comparison of Tautavel man with *Homo erectus* and early *Homo sapiens* in China. In "L'*Homo erectus* et la Place de l'homme de Tautavel Parmi les Hominidés Fossiles, Vol 2." Nice: Louis-Jean Scientific and Literary Publications, pp 605–616.

Wu Xinzhi (1961): Study on the Upper Cave man of Choukoutien. Vertebrata PalAsiatica 3:202–211.

Wu Xinzhi (1981): A well-preserved cranium of an archaic type of early *Homo sapiens* from Dali, China. Sci Sinica 24(4):530–41.

Wu Xinzhi, Zhang Yinyun (1978): Fossil man in China. In IVPP (eds): "Symposium on the Origin of Man." Beijing: Science Press, pp 28–42.

Xia Ming (1982): Uranium-series dating of fossil bones from Peking Man Cave—Mixing model. Acta Anthropol Sinica 1:191–196.

Yi Seonbok, Clark GA (1983): Observations on the lower Paleolithic of northeast Asia. Curr Anthropol 24:181–202.

Zhou Guoxing, Hu Chengchih (1979): Supplementary notes on the teeth of Yuanmou man with discussions on morphological evolution of mesial upper incisors in the hominoids. Vertebrata PalAsiatica 17:149–162.

Zhou Mingzhen, Li Yanxian, Wang Linghong (1982): Chronology of the Chinese fossil Hominids. In 'L'*Homo erectus* et la Place de l'Homme de Tautavel Parmi les Hominidés Fossiles, Vol 2." Nice: Louis-Jean Scientific and Literary Publications, pp 593–604.

The Origins of Modern Humans: A World Survey of the Fossil Evidence, pages 485–516

Prehistoric and Modern Tooth Size in China

C.L. Brace, Shao Xiang-qing, and Zhang Zhen-biao

Museum of Anthropology, University of Michigan, Ann Arbor, Michigan 48109
(C.L.B.), Anthropology Section, Department of Biology, Fudan University,
Shanghai, People's Republic of China (S.X-q.) and Institute of Vertebrate
Paleontology and Paleoanthropology, Beijing (Z.Z.-b.), People's Republic of China

INTRODUCTION

The emergence of "modern" human form is a phenomenon of the last forty thousand years or less in those parts of the world where there is actually enough evidence to put the question to a test. Prior to that, while it is agreed that the designation *Homo sapiens* is technically correct, most observers recognize that the less-than-modern degree of observed skeletal robustness at least deserves recognition by the use of the term "archaic modern." Some find it of value to use the concept of a "Neandertal Stage" to refer to the general level of skeletal robustness observable before the subsequent reductions produced the configuration that we call "modern."

In Australia, aspects of archaic form survived in clearly visible fashion up to historic contact, and it is clear that the process of reduction was late in beginning [Brace, 1980; Thorne and Wolpoff, 1981; Wolpoff et al, this volume].

On the Asian mainland, specifically in China, premodern levels of robustness are obvious in the Dali (Fig. 1) and Maba (Mapa) (Fig. 2) skulls, which can be considered as representatives of a Chinese Neandertal Stage [Woo and Peng, 1959; Wu and You, 1979; Wu, 1981]. Dali may in fact be older than the European Neandertals and therefore, like the Solo collection from Java, near that *erectus*-to-*sapiens* transition that makes precise categorization impossible. In any case, it is obvious that archaic form preceded modern form in China just as it has elsewhere.

Right at the end of the Pleistocene, the specimens from the Upper Cave at Zhou Kou Dian (Fig. 3) still preserve traces of archaic form that have largely disappeared in modern Chinese.

Elsewhere in the world, Late Pleistocene and post-Pleistocene reductions in human skeletal robustness are clearly apparent. The region where this has been most convincingly demonstrated is western Europe, and it is

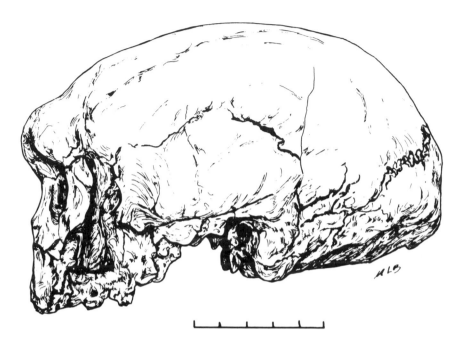

Fig. 1. The Dali skull, found in 1978. A candidate for a representative of the beginning of the Neandertal Stage in China. Drawn by M.L. Brace from a photograph taken in Beijing in 1980 with the permission of Professor Wu Xin-zhi.

displayed in most unequivocal fashion by human tooth measurements [Brace and Mahler, 1971; Brace, 1979a]. The association of structural reduction with conditions of selection relaxation may be only circumstantial, as is always the case when historic or prehistoric sequences are the subject of consideration, but the repeated observation of similar sequences in different parts of the world increases the probability that the associations observed are indeed causally related. Major changes in human subsistence strategies have been closely followed by changes in human physique. The development of such items as traps, nets, and fish hooks and the concentration on plant food resources that all come with the Mesolithic are followed by marked reductions in robustness, especially of the male physique [Brace, 1980; Brace and Ryan, 1980]. In spite of efforts by those who have tried to show that it really was the result of natural selection, there is no convincing demonstration that the reductions in bone and muscularity were of any particularly adaptive value to their possessors. On the contrary, these reductions occur just after the time when survival became possible *without* the previously normal levels of robustness. The suggestion has been offered

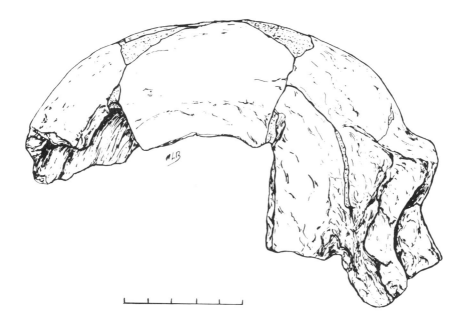

Fig. 2. The Maba skull, a Chinese Neandertal found in 1958. Drawn by M.L. Brace from a cast in Beijing in 1980.

that this was the consequence of mutations alone, accumulating after the forces of selection had been reduced—a phenomenon referred to as the Probable Mutation Effect (PME) [Brace, 1963, 1979b]. The logic is essentially the same as that used by students of biochemical evolution to account for the change in the newly superfluous sections of duplicated nucleic acid segments that code for the structure of proteins [Ohno, 1970; Ohta, 1980; Doolittle, 1981].

While the evidence for reduction in the recent past is clearly supported by an appraisal of the few crania available, the evidence is sparse and crude, and only large and dramatic changes can be demonstrated by these means. The continuing trends of regionally identifiable post-Pleistocene change are more easily treated by the study of material that exists in greater quantities and that can be assessed in relation to its response to selective force change.

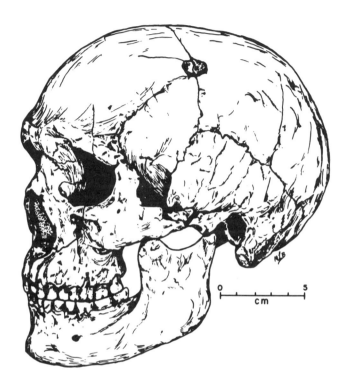

Fig. 3. The male skull from the Upper Cave, Zhou Kou Dian (Choukoutien), a late Pleistocene specimen from the Upper Paleolithic (?)/Mesolithic (?). Drawn from a cast by M.L. Brace.

We are referring to teeth. For the remainder of this paper, then, our attention will be focused on the dentition—specifically the metrics that can be used for quantitative comparisons.

From time to time, those who work in the realm of dental anthropology have turned their attention to various nondietary uses of the human dentition—for example, the use of teeth as tools, or as auxiliary manipulatory devices [Brace, 1962; Merbs, 1968; Molnar, 1972; Barrett, 1977; Tindale, 1977; Smith, 1983]. Despite this, we should not lose track of the fact that the primary role of the dentition from the beginning has been played in the processing of food. Until fairly recently, the loss of one's teeth constituted a sentence of certain starvation.

Uniquely among the mammals, human beings have developed nondental ways of processing food that have reduced this absolute dependence upon the possession of a functional dentition to the point where it no longer matters whether one has any teeth or not. It has been suggested that one of the keys to human survival in the periglacial portions of the Old World, as the last glacial episode tightened its grip on the northern hemisphere, was the use of deliberately applied heat to thaw foods that otherwise would have been inedible by virtue of being frozen solid. One of the unexpected consequences of this innovation was a reduction in the intensity of the forces of selection that had previously maintained the human dentition at a relatively uniform level for the previous million years or so. The reductions that then ensued produced recognizably modern faces out of a previous configuration that we term Neandertal form [Brace, 1977, 1978, 1979a].

If the key to human survival has been in that complex of customs, traditions, and learning called culture, reflective people have used this in part or in whole to illustrate a class of proposed dualities to define the human condition. Particular forms of behavior are said to characterize the civilized as opposed to the uncivilized, the human versus the nonhuman, and ultimately man versus nature.

Not only has this imposed duality been of importance in man's image of himself, but it has been extended in application to the treatment of the foodstuffs that at bottom are important solely for the extent to which they sustain human life [cf Elias, 1978]. Food is considered fit for human consumption if it is properly cooked. The nonprocessed, or raw, is indicative of barbarism or worse. It is perhaps not surprising that these views have been proposed with particular clarity from within that contemporary culture which is especially known for its self-declared culinary excellence, namely France [Lévi-Strauss, 1964, 1965, 1967, 1968]. If cooking is indeed a language by which a society reveals the bases of its nature, including the regularities and the contradictions, as Lévi-Strauss has suggested, then surely the most intricate and remarkable society must be that of China. The myriad of nested dualities by which Chinese gastronomy characterizes the edible is unmatched in any other culture [compare Brillat-Savarin, 1960, and Lin and Lin, 1972]. It is evidently the source for the gastronomic fastidiousness apparent in the *Analects* of Confucius 500 years before Christ [Legge, 1893, pp 232–233; Chang, 1976, p 133] and is part of a culinary tradition that extends by written record back to the Bronze Age and continues in full force today [Sakamoto, 1977]. The roots of the Chinese Bronze Age are clearly in situ in the preceding Neolithic [Chang, 1980, pp 148, 340, 344–348], and it is obvious that the span of time represented is of sufficient magnitude that measurable biological change could be discerned if those archeologically recorded pieces of evidence for culinary practices really do constitute changes in the nature of the selective forces that have influenced the maintenance of human form.

Now, what occidental or oriental commentators on matters from philosophy to gastronomy *say* is of basic importance may not in fact have anything at all to do with the real bases of human survival. On the other hand, it would be extraordinary if the long and unbroken record of cultural elaboration that is preserved in China did *not* contain some aspects that changed the nature and intensity of the selective forces that have influenced the possibilities for human survival; and it would be predictable that these would be in just those dimensions where that cultural record has attained the most extraordinary levels of elaboration. Since one of these is in the gastronomic, there is more than a little reason for us to anticipate that we should expect to discover changes in human form since the Neolithic that may be related.

The suggestion was offered that tooth size change began when food preparation practices altered the nature of the selective forces that had previously been in operation [Brace, 1977, 1979a]. Contrary to usual expectation, however, the significant food preparation practices were seen to have their beginnings, not with the Neolithic, but back in the Mousterian with the adoption of earth oven cookery. In line with this, it was noted that modern populations with the longest prehistoric-historic association with developed food preparation techniques should show the greatest degree of dental reduction, while those who had most recently acquired such techniques should possess teeth that retain a more Pleistocene level of development. The extreme case of the latter instance should occur in aboriginal Australia, and the available data support such an expectation [Brace, 1980].

On the Asian mainland, the archeological evidence indicates a much greater time depth for the use of elaborated food preparation techniques, and the published data show that dental reduction was indeed evident to a degree not manifest as one proceeds towards Australia [Brace, 1978, 1980; Brace and Hinton, 1981; Brace and Vitzthum, in press]. It is not surprising, then, that Asian tooth size is particularly small where the evident antiquity of food preparation practices is well documented and where culinary sophistication has attained levels unmatched anywhere in the world, namely the Chinese mainland [Keys, 1963; Lin, 1972; Chung, 1966; Chang, 1973, 1976, 1977].

Of course, as has been noted [Shao, 1981], China is very large, and one measured group is a scarcely sufficient basis on which to generalize about the country as a whole. Furthermore, the old question of whether the sources of Chinese civilization were to be located in the North or the South, or even within the bounds of eastern Asia at all, remains a matter for consideration [Arne, 1925; Li, 1976; Ho, 1976; Chang, 1977] as does that of the role of Southeast Asia vis-à-vis that of China proper [Solheim, 1967, 1969, 1970, 1972; Chang, 1977, p 142].

With that in mind, we present dental metrics for one Mesolithic and a number of modern and Neolithic Chinese samples. Although these data do not provide a definitive answer to the questions mentioned above, they do tend to confirm the view that post-Pleistocene dental reduction in Asia has proceeded at a rate entirely comparable with that in the western portions of the Old World. Furthermore, there is some reason to expect that selective force change has been in operation longer in southeastern China and adjacent Southeast Asia than elsewhere on the Asian continent.

MATERIALS AND METHODS

One Mesolithic, three Neolithic, and six modern samples were used for our analysis. One of the modern samples has been previously reported, ie, the material from Hong Kong (Xiang Gang) [Brace, 1978; Brace and Ryan, 1980]. The remaining material was measured in the People's Republic of China by the senior author in 1980.

The Shanghai sample consists of extractions that were collected at the Shanghai Dentistry Center from individuals of known sex. These are housed in the Anthropology Section of the Department of Biology, Fudan University, Shanghai. To these were added measurements made on crania in the collections at Fudan University. These crania came from post-Quing (Ching) Dynasty burial sites that were preempted by the expansion of the modern city of Shanghai. Assessment of sex was made as a result of consultation between the first two authors.

The Yunnan sample consists of modern crania from the Province of Yunnam stored in the collections of the Institute of Vertebrate Paleontology and Paleoanthropology in Beijing (Peking). Assessment of sex was made by the senior author.

The Nanjing (Nanking) sample consists of crania stored in the Department of Biology, Fudan University, and in the Department of Anatomy at the Nanjing Railway Medical College. Again, assessment of sex was made as a result of consultation between the first two authors.

The sample designated Beijing is principally housed in the collections of the Institute of Vertebrate Paleontology and Paleoanthropology in Beijing. Most of it was assembled from the dissecting rooms of the former Peking Union Medical College by Davidson Black. The specimens we included in our Beijing category were either individuals from Hebei Province or of unspecified provincial origin but presumed to be north Chinese. Sexual identity was known from the dissecting room records. A few more individuals were included from the collections of the Beijing First Medical College and the Beijing Second Medical College.

The sample designated Harbin was measured at the Traditional Chinese Medical College in Harbin, the Harbin Railway Health School, and the Heilongjiang Provincial Health School in Harbin. The crania came from post-Qing Dynasty burial sites that had yielded to the expansion of the modern city of Harbin. Assessment of sex was made by the senior author.

The Yang Shao sample was excavated by J. Gunnar Andersson between 1921 and 1924 at the Yang Shao type site of Xia Guo Tun (Sha Kuo T'un) in Henan (Honan) and in Gansu (Kansu) [Andersson, 1925; Black, 1925, 1928]. It is no longer possible to be certain which material comes from the two different provinces, and since the provisionally separated samples were not statistically distinguishable, they were combined in the present analysis and regarded as belonging to the fifth millennium B.C. Chinese Neolithic of the middle reaches of the Huang He (Yellow River) drainage area [Chang, 1977, p 85]. The specimens are housed in the storage facilities of the Institute of Vertebrate Paleontology and Paleoanthropology in Beijing. Assessment of sex was estimated by the senior author.

The Liu Lin material was excavated in the northern part of Jiangsu (Kiangsu) Province in 1960 and 1964 by the excavation team from the Nanjing Museum [Yin and Zhang, 1962; Qu, 1965]. It dates from the Chinese Neolithic of the mid- to late fourth millennium B.C. just before the Long-shan (Lung-shan) [Chang, 1977, p 133]. The specimens are housed in the collections of the Anthropology Section of the Department of Biology, Fudan University, Shanghai. Assessment of sex was made by the first two authors using both cranial and postcranial material.

The material from the Xi Chang site was excavated in 1971 and 1972 by the Henan Provincial Museum in the village of Xia Wang Gang just south of Xi Chuan in Henan and is housed in the Institute of Vertebrate Paleontology and Paleoanthropology in Beijing. The age is 5 to 6 ky BP and cultural affiliations are with the Yang Shao Neolithic. Assessment of sex for half of the population was made by the first and third authors in collaboration. The remainder was done by the third author using the same criteria.

The Mesolithic sample consists of the three crania found in the Upper Cave at Zhou Kou Dian (Choukoutien). Tooth measurements were not recorded in the descriptive publication [Weidenreich, 1938–1939], and, since the original specimens were lost in 1941 during the Sino-Japanese War, the measurements reported here were made on the excellent casts retained in the Institute of Vertebrate Paleontology and Paleoanthropology in Beijing [see also Brace and Vitzthum, in press].

The tooth measurements taken were the standard mesial-distal (MD) and buccal-lingual (BL) ones used in dental anthropology [see Brace, 1979a, p 529]. The measurements used were all made by the senior author except for those on about half of the Xi Chang specimens. These were made by the third author after techniques were checked so that remeasuring of a sample of specimens by both workers yielded results that did not differ by more than plus or minus 0.1 mm, the acceptable standard of measuring error for a single worker.

RESULTS

The mean measurement for each tooth category, its standard deviation and N for both sexes of the Neolithic and modern samples measured in 1980 are presented in Tables I–VIII. The Neolithic data are contained in Tables I–III, and the modern ones are to be found in Tables IV–VIII. As is so frequently the case, sample size for single rooted teeth is often quite small, reflecting the fact that they fall out so easily either in the burial ground or in storage. Enough are present, however, so that something of an appraisal can be made. Even so, the results of our comparisons must be viewed with considerable caution.

TABLE I. Mesial-Distal and Buccal-Lingual Measurements, Standard Deviations, and N for Female and Male Human Teeth From Yang Shao Levels in Henan and Gansu

	Female					Male				
	MD	sd	BL	sd	N	MD	sd	BL	sd	N
I^1	8.6	0.5	7.5	0.5	7	8.7	0.4	7.3	0.2	10
I^2	7.5	0.6	6.6	0.5	8	7.3	0.4	6.6	0.4	11
C	8.0	0.4	8.3	0.5	12	8.0	0.4	8.5	0.4	16
P^1	7.6	0.4	9.7	0.7	13	7.4	0.5	9.5	0.6	15
P^2	6.8	0.5	9.4	0.6	16	6.7	0.5	9.2	0.5	20
M^1	10.3	0.6	11.5	0.6	19	10.3	0.6	11.6	0.5	22
M^2	9.8	0.5	11.4	0.8	20	9.8	0.5	11.6	0.7	21
M_3	8.8	0.9	10.7	0.9	8	9.3	0.9	11.2	0.9	12
I_1	5.3	0.3	5.8	0.4	7	5.5	0.4	6.0	0.3	8
I_2	6.0	0.3	6.4	0.5	12	6.1	0.5	6.3	0.4	13
C	6.9	0.3	7.5	0.5	14	7.1	0.4	8.1	0.5	21
P_1	7.1	0.5	7.9	0.5	21	7.1	0.4	8.3	0.5	29
P_2	6.9	0.5	8.4	0.6	18	7.1	0.4	8.5	0.5	30
M_1	11.1	0.6	10.9	0.6	24	11.2	0.5	11.2	0.5	34
M_2	11.0	0.7	10.6	0.6	25	11.0	0.6	10.8	0.5	31
M_3	10.8	0.7	10.2	0.5	10	11.0	1.1	10.6	0.7	28

TABLE II. Mesial-Distal and Buccal-Lingual Measurements, Standard
Deviations, and N for Female and Male Human Teeth From the Liu Lin
Neolithic Site, Jiangsu

	Female					Male				
	MD	sd	BL	sd	N	MD	sd	BL	sd	N
I^1	8.5	0.6	7.1	0.4	18	8.7	0.5	7.4	0.4	9
I^2	6.8	0.5	6.5	0.5	11	7.2	0.5	6.6	0.5	10
C	7.9	0.4	8.3	0.4	24	7.9	0.6	8.5	0.6	14
P^1	7.2	0.4	9.4	0.4	28	7.4	0.5	9.7	0.4	9
P^2	6.8	0.4	9.3	0.4	28	6.8	0.3	9.6	0.4	11
M^1	10.2	0.6	11.4	0.4	29	10.6	0.5	12.0	0.5	11
M^2	10.0	0.5	11.2	0.5	29	10.3	0.6	12.0	0.4	12
M^3	8.7	0.8	10.3	0.7	18	9.2	0.8	11.2	0.8	9
I_1	5.4	0.4	5.9	0.4	22	5.6	0.4	6.1	0.3	12
I_2	5.9	0.3	6.4	0.4	27	5.9	0.5	6.5	0.3	16
C	6.9	0.4	7.7	0.4	36	7.0	0.6	8.0	0.5	20
P_1	7.0	0.5	8.0	0.4	38	7.1	0.5	8.4	0.3	18
P_2	6.9	0.5	8.3	0.5	40	7.0	0.5	8.6	0.5	20
M_1	11.0	0.5	11.0	0.5	39	11.3	0.6	11.4	0.6	19
M_2	10.6	0.6	10.4	0.5	39	11.1	0.6	10.9	0.5	22
M_3	10.3	0.8	10.3	0.6	28	10.7	1.2	10.2	1.2	21

To facilitate comparisons, we have taken several steps to combine and
simplify the data in Tables I–VIII. First, the separate mesial-distal and
buccal-lingual measurements were multiplied to produce approximations of
cross-sectional areas. Then, for each tooth, the female and male cross-
sectional areas were added together and in each instance the sum was
divided by two to give a population average cross-sectional area. Then the
upper and lower cross-sectional areas were summed for each tooth category.
The resulting composite cross-sectional areas for the six modern samples are
recorded in Table IX. Similarly, Table X contains the composite cross-
sectional areas for the terminal Pleistocene Zhou Kou Dian Upper Cave
individuals and the three north Chinese Neolithic samples.

The last column in Table IX was constructed by averaging the figures in
the other columns to produce a mean Chinese composite tooth size representation. In the same fashion, the last column in Table X was constructed
by averaging the Neolithic figures in the preceding columns.

The bottom row in both Tables IX and X is made up of the sums of each
column. Each figure in the final row is the summary tooth size figure (TS)
for the sample indicated. This is the most economical way to represent
tooth size for a given group and makes visual comparisons quick and easy
even if there are problems in calculating a usable variance [Brace, 1980].

TABLE III. Mesial-Distal and Buccal-Lingual Measurements, Standard
Deviations, and N for Female and Male Human Teeth From the Xi
Chang Neolithic Site, Henan

	Female					Male				
	MD	sd	BL	sd	N	MD	sd	BL	sd	N
I^1	8.7	0.5	7.1	0.3	10	8.6	0.5	7.2	0.4	12
I^2	6.7	0.7	6.4	0.6	12	7.2	0.7	6.7	0.7	20
C	7.8	0.5	8.2	0.6	16	8.0	0.4	8.5	0.5	23
P^1	7.2	0.4	9.4	0.4	19	7.4	0.4	9.7	0.8	31
P^2	6.8	0.5	9.2	0.4	19	6.9	0.4	9.6	0.6	31
M^1	10.2	0.5	11.2	0.4	21	10.5	0.6	11.6	0.6	30
M^2	9.8	0.4	11.1	0.4	20	9.9	0.6	11.4	0.8	30
M^3	9.0	0.9	10.4	0.9	13	9.2	0.8	11.0	0.7	20
I_1	5.6	0.8	5.8	0.7	11	5.6	0.4	5.9	0.4	19
I_2	6.1	0.4	6.1	0.3	21	6.2	0.4	6.3	0.4	43
C	6.8	0.4	7.5	0.4	39	7.1	0.4	8.0	0.5	63
P_1	7.0	0.4	7.9	0.4	46	7.2	0.5	8.3	0.5	71
P_2	6.9	0.6	8.2	0.5	55	7.0	0.5	8.5	0.6	80
M_1	11.1	0.5	11.0	0.4	71	11.4	0.6	11.3	0.5	86
M_2	10.6	0.6	10.5	0.5	70	10.9	0.6	10.8	0.5	77
M_3	10.6	0.8	10.2	0.7	53	10.7	0.8	10.4	0.6	68

The columns of cross-sectional areas, however, can be used for statistical comparisons even in the absence of usable variance estimates. The nonparametric Wilcoxon matched-pairs signed-ranks test can be used to determine whether there is any significance in the differences between any two such columns [see Siegel, 1956, pp 75–83].

A glance at Table XI will show that there is no good reason to regard the Neolithic samples as significantly different from each other. Strictly speaking, the difference between the Xi Chang and Yang Shao samples borders on significance, but the TS difference is less than 20 mm^2, and this is not in the realm of magnitude that other studies [cf Brace, 1980] have suggested are large enough to indicate some interpretable meaning. This, then, is why we have combined the Neolithic data into a single set of composite cross-sectional areas, shown in the last column of Table X, and used them as the basis for our subsequent comparisons.

When we look at Table XII, however, it is clear that there is much less justification for using a single combined set of data, for example, the last column of Table IX, to represent a picture of modern China. Even though we used two-tailed expectations in deriving our probability levels (and, given the nature of our prediction, the use of one-tailed logic could easily be justified), it is clear that our samples from the southern part of China

TABLE IV. Mesial-Distal and Buccal-Lingual Measurements, Standard
Deviations, and N for Modern Female and Male Human
Teeth From Shanghai

	Female					Male				
	MD	sd	BL	sd	N	MD	sd	BL	sd	N
I^1	8.4	0.6	7.0	0.5	57	8.7	0.5	7.3	0.4	70
I^2	6.7	0.6	6.2	0.5	65	7.0	0.8	6.5	0.7	62
C	7.7	0.4	8.0	0.5	75	7.8	0.5	8.2	0.6	62
P^1	7.1	0.4	9.3	0.4	36	7.2	0.4	9.6	0.6	34
P^2	6.6	0.4	9.0	0.5	40	6.7	0.4	9.2	0.6	37
M^1	10.2	0.5	11.1	0.5	130	10.4	0.5	11.5	0.5	104
M^2	9.7	0.6	11.1	0.6	97	9.9	0.6	11.5	0.7	79
M^3	9.0	0.9	10.3	1.0	55	8.8	1.0	10.6	1.3	95
I_1	5.4	0.4	5.7	0.3	58	5.5	0.3	5.9	0.7	63
I_2	6.1	0.8	6.3	0.8	58	6.1	0.4	6.3	0.4	61
C	6.9	0.5	7.6	0.5	59	7.1	0.5	8.0	0.5	43
P_1	6.9	0.3	7.7	0.5	11	7.1	0.5	8.2	0.6	20
P_2	7.0	0.4	8.0	0.4	12	7.0	0.6	8.4	0.6	14
M_1	11.2	0.5	10.7	0.4	29	11.4	0.5	10.9	0.5	42
M_2	10.7	0.5	10.1	0.4	22	11.0	0.7	10.5	0.6	33
M_3	10.3	1.1	9.8	1.0	50	10.6	1.2	10.1	1.0	53

have teeth that are significantly smaller than those from northern China. If
we were to use the last column of Table IX to represent modern Chinese,
we could note that they had significantly smaller teeth than did the people
of the Chinese Neolithic as recorded in the last column of Table X.

However, as shown in Table XIII, if we make separate groups north and
south of the Chang Jiang (Yangtze) as would be justified by a reading of
Table XII, then it is clear that the southern group is significantly smaller
than the northern group as a whole, but the northern group is not significantly distinguishable from the Neolithic. In fact, the summary tooth size
for the modern north Chinese is identical with that of the north Chinese
Neolithic.

Both columns of Table XIII are significantly different (.01) from the last
column in Table IX and also from each other. In fact, Table IX shows that
tooth size in modern China is distributed in the form of a clear-cut north-
south cline. At the northern end, teeth are fully Neolithic in size, while at
the southern end, a significant amount of reduction has occurred. If we take
the minimal north-south distinction as represented in Table XIII, our south
Chinese tooth size is 44 mm^2, or 4% less than our north Chinese figure,
and this is close to the level that has previously been used to signify some
basic biological meaning [Brace, 1979a, 1980; Brace and Hinton, 1981; Brace

TABLE V. Mesial-Distal and Buccal-Lingual Measurements, Standard
Deviations, and N for Modern Female and Male Human Teeth
From Yunnan

	Female					Male				
	MD	sd	BL	sd	N	MD	sd	BL	sd	N
I^1	8.8	0.4	7.4	1.1	2	8.3	0.8	6.7	0.3	4
I^2	7.0	0.7	6.4	0.7	6	7.2	0.5	6.4	0.8	4
C	7.7	0.4	8.2	0.6	19	8.0	0.3	8.6	0.5	15
P^1	7.2	0.4	9.2	0.7	37	7.4	0.5	9.5	0.7	38
P^2	6.7	0.4	9.3	0.6	44	6.9	0.5	9.4	0.7	30
M^1	10.3	0.6	11.3	0.5	98	10.4	0.5	11.5	0.6	70
M^2	9.6	0.5	11.2	0.6	83	9.8	0.6	11.5	0.8	66
M^3	8.7	0.8	10.8	0.7	26	9.0	0.7	11.0	0.8	30
I$_2$	5.2	0.4	5.6	0.6	5	5.6	0.1	5.6	0.6	2
I	5.6	0.6	5.8	0.5	8	5.9	0.4	6.0	0.3	9
C	6.9	0.6	7.8	0.7	13	7.1	0.4	7.9	0.5	19
P$_1$	7.2	0.4	8.0	0.6	21	7.2	0.5	8.1	0.7	33
P$_2$	7.0	0.4	8.2	0.6	36	7.0	0.5	8.4	0.5	31
M$_1$	11.1	0.6	10.7	0.5	50	11.2	0.6	10.8	0.6	59
M$_2$	10.8	0.7	10.3	0.5	44	10.8	0.6	10.4	0.6	53
M$_3$	10.4	0.7	10.0	0.7	38	10.5	0.7	10.2	0.7	41

and Nagai, 1982]. Certainly that level is attained if one compares Harbin
with Shanghai or, even more clearly, Harbin with Hong Kong.

DISCUSSION

The data we present demonstrate that there is a rough tooth size cline in
modern China running from Neolithic-sized teeth in the north to southern
levels that are reduced by 4 to 7%, depending on which samples are used to
stand for southern Chinese. We suggest that this really does represent a
picture of differential dental reduction that has taken place in the recent
past, but there are a couple of potentially confusing factors that we should
bring up before the reader can take this as having been demonstrated.

First, we do not have the data to test for the nature of east-west variation.
There are some hints, however, that there is also an east-west or coastal-
island cline. The composite tooth-size figures for Hong Kong are significantly smaller than those for Yunnan, although Yunnan is in fact just as far
south as Hong Kong. Yunnan, however, is inland rather than on the coast.
Further, there is always the question of how much effect was produced by
the spreads of Han Chinese from the north into Yunnan during historic
times [Weins, 1952, pp 101–104, 107;1967, Chapter 5].

TABLE VI. Mesial-Distal and Buccal-Lingual Measurements, Standard
Deviations, and N for Modern Female and Male Human Teeth
From Nanjing

	Female					Male				
	MD	sd	BL	sd	N	MD	sd	BL	sd	N
I^1	8.6	0.1	7.6	0.5	2	8.4	0.2	7.5	0.4	3
I^2	6.7	0.5	6.3	0.4	4	6.9	0.3	7.0	0.4	3
C	7.7	0.2	8.2	0.6	9	7.8	0.6	8.7	0.5	10
P^1	7.2	0.3	9.2	0.6	27	7.2	0.5	9.4	0.6	29
P^2	6.6	0.6	9.1	0.6	29	6.6	0.5	9.3	0.6	26
M^1	10.2	0.5	11.3	0.6	78	10.3	0.5	11.4	0.4	65
M^2	9.6	0.6	11.2	0.6	69	9.9	0.6	11.5	0.6	55
M^3	8.7	0.9	10.8	0.7	24	9.1	0.7	11.3	0.8	24
I_1	5.2	—	5.6	—	1	5.0	0.3	5.9	0.5	4
I_2	5.6	0.6	6.2	0.4	5	5.8	0.6	6.6	0.2	3
C	6.9	0.4	7.8	0.5	26	7.3	1.0	8.3	0.8	21
P_1	7.0	0.5	7.9	0.5	47	7.1	0.4	8.1	0.6	34
P_2	7.0	0.9	8.1	0.5	45	7.0	0.5	8.2	0.6	47
M_1	11.1	0.6	10.6	0.5	132	11.3	0.5	11.0	0.5	104
M_2	10.6	0.8	10.3	0.6	122	10.9	0.6	10.6	0.5	83
M_3	10.5	0.8	9.9	0.7	82	10.9	0.7	10.3	0.7	76

The other tentative east-west datum is from farther north. There are some specimens from the northern coastal province of Shandong (Shantung) in the Davidson Black collection in the Institute of Vertebrate Paleontology and Paleoanthropology in Beijing. A complete tooth size figure is only available for a sample of males. This is 1,190, with an average N of 9 and a range of N from 4 to 13. This is well below the TS of 1,259 for our male Beijing sample and even the 1,230 figure for our Shanghai sample. In fact, it is only slightly above the figure of 1,181 for male Hong Kong Chinese and just under the An Yang Bronze Age (also mostly male [cf Yang, 1966, p 5]) figure of 1,196 [Brace and Ryan, 1980; Brace, 1978].

A visual appraisal of specimens in the collections of the Institute of Archeology in Beijing has suggested that there was an east-west cline of cranio-facial robustness that was tentatively evident in the Neolithic and somewhat more noticeable in the Bronze Age. At the moment, however, this is just an impression and it requires proper measuring and testing. Despite the fact that we cannot produce quantitative evidence in its support, our overview of the available material suggests to us that the greater cranio-facial robustness by which the modern north Chinese are distinguished from modern south Chinese enters from the western part of north China, becoming more evident as the Neolithic gives way to the Bronze Age. If this

TABLE VII. Mesial-Diatal and Buccal-Lingual Measurements, Standard
Deviations, and N for Modern Female and Male Human Teeth
From Beijing

	Female					Male				
	MD	sd	BL	sd	N	MD	sd	BL	sd	N
I^1	8.6	0.5	7.3	0.4	7	8.6	0.6	7.4	0.5	34
I^2	6.9	0.4	6.3	0.4	8	6.9	0.7	6.6	0.6	50
C	7.6	0.6	8.2	0.6	13	7.9	0.4	8.6	0.5	78
P^1	7.2	0.4	9.2	0.5	16	7.3	0.4	9.6	0.6	93
P^2	6.8	0.4	9.2	0.7	15	6.7	0.5	9.4	0.6	89
M^1	10.2	0.5	11.2	0.5	29	10.3	0.5	11.5	0.6	103
M^2	9.6	0.6	11.3	0.6	25	9.9	0.6	11.7	0.6	99
M^3	8.9	0.5	10.7	0.6	15	9.0	1.0	11.0	1.4	66
I_1	5.4	0.2	5.8	0.4	5	5.6	0.3	6.0	0.3	41
I_2	6.0	0.2	6.2	0.3	8	6.2	0.4	6.4	0.3	62
C	6.8	0.2	7.7	0.5	10	7.2	0.4	8.1	0.5	80
P_1	6.9	0.4	7.9	0.5	15	7.2	0.4	8.2	0.5	95
P_2	6.9	0.6	8.0	0.6	13	7.1	0.6	8.4	0.5	96
M_1	11.0	0.5	10.7	0.6	23	11.4	0.6	11.0	0.5	107
M_2	10.8	0.6	10.3	0.6	21	11.1	0.7	10.8	0.6	94
M_3	10.7	1.0	10.2	0.8	17	11.0	1.0	10.4	0.7	80

can be sustained by a proper consideration of the evidence, then it may just
indicate that the historic picture of repeated invasions of China from the
northwest can be traced back beyond the dawn of recorded history into the
Neolithic. This is a problem for future work and testing.

There are a couple of other matters that must be mentioned before we
can offer our tentative conclusions. First, the modern north Chinese have
teeth that are virtually the same size as those of the Neolithic, and we have
to consider the possibility that there may have been no tooth size reduction
in China over the past 6 or 7 ky even though this is counter to what can
be shown for virtually every other part of the world where the evidence has
been considered, namely Europe [Brace, 1979a], North America [Brace and
Mahler, 1972], Southeast Asia [Brace and Vitzthum, in press], and even
Australia [Brace, 1980]. There is some reason for us to suspect, however,
that the modern north Chinese are not the unmixed descendants of the
north Chinese Neolithic. This is not something that we can show from our
tooth measurements alone, however, and it is an illustration of the fact that
more than one dimension must be taken into account when the biological
status of a population is in question.

For one thing, our suspicion is based on the observation that the modern
north Chinese just do not look like unmixed in situ descendants of the

TABLE VIII. Mesial-Distal and Bucco-Lingual Measurements, Standard Deviations, and N for Modern Female and Male Human Teeth From Harbin

	Female					Male				
	MD	sd	BL	sd	N	MD	sd	BL	sd	N
I^1	8.4	0.4	7.0	0.2	6	8.7	0.3	7.3	0.5	7
I^2	6.8	0.5	6.3	0.5	32	7.0	0.8	6.6	0.5	17
C	7.7	0.4	8.0	0.5	70	8.1	0.4	8.7	0.7	55
P^1	7.3	0.4	9.4	0.5	82	7.5	0.4	9.8	0.7	64
P^2	6.8	0.4	9.2	0.5	69	7.0	0.5	9.5	0.6	58
M^1	10.3	0.5	11.1	0.5	134	10.6	0.6	11.7	0.5	88
M^2	9.8	0.6	11.2	0.6	118	10.1	0.6	11.9	0.6	85
M^3	8.7	1.0	10.6	1.2	40	9.4	0.6	11.6	0.8	42
I_1	5.3	—	5.4	—	1	—	—	—	—	—
I_2	6.5	0.4	6.8	1.3	2	6.0	0.1	6.5	0.3	2
C	7.2	0.3	7.6	0.2	3	7.2	0.4	8.5	0.5	5
P_1	7.1	0.5	7.9	0.7	12	7.4	0.5	8.1	0.4	14
P_2	7.0	0.4	8.2	0.6	13	7.4	0.6	8.5	0.4	18
M_1	11.3	0.6	10.8	0.5	82	11.3	0.5	11.0	0.5	78
M_2	11.2	0.7	10.0	0.7	61	11.2	0.6	10.8	0.5	65
M_3	10.7	0.8	10.0	0.7	50	11.0	0.9	10.4	0.6	57

TABLE IX. Composite Cross-Sectional Areas and Summary Tooth Size Figures (TS) for Six Modern Chinese Samples

	Hong Kong	Shanghai	Yunnan	Nanjing	Beijing	Harbin	Mean
I1	89.5	93.6	90.8	94.0	95.8	91.9	92.6
I2	79.6	82.8	79.6	82.6	83.4	86.5	82.4
C	114.5	117.2	121.7	123.6	120.6	124.0	120.3
P1	120.5	123.5	126.6	123.6	125.1	129.6	124.8
P2	118.3	118.8	122.5	118.1	120.7	125.2	120.6
M1	228.9	239.0	238.5	238.2	238.5	242.3	237.6
M2	214.5	222.8	222.4	224.0	229.0	235.0	224.6
M3	188.1	199.2	203.2	207.3	210.0	213.0	203.5
TS	1,154	1,197	1,206	1,211	1,223	1,248	1,206

Chinese Neolithic. This has been mentioned previously in rather general terms [Li, 1976]. Brow ridges are heavier, nuchal muscle markings are more pronounced, and the nasal skeleton is higher and more elongated in modern north Chinese. In these aspects, the configuration visible in the Chinese Neolithic simply looks more like a robust version of what is now visible among the modern *south* Chinese. This is what leads us to voice the suspicion that the appearance of modern north Chinese is influenced by gene flow from outside the area inhabited by the denizens of Neolithic

TABLE X. Composite Cross-Sectional Areas and Summary Tooth Size
Figures (TS) for a Mesolithic and Three North Chinese Neolithic
Samples

	Mesolithic	Xi Chang	Liu Lin	Yang Shao	Mean Neolithic
I1	94.3	95.2	95.2	97.5	96.0
I2	94.0	83.9	83.8	87.6	85.1
C	144.2	120.4	121.1	122.1	121.2
P1	127.4	127.6	127.2	129.7	128.2
P2	124.7	123.4	123.1	121.8	122.8
M1	260.3	243.2	246.7	244.0	244.6
M2	253.9	226.4	233.3	230.7	230.1
M3	206.0	208.4	203.4	213.6	208.5
TS	1,305	1,228	1,234	1,247	1,236

TABLE XI. Probability Levels for Three Neolithic Samples
Compared by the Wilcoxon Matched-Pair Signed Rank Test

	Yang Shao	Liu Lin
Liu Lin	—	
Xi Chang	< .02	—

TABLE XII. Probability for a Combined Northern Neolithic and Six
Modern Chinese Samples Compared by the Wilcoxon Matched-Pair
Signed-Rank Test

	Hong Kong	Shanghai	Yunnan	Nanjing	Beijing	Harbin
Shanghai	.01					
Yunnan	.01	—				
Nanjing	.01	< .02	—			
Beijing	.01	< .02	—	< .05		
Harbin	.01	< .02	.01	< .05	< .05	
Neolithic	.01	.01	.01	< .05	< .02	—

China. Figures 4–7 show, respectively, a male and female north Chinese
and a male and female south Chinese skull. The contrast in nasal elevation
and mid-facial development is visually obvious. As always in such cases,
this is much more evident when the male specimens are compared. In
populations where nasal development is stressed, it is always more pronounced among males. We did not have this point in mind when the
drawings were made. In fact, when we chose those particular skulls for
illustrative purposes, our principal object was to get the most complete and
undamaged specimens of known sex from samples where, among other

TABLE XIII. Composite Cross-Sectional Areas and Summary Tooth Size
Figures (TS) for Modern Chinese Samples From South and North of the
Yangtze River

	South Chinese	North Chinese
I1	92.0	93.8
I2	81.2	84.9
C	119.2	122.3
P1	123.5	127.3
P2	119.4	122.9
M1	236.2	240.4
M2	220.9	232.0
M3	199.4	211.5
TS	1,192	1,236

imperfections, at least some teeth were missing from almost all of the
specimens available.

The other thing we have to consider is whether the north-south tooth
size cline displayed in our data may not simply be a reflection of differences
in body size. There is a low (0.155) but significant correlation between
human tooth size and body size [Garn et al, 1968a; Garn et al, 1968b], and
differences in body size and robustness between north and south Chinese
are a matter of common observation [Lin, 1935, p 301; Snow, 1938, p 319]
and anthropological record [Koganei, 1902, 1903; Li, 1928; Woo and Morant, 1932; Liu, 1937; von Eickstedt, 1944; Howells, 1983]. We cannot solve
this problem directly. For one thing, there were no postcranial skeletons
associated with the crania that furnished us with the dental data for most
of our modern Chinese samples. For another, it is clear that major differences in tooth size exist between living human populations that do not
differ significantly in body size, and it has yet to be seen whether the tooth-
size body-size regression established in one group can be legitimately used
in another. In any case, the percent difference in stature between north and
south China, whether it is based on the male data used by Koganei [1903]
or calculated from the measurements reported in Li [1928, p 33], is equal to
or less than the percent tooth size difference. Since the allometric relation
between the two variables is such that large body size changes are needed
to produce smaller tooth size changes [Kurtèn, 1967; Gould, 1975; Gingerich
et al, 1982], we would defend the view that the tooth size differences
between north and south China are more than can be accounted for by
differences in body size alone.

To put our Chinese data in perspective, Figure 8 shows a bar graph
superimposed upon a map of eastern Asia. The bars for China are derived

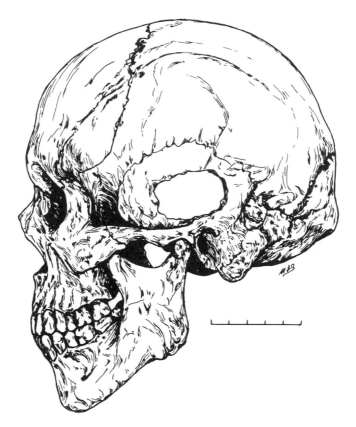

Fig. 4. Modern male from Beijing. Drawn by M.L. Brace from a specimen in the Davidson Black collection at the Institute of Vertebrate Paleontology and Paleoanthropology in Beijing.

from the data used to make Tables IX, X, and XIII of the present paper; the bars for Japan and Korea are from Brace and Nagai [1982], and the bars for Vietnam, Laos, Sarawak, Thailand, and Java are from data in Brace and Vitzthum [in press]. It is clear that tooth size reduces on the mainland from north to south China and adjacent Southeast Asia where it then begins to increase in size again out into Indonesia.

Tooth size for modern Japanese, as might be expected [eg, from the assessment of dental morphology by Turner, 1976, 1979; Turner and Hanihara, 1977], is indistinguishable from that in Korea and north China. The short bar visible in northern Japan is for the Ainu in whom tooth size shows even more reduction than that visible in south China. A full consid-

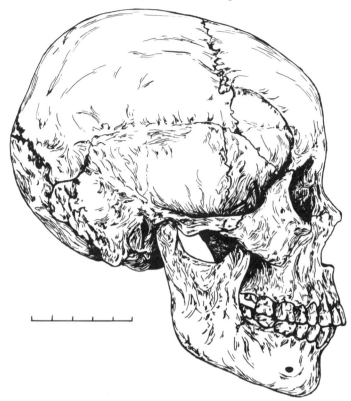

Fig. 5. Modern female from Beijing. Drawn by M.L. Brace from a specimen in the Davidson Black collection at the Institute of Vertebrate Paleontology and Paleoanthropology in Beijing.

eration of this interesting situation is developed elsewhere [Brace and Nagai, 1982].

If the gradients in tooth size are clear for the modern populations of eastern Asia, this is not the case for the Neolithic. Figure 9 shows a Neolithic bar graph superimposed upon a map of Asia. The Chinese bars represent the Neolithic samples before the Gansu and Henan Yang Shao material was combined. The Southeast Asian bars are based on material from Vietnam, Laos, Niah Cave in Sarawak, and Thailand [Brace and Vitzthum, in press]. These are not significantly different from each other when compared by the Wilcoxon matched-pairs signed-ranks test. Admittedly, we lack a good Neolithic sample from south China, but for the moment, our data do not show any significant tooth size differences between continental east Asian populations during the Neolithic.

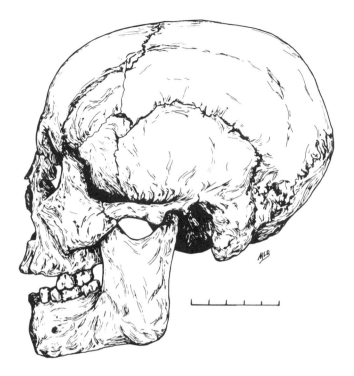

Fig. 6. Modern male from Yunnan. Drawn by M.L. Brace from a specimen in the collections of the Institute of Vertebrate Paleontology and Paleoanthropology in Beijing.

In Japan, the longer bar represents the Yayoi rice cultivators who moved to Kyushu and western Honshi from Korea in the third century B.C. Again, tooth size is indistinguishable from that of the mainland Neolithic [Brace and Nagai, 1982]. The short bar for Japan in Figure 9 represents the indigenous Jomon. It has been argued that this is the consequence of the long-term use of pottery in Jomon Japan, a development that pre-dates the appearance of pottery in China even though it was not associated with a food-producing mode of subsistence. The consequent long-term in situ dental reduction, it is suggested, may account for the small teeth of the modern Ainu shown in the short bar in Figure 8.

If we combine some of our separate Neolithic samples yet retain a Chinese and Southeast Asian distinction, we can present a picture of continental Asian tooth size for Mesolithic, Neolithic, and modern data in a single graph as shown in Figure 10. The bar for the Southeast Asian Neolithic was made by averaging the figures for Neolithic sites in Vietnam, Laos, and Thailand since it has been noted that these were not significantly

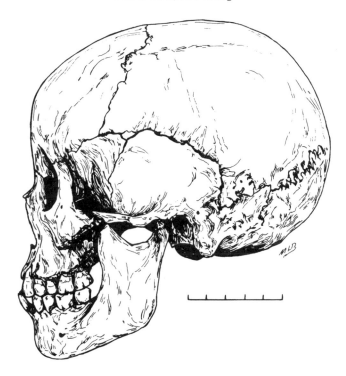

Fig. 7. Modern female from Yunnan. Drawn by M.L. Brace from a specimen in the collections of the Institute of Vertebrate Paleontology and Paleoanthropology in Beijing.

different from each other by the Wilcoxon matched-pairs, signed-ranks test. The bar for the Southeast Asian Mesolithic was constructed by averaging the figure from the Tam-Pong rock shelter in Laos [Brace and Vitzthum, in press, Table III] with the figure from the Hoabinhian site of Gua Kepah on the west coast of the Malay Peninsula [Brace and Hinton, 1981, Table I].

The question of whether either the Upper Cave of Zhou Kou Dian (Choukoutien) or the Hoabinhian of Southeast Asia is properly labelled Mesolithic has been argued by archeologists for years [Chang, 1968, pp 67–68; Aigner, 1972; Mathews, 1966, p 5; Solheim, 1969, p 129], and is relatively unimportant here. What matters for our purposes is that the samples precede those that authorities agree can be called Neolithic. In both of the above instances, there is no doubt that this is the case. In both instances also, even though the samples are relatively small, tooth size is clearly larger than is the case for more recent people in the same areas.

Although the Zhou Kou Dian sample is so small that no generalization based on it is very secure, it may be a good deal older than the Hoabinhian

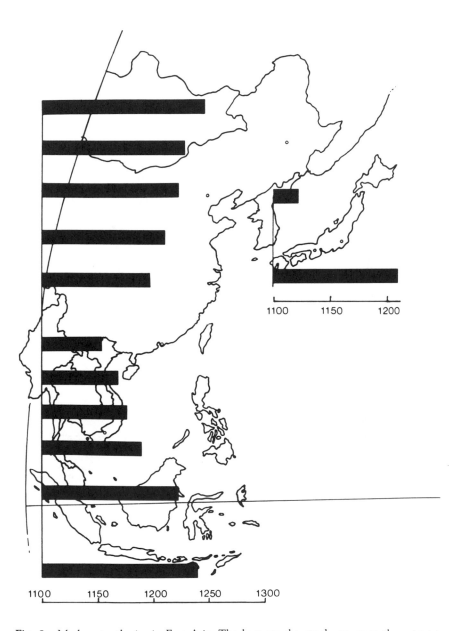

Fig. 8. Modern tooth size in East Asia. The bars on the graph represent the extent to which summary tooth size rises above 1,100 mm². The bar graph is superimposed on a map so that each bar represents tooth size in a population that currently lives approximately at the latitude where the bar is located. The modern Korean and Japanese bars represent data from Brace and Nagai [1982]; the Southeast Asian bars are from Brace and Vitzthum [in press]; and the Chinese bars are based on data in Tables IV–VIII.

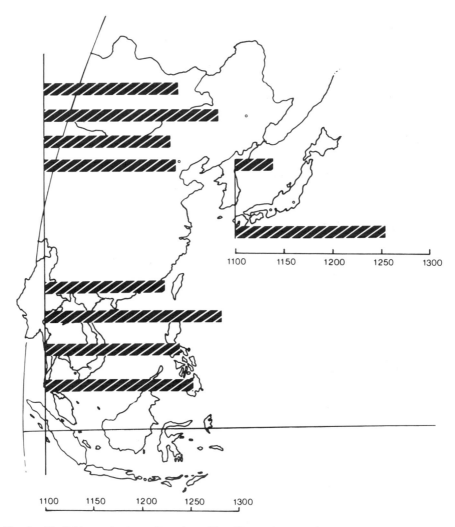

Fig. 9. Neolithic tooth size in East Asia. The Chinese bars are from the Liu Lin and Xi Chang data in Table X and the Gansu and Henan Yang Shao material before it was combined. The Southeast Asian bars are based on data from Vietnam (Lang Cuom), Laos (Tam-Hang), Sarawak (Niah Cave), and Thailand (Ban Kao) recorded in Brace and Vitzthum [in press].

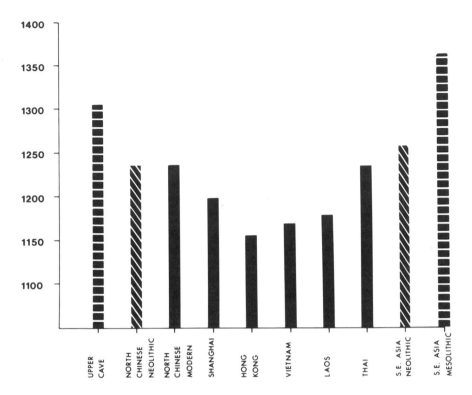

Fig. 10. Summary tooth size bars for Mesolithic, Neolithic, and Modern samples from China and Mainland Southeast Asia. The height of each bar is proportionate to the number of square millimeters of summary tooth size (TS). The Chinese Mesolithic (?) bar represents the Upper Cave at Zhou Kou Dian, the Neolithic bar is from the last column in Table X and the modern bars are based on the data in Table IX. The Southeast Asian Mesolithic bar is based on data combined from the Tam-Pong rock shelter in Laos and Gua Kepah from the Malay Penninsula. The Southeast Asian Neolithic bar is based on data combined from Vietnam, Laos, and Thailand. These and the modern Southeast Asian data are reported in Brace and Vitzthum [in press].

material from Southeast Asia, Sarawak, and Indonesia, and yet tooth size is smaller than its southern counterparts. This may indicate that in eastern Asia, as in Europe, late Pleistocene dental reduction began rather earlier in the northern areas of human habitation than it did farther south. The reasons for this have been previously suggested [Brace, 1977, 1978, 1979a].

Subsequent dental reduction as indicated by our Neolithic figures shows that there was no significant difference in tooth size when north China is compared with Southeast Asia. Actually, when the Southeast Asian Neolithic samples are combined to make the average used for Figure 10, the composite cross-sectional areas differ from those of north China, as shown, by nearly 20 mm^2, which has a probability value of .05, although this level is not achieved when the samples are individually compared with the north Chinese Neolithic. Admittedly, we lack data from south China. Even though this is geographically considered Southeast Asia [Solheim, 1969, pp 125–127;1979, p 74], the pattern of modern tooth size variation shown in Figure 10 suggests to us that the selective force alteration that led to dental reduction has been in operation in the south Chinese part of Southeast Asia for as long as anywhere on the Asian continent.

That harbinger of the Neolithic beloved to the archeologist, the potsherd, signals something of equal importance to the biological anthropologist—an alteration of major significance in the selective forces that previously maintained the jaws and teeth. Cooking vessels enable the possessors to process foods in such a manner that the retention of teeth is no longer of such crucial importance for sheer survival. It is no accident that Neolithic and subsequent burial grounds are characterized by the presence of individuals in some numbers who had been edentulous for many years prior to death. Nothing like this is visible in burials from earlier or pre-pottery time levels, although it is interesting to note that edentulous individuals occur in considerable numbers among the preagricultural inhabitants of Japan. In this case, however, the Jomon Japanese were the creators and hence the beneficiaries of the oldest pottery tradition in the world, and the consequences for their maintenance of usable tooth substance have been noted elsewhere [Brace and Nagai, 1982]. The descendants of those who first developed techniques for prolonging the life of the toothless—especially by the use of pottery cooking vessels—include those modern people with the most reduced jaws and teeth amongst the living people of the world. This we can guess is because of the relaxation of the forces of selection that had previously maintained the formerly necessary chewing machinery.

The pattern of modern Asian tooth size variation shown in Figure 10, where reduction is most obvious in south China and adjacent Southeast Asia and size increases as one goes both north and south of that area, suggests that this is where selective force relaxation has its greatest antiquity. Although the increase in tooth size in our northernmost samples may well reflect some covariance with the somewhat larger northern body size, this certainly cannot be a factor in the increase in tooth size as one goes south in peninsular Southeast Asia and out into Indonesia. Even in the absence of direct evidence, we could guess that the search for the earliest Asian Neolithic should focus on the area where dental reduction among the modern inhabitants is most evident. The recent report of a Neolithic date of nearly 9,000 B.C. in Jiangxi (Kiangsi) [Meachem, 1977, p 423] provides gratifying support for our expectations.

CONCLUSIONS

As the record of history shows us, people do not necessarily stay in the area where they were biologically shaped. Not only do they move out of the area characterized by the forces responsible for their shaping, but others shaped by different backgrounds can move into the area in question. In China, both events appear to have happened. For example, the low nasal bridge and relatively broad piriform aperture of the peoples of south China and adjacent Southeast Asia are appropriate reflections of in situ selective forces extending back into the Pleistocene (the rationale for this expectation is discussed by Carey and Steegmann [1981]). But the same kind of nasal form also characterizes the north Chinese Neolithic all the way west to Gansu where it is not an appropriate reflection of local selective forces. We can suspect that the people of the north Chinese Neolithic actually had their roots in the southeast and moved north and west from there. The relaxation of selection for the maintenance of preagricultural levels of tooth size, which was a product of their culinary practices, led to the reduction of the jaws and teeth of the descendants of the Chinese and adjacent Southeast Asian Neolithic. This, we suspect, is the meaning of the fact that the smallest dentitions on mainland Asia range from Hong Kong to Vietnam.

During the Neolithic, tooth size in north China and Southeast Asia was essentially the same and, in both instances, evidently reduced from Mesolithic levels. If no other factors had intervened, we would expect to find the modern descendants of the East Asian Neolithic to show similar degrees of dental reduction. In north China, however, modern teeth are exactly the same size as they were in the Neolithic. We suggest that the trend of dental reduction within the area influenced by the Neolithic was counteracted in the north by the influx of genes from areas beyond the reaches of the Neolithic, and such areas of course could only be to the west and north. There one would also expect that in situ selective forces had maintained a prominence of nasal skeleton not characteristic of the Neolithic. The movement by such northern and western people into the area of burgeoning Neolithic civilization and the contribution of their biological attributes to the amalgamated resultant population would produce exactly what is visible in northern China today.

In Southeast Asia, a similar amalgamation of people who were influenced by the selective force changes that date from the beginning of the Neolithic and those who had only later been affected should also be visible in the form of larger jaws and teeth as one proceeds south of the area of the earliest Neolithic development. In this case, however, the larger-toothed nonagriculturalists would not have had the more prominent nasal skeletons appropriate to a northern locale, and indeed the increase in tooth size from Thailand to Java and out towards New Guinea is not accompanied by nasal enlargement. In those areas south and east of the earliest Neolithic also it appears that the amalgamation was not so much the product of previously nonagricultural people moving into cultivated areas, as was the case historically and possibly prehistorically in north China, but rather of the expansion of people bearing an agricultural way of life into areas where the indigines previously had been foragers.

Although there are many possible influencing factors and we have had to temper our appraisal by considering an assessment of nasal form that for the moment is no more than impressionistic, nevertheless, we suggest that there is some reason to regard the modern inhabitants of continental Asia who have the smallest teeth to be just those who have been longest associated with the oldest and most sophisticated culinary traditions in the world—namely those of south China [Kee, 1963, p xx; T'ang, 1980, p 3], itself the northeastern corner of Southeast Asia. From the logic of our analysis, then, it is there that we would expect to find evidence for the oldest Neolithic cultures in Asia, and there that we would locate the original source of the impetus that has flowered into the development of Chinese civilization. However, only further data collection and analysis can determine whether our interpretation will be supported or rejected.

ACKNOWLEDGMENTS

The research on which this paper was based was made possible by support from the Committee on Scholarly Communication with the People's Republic of China and by a grant from the Horace H. Rackham School for Graduate Studies at the University of Michigan. The time for working up the data and shaping the text of the paper was provided by a sabbatical leave given to the senior author by the University of Michigan in 1982. Attempts to add to the poorly represented South Chinese samples were denied support by the National Science Foundation. For assistance with the specimens used, we are grateful to Wang Shao-jie in the division of Anthropology at Fudan University in Shanghai; Dr. Wang Si-jing, Director of the Faculty of Basic Medical Science at the Railway Medical College in Nanjing; Dong Xing-ren at the Institute of Vertebrate Paleontology and

Paleoanthropology in Beijing; Prof. Zhu Xin-ren, Chairman, Department of Anatomy, the Traditional Chinese Medical College in Harbin; Zhu Shao-hong at the Railway Health School in Harbin; and Yan Jia-hue at the Heilongjiang Provincial Health School in Harbin. For assistance with permissions, travel arrangements and translation for the senior author, we are indebted to Dr. Woo Ru-kang and Dr. Wu Xin-zhi of the Institute of Vertebrate Paleontology and Paleoanthropology in Beijing and to Chi Wen-rong of the Foreign Affairs Office of the Heilongjiang Provincial Science and Technology Committee in Harbin. For help with the extensive data processing, we thank V. Vitzthum, P. Bridges, and K. Rosenberg in the Museum of Anthropology at the University of Michigan. Finally, we wish to record our gratitude to Cao Shen without whose benign and creative influence none of this would have been possible.

LITERATURE CITED

Aigner JS (1972): Relative dating of North Chinese faunal and cultural complexes. Arctic Anthropol 9(2):36–79.

Andersson JG (1925): Preliminary report on archeological research in Kansu. Mem Geol Survey China Ser A 5:1–51.

Arne TJ (1925): Painted stone age pottery from the province of Honan, China. Paleontol Sinica Ser D 1[Fasc 2]:1–34.

Barrett MJ (1977): Masticatory and nonmasticatory uses of teeth. In Wright RVS (ed): "Stone Tools as Cultural Markers: Change, Evolution, and Complexity." Atlantic Highlands, NJ: Humanities Press, pp 18–23.

Black D (1925): The human skeletal remains from Sha Kuo T'un cave deposit in comparison with those from Yang Shao Tsun and with recent North China material. Palaeontol Sinica Ser D, 1[Fasc 3]:1–148.

Black D (1928): A study of Kansu and Honan aeneolithic skulls and specimens from later Kansu prehistoric sites in comparison with North China and other recent crania. Part 1. On measurement and identification. Palaeontol Sinica Ser D, 6:[Fasc 1]:1–83.

Brace CL (1962): Cultural factors in the evolution of the human dentition. In Montagu MFA (ed): "Culture and the Evolution of Man." New York: Oxford University Press (Galaxy Book), pp 343–354.

Brace CL (1963): Structural reduction in evolution. Am Naturalist 97:39–49.

Brace CL (1977): Occlusion to the anthropological eye. In McNamara JA, Jr (ed): "The Biology of Occlusal Development." Ann Arbor: Center for Human Growth and Development, Craniofacial Growth Series Monograph No. 7, pp 79–209.

Brace CL (1978): Tooth reduction in the Orient. Asian Perspect 19(2):203–219.

Brace CL (1979a): Krapina, "classic" Neanderthals, and the evolution of the European face. J Hum Evol 8(5):527–550.

Brace CL (1979b): "The Stages of Human Evolution, 2nd ed." Englewood Cliffs, New Jersey: Prentice-Hall.

Brace CL (1980): Australian tooth-size clines and the death of a stereotype. Current Anthropology 21(2):141–164.

Brace CL, Hinton RJ (1981): Oceanic tooth-size variation as a reflection of biological and cultural mixing. Current Anthropology 22(5):549–569.

Brace CL, Mahler PE (1971): Post-Pleistocene changes in the human dentition. American Journal of Physical Anthropology 34(2):191–204.

Brace CL, Nagai M (1982): Japanese tooth size, past and present. American Journal of Physical Anthropology 59(4):399–411.

Brace CL, Ryan AS (1980): Sexual dimorphism and human tooth size differences. Journal of Human Evolution 9(5):417–435.

Brace CL, Vitzthum V (in press): Human tooth size at Mesolithic, Neolithic, and modern levels at Niah Cave, Sarawak: Comparisons with other Asian populations. Sarawak Museum.

Brillat-Savarin JA (1960): "The Physiology of Taste, or Meditations on Transcendental Gastronomy." (Introduction by Arthur Machen.) New York: Dover.

Carey JW, Steegmann AT, Jr (1981): Human nasal protrusion, latitude, and climate. Am J Phy Antropol 56(3):513–519.

Chang K-c (1968): "The Archaeology of Ancient China, 2nd ed." New Haven: Yale University Press.

Chang K-c (1973): Food and food vessels in ancient China. Trans Acad Sci Ser 2 35(6):495–520.

Chang K-c (1976): "Early Chinese Civilization: Anthropological Perspectives." Cambridge: Harvard-Yenching Institute Monograph Series, Vol 23.

Chang K-c (1977a): "The Archaeology of Ancient China, 3rd ed." New Haven: Yale University Press.

Chang K-c (ed) (1977b): "Food in Chinese Culture: Anthropological and Historical Perspectives." New Haven: Yale University Press.

Chang K-c (1980): "Shang Civilization." New Haven: Yale University Press.

Chung S (1966): "Court Dishes of China, the Cuisine of the Ch'ing Dynasty." Translated by Lucille Davis. Rutland, VT: Tuttle.

Doolittle RF (1981): Similar amino acid sequences: Chance or common ancestry? Science 214:149–159.

Eickstedt E von (1944): "Rassendynamik von Ostasien." Berlin: de Gruyter.

Elias N (1978): "The Civilizing Process; The Development of Manners." New York: Urizen.

Garn SM, Lewis A, Kerewsky R (1968a): The magnitude and implications of the relationship between tooth size and body size. Arch Oral Biol 13(1):129–131.

Garn SM, Lewis A, Walenga AJ (1968b): Two-generation confirmation of crown-size body-size relationships in human beings. J Dent Res 47(6):1197.

Gingerich PD, Smith BH, Rosenberg K (1982): Allometric scaling in the dentition of primates and prediction of body weight from tooth size in fossils. Am Phys Anthropol 58(1):81–100.

Gould SJ (1975): On the scaling of tooth size in mammals. Am Zool 15:351–362.

Ho P-t (1976): "The Cradle of the East: An Inquiry Into the Indigenous Origins of Techniques and Ideas of Neolithic and Early Historic China, 5000–1000 B.C." Hong Kong: Chinese University of Hong Kong.

Howells WW (1983): Origin of the Chinese peoples: Interpretations of the recent evidence. In Keightley DN (ed): "The Origins of Chinese Civilization." Berkeley: University of California Press, pp 297–319.

Kee J (1963): Introduction. In Keys J: "Food for the Emperor." New York: Gramercy, pp x–xx.

Keys J (1963): "Food for the Emperor: Recipes of Imperial China With a Dictionary of Chinese Cuisine." New York: Gramercy.

Koganei Y (1902): Kurze Mitteilung über Messungen an männlichen Chnesen-Schädeln. Int Centralblatt Anthropol Verwandte Wissenchaft 7(3):129–133.

Koganei Y (1903): Messugen an chinesischen Soldaten. Mitt Med Facultat Kaiserlich Jpn Univ Tokyo 6:123–145.

Kurtèn B (1967): Some quantitative approaches to dental microevolution. J Dent Res 46:[Suppl]:817–828.

Legge J (1893): "The Chinese Classics, Vol 1: Confucian Analects, the Great Learning and the Doctrine of the Mean." Oxford: Clarendon.

Lévi-Strauss C (1964): "Mythologiques, Vol 1: Le Cru et le Cuit." Paris: Librairie Plon.

Lévi-Strauss C (1965): Le triangle culinaire. Arc 26:19–29.

Lévi-Strauss C (1967): "Mythologiques, Vol 2: Du Miel aux Cendres." Paris: Librairie Plon.

Lévi-Strauss C (1968): "Mythologiques, Vol 3: L'Origine des Manieres de Table." Paris: Librairie Plon.

Li C (1928): "The Formation of the Chinese People: An Anthropological Inquiry." New York: Russell and Russell.

Li K-c (1976): The beginning of millet farming in prehistoric China, and a short review of Chinese prehistory. Bull Dept Archaeol Anthropol Nat Taiwan Univ Taipei 39-40:116–139.

Lin JH, Lin TL (1972): "Chinese Gastronomy." New York: Pyramid.

Lin Y (1935): "My Country and My People." New York: Day.

Lin Y (1972): Forward. In Lin HJ, Lin TL: "Chinese Gastronomy." New York: Pyramid, pp 7–8.

Liu CH (1937): A tentative classification of the races of China. Z Rassenkunde 6(2):129–150.

Mathews JM (1966): The Hoabinhian affinities of some Australian assemblages. Archeol Phys Anthropol Oceania 1(1):5–22.

Meacham W (1977): Continuity and local evolution in the Neolithic of South China: A non-nuclear approach. Curr Anthropol 18(3):419–440.

Merbs CF (1968): Anterior tooth loss in arctic populations. Southwest J Anthropol 24(1):20–32.

Molnar S (1972): Tooth wear and culture: A survey of tooth functions among some prehistoric populations. Curr Anthropol 13(5):511–526.

Ohno S (1970): "Evolution by Gene Duplication." New York: Springer.

Ohta T (1980): "Evolution and Variation of Multigene Families." New York: Springer.

Qi Z-q (1965): Excavations (second season) of the Neolithic site of Liulin, P'i Hsien, Kiangsu Province. Acta Archaeol Sinica 2:9–47.

Sakamoto N (1977): "The People's Republic of China Cookbook." New York: Random House.

Shao X-q (1981): Comment on Oceanic tooth-size variation as a reflection of biological and cultural mixing. Curr Anthropol 22(5):560–561.

Siegel S (1956): "Nonparametric Statistics for the Behavioral Sciences." New York: McGraw-Hill.

Smith BH (1983): "Dental Attrition in Hunter-Gatherers and Agriculturalists." PhD Dissertation. Ann Arbor, Michigan: University of Michigan.

Snow E (1938): "Red Star Over China." New York: Random House.

Solheim WG II (1967): Southeast Asia and the west. Science 157:896–902.

Solheim WG II (1969): Reworking Southeast Asian prehistory. Paideuma 15:125–139.

Solheim WG II (1970): Northern Thailand, Southeast Asia, and world prehistory. Asian Perspect 13:145–162.

Solheim WG II (1972): An earlier agricultural revolution. Sci Am 226(4):34–41.

Solheim WG II (1979): New data on late Southeast Asian prehistory and their interpretation (excerpts). J Hong Kong Archaeol Soc 8:73–87.

T'ang LH (1980): Introduction. In Yeh C-H (ed): " Chinese Cuisine (II), Wei-Chuan's Cook Book." (translated by Simonds, N), Taipei: Wei-Chuan, pp 3–4.

Thorne AG, Wolpoff MH (1981): Regional continuity in Australasian Pleistocene hominid evolution. Am J Phys Anthropol 65(3):337–349.

Tindale NB (1977): Further report on the Kaiadilt people of Bentinck Island, Gulf of Carpentaria, Queensland. In Allen J, Golson J, Jones R (eds): "Sunda and Sahul." London: Academic Press, pp 247–273.

Turner CG II (1976): Dental evidence on the origins of the Ainu and Japanese. Science 193:911–913.

Turner CG II (1979): Dental anthropological indications of agriculture among the Jomon people of central Japan. Am J Phys Anthropol 51(4):619–636.

Turner CG II, Hanihara K (1977): Additional features of Ainu dentition. V. Peopling of the Pacific. Am J Phys Anthropol 46(1):13–24.

Weidenreich F (1938–1939): On the earliest representatives of modern mankind recovered on the soil of East Asia. Peking Nat Hist Bull 13(3):161–174.

Wiens HJ (1952): "China's March Into the Tropics." Washington, DC: Office of Naval Research.

Wiens HJ (1967): "Han Chinese Expansion in South China." Hamden, CN: Shoe String.

Woo JK, Peng RC (1959): Fossil human skull of early paleoanthropic stage found at Mapa, Shauquan, Kwangtung province. Vertebrata Palasiatica 3(4):176–182.

Woo TL, Morant GM (1932): A preliminary classification of Asiatic races based on cranial measurements. Biometrika 24(1 and 2):108–134.

Wu X-z, You Y-z (1979): A preliminary observation of Dali man site. Vertebrata Palasiatica 17(4):294–303.

Wu X-z (1981): A well-preserved cranium of an archaic type of early Homo sapiens from Dali, China. Sci Sinica 24(4):530–539.

Yang H-m (1966): A preliminary report of human crania excavated from the Hou-chia-chuang and other Shang Dynasty sites at An-yang, Honan, North China. Ann Bull China Council East Asian Stud Taipei 5:1–13.

Yin H-z, Zhang Z-x (1962): Excavations at the Neolithic site of Liu Lin, P'ei Hsien, Kiangsu (1960). Acta Archaeol Sinica 1:81–100 (English summary, pp 101–102).

The Origins of Modern Humans: A World Survey of the Fossil Evidence, pages 517–563
© 1984 Alan R. Liss, Inc., 150 Fifth Avenue, New York, NY 10011

The Americas: The Case Against an Ice-Age Human Population

Roger C. Owen

Department of Anthropology, Queens College of the City University of New York,
Flushing, New York 11367

> As a general rule, the rarer things are the more they are esteemed. In anthropology the rarest things are the finds, both cultural and physical, relating to the most ancient human beings. If the oldest find on record is, say, 20,000 years old, then anyone who finds something older than this is assured of good publicity, because of the fascination which the history of the human race holds for scientist and layman alike. On the other hand, the circumstances of each claim had better be well documented for otherwise they will surely be challenged [Stewart, 1973, p 155].

INTRODUCTION

Intellectual interest in the origin and antiquity of human presence in the Americas arose shortly after the arrival of Columbus in the Western Hemisphere and continues to the present [Wilmsen, 1965]. Today, although no scholars believe early hominid evolution to have occurred in the Western Hemisphere, considerable difference of opinion exists as to when the first humans arrived.[1]

[1]Research for this paper was done while writing a general book on the anthropology of native North America. As an ethnologist I had assumed, as I had earlier so written [Owen et al, 1967, pp 3–4] and as I had regularly taught in my classes, that the reality of a pre-Clovis occupation of the Western Hemisphere had been proven. As I searched for the evidence upon which proof for this proposition might rest, I gradually reached the negative opinion expressed herein. For comments on and/or contributions to the development of the manuscript, I am grateful to Alan L. Bryan, Dina F. Dincauze, Ruth Gruhn, C. Vance Haynes, Clement W. Meighan, and R.E. Taylor. I would also like to thank Thomas D. Dillehay, and Thomas Stafford for providing unpublished materials. I would like to extend special gratitude to George F. Carter for his detailed and helpful comments on a manuscript in which he could find little to agree with. For exceptional patience in listening to my interminable vocal revisions as well as for keen criticism I wish to thank my colleagues: Lynn Ceci, Warren DeBoer, James Moore, Frank Spencer, and Walter S. Newman.

At one end of the spectrum of opinion are a small group of archaeologists, specialists in the study of "early" early Man in the Americas, who believe that the first Americans to have entered the Hemisphere did so in Mid- or Early Wisconsin times, 40 ky or more ago, perhaps even as long ago as 200+ky. In open opposition to those of this opinion have been, until recently, only two scholars: C.V. Haynes and P.S. Martin. Haynes and Martin insist that the evidence does not support an entrance date of greater than 12 ky ago. The vast majority of Americanist archaeologists, however, have either ignored the question or accepted as probable, perhaps proven, that humans were in the Americas during the Late Pleistocene, 20 ky ago or more.

The roots of the present controversy are entwined with the history of Americanist archaeology of the past 100 years. During the last portion of the 19th and the early part of the 20th centuries, William Henry Holmes and Aleš Hrdlička of the United States National Museum fought a continuous and often acrimonious battle with advocates of autochthonous New World hominid evolution as well as with many archaeologists and antiquarians who believed in great age for the original native occupation of the Americas [Willey and Sabloff, 1974, pp 52–56]. In 1915, Hrdlička observed:

> ... Man could not have originated here, for there was nothing to originate from! He, therefore, must have come from elsewhere. But before coming, he must have peopled that elsewhere, which comprised all of the more habitable parts of the Old World. Such peopling, according to ample evidence, has not taken place before the Paleolithic, or, more likely, the earlier parts of the Neolithic period, or during and after the final phase of the last ice invasion of Europe [Hrdlička, 1915].

Contesting with Holmes and Hrdlička were numerous antiquarians and archaeologists who believed that evidence existed that could prove the Pleistocene presence of humans in the Western Hemisphere. Fossil pig teeth and bones, a skull of questionable provenience (Calaveras), a cement giant (Cardiff), the Trenton Gravels, naturally fractured stones, as well as apparently ancient hearths and sites were brought forward only to be rejected by Hrdlička and Holmes as ineffective evidence of a Pleistocene presence of humans in the Americas.

In 1926, J.D. Figgins, while excavating in a fossil bone bed near Folsom, New Mexico, found an irrefutable association between carefully flaked stone dart or spear points and the now extinct *Bison antiquus*. In 1932, near Clovis, New Mexico, archaeologists established the existence of similar stone points in association with mammoth remains. The Clovis find led to the definition of a hemisphere-wide cultural complex associated with elephant hunters [Haynes, 1966]. Widespread evidences of the Clovis Complex have permitted highly accurate dates: The Clovis Complex dates between 11.5 ky and 11 ky ago [Haynes, 1969a, 1980b, 1982a]. Folsom is often associated with extinct bison remains rather than elephant and is generally younger than Clovis. In turn, Folsom is followed in time by the Plano Complex (9.5–7.5 ky ago), a set of widespread but regional adaptations by paleo-hunters also characterized by large lithic points but ones without the fluting of Clovis and Folsom.

Since the 1930s, when the presence of humans south of the North American ice sheets at the end of the Pleistocene was proven, the search for a yet earlier human occupation of the Americas has continued unabated. Today, in a way, professional Americanist archaeological attitudes are reminiscent of that earlier time in the century when Hrdlička attempted to apply natural science criteria to data put foreward as evidence for Man's ancient presence in the Americas. C.W. Meighan has suggested that contemporary scholarly opinion on the early peopling of the Americas falls into three camps: the Believers, the Agnostics, and the Skeptics [Meighan, 1983a, pp 445–448].

The Believers

Some Americanist archaeologists today believe that "greater-than-Clovis" age has been proven for perhaps as many as 100 sites or other archaeological evidences such as human bones, tools made of animal bone, apparently flaked stones, "hearths," and others.

In 1964, A. Krieger published a list of over 20 locations in North America alone from which he believed materials older than 12 ky had come [Krieger, 1964]. In 1976 and 1978, R.S. MacNeish noted approximately 75 sites (several purportedly 50 ky old or more) where he believed pre-Clovis materials had been found [MacNeish, 1976, 1978]. In the same year, A.L. Bryan edited "Early Man in America: From a Circum-Pacific Perspective," a volume that generally accepts and expands upon the dimension of the proposed pre-Clovis occupation suggested by Krieger and MacNeish [Bryan, 1978]. In 1979, R.L. Humphrey and D. Stanford edited "Pre-Llano Cultures of the Americas: paradoxes and possibilities," adding yet more data [Humphrey and Stanford, 1979]. In 1981, another volume appeared that reaffirmed several existing pre-Clovis candidates and proposing some new ones [Bryan, 1981], as did *Peopling of the New World* [Ericson et al, 1982].

Of some of the sites proposed as most ancient in the Americas, Calico Mountain, Texas Street, and Santa Rosa Island, all in California, Bryan has stated:

> In my opinion, the only valid reason why this evidence has not been accepted is because no one has yet presented a carefully detailed descriptive report based upon stratigraphically controlled excavations. So far, the reports on these sites are either preliminary or they contain vague generalizations leading to apparently significant conclusions about flaked stone artifacts, hearths, and their associations; but as the exact provenience and precise descriptions of these features and associations, as well as technical descriptions of the specific artifacts, have never been detailed, the reader is left confused and unconvinced [Bryan, 1978, p 312].

Dincauze, in an analysis of the logic and methods of several pre-Clovis proponents (Carter, Bryan, MacNeish, Morlan, and Adovasio) concludes that there is no "school" of pre-Clovis advocates nor even a consistent point of view:

> . . . refusing to stand together, the Pre-Clovis proponents challenge the skeptics to deal with each of them separately—a formidable and thankless task which has been undertaken by only a dauntless few. The strength of numbers, therefore, is with the proponents [Dincauze, 1984, p 293].

In recent years, there have been several symposia and/or publications that have been participated in almost exclusively by proponents of the pre-Clovis position [eg, Bryan, 1978, 1981; Humphrey and Stanford, 1979; Ericson et al, 1982; Shutler, in press]. Generally in these outlets, unitary phenomena such as an artifact (the Old Crow "flesher" dated to 27 ky), or a bone (Sunnyvale "dated" at 70 ky) or a "site" (Calico Mountain estimated to be 200 ky) are the weapons entered into the battle with an archaeological "establishment" perceived as hostile to the pre-Clovis position. New evidences are constantly being nominated as proofs but, simultaneously, old data are constantly questioned or discarded as new analytic techniques are applied.

With rare exceptions, no general anthropological formulations attempt to weave the phenomena presented as evidence to the pre-Clovis contention into the generally acknowledged fabric of world prehistory. Little consensus exists among pre-Clovis specialists as to which of the evidences proposed is valid and which is not. But, generally, an assumption is evident that, as more data are entered into the discussion, the more likely it becomes that, by sheer quantity, the "proofs" acquire greater validity. Dincauze refers to this as the "probablist" fallacy [Dincauze, 1984, p 282], which, in colloquial terms might be expressed, "where many colleagues believe they see tiny wisps of smoke, there must be a big fire somewhere."

Some of the leading advocates of very ancient human occupation of the Americas occasionally adopt an attitude of persecuted prophets crying the truth in the wilderness, but going unheard due to the closed minds of the majority of "establishment" Americanists [cf Davis, 1982, p 203]. Others decry how some pettifogging, inflexible archaeologists or geologists may obscure the "scientific" approach by demanding:

> . . . proof, testimony, evidence, and the like... Under threat of the ostracism in which all of us find ourselves who were unfortunate enough to have discovered remains that went beyond the 14,000 cut-off point . . . [Lorenzo, 1978, p 29].

But, because most of today's archaeologists belong to Meighan's second category, and are sufficiently agnostic to at least passively accept a pre-Clovis horizon, the persecution complex is clearly unwarranted. G. Willey, certainly as "establishment" as it is possible to become, spoke for this position when he wrote:

> . . . I think it likely that the "pre-projectile point horizon is a reality and that man first crossed into America as far back as 40,000 to 20,000 B.C. [Willey, 1966, p 37].

The Agnostics

The vast majority of archaeologists, if they do not believe the point to be yet proven, accord a high probability to the presence of humans in the Americas during Wisconsinan times and perhaps earlier. An appreciation of the widespread acceptance of the pre-Clovis arrival of humans in the Western Hemisphere can be had by examining the topic in some of the textbooks in wide use during the past decades. In 1947, Martin, Quimby, and Collier proposed no more than 20 ky [Martin et al, 1947, p 79]; in 1949, Wormington accepted "more than" 25 ky [Wormington, 1949, p 160]; in 1966, Willey allowed perhaps 30 ky [Willey, 1966, p 29]; in 1967, Owen, Deetz, and Fisher were comfortable with 15 ky proven and 20–30 ky possible [Owen et al, 1967, p 3] and most recently, Snow favors an entry date for the first Americans of at least 27 ky [1976, pp 20–23]. Again speaking for the "establishment":

> . . . but the consensus of Americanist opinion as we go into the 1970s is that Levallois-Mousteroid techniques, by whatever slow and circui-

tous diffusion, were transferred from Asia to the Americas and may well have provided the bifacial-flaked blade technology which, eventually, gave rise to the early American Clovis and related industries [Willey and Sabloff, 1974, p 172; see also Muller-Beck, 1966].

The Skeptics

One might conclude that the issue has now been settled—humans were in the Americas more than 12 ky ago. But, it is not so—the case is far from settled. Even as formal protests against assumption of a pre-Clovis occupation were being lodged in the 1960s, J. D. Jennings (who would later join the Believers—[Jennings, 1978, p 1]), perhaps spoke for numerous silent critics of the "Lower Lithic," the "Pre-projectile Horizon," the "American Paleolithic," the "Percussion Stage," the "Protolithic," the "Biface Horizon," (all names applied to the proposed pre-Clovis epoch) when, in 1968, he wrote:

> Most archaeologists reject or even ignore the entire matter. Some base their rejection on the lack of controls; others appear to reject the idea of a mid-Wisconsin emigration, to say nothing of a pre-Wisconsin dating, as the typology of the implements would require. Some, like Jennings, express a blind faith that all the claims could be true and may one day be substantiated, although this faith seems to weaken as years pass without the emergence of any convincing proof [Jennings, 1968, p 68].

The first to express publicly their growing doubt regarding pre-Clovis American hominids were two University of Arizona scholars, C.V. Haynes, a geochronologist and archaeologist, and P.S. Martin, a geochronologist and palynologist. Haynes divided the Paleo-American epoch into three subperiods: Early (pre-28 ky); Middle (28–12 ky); and Late (12–7 ky).[2] He concluded that the reality of only the last had been so far proven. Although not rejecting completely the possibility that further research might uncover satisfactory evidence of a pre-Clovis occupation he noted that none had so far done so [Haynes, 1966, 1969a]. Of the data so far available, Haynes writes:

---

[2]I have modified "Paleo-Indian" to "Paleo-American," because the latter is more accurate and descriptive and also preferred by native Americans themselves. That the distinction is not trivial is emphasized in Berkhofer [1978].

"There is no one place where the evidence is so compelling that if you look at it in a court of law you would want to be tried on the basis of that evidence . . . I think if you were to dig anywhere where there are Pleistocene sediments of this age range, that given the right combination of the morphologies and fossil deposits, you would find something that would be interpreted as artifacts. In other words, there is a sort of background noise in the buried record of things that can be taken as artifacts . . . [Haynes, 1979a].

Compared to Haynes, P. S. Martin takes a much stronger opposition to any pre-Clovis occupation of the Americas. Martin rejects all suggestions that humans had occupied the Americas prior to those who left the earliest fluted points. As a testable hypothesis, Martin offered an elegant scheme that attributes the extinction of several dozen large animal species to predatory activities of the first humans into the hemisphere, commencing about 12 ky ago [Martin, 1973, 1982; Mosimann and Martin, 1975].

More recently, other Americanist archaeologists have come to question the validity of a pre-Clovis human presence: J.B. Griffin [1979], F.H. West [1982], T. Lynch [1983], L.A. Payen [1983a,b], C.W. Meighan [1983a], and D.F. Dincauze [1984] are some. For example, Griffin has remarked that: ". . . an age of about 20,000 years for man in the New World . . . is not unreasonable but, also, is not yet satisfactorily proved [1979, p 47]." Similarly, Lynch, in a summary of early archaeological materials from South America, concludes that, taken as a whole, the evidence will not support the presence of humans in South America prior to terminal Pleistocene times, or "14,000 years ago [Lynch, 1983, p 95]." Also, Payen, after examination of the published literature on 70 alleged North American pre-Clovis sites and complexes, as well as analysis of museum collections, re-examination of sites and some new excavation, concluded that there is no compelling evidence of humans in the Americas more than 12 ky ago [Payen, 1983a,b]. Dincauze, in review of the logic, methods, and evidence employed in the pre-Clovis debate concludes:

. . . it is premature, if not unwise, for Americanist prehistorians to operate under the presumption that there were people in this hemisphere before about 12,000 years ago [Dincauze, 1984, p. 311].

This, then, is the dimension of disagreement among competent scholars: either there may be a hundred or so sites or other evidences that indicate human presence in the Americas during the Pleistocene, or there is none! It is the conclusion of this reviewer that, at this time, there is no substantial

evidence that requires belief that there were hominids in the Americas pre-
12 ky ago. The apparent absence of any ancient human fossil skeletal
material from the Pleistocene, and the continuing ambiguity of pre-Holo-
cene archaeological findings, lend support to the growing group of scholars
who propose the Clovis hunters as the earliest Americans. But an examination of the evidence is in order.

EVIDENCES OF A PRE-12 ky OCCUPATION OF THE AMERICAS

If one were to begin inquiry regarding a pre-Clovis American population
from anthropological reasoning other than the archaeological, from linguis-
tics or physical anthropology, for example, the controversy might take a
different form. Neither subdiscipline provides much comfort to those who
propose great time depth for native American cultural development in the
Western Hemisphere.

Linguistic Evidence

Linguists have catalogued the great diversity of native American languages into dozens of stocks and hundreds of languages. But Edward Sapir
and his student Morris Swadesh, believed the diversity could be more
evident than real. Sapir suggested a reduction of North American stocks to
but six [Sapir, 1929]; Swadesh went much further. He believed the great
bulk of native American languages formed a single phylum, going back
perhaps 15 ky [Swadesh, 1964]. In the same volume wherein Krieger published his original enumeration of pre-Clovis evidence, Swadesh provided a
linguistic counter-argument. Although not prepared to dispute the archaeologists' contention of a possible great age for American culture, Swadesh
denied that linguistic diversity in the Americas need have taken so long to
develop:

> The archeological date of man's first appearance in the New World has
> been pushed back, possibly to something over 30,000 years ago.
> Linguistic differentiation in America is great, but probably is not great
> enough to require more than half the time given by the foregoing
> archeological date [Swadesh, 1964, p 529].

Swadesh hypothesized three "waves" of migrants into the Americas: 1)
carriers of an ancient proto-Native American language whose linguistic
descendents now occupy all of South and Central America as well as most
of North America, 2) ancestors of today's Nadene speakers, and 3) carriers

of proto-Eskimo-Aleut. He believed these migrations might have begun around 15 ky ago and have been complete by about 8 ky ago.

More recently, a three-"wave" origin of native Americans is also supported by conclusions reached by J. Greenberg. Greenberg's research among the more than 1,500 native American languages has led him to conclude, as did Swadesh, that they all belong to one of three groups: 1) the first to arrive, or "Amerind," 2) ancestral Nadene speakers, and 3) Eskimo-Aleut. Greenberg proposes a migrational chronology even more conservative than that of Swadesh: 12, 6, and 4 ky, respectively, for the three entering language groups [Greenberg, 1983].

Physical Anthropological Evidence

Morphologically, native Americans are collectively more alike than any other of the world's continental populations, those of Africa or Europe, for example. Native Americans are phenotypically and genetically very similar to their east Asian ancestors and "cousins." Australian aborigines, on the other hand, isolated and evolving on their island for 30 ky or more, are neither. They do not look like nor are they genetically close to populations from which they must have come [Laughlin et al, 1979, pp 102–103]. Discussing the homogeneity of native Americans, T.D. Stewart remarked:

> . . . such homogeneity is not consistent with the passage of a long period of time following the establishment of the first beachhead because the hemisphere offered ideal conditions for the action of selection and drift . . . [Stewart, 1960, pp 269–270].

A.B. Harper, examining gene ratios among Indians, Eskimos, and Asians and comparing these to known archaeological, biological, and linguistic data suggests:

> All evidence, therefore points to an independent separation of American Indians from Siberian populations some 15,000 years ago . . . the concordance of archaeologic, genetic, biological and linguistic data strongly commend this model [Harper, 1980, p 553].

A.M. Brues, in review of the process of racial differentiation in the Americas, sees it as "only half-finished," with a possible 15 ky or so as simply not enough time to permit the processes of raciation to produce the variability known from other continents [Brues, 1977, p 30]. In pigmentation, hair form, many blood types, incisor patterns, and in many other ways, native Americans are remarkably homogeneous.

In *The First Americans: Origins, Affinities, and Adaptation*, W.S. Laughlin and his associates generally conclude that the morphological features and genetic components shared by native Americans, including Eskimos, are very close to those of Siberians and other east Asians [Laughlin and Harper, 1979]. They conclude that these facts are consonant with a fairly recent common origin and perhaps a maximum time depth of 14–20 ky [Laughlin and Wolf, 1979; Szathmary, 1979].

Research on native American teeth also supports a fairly recent human arrival in the Western Hemisphere. Native American tooth patterns both ancient and modern are very similar to each other and to contemporary northeast Asians ones as well [Turner and Bird, 1981]. Furthermore, research on tooth patterns throughout the Western Hemisphere, strongly supports the three-"wave" linguistic hypothesis proposed by Swadesh and Greenberg. C. Turner believes that, sometime before 15 ky ago, foragers with a distinctive tooth pattern lived in the Lena River basin in Siberia. From these the first ancestral American immigrants may have come. Between 14 and 12 ky ago, a second group moved into Beringia—the Nadene, followed thereafter, around 9 ky ago, by the ancestral Eskimo-Aleut [Folsom and Folsom, 1982].

Human Skeletal Evidence

Human biological evolution, worldwide, has been documented by numerous discoveries and subsequent analysis of hominid bone. One would suppose that human presence in the Western Hemisphere would be no less well documented by surviving skeletal material than has been recovered from Africa, Asia, Europe, and Australia. This is not the case, however, in the Americas. At the present time, only seven finds of skeletal material are regarded by some as being of possible Early or Middle Paleo-American age, that is, more than 12 ky old [Smith, 1976; Protsch, 1978][3] (see Table I). The age of each of these has been challenged on one ground or another by various authorities.

In the opinion of T.D. Stewart, all seven specimens are, morphologically, comfortably within the range of fully modern *Homo sapiens* [Stewart, 1960, 1973, pp 169–170, 1981; Smith, 1977]; neither are there morphological nor cultural grounds for excluding any from a Holocene age. The attribution to all these putatively ancient bones of a Paleo-American age (pre-12 ky ago) is

---

[3]All data on American skeletal material unless otherwise attributed is from either Smith, 1976, or Protsch, 1978.

a result of either chronometric assessments or is based upon geological or paleontological criteria.

Four of the most ancient dates on American skeletal finds (Sunnyvale, Del Mar, Los Angeles, and Yuha) have been obtained through use of amino acid racemization (AAR), a technique utilized by J. Bada and his associates at the University of California, La Jolla [Bada et al, 1974; Bada and Masters, 1982]. The use of the AAR technique for dating ancient human bone has been sharply criticized, since its validity depends largely upon knowledge of environmental temperature throughout the history of the specimen studied. For ancient human bone, especially that found in zones of great temperature fluctuations, eg, southern California, such control is impossible [Lajoie et al, 1980]. Critics of AAR dates have suggested that heat contamination of the specimens either before or after collection could account for their apparent great age [Smith, 1976, pp 133–136; Von Endt, 1978]. Furthermore, since the AAR procedure for southern California specimens was calibrated using a questioned C-14 date obtained from the Laguna bones, further reservations may be in order. Nonetheless, Bada and his associates retain their faith in the validity of the technique [Bada and Masters, 1982].

The Early Paleo-American Period (pre-28 ky)

Sunnyvale. The Sunnyvale bones, recovered from near San Francisco, California, in 1972, were apparently interred in a well-defined grave and associated in the ground with sea shells, a common burial inclusion during the Archaic Period in California (less than 8 ky ago). The specimen provided an AAR date of 70 ky. Despite this very ancient date, several others suggest a much younger age. The shells, stratigraphically older than the burial yielded C-14 dates of just over 10 ky [Haynes, 1984]; the bones themselves when subjected recently to uranium series analysis yielded an age of only 8.3 ky [Bischoff and Rosenbauer, 1981; for a counter opinion see Bada and Finkel, 1982]. Most recently, four C-14 determinations from this nearly complete skeleton strongly support a middle Holocene age in the range of 3.5 to 5 ky ago. This estimation agrees well with other dates derived from associated shell as well as with general geologic, anthropometric, and archaeological contexts [Taylor et al, 1983]. The only discordant date for Sunnyvale, and the only one that suggests great age, is from AAR. Furthermore, on the basis of the evidence he has seen, Carter also concludes Sunnyvale is recent [Carter, 1984].

The Del Mar bones. The Del Mar specimen, also known as the "Scripps Estate," or "La Jolla," or "San Diego Man," is comprised of the remains of

TABLE I. Human Skeletal Evidence of Possible Paleo-American Age

Specimen	Date found /by whom	Sex	Age	Maximum estimate	Technique	Consensus of critics
a) Sunnyvale, CA	1972	—	—	70 ky	AAR	5–3.5 ky
b) Del Mar, CA	1920's–30's, M.J. Rogers	Several indiv.	—	48 ky	AAR	Holocene
c) Taber, AB, Canada	1961–1962, A. Stalker	—	Infant	37 ky	Geologic	Holocene
d) Otovalo, Ecuador				36 ky	C-14	2.67–2.3 ky
e) Los Angeles, CA	1936, I.A. Lopatin	F	Adult	26 ky	AAR/C-14	6 ky
f) Yuha, CA	1971, W.M. Childers	M	17–20	23.6 ky	C-14	5 ky
g) Midland, TX	1953, K. Glasscock	F	25–30	20 ky	Various	7.1 ky
h) Laguna Beach, CA	1933, Wilson and Marriner	?	Adult	17.15 ky	C-14	
i) Brown's Valley, MN	1933, A.E. Jenks	M	25–40	12 ky	Geological/ typological	8.5 ky
j) Tepexpan, Mexico	1947, H. de Terra	F?	Adult	11.003 ky	C-14	Recent
k) Minnesota, MN	1931, P.F. Stary	F	ca 15	11 ky	Geological	6 ky
l) Marmes, WA	1965, Fryxell and Marmes	?	Adult	10.678 ky	C-14	Holocene
m¹) Melbourne, FL	1925, Loomis, Gidley, and Singleton	F	Adult	10 ky[a]	Geologic/ fluorine	Recent?
m²) Vero Beach, FL	1916, H. Sellards	F	Adult	10 ky[a]	Geologic	Recent?
n) Natchez, MS	1846, M.W. Dickson	M	Adult	10 ky[a]	Geologic/ fluorine	Recent
o) Lagoa Santa, Brazil	1830s, P.W. Lund	Several indiv.	—	10 ky[a]	Various	10 ky
p) Arlington Springs, CA	1959, P. Orr	?	Adult	10 ky	C-14	10 ky

[a]Early Holocene.

several individuals excavated during the 1930s and 1940s from southwest California. Several AAR dates were obtained, the most ancient of which is 48 ky. But sea shells from the same deposits provided C-14 dates of from 5 to 7 ky, [Protsch, 1978, p 59], and a recently done uranium series indicates 11 ky [Bischoff and Rosenbauer, 1981]. Protsch urges that AAR dates on the Del Mar bones be regarded with "utmost scepticism" [Protsch, 1978, p 60]."

The Taber child. This find consists of several bones of an infant recovered from an eroding bank in 1961 in the province of Alberta, Canada. The geological field party that found the specimen did not identify it as human and the bones were shipped to the National Museum of Canada. However, when their possible significance was learned, A.M. Stalker returned and attempted to locate exactly from whence the bones had come. He believes he succeeded with a possible error of only 5 feet or so. Based upon geological considerations, Stalker estimated the bones to be at least 32 ky old and possibly more than 37 ky [Stalker, 1969].

More recently, analysis of protein in the bones suggests an age of between 5.5 and 2.8 ky; while C-14 on them provided only 3.53 ky, and reexamination of the sedimentological context also points to a Holocene age [Dincauze, 1984, p 308]. In the face of these consistent late dates, the original geological age estimation, especially with the precise provenience of the bone in substantial doubt, is not compelling.

The Otovalo skeleton. Bones found on a museum shelf in Quito, Ecuador, were traced to a deposit of limonitic calcareous conglomerate approximately 30 feet below the surface of a contemporary terrace near Otavalo. It appeared as if the individual had fallen into a fissure. A number of dates have been obtained: 36–29 ky from aragonite filling the bone cavities; 28 ky from bone infused with interstitial calcium carbonate; 28 ky from AAR from the skull while C-14 analysis of bone collagen, often regarded as the best fraction to date, have yielded dates of 2.67 and 2.3 ky [Davies, 1978, p 273]. Obviously, it is premature to draw conclusions in the face of such contradictory dates.

These four specimens of human bone are all that have been so far offered as proof of an Early Paleo-American (pre-28 ky) occupation.

The Middle Paleo-American Period (28–12 ky)

The Los Angeles bones. In 1936, workmen near Los Angeles, California, encountered at a depth of 12 feet portions of a cranium and a humerus, both heavily mineralized. Later, remains of a mammoth were recovered at a similar depth from the same area. Fluorine testing indicated the human and the mammoth bones might be of approximately the same age. When subjected to C-14 and AAR dating procedures, the human bones yielded "more than" 23.6 ky and 26 ky, respectively. The specimens are highly mineralized, and for C-14 dating, only a minute portion was taken. The "more than 23.6 ky" date could mean that no significant radioactivity was present. Protsch apparently believes the specimen to be roughly 6 ky old [Smith, 1976, p 133]. But if truly associated with mammoth bones, the skeletal material should possess somewhat greater age, but not necessarily greater than 10 ky.

The Yuha burial. The Yuha burial was discovered in 1971 in the Yuha Desert of southern California by W.M. Childers [Childers, 1974; Rogers, 1977]. Radiocarbon tests on crusted caliche (calcium carbonate) scraped from the bones indicated an age of between 23.6 and 21.5 ky; an AAR date of 23.6 ky [Bischoff and Childers, 1979] and a thorium date of 19 ky were also obtained [Bischoff et al, 1976]. Although one might suppose that the find is well dated, this is not the case.

The use of caliche to date objects coated with it has been questioned, in part due to the potential for contamination from older "fossil" caliche [Payen et al, 1978]. Use of AAR dating on the Yuha find, in particular, has been criticized by J. Bada, the originator of the technique whose own research led him to believe Yuha to be about 5 ky old, an age later supported by a uranium series as well [West, 1983, pp 32–33]. Furthermore, the general conformation of the Yuha burial is similar to a well-known group of cairn burials from the southern California desert, burials which regularly date no older than 5 ky [Wilke, 1978].

In 1980, most of the skeleton mysteriously disappeared and is believed to have been stolen. Meanwhile, a surviving piece of the Yuha specimen (approximately 30 gm) has been dated at the University of Arizona. The results, to be published in *Nature* (London), were obtained from isotopes of six chemical fractions, and all yielded dates for the bone of less than 4 ky. The researchers conclude the specimen to be of Holocene age [West, 1983; Stafford et al, 1984].

The Midland Bone. The Midland find (aka "Scharbauer"), highly mineralized and fragmentary, was discovered underlying Folsom material near Midland, Texas. Dates of from 20–4 ky have been obtained using various techniques on remains associated with the human bones. Estimated ages range from 7.1 to 9.27 ky but, because the bones may be contemporaneous with Pleistocene fauna, the specimen might be slightly older [Smith, 1976, p 126].

The Laguna Beach bones. "Laguna Man," the final candidate to be a remnant of a pre-Clovis American population, consists of two heavily mineralized bones found in 1931 on a construction site in southern California. The bones had apparently washed down from a higher elevation and were located between two strata, both of which date to less than 9 ky ago [Meighan, 1983b]. After discovery, the bones were sent to Europe and to Africa and there has been some question, beyond resolution now, as to whether or not the original material was ever returned [Bada and Masters, 1982, p 175]. When shown the bones in 1967, L.S.B. Leakey decided to have them dated using the then new bone collagen C-14 technique. Dates obtained were 17.15 and 14.8 ky [Protsch, 1978, p 33].

Because of questions regarding their actual identity, the fact of their secondary deposition and possible contamination, as well as the completely modern appearance of these bones, the Laguna specimen is regarded skeptically by critics. Their C-14 dates, however, take on additional importance because Bada used them to calibrate his AAR procedures for dating of other southern California specimens. If the C-14 dates for the Laguna specimen are inaccurate, or if the bones are not as represented, then the validity of all the southern California AAR dates is in jeopardy.

But, on the other hand, if the AAR dates for these southern California specimens are correct, then the earliest of them represent the first known modern *Homo sapiens* from any continent. Such an eventuality would force a major revision in the currently accepted understandings of recent human evolution. This possibility, widely publicized by the popular media [Sullivan, 1975; Rensberger, 1976, 1977] led one opportunistic author with anthropological credentials that have been questioned [Goodman, 1981; Feder, 1983], to write a book in which he proposed southern California as the cradle of modern Man! Dennis Stanford, in his review of the work, found he could not recommend it to anyone for serious reading, but especially not to laymen [Goodman, 1981; Stanford, 1981].

Summarizing, it is these seven specimens of bone (Sunnyvale, Del Mar, Taber, Otovalo, Los Angeles, Yuha, Laguna Beach) from which support for a Pleistocene occupation of the Americas must be drawn. Even if the offered evidence were less ambiguous, it would seem scarcely credible that any substantial occupation of the Americas in pre-Clovis times could leave behind so few of its kind for our study.

The Late Paleo-American Period (12–7 ky). We have somewhat better skeletal documentation for the early portion of the Late Paleo-American Period (12–10 ky) but even from this quite recent period there are only ten candidates for possible inclusion [Smith, 1976; Protsch, 1978]. Several of these specimens have, in the past, figured as possible pre-Clovis remains, but further research has placed them all into this more recent category.

Brown's Valley. This specimen, red-ocher covered, was found in Minnesota in 1933. The gravel stratum where the remains were found was dated geologically to about 12 ky ago; while associated artifacts have been typologically dated to about 8.5 ky.

Tepexpan. The nearly complete Tepexpan skeleton was discovered in Mexico in the late 1940s by H. De Terra. Apparently associated with mammoth bones, the find was dated to 11.003 ky using C-14 on peat from the same stratum. Heizer and Cook, viewing the results of fluorine analysis, suggest the human bones are younger than those of the mammoth. Krieger and others concluded the bones represented a relatively recent intrusive burial [Smith, 1976, pp 130–132].

The Minnesota girl. The Minnesota find (aka "Pelican Rapids") is a nearly complete skeleton of a juvenile girl bulldozed out of the ground in 1931. First thought to be as ancient as 18–25 ky on geological criteria, revision of this estimate to 11 ky has been accepted as reasonable by many experts. Others, however, believe the find to represent a much more recent (perhaps 6 ky) intrusive burial. This belief is based upon two Archaic Period artifacts found with the bones: an antler "dagger" and a pendant made from a Gulf of Mexico species of conch shell.

The Marmes skull. The Marmes skull, found in situ in the state of Washington in 1965, was dated through shells located in the stratum above the bones (10.678 ky). Some specialists question the validity of the date for the skull because shell tends to give dates that are too old, but unpublished C-14 dates of between 10–11 ky tend to support the original estimation [Haynes, 1969b, p 353; 1984].

The Melbourne and Vero Beach bones. The Melbourne and Vero Beach bones are of more interest in the study of American archaeology than they are as putative fossils. Found in 1916 and 1925, respectively, and claimed at the time to be associated with Pleistocene fauna, they are judged by most experts to be of fairly recent origin [Stewart, 1973, pp 155–158; Meltzer, 1983, pp 30–34].

The Natchez pelvis. The Natchez pelvis, first described in 1846, was widely discussed and examined during the 19th century; but, because of its uncertain provenience and its lack of cultural or other associations, it is now accorded little significance. Fluorine tests made in the late 19th century indicated that it was contemporaneous with some now extinct animals such as mastodon, horse, and others. However, too little is known of it to make a judgement as to its age [Stewart, 1973, pp 153–158].

Lagoa Santa bones. Finds made in caves near Lagoa Santa, Brazil, between 1835 and 1844, were among the first scientifically credible data recovered bearing on the antiquity of America's native people. In these caves bones of humans co-occurred with Pleistocene animals in such a way as to suggest contemporaneity. "Confins Man," reported in 1937, appears to represent a deliberate burial in a stratum that also included the bones of mastodon and horse shown by fluorine tests to be contemporaneous with the human bone. Later research by W.R. Hurt in other caves in the Lagoa Santa region failed to sustain any widespread association of human and extinct mammal bones but humans do appear to have been in and around these caves for 10 ky or so [Mattos, 1961].

A.L. Bryan, in 1970, while examining a paleontological collection apparently from near Belo Horizonte in Minas Gerais, discovered a highly mineralized calotte that, although similar to others in the known Lagoa Santa population in most ways, is possessed of exceptional thickness and very heavy brow ridges. Bryan believes this bone may represent, along with another found earlier by Lund in the same area, a distinct, perhaps earlier predecessor of the abundant Lagoa Santa type. At least one "early" early man specialist urges consideration of these two specimens as possible paleoanthropine Americans of the *erectus* variety [Davis, 1980, pp 60–65]. Unfortunately, the calotte was lost before much analysis could be done [Bryan, 1978, pp 318–321].

Arlington man. In 1959, when first noticed protruding from a canyon wall, the Arlington find, from Santa Rosa Island, California, was considered by its finders to be possibly 35 ky old because of its apparent location in a stratum known to be of that age. Meticulous excavation, in the presence of many distinguished paleo-Americans specialists, revealed the bone to be intrusive in the stratum, notwithstanding its great depth, 35 feet or more, from the present canyon edge. A date of 10 ky has been obtained from both bone collagen and from charcoal recovered near the bone. Both the bone and charcoal are presumed to have washed into their present location.

Archaeological Evidences

In addition to the skeletal evidence to an Early or Middle Paleo-American presence, several dozen archaeological sites and isolated artifacts possess either dates, typological characteristics, or ancient associations that would place them into the Early or Middle Paleo-American Periods (pre-12 ky BP) (see Table II). Because of their great number, only some of the more prominent and/or promising data will be discussed here. The general dimensions of the controversy can be extracted from these.

Exactly what constitutes sufficient evidence to demonstrate an archaeological assertion has rarely been specified. But with respect to "early" early

Man in the Americas, J.B. Griffin has outlined a set of evidentiary criteria that he believes should be met by archaeological phenomena if they are to be accepted as valid [Griffin, 1979, p 44]. Although criticized by some as "arbitrary standards" [Morlan and Cinq-mars, 1982, p 355], their reasonableness is attested to by the fact that the past million years or so of our bio-cultural evolution has been documented by abundant data that do meet them. If there is some reason why the basic evidentiary standards of scientific archaeology should be suspended in the study of ancient American prehistory, it is not readily apparent. Furthermore, as the Clovis Complex easily satisfies these criteria [Haynes, 1982a] why should lesser requirements be applied to materials purportedly a few thousand years older?

Griffin's proposed evidentiary standards are 1) a clearly identifiable geologic context, agreed upon by competent authority; 2) recovery of a sample of material-cultural items adequate to define a society's material culture; 3) faunal and floral materials sufficient to define environmental and cultural relations; 4) human skeletal material to support the cultural and temporal conclusions drawn from other evidence; 5) radiometric and other dates to support geologic, cultural, and other conclusions as to the age of the evidence.

The Early Paleo-American Period (pre-28 ky)

The Calico Mountain sites (aka "Calico Hills"). R.D. Simpson has been exploring in the Calico Mountains region of the Mohave Desert in southern California since the 1940s. Joined in 1964 by L.S.B. Leakey, Simpson uncovered "artifacts" that yielded dates, derived from carbonate covering them, of as old as 202 ky [Simpson et al, 1981; Shlemon and Bischoff, 1981]. By 1978, over one thousand "artifacts" had been identified including hand axes, chopping tools, scrapers, burins, gravers, and others [Simpson, 1978, 1982]. Work continues at the site.

Haynes and others, including Jennings, question both the dating and artifactual identity of the material from Calico; they conclude that the recovered materials are "geofacts," may be from levels 200–500 ky old, and are not of human manufacture [Haynes, 1973; Jennings, 1978, pp 22–23]. More recently, MacNeish, generally accepting of an Early Paleo-American presence, admitted he was unconvinced that Calico possessed indisputable cultural materials [MacNeish, 1982, p 313].

In an important study, Payen applied the Barnes Test to purported artifacts from Calico. This test, which examines the angle of fracture of chipped stone objects, was developed for use with Old World specimens to distinguish human chipped stones from those made by nature. Payen, on the basis of his test, concluded that the Calico tools more closely approximate samples of naturally fractured rock than they do unquestioned samples of Paleo-American handicrafts, those from Clovis, for example [Payen, 1982; Taylor and Payen, 1979; Payen, 1983b].

The Texas Street site. G.F. Carter, geographer and archaeologist, has long championed not only a pre-Clovis American population but one even during or before the Third Interglacial. The cornerstone of his belief are burn areas or "hearths" and apparently associated "flakes" and "cores" found on a series of geological terraces near San Diego, California, that he believes may be 200 ky or more old [Carter, 1957, 1978, 1980; Minshall, 1976, pp 5–15; Reeves, 1981, p 47].

Although the data from Texas Street are dismissed by most archaeologists as the result of natural fires and normal geological processes [Krieger, 1958], their cultural character has been accepted by Bryan [Bryan, 1978, p 312]. Examination of the angle of fracture of Texas Street specimens by means of the Barnes Test suggest that they, too, along with those of Calico Mountain provenience, are also geofacts [Taylor and Payen, 1979, p 274].

Yuha Pinto Wash. In 1980, Childers and Minshall reported finding 80 tools ("choppers and chopping tools, scrapers and utilized flakes") as well as three pieces of possible human bone exposed from under 21 m of alluvium by storm-caused outwash in the Yuha Pinto Wash, in southwestern California [Childers and Minshall, 1980]. The age of the finds, approximately 50 ky old, was estimated from geological criteria. No full report is yet available.

Old Crow Basin. One of the most exciting archaeological discoveries in the Americas in recent years was made along the Old Crow River in Canada's Yukon Territory. Here, a Kutchin Indian serving as a paleontological assistant picked up a caribou tibia serrated at one end, in a fashion similar to the hide scrapers until recently used by Indians of the area [Irving and Harington, 1973; Harington et al, 1975; Canby, 1979, p 344]. This flesher revealed a C-14 age of 27 ky and with that, apparently, the pre-Clovis question was answered, at least for Beringia. But because the date on the flesher was obtained from the mineral fraction apatite now believed to be unreliable, the age of the implement is still uncertain [Irving, 1982, p 69].

Interdisciplinary work of the highest caliber in the Old Crow region has recovered, as of 1980, over 251 specimens of fractured or apparently cut, grooved, faceted, scraped, polished, or butchered elephant bone and ivory believed to have been worked by ancient humans [Morlan, 1978, 1980; Morlan and Cinq-Mars, 1982, p 360; Irving, 1978; Bonnichsen, 1978; Bon-

TABLE II. Archaeological Evidences of Paleo-American Period Occupation

Name/location	Earliest date(s)	Method(s)	Cultural materials	Critics	Status
The Early Paleo-American Period Locations (28 ky ago and more)					
a) Calico Mts. Site, CA	202 ky	Geologic	Flakes/cores	Haynes and others	Questioned[a]
b) Texas Street, San Diego, CA	200 ky+	Geologic	Flakes/cores/hearths	Kreiger	Questioned[c]
c) Yuha Pinto Wash, CA	50 ky	Geologic/C-14	Choppers/scrapers	—	Uncertain[f]
d) Old Crow Basin, Yukon Terr, Canada	Mid-Pleistocene 27 ky	C-14 Geologic	Fractured bone "tools"	—	Questioned[c]
e) China Lake, CA	42.35 ky	Uranium Series	Flakes	MacNeish	Questioned[c]
f) El Bosque, Nicaragua	30 ky	C-14 and others	Flakes	—	Uncertain[f]
g) El Cedral, Mexico	33.3 ky	C-14	Scrapers/modified bones	—	Uncertain[f]
h) Santa Rosa Island, CA	40 ky	C-14	Mammoth bones/scraper	—	Uncertain[f]
i) Lewisville, TX	38 ky+	C-14	Hearths/stone tools—Clovis pt.	Heizer/Brooks/Haynes	Rejected
j) Tule Springs, NV	32 ky	C-14	Obsidian flake/biface chopper	Shutler and others	Rejected[d]
k) Tlapacoya, Mexico	24 ky	C-14 Geologic	Obsidian pt./hearths	Haynes	Questioned[d]
l) Pikimachay Cave, Peru	23 ky	C-14	Stone "tools"	Haynes, Lynch	Uncertain[c]
m) Valsequillo, Mexico	21.85 ky	C-14	Points/scrapers	—	Uncertain[c,d]
n) Meadowcroft Rockshelter, PA	19 ky	C-14	Tool kit/hearths	Haynes	Uncertain[b]
o) Timlin Site, NY	16 ky	C-14	Cores/flakes	—	Uncertain[f]
p) Selby, Dutton Sites, CO	15 ky	Geologic	Bone tools/flakes	—	Rejected?[f,e,c]
q) Wilson Butte Cave, ID	14.55 ky	C-14	Biface/blade	Haynes/Aikens	Uncertain[d,e]
r) El Jobo Complex, Venezuela	14.2 ky	C-14	El Jobo pts.	Haynes/Lynch	Questioned[d]

s) Blue Fish Caves, Yukon Terr, Canada	14 ky	C-14	Flakes	—	Uncertain[f]
t) Monte Verde, Chile	14 ky	C-14	Flakes/modified bones	—	Uncertain[f]
u) The Levi Site, TX	13.75 ky	C-14	Flakes	—	Uncertain[f]
v) Fort Rock Cave, OR	13.2 ky	C-14	Pts./scrapers/milling stones	—	Possible[d]
w) Los Toldos, Argentina	12.65 ky	C-14	Scrapers/flakes	Lynch	Questioned[d,e]
x) Malakoff Heads, TX and Mexico	Pleistocene	Geologic	Modified boulders	Agogino	Uncertain[b]
The Late Paleo-American Period (12 ky ago)					
a) The Clovis Complex (dozens of locations throughout hemisphere)	12 ky	All methods	Fluted pts./complete tool kit	None	Demonstrated
b) The Folsom Complex	11 ky	All methods	Fluted pts./complete tool kit	None	Demonstrated
c) The Plano Complex (hundreds of locations)	9.5 ky	All methods	Large blades/complete tool kit	None	Demonstrated

[a]Poor dates, human presence questioned.
[b]Poor dates, human presence unquestioned.
[c]Good dates, human presence questioned.
[d]Good dates, human presence unquestioned, but association between dates and human presence questioned.
[e]Good dates, but cultural evidence too sparse to permit conclusions.
[f]Conclusions too recent to permit careful assessment or too little evidence so far published to permit careful evaluation.

nichsen and Young, 1980]. The researchers believe these apparently modified bones to comprise an ancient tool kit, perhaps of mid-Pleistocene age,
but one totally lacking stone tools. As one investigator expressed: ". . . data
for man's presence found in these beds of Illinoian age are meager in
number, but to our minds compelling" [Irving, 1982, p 78]. Large numbers
of microflakes of cherts and quartzites have been found in samples of soil
and sand; Morlan proposes that they might be "microdebitage" from an as
yet undiscovered lithic technology [Morlan, 1981, pp 13–16].

A date of 29.3 ky has been obtained from a single apparently modified
bone, but the remainder of Old Crow dates are based on stratigraphic
calculations and dates derived from testing of geological phenomena [Mor-
lan, 1981, pp 9–11]. Apparent associations between bone possibly made into
tools and geologic strata, if correct, may place the earliest human occupation
of Beringia a quarter of a million or more years ago [Jopling et al, 1981].

Interpretation of fractured and abraided bones as sophisticated tool kits
obviating the need for stone has been criticized. L.S. Binford, for one, has
derided such interpretations "as having produced a series of modern myths"
[Binford, 1981, p 42]. He notes that it is only with great difficulty, if at all,
that human modification of bone, antler, and other organic material may
be distinguished from the work of other factors in nature. Furthermore,
perhaps heeding Haynes' suggestion to test the "background noise" in the
earth itself, a group of paleontologists excavated several Nebraska sites
known to be millions of years old. These excavations uncovered modified
bones to match the proposed tool kit from Old Crow, a result that should
force caution in the future before fractured or flaked bones are identified as
human "artifacts" [Myers et al, 1980; see also G. Haynes, 1983].

Further complications in the Old Crow region stem from an extremely
complex geological history caused by great amounts of soil movement
through the millenia. Most of the objects so far identified as of human
manufacture have been redeposited, perhaps many times, thus destroying
whatever associations they may once have had.

Despite the evident confidence of the numerous reporters of work in the
Old Crow region, if bone apatite gives inaccurate C-14 dates and if human
modification of bone cannot be distinguished from the work of nonhuman
predators or of the results of various other natural processes, then no
compelling evidence of an Early or Middle Paleo-American human presence
in Beringia so far exists.

China Lake. At China Lake, in southeastern California, E.L. Davis and
her co-workers have uncovered Clovis artifacts as well as some believed by
them to be earlier [Davis et al, 1980]. From a mammoth tooth, described as

in "immediate contact" with two "finishing flakes of obsidian and chert," enamel was taken which yielded a uranium series date of 42.35 ky [Davis et al, 1981, p 36; Davis, 1982]. To believe that three objects resting together in the 20th century have remained so during the preceeding 40 ky of geological and climatological chaos that Davis' writing ably discusses reveals a sanguine nature. Of Davis' conclusions, associating humans with mastodons 42 ky ago, MacNeish, generally accepting of much pre-Clovis data, commented: "I would love to believe Davey, but I need some more valid contextual evidence [MacNeish, 1982, p 314]."

El Bosque. Bone apatite dates, carbonate nodule dates, stratigraphic and geomorphic information indicate the age of this site, in Nicaragua, to be greater than 30 ky [Gruhn, 1978, p 257]. The cultural material consists of crudely flaked pieces of jasper:

> . . . at least one of which is believed by several archaeologists who have worked at the site (J. Espinosa, A. Bryan, R. Gruhn, W. Irving, and R. Morlan) probably to have been flaked by man (ibid.)

The pieces of jasper are in association with numerous bones of large mammals such as giant ground sloth, horse, and others. One authority concluded that a bone from El Bosque was so mineralized as to be perhaps pre-Wisconsinin in age, perhaps 150 ky old [Page, 1978, p 254]. Too little is known as yet regarding cultural materials at El Bosque to lend confidence in its interpretation as an archaeological site.

El Cedral ("La Amapola"). In the late 1970s, an interdisciplinary team in Mexico excavated "La Amapola," the site of an ancient spring, now dry. Among the items recovered were ten possibly modified bones. One of these provided a C-14 date of 21.96 ky. Stones judged to be artifacts were also recovered, one of which, a circular scraper, was recovered from a stratum 33.3 ky old [Lorenzo and Mirambell, 1981, pp 112–124]. The validity of this date must await further tests, but swampy soil around springs, constantly stirred by incoming animals, including humans, often provides strange bedfellows.

Santa Rosa Island. Extensive research by P.C. Orr demonstrated a human occupation of Santa Rosa Island, California, for perhaps 10 ky prior to the arrival of the Spanish [Orr, 1968]. His work, now continued by R. Burger and others, led him to believe that humans had been on the island much longer, perhaps as long as 40 ky or more. The principle data his conjecture rested upon were numerous "fire" areas within which bones of a type of dwarf mammoth were sometimes found. Orr believed some of the burn areas to be "barbecue pits" wherein humans anciently may have cooked the small mammoths. Never, however, did he encounter any burn areas that contained stone tools. Burger, in the late 1970s, excavated one such, known as the Woolley Mammoth site, and uncovered not only mammoth bones but what he believes may prove to be the long-sought stone tools. Charcoal from the site has provided a date of greater than 40 ky [Berger, 1978, 1981, 1982]. No detailed report of this find is available.

With respect to the burn areas, or "barbecue pits," M. Wendorf has proposed that most, if not all the small reddened areas on the Channel Islands and on the mainland as well, are a result of root and stump fires that can leave hearthlike remains within which collections of stone and small animal bone are sometimes found [Wendorf, 1982]. That the small mammoth of the island sometimes blundered into such a burning pit may stretch credulity, but no more, perhaps, than the proposal that predatory humans and small mammoth coexisted on the confined space of the island for nearly 30 ky. Nor would it be surprising to discover mammoth bones and artifacts in such burn pits [cf Meighan, 1983a, p 456]. Mammoth may have lived on the island for at least 20 ky and must have left great numbers of bones. Later, we know humans to have been on Santa Rosa Island for perhaps 10 ky and to have left behind abundant cultural remains. Given the degree of soil movement on the island, recalling Arlington Man for a moment (discussed above), it is not unlikely that elephant bones, burn pits, and Archaic Period tools would come to co-occur fortuitously.

Lewisville. Two dates of greater than 37 ky ago on "charcoal" from a hearth held this site, located in Texas, in contention as a pre-Clovis candidate for many years despite the presence as well on the site of a Clovis point. Krieger believed the point might be a "plant," but reexamination of the site has shown that the original ancient dates were derived from lignite, not charcoal as had been believed [Crook and Harris, 1958; Krieger, 1964; Chedd, 1980, p 49]. The cultural material at the site apparently dates to about 11 ky ago.

Tule Springs. Tule Springs is a famous paleontological zone with remains of camel, bison, horse, and other extinct fauna being found here. Disarticulated camel bones, some of them burned, imply human presence as do associated stone tools. In 1954, charcoal samples provided dates of from 23.8 to 32 ky. Additional work, however, has demonstrated that the maximum age of scant human activity at the site is no more than 10–11 ky [Shutler et al, 1967].

The Middle Paleo-American Period (28–12 ky). Most anthropologists who have published an opinion have accepted the probability, if not the certainty, that humans were present in the Western Hemisphere during all or most of the time between 28 and 12 ky ago. Among them are Bordes, Chard, Comas, Dummond, Krieger, Laughlin, MacNeish, Meighan, Rouse, F. Wendorf, Wormington, and Willey. Nonetheless, a growing number of critics are coming to believe that neither the Early nor the Middle Paleo-American Periods have been unequivocally demonstrated and may never be. In 1983, D. Stanford summed up what is still needed to prove a pre-Clovis presence:

> If I could find one clearly stratified site with some busted mammoth bones, a couple of crude flake tools, and a single human bone, all in unquestionable association with a charcoal hearth date 19,500 years ago—I'd have my dream [in Selig, 1983, p 1].

Several dozen locations have been proposed as evidences of a Middle Paleo-American population. Many possess an early date with doubtful tools, or a date with a tool or two associated but with the possibility that the tools belong to a later horizon and have drifted downward, as Stanford has discovered at the Dutton Site.

Tlapacoya. Eighteen localities have been excavated in the environs of the Cerro de Tlapacoya, Mexico, between 1965 and 1973. Interdisciplinary teams believe they have defined a cultural sequence evidenced by an obsidian projectile point, obsidian flakes, and three hearths associated with a sequence of C-14 dates from 24 to 15 ky [Mirambell, 1978]. Haynes believes the C-14 dates to be valid for the site but is concerned about the association of the dated material, a log, and a single obsidian blade [Haynes, 1969b, p 353].

Pikimachay Cave. R.S. MacNeish, in 1967, sought preceramic sites in highland Peru, near Ayachucho:

> The long hill we were descending had a number of bad hairpin curves in it, so I was paying attention to the driving. In fact, I didn't look up until the road wound to the north away from the basin. When I looked up, there it was! Halfway up a large rounded hill full of ancient terraces was a huge dark cave, highlighted by the morning sun shining on the surrounding cliffs [MacNeish, 1979, p 3].

Thus was discovered one of the most significant Paleo-American sites in the Americas: Pikimachay Cave (piki = flea, machay = cave). Herein, MacNeish has delineated 13 superimposed zones that appear to contain 43 definable phases of occupancy extending in time from Spanish days to perhaps 23 ky ago. The levels of greatest interest are Pacacaisa (23 ky) and Ayacucho (14 ky). He has recovered scrapers, spoke shaves, worked flakes, and other apparent tools [MacNeish, 1979]. Of these, MacNeish remarked: "The material is extremely crude with the majority made from volcanic tufa (sic)—(actually tuff), that does not show evidence of man's work very clearly [MacNeish, 1978, p 476]." Of the same "tools" Haynes wrote:

> At Pikimachay Cave there are now several C-14 dates above and below the 14,000-BP level which are stratigraphically consistent but all dates are from bone, a material notorious for being contaminated. I do not believe the lower levels to be cultural but the 14 ky level cannot be dismissed entirely [Haynes, 1984].

In review of the earliest level in Pikimachay Cave, Lynch calls attention to a disagreement MacNeish has had with geologists as to which strata in the cave represent glacial advance or retreat, a controversy that, if Mac-Neish is correct, must be resolved by concluding that the glacial sequence around the cave is different from that of Europe and North America [Lynch, 1983, p 93]. As already noted, Lynch has concluded that no solid evidence exists in South America of a pre-14 ky human occupation.

Valsequillo (Hueyatlaco). In 1962, C. Irwin-Williams and J.A. Camacho began excavation around the Valsequillo Reservoir, near Puebla, Mexico, where large quantities of paleontological and archaeological materials had been recovered. Bifacially chipped projectile points, blades, and a "psuedo-fluted" point resemble specimens from El Inga, Ecuador, and a bipointed projectile point is similar to some Plano points (Late Paleo-American, 9.5–7.5 ky) such as Lerma, Cascade, and others. These artifacts were associated stratigraphically with the bones of horse, camel, elephants, and of other extinct animals. Another group of artifacts, "crudely edge-worked points," reminds Irwin-Williams not of any other New World tool complex but rather of Old World Upper Paleolithic materials [Irwin-Williams, 1967, p 343–346].

Dates of 21.85 and 9.15 ky from C-14 on fresh-water shell have been obtained, a dating ambiguity that leaves the precise date of the site in substantial doubt [Haynes, 1969a, p 712]. An absence of charcoal, heavy mineralization of the bones, and very complex "stream bed" stratigraphy composed of gravels that might be as much as 200 ky old, combine to defy easy confirmation of a pre-Clovis age. Haynes believes Valsequillo possesses some of the best evidence of a late Wisconsin human presence in the Americas but, on the other hand, he notes that, apart from the dates, the sites illustrate no cultural materials that could not be quite recent [Haynes, 1969b, 1979b].

Meadowcroft Rockshelter. Meadowcroft Rockshelter, Pennsylvania, which since 1973 has been meticulously excavated by a team directed by J.M. Adovasio, is a deeply stratified, multicomponent site located in southwestern Pennsylvania. Eleven strata have been defined within which firepits, ash and charcoal lenses, large burned areas, as well as refuse and storage areas have been identified. Adovasio believes the site to have been first occupied 19 ky ago, then intermittently until historic times. Radiocarbon dates for lower Stratum IIa, which Adovasio believes to be the oldest, range from 10.85 to 17.65 ky. Over 400 items have been discovered from this stratum and there is no question as to their cultural nature [Adovasio et al, 1979–1980a,b, 1982].

Dates in the cave for the earliest levels have been derived from charcoal using C-14. Haynes believes that this charcoal may have been contaminated by older carbon brought into the lower levels by ground water. He and others have called attention to the absence in the cave of any evidence of other than a Holocene flora or fauna, a condition regarded as unlikely for a site that, if correctly dated at 19 ky ago, would have been only a short distance from the leading edge of the continental ice mass [Haynes, 1980a]. Fossil pollens from a source near Meadowcroft indicate that the area was dominated by tundra with spruce forests 19–12 ky ago but no evidence of such a flora is found in the Meadowcroft deposits [Mead, 1980]. Furthermore, all of the Meadowcroft cultural assemblage falls within an acceptable Clovis pattern. The only real indicator of great age in the cave is in the C-14 dates from sediment 70 cm or less in thickness. Haynes believes Meadowcroft to be no older than Clovis, if as old [Haynes, 1980a, p 582].

Despite these criticisms, Adovasio and his colleagues stand their ground and insist that Meadowcroft constitutes the best evidence recovered so far for a pre-Clovis occupation anywhere in the New World [Adovasio et al, 1980c, p 595].

The Timlin Site. Claims have been made since the early 1970s that early "Paleolithic" tools were to be found in the Catskill Mountains at the Timlin Site, near Cobleskill, New York [Raemsch and Vernon, 1977]. Dismissed by a number of authorities as preposterous, the claims have generally been taken lightly [Cole et al, 1977]. But further excavation at the site by A.L. Bryan in 1979, although not supportive of earlier claims, did find cores and flakes at a level that might be 16 ky old although according to Bryan, dates from the site are contradictory [Bryan et al, 1980]. No detailed report is yet available on this site.

Pollen profiles suggest a date of between 14 and 13 ky for the level in which the cultural material was found. Bracketing C-14 dates on mammal bone support approximately this age also: 12.9 and 15 ky.

The Selby and Dutton sites. Extensive excavation in 1979 of these two Pleistocene fossil beds in eastern Colorado by D. Stanford, discovered what to him appeared to be four classes of evidence of human presence among the abundant remains of 24 or more extinct Pleistocene vertebrate species. These classes of evidence were: 1) bone expediency tools, 2) flaked bone, 3) bone processed to extract marrow, and 4) at Dutton, seven tiny stone flakes [Stanford, 1979; Stanford et al, 1981]. Geological considerations suggest an age of approximately 15 ky ago for the materials.

But because the existence of any bone "tool kits" in the Americas has been seriously challenged, further proof of the existence of one at these locations is required. Furthermore, since the Dutton site possesses a Clovis occupational level, and the seven tiny stone flakes underlay that, there exists the possibility that they are of Clovis origin. Also, at Dutton, Stanford believed he might have found an artifact associated with broken camel bone in a stratum 16 ky old but further work proved the tool to be lying at the bottom of a gopher hole. Stanford appears to be no longer sure that the recovered fractured bones form part of a human tool kit [Selig, 1983]. Perhaps the Selby and Dutton sites warrant removal from the ranks of pre-Clovis candidates.

Wilson Butte Cave. Wilson Butte Cave on the Snake River in southern Idaho was reported by R. Gruhn in 1961. Three artifacts—a biface, a blade, and a "burinated" flake were recovered from the lowest level, Stratum C, a water-laid sand, a level that yielded C-14 dates to 14.55 ky, obtained from numerous bones of small animals found in the layer [Gruhn, 1961]. Haynes suggests that the artifacts might have been carried into the lower level by rodent activity within the cave [Haynes, 1969b, pp 353-354]. C.M. Aikens, summarizing the archaeology of the region, regards the dates from Wilson Butte Cave as "controversial"; the artifact group is too limited to be meaningfully assessed [Aikens, 1978].

The El Jobo Complex (Taima-Taima). The El Jobo Complex of coastal Venezuela was first defined in 1956 by Cruxent and Rouse from surface-collected material. The "type" tool is a long lanceolate projectile point with a thick cylindrical cross section [Cruxent and Rouse, 1956]. El Jobo points have been found associated with bones of mastodon, glyptodon, and other extinct forms at a waterhole site at Muaco, and elsewhere. Burned bone from Muaco was dated to 14-16 ky but because modern glass co-occurs in the deposit, there is suspicion that it has been seriously mixed. There is no question regarding the cultural character of El Jobo artifacts. If it were not for their association with Pleistocene fauna and with pre-Clovis dates, the El Jobo tools could fall within the typological range of post-Clovis blade complexes, referred to as "Plano" (see page 549).

Taima-Taima, also a waterhole site, was excavated by Bryan and Gruhn in 1976 in order to confirm previously reported associations between extinct fauna and El Jobo artifacts. They uncovered the semi-articulated remains of a young mastodon with an El Jobo point midsection in its pubic cavity, apparently killed and butchered by humans [Bryan et al, 1978]. In the vicinity of the mastodon skeleton, and presumed by Gruhn and Bryan to be the intestinal contents of the animal, were fragments of sheared wood twigs, twigs which have yielded four C-14 dates ranging from 14.2 to 12.98 ky [Gruhn and Bryan, 1981].

If the interpretations of Gruhn and Bryan are correct, there must have been an elephant hunting, non-Clovis point-using adaptation in northern Venezuela 13 ky ago. But mucky earth—past, intermittent, or present—can lead to movement in soils and their contents, the extent of which would be impossible to know. Plodding large animals in search of drink, ravenous predators attacking game, humans digging for water (or archaeological workers in search of evidences of Paleo-hunters), burrowing animals, or simply the natural movement associated with bog formation will result in displacement of soil contents beyond the ability of even the most meticulous workers to recapitulate. Haynes, Lynch, and others have criticized these and other dates and associations derived from materials obtained from waterholes [Haynes, 1979b; Lynch, 1974, 1978, p 469].

Dincauze questions the association of the fragment of the El Jobo point with the mastodon bones on another basis. She reasons that, as both are found together on an impermeable floor at the base of old spring deposits, they could have moved down separately to become co-residents of the bottom at different times, up until about 10 ky ago when a soil formed which seals the deposit below it [Dincauze, 1984, pp 289–290]. Thus, neither their relationship to each other nor the age of either is really certain.

The Blue Fish Caves. Two small caves, in northeast Yukon Territory, Canada, were test excavated by Cinq-mars from 1978 to 1981, and revealed late Pleistocene faunal remains. Lithic cultural evidence consists of an angle burin made on a retouched chert blade, flakes, and a possible micro-blade fragment [Morlan and Cinq-mars, 1982]. In addition to these stone items, all made of rock exotic to the caves, two mammal tibiae, polished primarily at one end, were found, as well as other bones which appeared to show human modification. Distribution of faunal remains within the cave also suggests human agency to Cinq-mars [Morlan and Cinq-mars, 1982, pp 366–372].

It is surprising that so little evidence of human presence in Beringia exists. If Clovis hunters prove to be the first Americans it would be reasonable to expect that their ancestors were in Beringia by 15 ky ago. But the Blue Fish Cave materials scarcely prove it. The data are too modest, the evidence for mixing in the deposits substantial, the conclusions regarding a human presence perhaps premature.

Monte Verde. Several seasons of multidisciplinary research at the Monte Verde site in south-central Chile have convinced T. Dillehay and his collaborators that they have recovered a tool kit associated with mastodon-hunting people of 14–12 ky ago [Dillehay et al, 1982; Collins, 1981]. Archaeological materials include fragmentary remains of perhaps six mastodons of various ages, with some bones possibly modified; wood and plant remains; one hundred or more lithic items most showing minimal, if any, modification; charcoal; plus a possible feature. A C-14 date obtained from bone (12.35 ky) and one from wood (13.03 ky) are believed to date human presence.

Due to the quality of excavation and research at Monte Verde, and the complexity of the remains, this site must be regarded as a leading candidate for pre-Clovis age. But, all of the evidence was recovered around an old stream course, thus there is the possibility that all are fortuitously associated. Perhaps supporting such an interpretation are the mastodon remains. Dillehay believes they represent a cultural event. Just how likely it is that a group of primitive hunters perhaps lacking projectile points could drop a whole family of elephants in a small area is a matter for conjecture. An alternative explanation for this collection of bones might be that water-action or nonhuman predators deposited them into a back-water or lair. Although the flow of the stream today is slight, weather disruptions in the area, for example, exceptional quantities of snow and rain due to periodic shifts by the Pacific Ocean current, "El Nino," can bring massive water flow (as Americans have learned in 1983).

Another complicating factor at Monte Verde might be the materials used for dating, especially the wood. Meighan reports that in the Chilean desert he recovered charcoal that gave a date of 5 ky on a site otherwise known to be only 2 ky old. Apparently, the ancient people were burning fossil wood [Meighan, 1983a, p 454]. More work is underway at Monte Verde, so final interpretations are not yet available.

The Levi site. During the 1960s and again during the 1970s, H.L. Alexander excavated a long, narrow rock shelter near Austin, Texas [Alexander,

1982]. He uncovered five strata, from recent at the top to Clovis at the next to bottom. The bottom stratum, Zone I, contains crude flakes, polished bones, a small hammerstone, use flakes, and burned bones. Hackberry seeds from the Clovis level provide a C-14 date of 13.75 ky while from Zone I, beneath the Clovis level, a hackberry date of 12.83 ky and a C-14 date from bone collagen of 10.825 ky were obtained [Alexander, 1982, p 138]. There is evidence that the Pedernales River may have flooded the cave in times past and mixed its contents. So all of the material in Stratum I could be from the Clovis occupation.

Fort Rock Cave. Two projectile points, several scrapers and gravers, plus a milling stone and a fragment of a handstone or mano (common Archaic Period tools), were found resting on Pleistocene lake gravels in Fort Rock Cave, Oregon. Nearby, also on ancient lake gravels, was a concentration of charcoal from which a date of 13.2 ky was obtained. In question is the degree of association between the dated charcoal and the artifacts; corroborating evidence is needed [Bedwell, 1973; Bryan, 1969, p 340].

Los Toldos. The stratified Los Toldos caves, located in southern Patagonia, Argentina, have been excavated since the early 1950s. A date from hearth charcoal of 12.65 ky appears to date an industry at the bottom of Cave 3, an industry that consists mostly of scrapers and flakes but that may include a unifacial point. Lynch believes the early date to be valid but the apparently associated artifacts to be too few in number to permit sound interpretation [Cardich, 1978; Lynch, 1974, p 369; 1978, pp 549-560].

The Malakoff Heads. Five apparently modified boulders, ranging in weight from 31 to 135 pounds, have been recovered by various people at various times in south Texas and in nearby Mexico. None was found in situ by professional investigators. The possibility apparently exists that the first three found may have been in association with Pleistocene fauna in a terrace that dates from 200 ky ago to as recent as 5–7 ky [Agogino, 1966]. Although undatable, and as likely of Archaic as Paleo-American age, Krieger included them among his evidence of pre-Clovis Americans [Krieger, 1964, p 47].

The Late Paleo-American Period (12–7.5 ky)

The Clovis Complex (12–11 ky). Evidences of human presence south of the retreating glacial masses are abundant and unambiguous beginning around 11.5 ky ago. Dozens of excavated sites in Canada, in the United States, and a few in Latin America testify to the presence of people hunting mammoths and other now extinct animals throughout the hemisphere at that time [Jennings, 1978, p 13]. They possessed a complex tool kit marked by a fluted, lanceolate stone projectile point that has come to be known as a Clovis point, a projectile point type that Bordes once labeled "the first American patent." At places such as Blackwater Draw, at Lehner and Naco, perhaps at Sandia Cave, and at many other locations, abundant Clovis fluted projectile points, scrapers, chopping tools, gravers, knives, and hammerstones, as well as fragments of bone tools provide testimony to the existence of a technology and culture comparable in general contour to those known for Old World humans of the Upper Paleolithic.

Haynes and Martin, along with many other scholars, believe the progenitors of the Clovis adaptation to have come from Asia via Beringia. In Asia, their cultural antecedents may have been the Dyuktai Complex or, as Haynes believes more likely, the Mal'ta-Afontova, either of which possessed complex lithics, including bifacially flaked stone blades [Haynes, 1982; Martin, 1982]. They entered Beringia perhaps 20 ky ago in pursuit of game animals during a time when steppe-tundra united Siberia and Alaska across Beringia. As the steppe-tundra and the Pleistocene megafauna began to decline 15 ky or so ago, these ancient hunters were led further to the southeast away from Beringia until, ultimately, 12 ky or so ago, they passed south of the ice sheets through the MacKenzie Corridor and burst into the northern plains of North America. Here they would encounter abundant and relatively easy prey. A rapid spread and successful adaptation for the Clovis people is indicated by a date for a fluted point from Fell's Cave at the tip of South America of approximately 11 ky, only .5 ky after the earliest known date for Clovis to the north. In North America, from Borax Lake in California to the Shoop Site in Pennsylvania, a number of stratified sites containing Clovis material provide an abundant, if incomplete basis for reconstructing their ancient way of life, as C.V. Haynes and others have attempted [Haynes, 1966, 1980b, 1982; Saunders, 1980; Johnson et al, 1980].

The appearance of Clovis kill sites in the archaeological record is dramatic and relatively abrupt. It is as if humans and elephants suddenly confronted each other for the first time; the result of the confrontation was to be the extinction of this whale of the land by 11 ky, little more than .5-1 ky after the first known kill. Within 2 ky or so of their postulated arrival, the Clovis-descendent people witnessed, and perhaps precipitated, the extinction of 31 genera of animals; some 90% of the hemisphere's total of animals by body weight [Martin and Wright, 1967; Martin, 1973, 1982; Mosimann and Martin, 1975]. Whatever its cause, as the "Pleistocene extinction" proceeded, new adaptations become evident in the archaeological record.

The Folsom Complex (11–9 ky). The Folsom Complex, primarily associated with the giant wide-horned bison (*Bison antiquus*), is characterized by a fluted point, slightly smaller than the typical Clovis, exhibiting a high degree of technical competence in its careful flaking around the deep channel bifacial flutes. In general, except for the difference in the projectile points, the Folsom tool kit closely resembles that of Clovis. Folsom sites tend to be larger than those of Clovis and, although originally defined in the high Plains to the east of the Rocky Mountains, a number of fluted point sites of approximately the same age are known from eastern North America as well (Bull Brook, MA; Debert, Nova Scotia; Port Mobil, NY; The Quad Site, VA; and others).

The Plano adaptation (9.5–7.5 ky). As the retreat of the glaciers continued, and as the post-Pleistocene biotic community underwent change in the direction of its modern composition, regional Paleo-American adaptations become apparent. The diagnostic tool of the different Plano cultures continues to be large projectile points with regionally variable shapes, often lanceolot but never fluted. Stone-chipping technology remains excellent, tool kits complex, and sites large.

Shifts in climate and in the character of the faunal community forced upon these descendents of the earlier Paleo-Americans an ever more archaeologically apparent reliance upon plant food resources. As the Paleo-American epoch ends, and the Archaic Period begins, culture adaptations are detectable in the archaeological record that will still be practiced by the native American people of the 16th century when Europeans begin to describe the pattern and variety of native American cultures.

DISCUSSION

Americanist archaeology has faced a dilemma since the 1960s, when the tacitly accepted presence of humans in the Americas prior to the Clovis hunters was challenged by the questions and possible alternative answers offered by C.V. Haynes and P.S. Martin. Professional response to this dilemma has been bidirectional: 1) proponents of the pre-Clovis position have intensified their activities or, at least, their publication rates, and 2) critics of the pre-Clovis proponents have become more vocal and direct in rejection of data put forward as evidence. In the effort to identify the earliest Americans, the issues are becoming clearer than at any time since that of Hrdlička.

The controversy as to who and when were the first Americans has reached what Kuhn has called the "paradigm debate" state in the development of a science [Kuhn, 1970, p 110]; a time when conflicting points of view, or paradigms, should be examined in light of available data in order to develop lines of reasoning which will link data into bodies of evidence that support one of the contending points of view or "paradigms." If the notion of paradigm can be overdone [see Pollie, 1983], it has useful applications in archaeology, a discipline that rarely has made explicit the assumptions or procedures that have led to the conclusions reached [cf Owen, 1967]. But, as the "new archaeology" with its emphasis upon logical positivism has come to dominate the field, examination of underlying paradigms has become more common [cf Binford and Sabloff, 1982; Ceci, 1982][4].

It is possible to detect at least four paradigms, or points of view, being employed in the analyses and conclusions drawn from existing data on Early Man in the Americas.

Paradigm 1: Humans of a premodern H sapiens or even H erectus variety spread across Beringia and into North America more than 40 ky ago utilizing a simple percussion-flaked chopper/scraper stone technology.

Bryan, Davis, Carter, Gruhn, Leakey, Minshall, MacNeish, Simpson, and others (including most or all the reporters of work done at Old Crow Basin), who accept as valid Calico Mountain, Texas Street, the Yuha site, and any other data of purported Sangamonin age, employ this paradigm. It is by no means a new point of view. Quite the contrary. F.W. Putnam, in the last two decades of the 19th century, while Curator of the Peabody Museum, had a great interest in proving that humans had occupied the Americas during the Pleistocene as Boucher de Perthes had done for Europe. Belief by some in an early Pleistocene American paleolithic has continued since Putnam's time.

To accept Paradigm 1, it is necessary to work from the following assumptions: 1) early H sapiens or earlier forms of the genus Homo entered the Americas, 2) while adaptively radiating into the Western Hemisphere they lost the capacity or will to practice the systematic chipping of stone that would produce a recognizable tool kit, 3) they did not utilize caves, 4) they did not bury their dead, 5) they did not systematically use hearths, etc.

With the understanding of worldwide cultural evolution developed by anthropologists during the past half century, it is difficult to believe that an ancient American society and culture born of the Paleolithic of the Old World would lack its parents' most common archaeological evidence—stone tools. This is especially true in light of archaeological research in China that has readily provided satisfactory evidence of human presence there for the past half million years or more [Aigner, 1978; Ikawa-Smith, 1982].

---

[4]I am grateful to Lynn Ceci for calling Kuhn's book to my attention.

It is a fact that some humans have employed relatively little rock in the face of its relative absence, as in the rainforests, but no such scarcity would afflict Beringians. Perhaps adaptive radiation into and across Beringia brought about cultural and social chaos of a kind we shall never understand. But if we may judge from known Paleo-arctic cultures, such as Akmak, with its skilled use of stone [Dumond, 1977, pp 36–46], it is clear that existence in arctic or subarctic ecological contexts does not mandate cessation of the use of stone as a raw material nor diminution of the technical skill with which it is worked.

At present, there is little reason to employ Paradigm 1. Not a single datum exists that requires assumption of any of these parameters. As MacNeish remarked regarding purported 40 ky dates for mammoth bone, charcoal, and putative human tools on Santa Rosa Island:

> We seem to be coming close to getting some good dates on early Early Man, but knowing exactly what his artifact assemblages were or how they fit into any stage or model eludes us for the most part [MacNeish, 1982, p 314].

If one must make the type of assumptions listed above (obviously without regard for the peaceful rest—wherever he may be—of William of Occam), then one has left behind concern for scientific parsimony and probability and entered another interpretative realm. Paradigm 1, despite its endurance through the years and notwithstanding the intellectual perserverance of its proponents, requires too many assumptions poorly supported by existing evidence and is, consequently, of little explanatory value.

Paradigm 2: Anatomically modern *H sapiens* spread across Beringia from Asia and into the Americas during early or middle Wisconsin times, perhaps 30 ky or more ago but perhaps as recently as 15–20 ky; were possessed of a simple tool kit perhaps lacking lithic elements; and either died out or became ancestors of later hunters including the Clovis people.

Adovasio, Bada, Berger, Childers, and Orr are some whose interpretations of their materials are based upon this second interpretative paradigm. Use of Paradigm 2 does not require the assumption of a premodern *sapiens* morphological type, so the absence of other than fully modern human bones in the Americas is not an embarassment. But, otherwise, it requires many of the same assumptions as does Paradigm 1. Because nowhere in the Americas has a flaked stone tool kit of middle or upper Paleolithic quality been found that dates to the pre-Clovis time, some explanation of its absence must be made. This absence is a particular embarassment because, in northeast Asia, the earliest Diuktai sites (18 ky or older) provide bifacially worked oval knives, subprismatic pebble cores, Levallois tortoise cores, scrapers, burins, and other artifact types associated with abundant remains of mammoth, wooly rhinoceros, bison, and other large mammals—thus meeting evidentiary criteria for establishing the presence of humans there and then [Mochanov, 1978a,b]. Furthermore, Australia, apparently peopled for the first time 30 ky or more ago, provides adequate artifacts and associations to permit a clear definition of a human presence there as well [White and O'Connell, 1979].

The confidence of proponents of Paradigm 2 rests heavily upon acceptance of dates derived from techniques in which more conservative specialists have little confidence: AAR; C-14 testing of calcium carbonate or the apatite fraction, for example; or upon trust that once deposited in the ground, cultural materials will not move from the location and associations that obtained millenia ago. If AAR dates are wildly wrong, or if dates taken on bone apatite cannot be trusted, or if some archaeological deposits in places such as Pikimachay, Meadowcroft, and Wilson Butte Cave have undergone even slight mixing of strata, then support for either Paradigm 1 or 2 becomes very tenuous. In no case are the data sufficient to approximate the evidentiary requirements met by archaeological complexes of much greater age elsewhere in the world. Use of Paradigm 2 is based more upon faith than upon evidence.

Paradigm 3: Sometime after 20 ky but before 12 ky ago, humans, possessed of a tool kit characterized by large nonfluted projectile points distinctive from, earlier, and perhaps ancestral to the Clovis complex, came south from Beringia and peopled many parts of the Western Hemisphere.

Some assumptions utilized by those who would employ Paradigm 3 are: 1) 12 ky ago, some force, as yet unspecified, muddles the American archaeological record; 2) humans, as predators, had little noticeable effect upon the biotic community until joined or succeeded by the Clovis hunters 11.5 ky ago; 3) they did not bury their dead; 4) they did not make hearths; 5) they did not systematically use caves, etc.

Bryan (although he appears to prefer Paradigm 1) believes there exists a "stemmed point tradition," prior to as well as contemporaneous with the Clovis hunters, a tradition that, if he is correct, would validate this paradigm. In the stemmed-point tradition Bryan includes Mount Moriah points from Smith Creek Cave, Nevada [Bryan, 1977, p 170]; Taima-Taima; Bird's "fishtailed points" from Patagonia; perhaps Lake Mohave stemmed points; Los Toldos; and others [Bryan, 1978, pp 306–310]. But none of these so far has been found in contexts unambiguously of Middle Paleo-American age.

Humans must have entered Beringia by 12–20 ky ago at least, consequently, it would not be surprising to find areas south of the ice-sheets occupied by them during this time period as well. But, surprisingly, not a single site provides substantial, unequivocal evidence of such a presence. The sites to which Bryan alludes as possible representatives of a cultural tradition that preceeds and is then contemporaneous with Clovis all provide either incomplete or controversial proof, at best. On the other hand, it is possible that all these sites are, in fact, later than Clovis. None possesses cultural, paleontological, or other attributes that force ascription of pre-Clovis dates.

Paleo-ecologists have pointed out that it is inconceivable that humans could be present in a habitat and not have a substantial and detectable impact upon the ecosystem. Although in a recent symposium a group of these specialists apparently accept the presence of humans in Beringia and perhaps throughout the hemisphere by at least 35 ky ago, they, inconsistently, could point to no such evident impact upon the Western Hemisphere's ecosystems before 10–12 ky ago [Schweger et al, 1982, p 437].

An enduring problem that may never be resolved for the period just preceeding 12 ky ago is that which apparently intruded into the interpretation of the Dutton site (see above): If there is a Late Paleo-American occupation at a location, it is almost certain that, due to small animal or human activity, or natural soil movement, or other disruptions, some cultural material will drift downward into what might otherwise be sterile, noncultural levels, or older paleontological beds, or even into genuinely older cultural strata. It is unlikely that, given the inherent lack of precision of C-14 dating, the dawn of Clovis could ever be separated through use of it from a cultural stratum one or two thousand years older, particularly if the only major differences are temporal not cultural.

Paradigm 4: The first Americans south of the continental glaciers were the Clovis hunters, approximately 12 ky ago.

To accept and employ Paradigm 4, the following assumptions must be made: All putative earlier Paleo-American data are either a result of "background noise," which naturally exists in the ground; or of faulty application of dating techniques; or of faulty association of cultural materials with properly dated noncultural materials; or faulty identification of ecofacts as artifacts; or, perhaps, occasionally, excessive zeal on the part of some pre-Clovis partisans prepared to believe, however thin the evidence. All of the assumptions that underly Paradigm 4 appear warranted and are being utilized by a growing number of Americanist archaeologists.

CONCLUSIONS

If Paradigm 4 is effective as an explanation of who and when were the first Americans, then 1) its application should constantly satisfy and agree with accepted tenets of scientific archaeology, 2) all new data should be explained within its assumptions, and 3) other appropriate lines of scientific reasoning, when applied to the same question, should reach the same conclusions. When any of these conditons are not met, the paradigm should be changed.

From the end of the 19th century, when natural science standards began to be applied to Americanist archaeology, it took approximately 30 years to demonstrate the reality of a Late Paleo-American cultural horizon. In the half century since, application of ever more rigorous and precise techniques and methods has resulted in the delineation of subsequent American pre-history concordant with that known from elsewhere in the world: Early foragers and hunters were followed by village-dwelling plant utilizers who, in their turn, were succeeded by urban, agricultural, "civilized," people. Success in solving the puzzle of American prehistory has rested directly upon the continued application of ever more rigorous techniques and methods in assessment of data. Today, although much of the content of native American prehistory remains to be learned, its general outline is well understood. There are no "Mysterious Moundbuilders," to be explained and, if "lost" cities may still be found, the pattern of urbanization in the prehistoric Americas is basically known. Archaeological fieldwork, where ever it is conducted in the Americas, employing universally recognized cannons of archaeological research, regularly brings further support to the general proposition. To date, scientific archaeology has not been able to demonstrate a pre-Clovis occupation of the Americas.

If an Early or Middle Paleo-American occupation is to take its place within the framework of existing scientific Americanist archaeology, then it is time to find answers to a number of specific questions [cf Meighan, 1983a, pp 456–457]: 1. If the hemisphere has been occupied for more than 12 ky, where are the cultural remains that worldwide archaeological knowledge would lead us to expect: stone tools and tool kits, skeletal material, living areas, etc. 2. Where are the American equivalents to Old World Middle and Upper Paleolithic sites? Why are there not lower levels at such likely places as Ventana Cave, Danger Cave, the Koster Site, or ancient human remains at the La Brea Tar Pits? 3. Where are the early Beringian sites? 4. If present, why did Early or Middle Paleo-Americans not have some detectable impact upon the Pleistocene fauna? 5. Why, as new evidences of a pre-Clovis occupation of the Americas are offered by proponents, are older ones eliminated? 6. Why cannot the Early and Middle Paleo-American archaeological record be expected to meet the worldwide evidentiary standards of scientific archaeology? 7. If the evidence of pre-Clovis people in the Americas were to be present in the hemisphere, would not scientific archaeology, practiced with the great intensity that has characterized research in North America for the past half century, have found it by now?

Failure to provide convincing answers to any of these questions is further evidence of the utility of Paradigm 4: Clovis hunters were the first Americans.

LITERATURE CITED

Adovasio JM, Gunn JD, Donahue J, Stuckenrath R, Guilday J, Lord K (1979–1980a): Meadowcroft Rockshelter—Retrospect 1977: Part I. North Am Arch 1:1:3–44.

Adovasio JM, Gunn JD, Donahue J, Stuckenrath R, Guilday J, Lord K, Volman K (1979–1980b): Meadowcroft Rockshelter–Retrospect 1977:Part II. North Am Arch 1:2:99–137.

Adovasio JM, Gunn JD, Donahue J, Stuckenrath R, Guilday J, Volman K (1980c): Yes Virginia, it really is that old: A reply to Haynes and Mead. Am Antiq 45:3:588–595.

Adovasio JM, Gunn JD, Donahue J, Stuckenrath R (1982): Meadowcroft Rockshelter, 1973–1977: A synopsis. In Ericson JE et al (eds): "Peopling of the New World." Los Altos, CA: Ballena Press, pp 97–131.

Agogino G (1966): The Malakoff heads. El Palacio 73:32–36.

Aigner J (1978): The paleolithic of China. In Bryan AL (ed): "Early Man in America: From a Circum-Pacific Perspective." Occas Papers 1, Dept of Anthro, Univ of Alberta, pp 25–41.

Aikens MC (1978): The far west. In Jennings JD (ed): "Ancient Native Americans." New York: Freeman, pp 131–182.

Alexander HL (1982): The pre-Clovis and Clovis occupations at the Levi site. In Ericson, et al (eds): pp 133–145.

Bada JL, Schroeder RA, Carter GF (1974): New evidence to the antiquity of man in North America deduced from aspartic acid racemization. Science 184:791–793.

Bada JL, Finkel R (1982): Uranium series ages of the Del Mar man and Sunnyvale skeletons. Science 217:755.

Bada JL, Masters PM (1982): Evidence for a 50,000 year antiquity of man in the Americas derived from amino acid racemization in human skeletons. In Ericson JE et al (eds): "Peopling of the New World." Los Altos, CA: Ballena Press, pp 171–179.

Bedwell SF (1973): "Fort Rock Basin Prehistory and Environment." Eugene: Univ of Oregon Press.

Berger R (1978): Thoughts on the first peopling of America and Australia. In Bryan AL (ed): "Early Man in America: From a Circum-Pacific Perspective." Occas Papers 1, Dept of Anthro, Univ of Alberta, pp 23–24.

Berger R (1981): Early man on the California Channel Islands. In Bryan AL (ed): "El Poblamiento de America: Evidencia Arqueologia de Ocupacion Humana en America Anterior a 11,500 Anos a.p." X Congreso, Union Internacional de Ciencias prehistorias y protohistoricas. Mexico, D.F., p 46.

Berger R (1982): The Wooley mammoth site, Santa Rosa Island, California. In Ericson JE et
al (eds): "Peopling of the New World." Los Altos, CA: Ballena Press, pp 163–170.
Berkhofer RF (1978): "The White Man's Indian, From Columbus to the Present." New York:
Vintage Books, Random House.
Binford LR (1981): "Bones: Ancient Men and Modern Myths." New York: Academic Press.
Binford LR, Sabloff JA (1982): Paradigms, systematics, and archaeology. J Anthropol Res
28:2:137–153.
Bischoff JL, Childers WM (1979): Temperature calibration of amino acid racemization: Age
implications for the Yuha skeleton. Earth Planetary Sci Letts 45:172–180.
Bischoff JL, Merriam R, Childers WM, Protsch R (1976): Antiquity of man in America
indicated by radiometric dates on the Yuha burial site. Nature 261:129–30.
Bischoff JL, Rosenbauer RJ (1981): Uranium series dating of human skeletal remains from
the Del Mar and Sunnyvale sites, California. Science 213:1003–1005.
Bonnichsen R (1978): Critical arguments for Pleistocene artifacts from the Old Crow Basin,
Yukon: A preliminary statement. In Bryan AL (ed): "Early Man in America: From a
Circum-Pacific Perspective." Occas Papers 1, Dept Anthro, Univ of Alberta, pp 102–118.
Bonnichsen R, Young D (1980): Early technological repertoires: Bone to stone. Can J
Anthropol 1:123–128.
Brues AM (1977): "People and Races." New York: Macmillan.
Bryan AL (1969): Early man in America and the late Pleistocene chronology of western
Canada and Alaska. Curr Anthropol 10:4:339–365.
Bryan AL (1977): "Smith Creek Cave." Nevada State Museum, Anthropol Papers 17.
Bryan AL (ed) (1978): "Early Man in America: From a Circum-Pacific Perspective." Occas
Papers 1, Dept of Anthro, Univ of Alberta.
Bryan AL (ed) (1981): "El Poblamiento de America: Evidencia Arqueológica de Ocupación
Humana en America Anterior a 11,500 Años a.p." X Congreso, Union Internacional de
Ciencias prehistoricas y protohistoricas. Mexico, D.F.
Bryan AL, Casamiquela RM, Cruxent JM, Gruhn R, Ochsenius C (1978): An El Jobo
mastodon kill at Taima-Taima, Venezuela. Science 200:1275–1277.
Bryan AL, Schnurrenberger D, Gruhn R (1980): An early post-glacial lithic industry in east-
central New York State. Am Quaternary Assoc, 6th Biennial Meeting Abstracts:52.
Canby TY (1979): The first Americans. Natl Geog 156:3:330–363.
Cardich A (1978): Recent excavations at Lauricocha (central Andes) and Los Toldos (Pata-
gonia). In Bryan AL (ed): "Early Man in America: From a Circum-Pacific Perspective."
Occas Papers 1, Dept of Anthro, Univ of Alberta, pp 296–300.
Carter GF (1957): "Pleistocene Man at San Diego." Baltimore: The Johns Hopkins Press.
Carter GF (1978): The American paleolithic. In Bryan AL (ed): "Early Man in America:
From a Circum-Pacific Perspective." Occas Papers 1, Dept of Anthro, Univ of Alberta,
pp 10–19.
Carter GF (1980): Earlier Than You Think: A Personal View of Man in America. Texas A
and M Univ Press.
Carter GF (1984): Personal communication.
Ceci L (1982): Method and theory in coastal New York archaeology. N Am Arch 3:1:5–36.
Chedd G (1980): On the trail of the first American. Science 80:March/April:44–51.
Childers MW (1974): Preliminary report on the Yuha burial, California. Anthropol J Can
12:1–9.
Childers MW, Minshall HL (1980): Evidence of early man exposed at Yuha Pinto Wash. Am
Antiq 45:2:297–308.

Cole JR, Godfrey LR, Funk RE, Kirkland JT, Starna WA (1977): On "some paleolithic tools from northeast North America."Curr Anthropol 78:3:541–546.

Collins MB (1981): The implications of the lithic assemblage from Monte Verde, Chile, for early man studies. In Bryan (ed): pp 62–65.

Crook WW Jr, Harris RK (1958): A Pleistocene campsite near Lewisville, Texas. Am Antiq 23:233–246.

Cruxent JM, Rouse I (1956): A lithic industry of paleo-Indian type in Venezuela. Am Antiq 22:172–179.

Davies DM (1978): Some observations on the Otavalo skeleton from Imbabura province, Ecuador. In Bryan AL (ed): "Early Man in America: From a Circum-Pacific Perspective." Occas Papers 1, Dept of Anthro, Univ of Alberta, pp 273–289.

Davis EL (1980): Evidence of early hominids in the New World: A South American example. In Davis EL et al (eds): "Evaluation of Early Human Activities and Remains in the California Desert." San Diego: Great Basin Foundation, pp 60–65.

Davis EL (1982): The geoarchaeology and history of China Lake, California. In Ericson JE et al (eds): "Peopling of the New World." Los Altos, CA: Ballena Press, pp 203–228.

Davis EL, Brown KH, Nichols J (eds) (1980): "Evaluation of Early Human Activities and Remains in the California Desert." San Diego: Great Basin Foundation.

Davis EL, Jefferson G, McKinney C (1981): Notes on a mid-Wisconsin date for man and mammoth, China Lake, California. In Bryan AL (ed): "El Poblamiento de America: Evidencia Arqueológica de Ocupación Humana en America Anterior a 11,500 Años a.p." X Congreso, Union Internacional de Ciencias Prehistoricas y Protohistoricas. Mexico, D.F., pp 36–39.

Dillehay TD, Pino QM, Davis EM, Valastro S Jr, Varel G, Casamiquela R (1982): Monte Verde: Radiocarbon dates from an early-man site in south-central Chile. J Field Arch 9:547–550.

Dincauze DF (1984): "An Archaeo-Logical Evaluation of the Case for Pre-Clovis Occupations." Vol. III, Advances in World Arch. New York: Academic Press, pp 275–323.

Dumond DE (1977): "The Eskimos and Aleuts." London: Thames and Hudson.

Ericson JE, Taylor RE, Berger R (eds) (1982): "Peopling of the New World." Los Altos, CA: Ballena Press.

Feder KL (1983): American disingenuous: Goodman's 'American Genesis'—a new Chapter in cult archaeology. Skeptical Inquirer VII:4:36–48.

Folsom F, Folsom M (1982): Sinodonty and Sundadonty: An argument with teeth in it for man's arrival in the New World. Early Man 4,2:16–21.

Goodman J (1981): "American Genesis." New York: Summit Books (Simon and Shuster).

Greenberg J (1983): cited in "A new wave to the New World," in Currents, Science 83, 14,10:7–8.

Griffin JB (1979): The origin and dispersion of American Indians in North America. In Laughlin WS, Harper AB (eds): "The First Americans: Origins, Affinities, and Adaptations." New York: G. Fisher, pp 43–55.

Gruhn R (1961): The archaeology of Wilson Butte Cave, south-central Idaho. Occas Papers of the Idaho State College Museum 6.

Gruhn R (1978): A note on excavations at El Bosque, Nicaragua, in 1975. In Bryan AL (ed): "Early Man in America: From a Circum-Pacific Perspective." Occas Papers 1, Dept of Anthro, Univ of Alberta, pp 261–262.

Gruhn R, Bryan AL (1981): A summary report and implications of the Taima-Taima mastodon kill site, northern Venezuela. In Bryan AL (ed): "El Poblamiento de America:

Evidencia Arqueológica de Ocupación Humana en America Anterior a 11,500 Años a.p." X Congreso, Union Internacional de Ciencias Prehistorias y Protohistoricas. Mexico, D.F., pp 48–49.

Harington CR, Bonnichsen R, Morlan RE (1975): Bones say man lived in Yukon 27,00 years ago. Can Geographical J 91:42–48.

Harper AB (1980): Origins and divergence of Aleuts, Eskimos and American Indians. Ann Human Biol 17:6:547–554.

Haynes CV (1966): Elephant-hunting in North America. Scientific American 214:104–112.

Haynes CV (1969a): The earliest Americans. Science 166:709–715.

Haynes CV (1969b): Comments. Curr Anthropol 10 (4): 353.

Haynes CV (1973): The Calico site: Artifacts or geofacts? Science 181:305–310.

Haynes CV (1979a): Quoted in Humphrey RL, Stanford D (eds): "Pre-Llano Cultures of the Americas: Paradoxes and Possibilities." Washington DC: Anthropol Soc of Washington, p viii.

Haynes CV (1979b): Personal communication.

Haynes CV (1980a): Paleoindian charcoal from Meadowcroft Rockshelter: Is contamination a problem? Am Antiq 45:3:582–587.

Haynes CV (1980b): The Clovis culture. Can J Anthropol 1:115–121.

Haynes CV (1982a): Were Clovis progenitors in Beringia? In Hopkins DM et al (eds): "Paleoecology of Beringia." New York: Academic Press, pp 383–398.

Haynes CV (1982b): Personal communication.

Haynes G (1983): Frequences of spiral and green-bone fractures on ungulate bones in modern surface assemblages. Am Antiq 48:1:102–114.

Haynes CV (1984): Personal communication.

Hopkins DM, Matthews JV Jr, Schweger CE, Young SB (1982): "Paleoecology of Beringia." New York: Academic Press.

Hrdlička A (1915): Letter to Theodore Roosevelt, March 4. Washington, DC: Aleš Hrdlička Papers, Natl Anthropol Archives, Smithsonian Institution.

Humphrey RL, Stanford D (eds) (1979): "Pre-Llano cultures of the Americas: Paradoxes and possibilities." Washington DC: Anthropol Soc of Wash.

Ikawa-Smith F (1982): The early prehistory of the Americas as seen from northeast Asia. In Ericson JE et al (eds): "Peopling of the New World." Los Altos, CA: Ballena Press, pp 15–33.

Irving WN (1978): Pleistocene archaeology in eastern Beringia. In Bryan AL (ed): "Early Man in America: From a Circum-Pacific Perspective." Occas Papers 1, Dept of Anthro, Univ of Alberta, pp 96–101.

Irving WN (1982): Pleistocene cultures in Old Crow Basin: Interim report. In Ericson JE et al (eds): "Peopling of the New World." Los Altos, CA: Ballena Press, pp 69–79.

Irving WN, Harington CR (1973): Upper Pleistocene radiocarbon-dated artifacts from the northern Yukon. Science 179:335–340.

Irwin-Williams C (1967): Associations of early man with horse, camel, and mastodon at Hueyatlaco, Valsequillo. In Martin PS, Wright HE Jr (eds): "Pleistocene Extinctions: The Search for a Cause." New Haven: Yale Univ Press, pp 337–347.

Jennings JD (1968): "Prehistory of North America." New York: McGraw-Hill.

Jennings JD (ed) (1978): "Ancient Native Americans." New York: W.H. Freeman.

Jennings JD (ed) (1983): "Ancient South Americans." New York: W.H. Freeman.

Jennings JD, Norbeck E (eds) (1964): "Prehistoric Man in the New World." Chicago: The University of Chicago Press.

Johnson DL, Kawano P, Ekker E (1980): Clovis strategies of hunting mammoth. Can J Anthropol 1:107–114.

Jopling AV, Irving WN, Beebe BF (1981): Stratigraphic, sedimentological and faunal evidence for the occurrence of pre-Sangamonian artifacts in northern Yukon. In Bryan AL (ed): "El Poblamiento de America: Evidencia Arqueológica de Ocupación Humana en America Anterior a 11,500 Años a.p." X Congreso, Union Internacional de Ciencias Prehistoricas y Protohistoricas. Mexico, D.F., p 27.

Krieger AD (1958): Review of: Pleistocene man at San Diego by G.F. Carter. Am Anthropol 60:974–978.

Krieger AD (1964): Early man in the New World. In Jennings JD, Norbeck E (eds): "Prehistoric Man in the New World." Chicago: Univ of Chicago Press, pp 23–81.

Kuhn TS (1970): "The Structure of Scientific Revolutions, 2nd ed." Chicago: Univ of Chicago Press.

Lajoie KR, Peterson E, Gerow B (1980): Amino acid bone dating: A feasibility study, south of San Francisco Bay region, Calif. In Hare PE, Hoering TC, King K Jr (eds): "Biogeochemistry of Amino Acids." New York: John Wiley and Sons.

Laughlin WS, Jorgensen JB, Frolich B (1979): Aleuts and Eskimos: Survivors of the Bering land bridge coast. In Laughlin WS, Harper AB (eds): "The First Americans: Origins, Affinities, and Adaptations." New York: Fischer, pp 91–104.

Laughlin WS, Harper AB (eds) (1979): "The First Americans: Origins, Affinities, and Adaptations." New York: Gustav Fisher, pp 1–11.

Laughlin WS, Wolf SI (1979): Introduction: The first Americans—origins, affinities, and adaptations. In Laughlin and Harper (eds): pp 1–11.

Lorenzo JL (1978): Early Man Research in the American Hemisphere: Appraisal and Perspectives. In Bryan AL (ed): "Early Man in America: From a Circum-Pacific Perspective." Occas Papers 1, Dept of Anthro, Univ of Alberta, pp 1–9.

Lorenzo JL, Mirambell L (1981): El Cedral, SLP, Mexico: Un sitio con presencia humana de mas de 30,000 a.p. In Bryan AL (ed): "El Poblamiento de America: Evidencia Arqueologica de Ocupación Humana en America Anterior a 11,500 Años a.p." X Congreso, Union Internaciónal de Ciencias Prehistoricas y Protohistoricas. Mexico, D.F., pp 112–124.

Lynch TF (1974): The antiquity of man in South America. Quaternary Res 4:356–377.

Lynch TF (1978): The South American paleo-Indians. In Jennings JD (ed): "Ancient Native Americans." New York: Freeman. pp 454–489.

Lynch TF (1983): The paleo-Indians. In Jennings JD (ed): "Ancient South Americans." New York: Freeman, pp 87–137.

MacNeish RS (1976): Early man in the New World. Am Sci 64:316–327.

MacNeish RS (1978): Late Pleistocene adaptations: A new look at early peopling of the New World as of 1976. J Anthropol Res 34:4:475–496.

MacNeish RS (1979): The early man remains from Pikimachay Cave, Ayacucho Basin, Highland, Peru. In Humphrey RL, Stanford D (eds): "Pre-Llano Cultures of the Americas: Paradoxes and Possibilities." Washington DC: Anthropol Soc. Washington, pp 1–47.

MacNeish RS (1982): A late comment on an early subject. In Ericson JE et al (eds): "The Peopling of the New World." Los Altos, CA: Ballena Press, pp 311–315.

Martin PS (1973): The discovery of America. Science 179:969–974.

Martin PS (1982): The pattern and meaning of holarctic mammoth extinction. In Hopkins DM et al (eds): "Paleoecology of Beringia." New York: Academic Press, pp 399–408.

Martin PS, Wright HE Jr, (eds) (1967): "Pleistocene Extinctions: The Search for a Cause." New Haven: Yale Univ Press.

Martin PS, Quimby GI, Collier D (1947): "Indians Before Columbus: Twenty Thousand Years of North American History Revealed by Archaeology." Chicago: Univ of Chicago Press.

Mattos A (1961): O homen das cavernas de Minas Gerais. Editora Itatiaia Limitada, Belo Horizonte.

Mead JI (1980): Is it really old? A comment about Meadowcroft Rockshelter "Overview." Am Antiq 45:3:579–582.

Meighan CW (1983a): "Early Man in the New World." In Masters PM, Fleming NC (eds): "Quaternary Coastlines and Marine Archaeology." Orlando: Academic Press.

Meighan CW (1983b): Personal communication.

Meltzer DJ (1983): The antiquity of man and the development of American archaeology. In: "Advances in Arch Method and Theory." New York: Academic Press, 6:1–51.

Minshall HL (1976): "The Broken Stones." San Diego: The Copley Press.

Mirambell L (1978): Tlapacoya: A late Pleistocene site in central Mexico. In Bryan AL (ed): "Early Man in America: From a Circum-Pacific Perspective." Occas Papers 1, Dept of Anthro, Univ of Alberta, pp 221–230.

Mochanov LA (1978a): Stratigraphy and absolute chronology of the paleolithic of northeast Asia, according to the work of 1963–1973. In Bryan AL (ed): "Early Man in America: From a Circum-Pacific Perspective." Occas Papers 1, Dept of Anthro, Univ of Alberta, pp 54–69.

Mochanov LA (1978b): The paleolithic of northeast Asia and the problem of the first peopling of America. In Bryan AL (ed): "Early Man in America: From a Circum-Pacific Perspective." Occas Papers 1, Dept of Anthro, Univ of Alberta, p 67.

Morlan RE (1978): Early man in northern Yukon Territory: Perspectives as of 1977. In Bryan AL (ed): "Early Man in America: From a Circum-Pacific Perspective." Occas Papers 1, Dept of Anthro, Univ of Alberta, pp 78–95.

Morlan RE (1980): Taphonomy and archaeology in the upper Pleistocene of the northern Yukon Territory: A glimpse of the peopling of the New World. National Mus. of Canada, Mercury Series, Arch. Survey of Canada Paper 94.

Morlan RE (1981): Big bones and tiny stones: Early evidences from the northern Yukon Territory. In Bryan AL (ed): "El Poblamiento de America: Evidencia Arqueológica de Ocupación Humana en America Anterior a 11,500 Años a.p." X Congreso, Union Internacional de Ciencias Prehistoricas y Protohistoricas. Mexico, D.F., pp 1–24.

Morlan RE, Cinq-mars J (1982): Ancient Beringians: Human occupation in the late Pleistocene of Alaska and the Yukon Territory. In Hopkins DM et al (eds): "Paleoecology of Beringia." New York: Academic Press, pp 353–381.

Mosimann JE, Martin PS (1975): Simulating overkill by paleoindians. Am Sci 63:304–313.

Muller-Beck H (1966): Paleohunters in America: Origins and diffusion. Science 152:1191–1210.

Myers TP, Voorhies MR, Conner RG (1980): Spiral fractures and bone psuedotools at paleontological sites. Am Antiq 45:483–490.

Orr PC (1968): "Prehistory of Santa Rosa Island." Santa Barbara Museum of Natural History.

Owen RC (1967): Assertions, assumptions, and Early Horizon (Oak Grove) settlement patterns in southern California: a rejoinder. Am Antiq 32:2:236–241.

Owen RC, Deetz JJF, Fisher AD (1967): "The North American Indians: A Sourcebook." New York: The Macmillan Company.

Page WD (1978): The geology of the El Bosque archaeological site, Nicaragua. In Bryan AL (ed): "Early Man in America: From a Circum-Pacific Perspective." Occas Papers 1, Dept of Anthro, Univ of Alberta, pp 231–260.

Payen LA (1982): Artifacts or geofacts at Calico: Application of the Barnes test. In Ericson JE et al (eds): "Peopling of the New World." Los Altos, CA: Ballena Press, pp 193–201.

Payen LA (1983a): The pre-Clovis of North America: Temporal and artifactual evidence. Dissertation Abstracts International 43:10.

Payen LA (1983b): The North American pre-Clovis: Chronological and archaeological evidence. SAS Research Reports 3.

Payen LA, Rector CH, Ritter E, Taylor RE, Ericson JE (1978): Comments on the Pleistocene age assignment and associations of a human burial from the Yuha Desert, California. Am Antiq 43:3:448-453.

Pollie R (1983): Brother, can you paradigm. Science 83:4:6:76-77.

Protsch RRR (1978): "Catalog of Fossil Hominids of North America." New York: Gustav Fischer.

Raemsch BE, Vernon WW (1977): Some paleolithic tools from northeast North America. Curr Anthropol. 18:1:97-99.

Reeves B (1981): Mission River and the Texas Street Question. In Bryan AL (ed): "El Poblamiento de America: Evidencia Arquelógica de Ocupación Humana en America Anterior a 11,500 Años a.p." X Congreso, Union Internacional Ciencias prehistoricas y protohistsoricas. Mexico, D.F., p 47.

Rensberger B (1976): Coast dig focuses on man's move to the New World. New York Times, Aug. 16.

Rensberger B (1977): Early, earlier, and earliest man. New York Times, June 26.

Rogers MJ (1966): "Ancient Hunters of the Far West." San Diego: Union-Tribune Publishing Co.

Rogers SL (1977): "An Early Human Fossil From the Yuha Desert of Southern California: Physical Characteristics." San Diego Mus. Papers 12. San Diego Mus. of Man.

Rutter NW, Schweger CE (eds) (1980): The ice-free corridor and peopling of the New World. Can J Anthropol 1:1.

Sapir E (1929): Central and North American Languages. Encyclopedia Brit. 14th ed. Reprinted in Selected Writing of E. Sapir, Mandelbaum D (ed) 1949:169-178. Univ. of Calif. Press.

Saunders JJ (1980): A model for man-mammoth relationships in late Pleistocene North America. Can J Anthropol 1:87-98.

Schiffer MB (ed) (1979): "Advances in Archaeological Method and Theory, 2." New York: Academic Press.

Schweger CE, Matthews JV Jr, Hopkins DM, Young SB (1982): Paleoecology of Beringia—a synthesis. In Hopkins et al (eds): pp 425-444.

Selig RO (1983): Bones and stones—or sheep? Anthropol Notes 5:1:1-3,14. Smithsonian InstT.

Shlemon RJ, Bischoff JL (1981): Soil-geomorphic and uranium-series dating of the Calico site, San Bernadino County, California. In Bryan AL (ed): "El Poblamiento de America: Evidencia Arqueológica de Ocupación Humana en America Anterior a 11,500 Años a.p." X Congreso, Union Internacional de Ciencias prehistoricas y protohistoricas. Mexico, D.F., pp 41-42.

Shutler R Jr (ed) (in press) "Early Man in the New World." Beverly Hills: Sage Publications (in press).

Shutler R Jr, Haynes CV Jr, Mawby JE, Mehringer PJ Jr, Bradley WG, Deacon JE (1967): Pleistocene studies in southern Nevada. Nevada State Mus. Anthro. Papers 13.

Simpson RD (1978): The Calico Mountains archaeological site. In Bryan AL (ed): "Early Man in America: From a Circum-Pacific Perspective." Occas Papers 1, Dept of Anthro, Univ of Alberta, pp 218-220.

Simpson RD (1982): The Calico Mountains archaeological project: A progress report. In Ericson JE et al (eds): "Peopling of the New World." Los Altos, CA: Ballena Press, pp 181–192.

Simpson RD, Patterson LW, Singer CA (1981): Early lithic technology of the Calico site, southern California. In Bryan AL (ed): "El Poblamiento de America: Evidencia Arqueológica de Ocupación Humana en America Anterior a 11,500 Años a.p." X Congreso, Union Internacional de Ciencias prehistoricas y protohistoricas. Mexico, D.F., pp 43–45.

Smith FH (1976): The skeletal remains of the earliest Americans: A survey. Tenn Anthropol 1:2:116–147.

Smith FH (1977): On the application of morphological "dating" to the hominid fossil record. J Anthropol Res 33:302–316.

Snow D (1976): "The Archaeology of North America." New York: Viking Press.

Stafford T et al (1984): Holocene age of Yuha burial: Direct radiocarbon determinations by accelerator mass spectometry. Nature (London). in press.

Stalker AM (1969): Geology and age of the early man site at Taber, Alberta. Am Antiq 34:4:425–428.

Stanford D (1979): The Selby and Dutton sites: Evidence for a possible pre-Clovis occupation of the high plains. In Humphrey RL, Stanford D (eds): "Pre-Llano Cultures of the Americas: Paradoxes and Possibilities." Washington DC: Anthropol Soc Washington, pp 101–123.

Stanford D (1981): "Who's on first," a review of J. Goodman's "American Genesis." Science 81:2:5:91–92.

Stanford D, Bonnichsen R, Morlan RE (1981): The Ginsberg experiment: Modern and prehistoric evidence of a bone-flaking technology. Science 212:438–440.

Stewart TD (1960): A physical anthropologist's view of the peopling of the New World. SWJ Anthropol 16:3:259–273.

Stewart TD (1973): "The People of America." New York: Charles Scribner's Sons.

Stewart TD (1981): The evolutionary status of the first Americans. Am J Phys Anthropol 56:461–466.

Sullivan W (1975): Settlements of most major land areas around Pacific put at 30,000 years old. New York Times, Sept. 3.

Swadesh M (1964): Linguistic overview. In Jennings JD, Norbeck E (eds): "Prehistoric Man in the New World." Chicago: Univ of Chicago Press, pp 527–556.

Szathmary EJE (1979): Blood groups of Siberians, Eskimos, subarctic and Northwest Coast Indians: The problem of origins and genetic relationships. In Laughlin WS, Harper AB (eds): "The First Americans: Origins, Affinities, and Adaptations." New York: Fischer, pp 185–209.

Taylor RE, Payen LA (1979): The role of archaeometry in American archaeology: Approaches to the evaluation of the antiquity of Homo sapiens in the California. In Schiffer MB (ed): "Advances in Archaeological Method and Theory, 2." New York: Academic Press, pp 239–283.

Taylor RE, Payen LA, Gerow B, Donahue DJ, Zabel TH, Tull AJT, Damon PE (1983): Middle holocene age of the Sunnyvale human skeleton. Science 220:1271–1273.

Turner CG II, Bird J (1981): Dentition of Chilean paleo-Indians and peopling of the americas. Science 212:1053–1054.

Von Endt DW (1979): Techniques of amino acid dating. In Humphrey RL, Stanford D (eds): "Pre-Llano Cultures of the Americas: Paradoxes and Possibilities." Washington DC: Anthropol Soc Washington, pp 71–100.

Wendorf M (1982): The fire areas of Santa Rosa Island: An interpretation. North Am Arch 3:2:173–180.

West FH (1982): The antiquity of man in North America. In Porter SC (ed): "Late Quaternary Environments of the United States, Vol. 1." Minneapolis: U of Minnesota Press.

West S (1983): Stolen bones. Science 83:4:1:28–35.

White JP, O'Connell JF (1979): Australian prehistory: New aspects of antiquity. Science 203:21–28.

Wilke PJ (1978): Cairn burials of the California deserts. Am Antiq 43:3:444–448.

Willey GR (1966): "An Introduction to American Archaeology, Vol. 1: North and Middle America." Engelwood, NJ: Prentice-Hall.

Willey GR, Sabloff JA (1974): "A History of American Archaeology." San Francisco: W.H. Freeman and Co.

Wilmsen EN (1965): An outline of early man studies in the United States. Am Antiq 131:(2) Part 1:172–197.

Wormington HM (1949): "Ancient Man in North America." Denver: Museum of Natural History Popular Series No. 4, 3rd ed.

Wormington HM (1957): "Ancient Man in North America." Denver: The Denver Museum of National History, Popular Series No. 4, 4th ed.

Author Index

Subject Index